T0211056

Lecture Notes in Artificial Intelligence   10369

Subseries of Lecture Notes in Computer Science

LNAI Series Editors

Randy Goebel
  University of Alberta, Edmonton, Canada
Yuzuru Tanaka
  Hokkaido University, Sapporo, Japan
Wolfgang Wahlster
  DFKI and Saarland University, Saarbrücken, Germany

LNAI Founding Series Editor

Joerg Siekmann
  DFKI and Saarland University, Saarbrücken, Germany

More information about this series at http://www.springer.com/series/1244

Alessandro Antonucci · Laurence Cholvy
Odile Papini (Eds.)

Symbolic and Quantitative Approaches to Reasoning with Uncertainty

14th European Conference, ECSQARU 2017
Lugano, Switzerland, July 10–14, 2017
Proceedings

 Springer

Editors
Alessandro Antonucci (iD)
IDSIA
Lugano
Switzerland

Odile Papini
Aix-Marseille University
Marseille
France

Laurence Cholvy
ONERA
Toulouse
France

ISSN 0302-9743   ISSN 1611-3349   (electronic)
Lecture Notes in Artificial Intelligence
ISBN 978-3-319-61580-6   ISBN 978-3-319-61581-3   (eBook)
DOI 10.1007/978-3-319-61581-3

Library of Congress Control Number: 2017943848

LNCS Sublibrary: SL7 – Artificial Intelligence

Printed on acid-free paper

This Springer imprint is published by Springer Nature
The registered company is Springer International Publishing AG
The registered company address is: Gewerbestrasse 11, 6330 Cham, Switzerland

Preface

The biennal ECSQARU conference is a major forum for advances in the theory and practice of reasoning under uncertainty. Contributions are provided by researchers in advancing the state of the art and practitioners using uncertainty techniques in applications. The scope of the conference includes, but is not limited to, fundamental and representation issues, reasoning, and decision-making in both qualitative and quantitative paradigms.

Previous ECSQARU conferences were held in Compiègne (2015), Utrecht (2013), Belfast (2011), Verona (2009), Hammamet (2007), Barcelona (2005), Aalborg (2003), Toulouse (2001), London (1999), Bonn (1997), Fribourg (1995), Granada (1993), and Marseille (1991).

The 14th European Conference on Symbolic and Quantitative Approaches to Reasoning with Uncertainty (ECSQARU 2017) was held in Lugano, Switzerland, during July 10–14, 2017. The event was co-located with the 10th International Symposium on Imprecise Probability: Theories and Applications (ISIPTA 2017).

A young researcher award granted by Springer for excellent research in the area of symbolic and quantitative approaches to reasoning with uncertainty was assigned to Nico Potyka.

The papers in this volume were selected from 63 submissions, after a strict single-blind review process by the members of the Program Committee. In addition, the volume contains the abstracts of five invited talks by outstanding researchers in the field: Leila Amgoud, Alessio Benavoli, Jim Berger, Didier Dubois, and Eyke Hüllermeier.

We would like to thank all the members of the Program Committee and the additional reviewers for their timely and valuable reviews. We also thank the members of the Organizing Committee for their work and contribution to the success of the conference.

We gratefully acknowledge operational support from IDSIA (Istituto Dalle Molle di Studi sull'Intelligenza Artificiale), USI (Università della Svizzera Italiana), and SUPSI (Scuola Universitaria Professionale della Svizzera Italiana) as well as financial support from ONERA.

July 2017

Alessandro Antonucci
Laurence Cholvy
Odile Papini

Organization

Executive Committee

Conference Chairs

Alessandro Antonucci	IDSIA, Switzerland
Laurence Cholvy	ONERA, France
Odile Papini	Aix-Marseille University, France

Organizing Committee

Alessandro Antonucci	IDSIA, Switzerland
Giorgio Corani	IDSIA, Switzerland

Program Committee

Leila Amgoud	IRIT, France
Nahla Ben Amor	LARODEC, Tunisia
Salem Benferhat	CRIL, France
Concha Bielza	Technical University of Madrid, Spain
Isabelle Bloch	ENST, France
Claudette Cayrol	IRIT, France
Giulianella Coletti	University of Perugia, Italy
Giorgio Corani	IDSIA, Switzerland
Inés Couso	University of Oviedo, Spain
Fabio G. Cozman	University of São Paulo, Brazil
Fabio Cuzzolin	Oxford Brookes University, UK
Luis De Campos	University of Granada, Spain
Thierry Denœux	UTC, France
Sébastien Destercke	UTC, France
Didier Dubois	IRIT, France
Florence Dupin de Saint-Cyr	IRIT, France
Zied Elouedi	LARODEC, Tunisia
Patricia Everaere	University of Lille 1, France
Alessandro Facchini	IDSIA, Switzerland
Hélène Fargier	IRIT, France
Laurent Garcia	University of Angers, France
Laura Giordano	University of Eastern Piedmont, Italy
Lluis Godo	CSIC-IIIA, Spain
Anthony Hunter	University College London, UK
Katsumi Inoue	National Institute of Informatics, Japan
Souhila Kaci	LIRMM, France

Gabriele Kern-Isberner   TU Dortmund University, Germany
Sébastien Konieczny   CRIL, France
Jérôme Lang   LAMSADE, France
Florence Le Ber   ENGEES, France
Philippe Leray   University of Nantes, France
Churn-Jung Liau   Academia Sinica, Taiwan
Weiru Liu   Queen's University Belfast, UK
Peter Lucas   Radboud University Nijmegen, The Netherlands
Francesca Mangili   IDSIA, Switzerland
Pierre Marquis   University of Artois, France
Maria Vanina Martinez   UNS, Argentina
Andrés Masegosa   NTNU, Norway
David Mercier   University of Artois, France
Enrique Miranda   University of Oviedo, Spain
Serafín Moral   University of Granada, Spain
Farid Nouioua   LSIS, France
Jose M. Peña   Linköping University, Sweden
Davide Petturiti   University of Perugia, Italy
Henri Prade   IRIT, France
Silja Renooij   Utrecht University, The Netherlands
Karim Tabia   CRIL, France
Choh Man Teng   IHMC, USA
Matthias Troffaes   Durham University, UK
Barbara Vantaggi   Sapienza University of Rome, Italy
Linda Van der Gaag   Utrecht University, The Netherlands
Leon van der Torre   University of Luxembourg, Luxembourg
Jiří Vomlel   UTIA, Czech Republic
Renata Wassermann   University of São Paulo, Brazil
Éric Würbel   LSIS, France

Additional Reviewers

Jérôme Delobelle   CRIL, France
Eduardo Fermé   University of Madeira, Portugal
Marcelo Finger   University of São Paulo, Brazil
Tommaso Flaminio   University of Insubria, Italy
Diogo Patrão   A.C. Camargo Cancer Center, Brazil
Martin Plajner   UTIA, Czech Republic
Ahmed Samet   IRISA, France
Nicolas Schwind   AIST, Japan
Gerardo. I. Simari   UNS, Argentina
Che-Ping Su   University of Melbourne, Australia
Sara Ugolini   University of Siena, Italy
Srdjan Vesic   CRIL, France
Chunlai Zhou   Renmin University of China, China

Sponsoring Institutions

Invited Talks

Evaluation Methods of Arguments: Current Trends and Challenges

Leila Amgoud

IRIT – CNRS, Toulouse, France

Argumentation is a reasoning process based on the justification of conclusions by arguments. Due to its explanatory power, it has become a hot topic in Artificial Intelligence. It is used for making decisions under uncertainty, learning rules, modeling different types of dialogs, and more importantly for reasoning about inconsistent information. Hence, an argument's conclusion may have different natures: a statement that is true or false, an action to do, a goal to pursue, etc. Furthermore, it has generally an *intrinsic strength*, which may represent different issues (the certainty degree of its reason, the importance of the value it promotes if any, the reliability of its source, ...). Whatever its intrinsic strength (strong or weak), an argument may be weakened by other arguments (called *attackers*), and may be strengthened by others (called *supporters*). The overall acceptability of arguments needs then to be evaluated. Several evaluation methods, called semantics, were proposed for that purpose. In this talk, we show that they can be partitioned into three classes (extension semantics, gradual semantics, ranking semantics), which answer respectively to following questions:

1. What are the coalitions of arguments?
2. What is the overall strength of an argument?
3. How arguments can be rank-ordered from the most to the least acceptable ones?

We analyze the three classes against a set of rationality principles, and show that extension semantics are fundamentally different from the two other classes. This means that in concrete applications, they lead to different results. Namely, in case of reasoning with inconsistent information, extension semantics follow the same line of research as well-known syntactic approaches for handling inconsistency, while the two other classes lead to novel and powerful ranking logics. We argue that there is no universal evaluation method. The choice of a suitable method depends on the application at hand. Finally, we point out some challenges ahead.

Bayes + Hilbert = Quantum Mechanics

Alessio Benavoli

IDSIA, Lugano, Switzerland

Quantum mechanics (QM) is based on four main axioms, which were derived after a long process of trial and error. The motivations for the axioms are not always clear and even to experts the basic axioms of QM often appear counter-intuitive. In a recent paper, we have shown that:

- It is possible to derive quantum mechanics from a single principle of self-consistency or, in other words, that QM laws of Nature are logically consistent;
- QM is just the Bayesian theory generalised to the complex Hilbert space.

In particular, we have considered the problem of gambling on a quantum experiment and enforced rational behaviour by a few rules. These rules yield, in the classical case, the Bayesian theory of probability via duality theorems. In our quantum setting, they yield the Bayesian theory generalised to the space of Hermitian matrices. This very theory is QM: in fact, we have derived all its four postulates from the generalised Bayesian theory. This implies that QM is self-consistent. It also leads us to reinterpret the main operations in quantum mechanics as probability rules: Bayes' rule (measurement), marginalisation (partial tracing), independence (tensor product). To say it with a slogan, we have obtained that quantum mechanics is the Bayesian theory in the complex numbers.

Encounters with Imprecise Probabilities

Jim Berger

Duke University, Durham, USA

Although I have not formally done research in imprecise probability over the last twenty years, imprecise probability was central to much of my research in other areas. This talk will review some of these encounters with imprecise probability, taking examples from four areas:

- Using probabilities of a "higher type" (I.J. Good's phrase), with an application to genome-wide association studies.
- Robust Bayesian bounds, with an application to conversion of p-values to odds.
- Importance (and non-importance) of dependencies in imprecise probabilities.
- Imprecise probabilities arising from model bias, with examples from both statistical and physical modeling.

Symbolic and Quantitative Representations of Uncertainty: An Overview

Didier Dubois

IRIT, CNRS and University of Toulouse, Toulouse, France

The distinction between aleatory and epistemic uncertainty is more and more acknowledged to-date, and the idea that they should not be handled in the same way becomes more and more accepted. Aleatory uncertainty refers to a summarized description of natural phenomena by means of frequencies of occurrence, which justifies a numerical approach based on probability theory. In contrast, epistemic uncertainty stems from a lack of information, and describes the state of knowledge of an agent. It seems to be basically qualitative, and is captured by sets of possible worlds of states of nature, one of which is the actual one. In other words, beliefs induced by aleatory uncertainty are naturally quantitative, while this is less obvious for beliefs stemming from epistemic uncertainty for which there are various approaches ranging from qualitative ones like three-valued logics and modal logics to quantitative ones like subjective probabilities. The qualitative approaches can be refined by considering degrees of beliefs on finite value scales or yet by means of confidence relations. Moreover aleatory and epistemic uncertainty may come together, and leads to the use of upper and lower probabilities.

In this talk, we review the various approaches to the representations of uncertainty, by showing similarities between quantitative and qualitative approaches. We give a general definition of an epistemic state or an information item, as defining a set of possible values, a set of plausible ones, a plausibility ordering on events. Moreover, epistemic states must be compared in terms of informativeness.

The basic mathematical tool for representing uncertainty is the monotonic set-function, called capacity of fuzzy measure. In the quantitative case, the most general model is based on convex probability sets, that is, capacities that stand for lower probabilities. In the qualitative case, the simplest non-Boolean approach is based on possibility and necessity measures. It is shown that possibility theory plays in the qualitative setting a role similar to the one of probability theory in the quantitative setting. Just as a numerical capacity can, under some conditions, encode a family of probability distributions, a qualitative capacity always encodes a family of possibility distributions. For decision purposes, Sugeno integral is similar to Choquet integral.

Logical reasoning under incomplete information can be achieved by means of a simplified version of epistemic logic whose semantics is in terms of possibility theory, in contrast with probabilistic reasoning. It can be extended to reasoning with degrees of beliefs using generalised possibilistic logic. Various ways of defining logics of uncertainty are outlined, absolute, comparative, or fuzzy.

Finally we discuss the issue of uncertainty due to conflicting items of information. In the numerical setting this is naturally captured by the theory of evidence, that essentially models unreliable testimonies and their fusion. A general approach to the fusion of information items is outlined, proposing merging axioms that apply to quantitative and qualitative items of information. Finally, we show that using Boolean valued capacities, we can faithfully represent conflicting information coming from several sources. In this setting, necessity functions represent incomplete information while possibility measures represent precise but conflicting pieces of information.

This talk owes much to works performed with M. Banerjee, D. Ciucci, L. Godo, W. Liu and J. Ma, H. Prade, A. Rico, S. Schockaert, among others.

References

1. Banerjee, M., Dubois, D.: A simple logic for reasoning about incomplete knowledge. Int. J. Approx. Reason. **55**, 639–653 (2014)
2. Ciucci, D., Dubois, D.: A two-tiered propositional framework for handling multisource inconsistent information these proceedings (2017)
3. Dubois, D.: Representation, propagation, and decision issues in risk analysis under incomplete probabilistic information. Risk Anal. **30**, 361–368 (2010)
4. Dubois, D., Godo, L., Prade, H.: Weighted logics for artificial intelligence - an introductory discussion. Int. J. Approx. Reason. **55**(9), 1819–1829 (2014)
5. Dubois, D., Liu, W., Ma, J., Prade, H.: The basic principles of uncertain information fusion. An organised review of merging rules in different representation frameworks. Inf. Fusion **32**, 12–39 (2016)
6. Dubois, D., Prade, H., Rico, A.: Representing qualitative capacities as families of possibility measures. Int. J. Approx. Reason. **58**, 3–24 (2015)
7. Dubois, D., Prade, H., Schockaert, S.: Reasoning about uncertainty and explicit ignorance in generalized possibilistic logic. In: Proceedings of the ECAI 2014, pp. 261–266 (2014)
8. Ferson, S., Ginzburg, L.R.: Different methods are needed to propagate ignorance and variability. Reliab. Eng. Syst. Saf. **54**, 133–144 (1996)
9. Flage, R., Dubois, D., Aven, T.: Combined analysis of unique and repetitive events in quantitative risk assessment. Int. J. Approx. Reason. **70**, 68–78 (2016)
10. Grabisch, M.: Set functions, Games and Capacities in Decision-Making. Springer (2016)
11. Walley, P., Fine, T.: Varieties of modal: classificatory and comparative probabilities. Synthese **41**, 321–374 (1979)

Learning from Imprecise Data

Eyke Hüllermeier

Paderborn University, Paderborn, Germany

This talk addresses the problem of learning from imprecise data. Although it has been studied in statistics and various other fields for quite a while, this problem received renewed interest in the realm of machine learning more recently. In particular, the framework of superset learning will be discussed, a generalization of standard supervised learning in which training instances are labeled with a superset of the actual outcomes. Thus, superset learning can be seen as a specific type of weakly supervised learning, in which training examples are imprecise or ambiguous. We introduce a generic approach to superset learning, which is motivated by the idea of performing model identification and "data disambiguation" simultaneously. This idea is realized by means of a generalized risk minimization approach, using an extended loss function that compares precise predictions with set-valued observations. Building on this approach, we furthermore elaborate on the idea of "data imprecisiation": By deliberately turning precise training data into imprecise data, it becomes possible to modulate the influence of individual examples on the process of model induction. In other words, data imprecisiation offers an alternative way of instance weighting. Interestingly, several existing machine learning methods, such as support vector regression or semi-supervised support vector classification, are recovered as special cases of this approach. Besides, promising new methods can be derived in a natural way, and examples of such methods will be shown for problems such as classification, regression, and label ranking.

Contents

Analogical Reasoning

Analogical Inequalities. 3
  Henri Prade and Gilles Richard

Boolean Analogical Proportions - Axiomatics and Algorithmic
Complexity Issues. 10
  Henri Prade and Gilles Richard

Argumentation

Evaluation of Arguments in Weighted Bipolar Graphs. 25
  Leila Amgoud and Jonathan Ben-Naim

Debate-Based Learning Game for Constructing Mathematical Proofs. 36
  Nadira Boudjani, Abdelkader Gouaich, and Souhila Kaci

Updating Probabilistic Epistemic States in Persuasion Dialogues. 46
  Anthony Hunter and Nico Potyka

From Structured to Abstract Argumentation: Assumption-Based
Acceptance via AF Reasoning 57
  Tuomo Lehtonen, Johannes P. Wallner, and Matti Järvisalo

On Relating Abstract and Structured Probabilistic Argumentation:
A Case Study. 69
  Henry Prakken

Bayesian Networks

Structure-Based Categorisation of Bayesian Network Parameters. 83
  Janneke H. Bolt and Silja Renooij

The Descriptive Complexity of Bayesian Network Specifications 93
  Fabio G. Cozman and Denis D. Mauá

Exploiting Stability for Compact Representation of Independency Models . . . 104
  Linda C. van der Gaag and Stavros Lopatatzidis

Parameter Learning Algorithms for Continuous Model Improvement
Using Operational Data. 115
  *Anders L. Madsen, Nicolaj Søndberg Jeppesen, Frank Jensen,
  Mohamed S. Sayed, Ulrich Moser, Luis Neto, Joao Reis, and Niels Lohse*

Monotonicity in Bayesian Networks for Computerized Adaptive Testing   125
    Martin Plajner and Jiří Vomlel

Expert Opinion Extraction from a Biomedical Database   135
    Ahmed Samet, Thomas Guyet, Benjamin Negrevergne, Tien-Tuan Dao,
    Tuan Nha Hoang, and Marie Christine Ho Ba Tho

Solving Trajectory Optimization Problems by Influence Diagrams   146
    Jiří Vomlel and Václav Kratochvíl

Belief Functions

Iterative Aggregation of Crowdsourced Tasks Within the Belief
Function Theory   159
    Lina Abassi and Imen Boukhris

A Clustering Approach for Collaborative Filtering Under the Belief
Function Framework   169
    Raoua Abdelkhalek, Imen Boukhris, and Zied Elouedi

A Generic Framework to Include Belief Functions in Preference
Handling for Multi-criteria Decision   179
    Sébastien Destercke

A Recourse Approach for the Capacitated Vehicle Routing Problem
with Evidential Demands   190
    Nathalie Helal, Frédéric Pichon, Daniel Porumbel, David Mercier,
    and Éric Lefèvre

Evidential k-NN for Link Prediction   201
    Sabrine Mallek, Imen Boukhris, Zied Elouedi, and Eric Lefevre

Ensemble Enhanced Evidential k-NN Classifier Through
Random Subspaces   212
    Asma Trabelsi, Zied Elouedi, and Eric Lefevre

Conditionals

Comparison of Inference Relations Defined over Different Sets
of Ranking Functions   225
    Christoph Beierle and Steven Kutsch

A Transformation System for Unique Minimal Normal Forms
of Conditional Knowledge Bases   236
    Christoph Beierle, Christian Eichhorn, and Gabriele Kern-Isberner

On Boolean Algebras of Conditionals and Their Logical Counterpart   246
    Tommaso Flaminio, Lluis Godo, and Hykel Hosni

A Semantics for Conditionals with Default Negation   257
Marco Wilhelm, Christian Eichhorn, Richard Niland,
and Gabriele Kern-Isberner

Credal Sets, Credal Networks

Incoherence Correction and Decision Making Based on Generalized
Credal Sets.   271
Andrey G. Bronevich and Igor N. Rozenberg

Reliable Knowledge-Based Adaptive Tests by Credal Networks   282
Francesca Mangili, Claudio Bonesana, and Alessandro Antonucci

Decision Theory, Decision Making and Reasoning Under Uncertainty

Algorithms for Multi-criteria Optimization in Possibilistic Decision Trees . . .   295
Nahla Ben Amor, Fatma Essghaier, and Hélène Fargier

Efficient Policies for Stationary Possibilistic Markov Decision Processes   306
Nahla Ben Amor, Zeineb EL khalfi, Hélène Fargier, and Régis Sabaddin

An Angel-Daemon Approach to Assess the Uncertainty in the Power
of a Collectivity to Act   318
Giulia Fragnito, Joaquim Gabarro, and Maria Serna

Decision Theory Meets Linear Optimization Beyond Computation   329
Christoph Jansen, Thomas Augustin, and Georg Schollmeyer

Axiomatization of an Importance Index for Generalized Additive
Independence Models   340
Mustapha Ridaoui, Michel Grabisch, and Christophe Labreuche

Fuzzy Sets, Fuzzy Logic

Probability Measures in Gödel$_\Delta$ Logic   353
Stefano Aguzzoli, Matteo Bianchi, Brunella Gerla, and Diego Valota

Fuzzy Weighted Attribute Combinations Based Similarity Measures   364
Giulianella Coletti, Davide Petturiti, and Barbara Vantaggi

Online Fuzzy Temporal Operators for Complex System Monitoring   375
Jean-Philippe Poli, Laurence Boudet, Bruno Espinosa,
and Laurence Cornez

Logics

Complexity of Model Checking for Cardinality-Based Belief
Revision Operators   387
  Nadia Creignou, Raïda Ktari, and Odile Papini

A Two-Tiered Propositional Framework for Handling Multisource
Inconsistent Information.   398
  Davide Ciucci and Didier Dubois

Reasoning in Description Logics with Typicalities and Probabilities
of Exceptions   409
  Gian Luca Pozzato

Orthopairs

Measuring Uncertainty in Orthopairs   423
  Andrea Campagner and Davide Ciucci

Possibilistic Networks

Possibilistic MDL: A New Possibilistic Likelihood Based Score
Function for Imprecise Data   435
  Maroua Haddad, Philippe Leray, and Nahla Ben Amor

Probabilistic Logics, Probabilistic Reasoning

The Complexity of Inferences and Explanations in Probabilistic
Logic Programming.   449
  Fabio G. Cozman and Denis D. Mauá

Count Queries in Probabilistic Spatio-Temporal Knowledge Bases
with Capacity Constraints.   459
  John Grant, Cristian Molinaro, and Francesco Parisi

RankPL: A Qualitative Probabilistic Programming Language   470
  Tjitze Rienstra

Generalized Probabilistic Modus Ponens   480
  Giuseppe Sanfilippo, Niki Pfeifer, and Angelo Gilio

A First-Order Logic for Reasoning About Higher-Order Upper
and Lower Probabilities   491
  Nenad Savić, Dragan Doder, and Zoran Ognjanović

Author Index   501

Analogical Reasoning

Analogical Inequalities

Henri Prade[1,2]($\boxtimes$) and Gilles Richard[1]

[1] IRIT, Toulouse University, Toulouse, France
{prade,richard}@irit.fr
[2] QCIS, University of Technology, Sydney, Australia

Abstract. Analogical proportions, i.e., statements of the form a is to b as c is to d, state that the way a and b possibly differ is the same as c and d differ. Thus, it expresses an equality (between differences). However expressing inequalities may be also of interest for stating, for instance, that the difference between a and b is smaller than the one between c and d. The logical modeling of analogical proportions, both in the Boolean case and in the multiple-valued case, has been developed in the last past years. This short paper provides a preliminary investigation of the logical modeling of so-called "analogical inequalities", which are introduced here, in relation with analogical proportions.

1 Introduction

Comparative thinking plays a key role in our assessment of reality. This has been recognized for a long time. Making comparison is closely related to similarity judgment [14] and analogy making [4]. Analogical proportions, i.e., statements of the form a is to b as c is to d provides a well-known way for expressing a comparative judgment between the two pairs (a, b) and (c, d) by suggesting that the comparison between the elements of each pair yields the same kind of result in terms of dissimilarity [12].

The interest of analogical proportions has been recently pointed out in classification in machine learning [1,2,8] and in visual multiple-class categorization tasks for handling pieces of knowledge about semantic relationships between classes. More precisely in this latter case, analogical proportions are used for expressing analogies between pairs of concrete objects in the same semantic universe and with similar abstraction level, and then this gives birth to constraints that serve regularization purposes [5]. Interestingly enough, constraints of the same kind but issued from pieces of knowledge stating *relative* comparisons between quadruplets of images, feature by feature, have been recently experienced with success [6,7]. These relative comparisons are inequalities between differences inside pairs rather than equalities. Moreover these comparisons involving quadruplets have been shown to be more useful in categorization tasks than comparisons involving only triplets, or pairs, of images.

Besides, it has been also recently noticed that similar relations in terms of comparison of pairs were also present in multiple criteria analysis for expressing, for instance, that the "difference" between two evaluation vectors on a criterion

© Springer International Publishing AG 2017
A. Antonucci et al. (Eds.): ECSQARU 2017, LNAI 10369, pp. 3–9, 2017.
DOI: 10.1007/978-3-319-61581-3_1

is smaller than (i.e., does not compensate) the "difference" between the vectors on the rest of the criteria [10].

This recent emergence of the interest for inequality constraints between pairs of items motivates this first formal study of "analogical inequalities", in relation with the Boolean and the multiple-valued modeling of analogical proportions. The paper first restates the necessary background on these proportions, before extending it in order to represent "analogical inequalities", both in the Boolean and in the multiple-valued settings.

2 Background on Analogical Proportions

We start with a reminder on analogical proportions, first in the case of binary attributes.

2.1 Boolean Case

Let us assume that four items a, b, c, d are represented by sets of binary features belonging to a universe U (i.e., an item is then viewed as the set of the binary features in U that it satisfies). Then, the dissimilarity between a and b can be appreciated in terms of $a \cap \bar{b}$ and/or $\bar{a} \cap b$, where $\bar{a}$ denotes the complement of a in U, while the similarity is estimated by means of $a \cap b$ and/or of $\bar{a} \cap \bar{b}$. Then, an analogical proportion between subsets is formally defined by the two conditions [9]:

$$a \cap \bar{b} = c \cap \bar{d} \text{ and } \bar{a} \cap b = \bar{c} \cap d \tag{1}$$

This expresses that "a differs from b as c differs from d" and that "b differs from a as d differs from c". It can be viewed as the expression of a co-variation. Analogical proportion has an easy counterpart in Boolean logic, where it is denoted by $a : b :: c : d$, a, b, c, d being now Boolean variables (supposed to refer to the value of the same attribute for 4 different items). In this logical setting, "equality" translates into "equivalence" ($\equiv$), $\bar{a}$ into the negation of a (i.e., $\neg a$), and $\cap$ is changed into a conjunction ($\wedge$), and we get the logical condition expressing that 4 Boolean variables make an analogical proportion [9]:

$$a : b :: c : d = (a \wedge \neg b \equiv c \wedge \neg d) \wedge (\neg a \wedge b \equiv \neg c \wedge d) \tag{2}$$

An analogical proportion is then a Boolean formula. The expression $a : b :: c : d$ takes the truth value "1" only for the 6 following patterns for $abcd$: $1111, 0000, 1100, 0011, 1010, 0101$. For the 10 other patterns of its truth table, it is false (i.e., equal to 0).

A worth noticing property, beyond reflexivity ($a : b :: a : b$), symmetry (if $a : b :: c : d$ then $c : d :: a : b$), and central permutation (if $a : b :: c : d$ then $a : c :: b : d$) is the fact that the analogical proportion remains true for the negation of the Boolean variables [11]. It expresses that the result does not depend on a positive or a negative encoding of the features:

$$\text{if } a : b :: c : d \text{ then } \neg a : \neg b :: \neg c : \neg d \qquad \text{(code independency).} \tag{3}$$

2.2   Multiple-Valued Case

Attributes or features are not necessarily Boolean, and a graded extension of analogical expression is needed. We assume that attributes are now valued in $[0, 1]$ (possibly after renormalization). The extension is obtained by replacing (i) the central $\wedge$ in (2) by min, (ii) the two $\equiv$ symbols by $\min(s \rightarrow_L t, t \rightarrow_L s) = 1 - |s - t|$, where $s \rightarrow_L t = \min(1, 1 - s + t)$ is Łukasiewicz implication, (iii) the four expressions of the form $s \wedge \neg t$ by the bounded difference $\max(0, s - t) = 1 - (s \rightarrow_L t)$, which is associated to Łukasiewicz implication, using $1 - (\cdot)$ as negation. The resulting expression [3] is then

$$a : b ::_L c : d = \begin{cases} 1 - |(a - b) - (c - d)|, \\ \quad \text{if } a \geq b \text{ and } c \geq d, \text{ or } a \leq b \text{ and } c \leq d \\ 1 - \max(|a - b|, |c - d|), \\ \quad \text{if } a \leq b \text{ and } c \geq d, \text{ or } a \geq b \text{ and } c \leq d \end{cases} \tag{4}$$

It coincides with $a : b :: c : d$ on $\{0, 1\}$. As can be seen, this expression is equal to 1 if and only if $(a - b) = (c - d)$, while $a : b ::_L c : d = 0$ if and only if (i) $a - b = 1$ and $c \leq d$, or if (ii) $b - a = 1$ and $d \leq c$, or if (iii) $a \leq b$ and $c - d = 1$, or if (iv) $b \leq a$ and $d - c = 1$. Thus, $a : b ::_L c : d = 0$ when the change inside one of the pairs (a, b) or (c, d) is maximal, while the other pair shows either no change or a change in the opposite direction. It can be also checked that code independency continue to hold under the form $a : b ::_L c : d = 1 - a : 1 - b ::_L 1 - c : 1 - d$.

Note that the algebraic difference between a and b equated with the difference between c and d, namely $a - b = c - d$, provides a constraint that is satisfied by the 6 patterns making true the analogical proportion $a : b :: c : d$ in the Boolean case, and by none of the 10 others. However, $a - b$ may not belong to $\{0, 1\}$ when $a, b \in \{0, 1\}$. While $|a - b| \in \{0, 1\}$, the constraint $|a - b| = |c - d|$ validates 8 patterns including 0110 and 1001. When considering the graded case, $a - b$ is not close in $[0, 1]$; moreover, the modeling of the analogical proportion by the constraint $a - b = c - d$ does not provide a graded evaluation of how far we are from satisfying it.

3   Inequalities

In the following, we propose a logical modeling for expressions of the form "a is to b at least as much as c to d", first in the Boolean case, and then in the multiple-valued case. We denote this expression by $a : b \ll c : d$.

3.1   Boolean Case

Starting from the Boolean expression (2) of the analogical proportion, we replace the two symbols $\equiv$ expressing sameness by two material implications $\rightarrow$ for modeling the fact that the result of the comparison of c and d is larger or equal to the result of the comparison of a and b. Namely, we obtain

$$a : b \ll c : d = ((a \wedge \neg b) \rightarrow (c \wedge \neg d)) \wedge ((\neg a \wedge b) \rightarrow (\neg c \wedge d)) \tag{5}$$

It can be checked from this definition that the following expected properties hold:

- $a : b \ll a : b$
- $a : b :: c : d \rightarrow a : b \ll c : d$
- $a : b :: c : d \equiv ((a : b \ll c : d) \wedge (c : d \ll a : b))$
- $(a : b \ll c : d) \equiv (\neg a : \neg b \ll \neg c : \neg d)$

Namely, $a : b \ll c : d$ is weaker than $a : b :: c : d$, while $a : b :: c : d$ holds if and only if both $a : b \ll c : d$ and $c : d \ll a : b$ hold; moreover, code independency is preserved.

The truth table of $a : b \ll c : d$ is given in Table 1. As can be seen $a : b \ll c : d$ holds true for the 6 patterns that makes analogical proportion true, plus the 4 patterns 0001, 0010, 1110, 1101. These latter patterns correspond to the 4 situations where $a \equiv b$ and $c \not\equiv d$. In these 4 situations a and b are indeed strictly closer than c and d, and these are the only cases in $\{0, 1\}$. Since the 4 situations where $a \equiv b$ and $c \equiv d$ are among the patterns making true $a : b :: c : d$, we have

$$a : b \ll c : d \equiv (a : b :: c : d) \vee (a \equiv b) \qquad (6)$$

It is also worth noticing that the central permutation property of analogical proportion now fails since 0010 and 1101 are true while 0100 and 1011 are false. This may be unexpected at first glance since the arithmetic proportion inequality, $a - b \leq c - d$, still satisfies central permutation in the numerical case; however it is made possible since $a - b \in \{-1, 0, 1\}$ and indeed $0 < 1 \Leftrightarrow -1 < 0$.

Table 1. Boolean valuations for $a : b \ll c : d$

a	b	c	d	$a : b \ll c : d$	a	b	c	d	$a : b \ll c : d$
0	0	0	0	1	1	0	0	0	0
0	0	0	1	1	1	0	0	1	0
0	0	1	0	1	1	0	1	0	1
0	0	1	1	1	1	0	1	1	0
0	1	0	0	0	1	1	0	0	1
0	1	0	1	1	1	1	0	1	1
0	1	1	0	0	1	1	1	0	1
0	1	1	1	0	1	1	1	1	1

Note that the quaternary relation $a : b \ll c : d$ induces a ternary relation (just as a continuous analogical proportion of the form $a : b :: b : c$ is a particular case of analogical proportion [13]). It can be seen that $a : b \ll b : c$ is true only for the four patterns 0000, 0001, 1110 and 1111, and false for the four other patterns. It expresses that the difference between b and c is greater or equal to the one between a and b.

3.2 Graded Case

The expression (5) can be extended to the multiple valued case, still keeping min for extending the central $\wedge$, $1- \mid s - t \mid$ for the $\equiv$ symbols, and the four expressions of the form $s \wedge \neg t$ as the bounded difference $\max(0, s - t)$. The resulting expression is then

$$
a : b \ll_L c : d = \begin{cases} \min(1, 1 - ((b - a) - (d - c))) & \text{if } a \leq b \text{ and } c \leq d \\ \min(1, 1 - ((a - b) - (c - d))) & \text{if } a \geq b \text{ and } c \geq d \\ 1 - (b - a) & \text{if } a \leq b \text{ and } c \geq d \\ 1 - (a - b) & \text{if } a \geq b \text{ and } c \leq d \end{cases} \tag{7}
$$

Thus $a : b \ll_L c : d$ can be read "c is more different from d than a is from b". It can be checked that the following expected properties still hold

- $a : b \ll_L c : d = a : b \ll c : d$ when $a, b, c, d \in \{0, 1\}$;
- $a : b \ll_L a : b = 1$;
- $a : b ::_L c : d \leq a : b \ll_L c : d$;
- $a : b ::_L c : d = \min((a : b \ll_L c : d), (c : d \ll_L a : b))$;
- $(a : b \ll_L c : d) = ((1 - a) : (1 - b) \ll_L (1 - c) : (1 - d))$.

In particular, $a : b \ll_L c : d = 1$ if and only if

- $a = b$, or
- $\mid b - a \mid \leq \mid d - c \mid$ if $a \leq b$ and $c \leq d$, or if $b \leq a$ and $d \leq c$.

Moreover $a : b \ll_L c : d = 0$ if and only if

- $\mid b - a \mid = 1$ and $\mid d - c \mid = 0$, or
- $b - a = 1$ and $c \geq d$, or
- $a - b = 1$ and $c \leq d$.

It is worth noticing that $a : b \ll_L c : d$ does not exactly amount at comparing absolute value distances, as in the constraint $\mid a - b \mid \leq \mid c - d \mid$. Indeed it can be checked that we may have $a : b \ll_L c : d = 0$, while $\mid a - b \mid \leq \mid c - d \mid$ holds (taking $a = d = 0$ and $b = c = 1$). Moreover $a : b \ll_L c : d$ provides a graded estimate of the extent to which the numerical constraint $a - b \leq c - d$ is satisfied.

Continuous analogical inequalities define the following graded comparative ternary relation:

$$
a : b \ll_L b : c = \begin{cases} \min(1, 1 + (a + c) - 2b) & \text{if } a \leq b \leq c \\ \min(1, 1 + 2b - (a + c)) & \text{if } a \geq b \geq c \\ 1 - (b - a) & \text{if } a \leq b \text{ and } b \geq c \\ 1 - (a - b) & \text{if } a \geq b \text{ and } b \leq c \end{cases} \tag{8}
$$

Note that $a : b \ll_L b : c = 1$ if and only if $a = b$, or if $b \leq (a + c)/2$ (resp. $b \geq (a + c)/2$) if $a \leq b \leq c$ (resp. $a \geq b \geq c$), i.e., if and only if b is closer (in the broad sense) to a than to c. It means that the difference between b and c is greater or equal to the one between a and b and the differences are oriented in the same way (when non zero).

Lastly, all the definitions considered in this paper apply to a single attribute. Just as in the case of the analogical proportion, the definitions straightforwardly extend to multiple attribute descriptions by applying them in a component-wise manner, attribute per attribute. If necessary, a global evaluation may be obtained by taking the average of the estimates obtained for each considered attribute.

4   Conclusion

The paper has provided a preliminary investigation of the idea of analogical inequality as a relaxation of the notion of analogical proportion, both in the Boolean case and in the multiple-valued case. It appears that this proper extension does not just amount at comparing differences (or distances) between the elements of two pairs, but, as in the case of the analogical proportion, it also takes into account the orientation of the variations when going from a to b, and from c to d. Moreover, it also provides a graded estimate of the extent to which "c is more different from d than a is from b". This enables us to turn such a statement into a soft constraint, where the threshold corresponding to the minimal amount to which the constraint should hold might be a matter of learning in practice.

References

1. Bayoudh, S., Miclet, L., Delhay, A.: Learning by analogy: a classification rule for binary and nominal data. In: Proceedings of 20th International Joint Conference on Artifical Intelligence (IJCAI 2007), Hyderabad, India, pp. 678–683 (2007)
2. Bounhas, M., Prade, H., Richard, G.: Analogical classification. A new way to deal with examples. In: Proceedings of ECAI 2014, pp. 135–140 (2014)
3. Dubois, D., Prade, H., Richard, G.: Multiple-valued extensions of analogical proportions. Fuzzy Sets Syst. **292**, 193–202 (2016)
4. Gentner, D., Holyoak, K.J., Kokinov, B.N. (eds.): The Analogical Mind: Perspectives from Cognitive Science. MIT Press, Cambridge (2001)
5. Hwang, S.J., Grauman, K., Sha, F.: Analogy-preserving semantic embedding for visual object categorization. In: Proceedings of International 30th Conference on Machine Learning (ICML), Atlanta, pp. 639–647 & JMLR: W&CP, 28 (1) 222–230 (2013)
6. Law, M.T., Thome, N., Cord, M.: Quadruplet-wise image similarity learning. In: Proceedings of IEEE International Conference on Computer Vision (ICCV) (2013)
7. Law, M.T., Thome, N., Cord, M.: Learning a distance metric from relative comparisons between quadruplets of images. Int. J. Comput. Vis. **121**(1), 65–94 (2017)
8. Miclet, L., Bayoudh, S., Delhay, A.: Analogical dissimilarity: definition, algorithms and two experiments in machine learning. J. Artif. Intell. Res. (JAIR) **32**, 793–824 (2008)

9. Miclet, L., Prade, H.: Handling analogical proportions in classical logic and fuzzy logics settings. In: Sossai, C., Chemello, G. (eds.) ECSQARU 2009. LNCS (LNAI), vol. 5590, pp. 638–650. Springer, Heidelberg (2009). doi:10.1007/978-3-642-02906-6_55

10. Pirlot, M., Prade, H., Richard, G.: Completing preferences by means of analogical proportions. In: Torra, V., Narukawa, Y., Navarro-Arribas, G., Yañez, C. (eds.) MDAI 2016. LNCS (LNAI), vol. 9880, pp. 135–147. Springer, Cham (2016). doi:10.1007/978-3-319-45656-0_12

11. Prade, H., Richard, G.: From analogical proportion to logical proportions. Logica Universalis **7**(4), 441–505 (2013)

12. Prade, H., Richard, G.: From the structures of opposition between similarity and dissimilarity indicators to logical proportions. A general representation setting for capturing homogeneity and heterogeneity. In: Dodig-Crnkovic, G., Giovagnoli, R. (eds.) Representation and Reality in Humans, Animals and Machines. Springer (2017)

13. Schockaert, S., Prade, H.: Completing rule bases using analogical proportions. In: Prade, H., Richard, G. (eds.) Computational Approaches to Analogical Reasoning - Current Trends, pp. 195–215. Springer, Heidelberg (2014)

14. Tversky, A.: Features of similarity. Psychol. Rev. **84**, 327–352 (1977)

Boolean Analogical Proportions - Axiomatics and Algorithmic Complexity Issues

Henri Prade[1,2(✉)] and Gilles Richard[1]

[1] IRIT, Toulouse University, Toulouse, France
{prade,richard}@irit.fr
[2] QCIS, University of Technology, Sydney, Australia

Abstract. Analogical proportions, i.e., statements of the form a is to b as c is to d, are supposed to obey 3 axioms expressing reflexivity, symmetry, and stability under central permutation. These axioms are not enough to determine a single Boolean model, if a minimality condition is not added. After an algebraic discussion of this minimal model and of related expressions, another justification of this model is given in terms of Kolmogorov complexity. It is shown that the 6 Boolean patterns that make an analogical proportion true have a minimal complexity with respect to an expression reflecting the intended meaning of the proportion.

1 Introduction

Despite the fact that conclusions obtained by analogical reasoning do no guarantee to be valid from a classical logic viewpoint, this kind of reasoning is considered as a valuable and often creative way to solve real life problems. Analogical proportions, i.e., relations between four items of the form a is to b as c is to d, constitute a key notion formalizing analogical inference and relying on the following principle: if such proportions hold on a noticeable subset of known features used for describing the four items, the proportion may still hold on other features as well, which may help guessing the unknown values of d on these other features from their values on a, b, and c [11,19]. It is only quite recently that a logical modeling of these proportions has been proposed [12,13], following several attempts at formalizing them in other settings [7,10]. This logical modeling makes clear that the analogical proportion holds if and only if a differs from b as c differs from to d and vice-versa.

The paper investigates two new justifications of the Boolean expression of an analogical proportion. First, starting from the core axioms supposed to be satisfied by an analogical proportion, and agreed by everybody for a long time, this paper exhibits the Boolean models compatible with these axioms. There are several ones, but the smallest model is the standard Boolean expression of an analogical proportion previously proposed. This smallest model is characterized by six possible Boolean patterns (among sixteen candidates). In the second part of the paper, we try to evaluate their cognitive significance in terms of algorithmic complexity (i.e. Kolmogorov complexity) and show that they are also minimal

A. Antonucci et al. (Eds.): ECSQARU 2017, LNAI 10369, pp. 10–21, 2017.
DOI: 10.1007/978-3-319-61581-3_2

among all Boolean patterns with respect to an algorithmic complexity-based definition of analogy. Indeed algorithmic complexity measures a kind of universal information content of a Boolean string. Despite its inherent uncomputability, there exist powerful tool for computing good approximations. Kolmogorov complexity has been proved to be of great value in diverse applications: for example, in distance measures [1] and classification methods, plagiarism detection, network intrusion detection [5], and in numerous other applications [9].

The paper is organized as follows. In Sect. 2, we provide a background on the definition of an analogical proportion and its basic properties in a Boolean setting. In Sect. 3, we start from the characteristic axioms of analogical proportions, and we investigate the different compatible Boolean models. In Sect. 4, we briefly review the main definition and theorems of Kolmogorov complexity. As we have the main tools, we are in a position to give a Kolmogorov complexity-based definition of analogy in Sect. 5. Section 6 is devoted to a set of experiments that empirically validate our definition. Finally, we conclude in Sect. 7.

2   Background on Boolean Analogical Proportion

At the time of Aristotle, the idea of analogical proportion originated from the notion of numerical proportion. In that respect the *arithmetic* proportion between 4 integers a, b, c, d, which holds if $a - b = c - d$, is a good prototype of the idea of analogical proportion, since we can read it as "a differs from b as c differs from d", which perfectly fits with "a is to b as c is to d", denoted by $a : b :: c : d$. When considering Boolean interpretation where $a, b, c, d \in \{0, 1\}$, it is tempting to carry on with the same definition as $\{0, 1\} \subset \mathbb{R}$, with the inevitable drawback that difference is not an internal operator in $\mathbb{B} = \{0, 1\}$. Nevertheless, if we draw the truth table (16 lines) corresponding to this definition, we get Table 1 highlighting that only 6 among 16 lines are valid proportions.

Table 1. Boolean valuations for $a : b :: c : d$

a	b	c	d	$a:b::c:d$	a	b	c	d	$a:b::c:d$
0	0	0	0	1	1	0	0	0	0
0	0	0	1	0	1	0	0	1	0
0	0	1	0	0	1	0	1	0	1
0	0	1	1	1	1	0	1	1	0
0	1	0	0	0	1	1	0	0	1
0	1	0	1	1	1	1	0	1	0
0	1	1	0	0	1	1	1	0	0
0	1	1	1	0	1	1	1	1	1

Boolean Definition. Looking for a purely logical definition of $a : b :: c : d$, we need to make use of the comparative indicators [14,15] that are naturally associated with a pair of variable (a, b):

– $a \wedge b$ and $\neg a \wedge \neg b$: they are *positive similarity* and *negative similarity* indicators respectively; $a \wedge b$ (resp. $\neg a \wedge \neg b$) is true iff only both a and b are true (resp. false);
– $a \wedge \neg b$ and $\neg a \wedge b$: they are *dissimilarity* indicators ; $a \wedge \neg b$ (resp. $\neg a \wedge b$) is true iff only a (resp. b) is true and b (resp. a) is false.

Then analogical proportion is defined by the two *logically equivalent* expressions [13]:

$$a : b :: c : d = (a \wedge \neg b \equiv c \wedge \neg d) \wedge (\neg a \wedge b \equiv \neg c \wedge d) \tag{1}$$

$$a : b :: c : d = ((a \wedge d) \equiv (b \wedge c)) \wedge ((\neg a \wedge \neg d) \equiv (\neg b \wedge \neg c)) \tag{2}$$

Expression (1) reads "a differs from b as c differs from d and b differs from a as d differs from c". This definition is equivalent to the previous one (it yields Table 1) with the advantage of being an internal definition inside $\mathbb{B}$. Expression (2) may be viewed as the logical counterpart of the well-known property of arithmetical proportions $a - b = c - d \Leftrightarrow a + d = b + c$. "$a$ is to b as c is to d" can now be read "what a and d have in common, b and c have it also (both positively and negatively)", which, however, is a less straightforward reading of the idea of analogy than the one associated with (1). As can be checked on Table 1, analogical proportions are *independent with respect to the positive or negative encoding* of properties: $(a : b :: c : d) \equiv (\neg a : \neg b :: \neg c : \neg d)$.

For representing objects one generally needs *vectors* of Boolean values, rather than single Boolean values, each component being the value of a binary attribute. The previous definition directly extends to Boolean vectors in $\mathbb{B}^n$ of the form $\overrightarrow{a} = (a_1, \cdots, a_n)$ as follows: $\overrightarrow{a} : \overrightarrow{b} :: \overrightarrow{c} : \overrightarrow{d}$ iff $\forall i \in [1, n]$, $a_i : b_i :: c_i : d_i$.

Equation and Creativity. The equation $a : b :: c : x$ has a unique solution $x = c \equiv (a \equiv b)$ provided that $(a \equiv b) \vee (a \equiv c)$ holds. Indeed neither $0 : 1 :: 1 : 0$ nor $1 : 0 :: 0 : 1$ holds true. This process can be extended componenwise to vectors. In that case, for instance, the following equation $010 : 100 : 011 : x$ has for unique solution the vector $(1, 0, 1)$ which is not among the 3 previous vectors $(0, 1, 0), (1, 0, 0), (0, 1, 1)$. Then analogical proportions for vectors are *creative* (an informal quality usually associated with the idea of analogy) as they may involve 4 distinct vectors.

A Previous View of Analogical Proportion. In [6], S. Klein suggests that an analogical proportion would hold as soon as a, b, c are completed by d taken as $d = c \equiv (a \equiv b)$. This amounts to define it as $A_K(a, b, c, d) \triangleq (a \equiv b) \equiv (c \equiv d)$. Then $0 : 1 :: 1 : 0$ and $1 : 0 :: 0 : 1$ become valid analogical proportions and leads to the model denoted Kl in the following section. The validity of such patterns may be advocated on the basis of a functional view of analogy where $a : f(a) :: b : f(b)$ sounds indeed valid, taking the negation in $\mathbb{B}$ for f. But, this is debatable since $A_K(a, b, c, d) \Leftrightarrow A_K(b, a, c, d)$ (which does not fit with intuition). It turns out that $a : b :: c : d \Rightarrow A_K(a, b, c, d)$.

Lower Approximations of Analogical Proportion. While $A_K(a, b, c, d)$ is an upper approximation of $a : b :: c : d$ true for 8 patterns, one may look for lower approximations that are true for 4 patterns only (taking into account code independency). There are 3 such approximations, given below, followed by the patterns they validate[1]:

$$(a \equiv b) \wedge (c \equiv d) \begin{array}{|cccc|} \hline 1&1&1&1 \\ \hline 0&0&0&0 \\ \hline 1&1&0&0 \\ \hline 0&0&1&1 \\ \hline \end{array}; \quad (a \equiv c) \wedge (b \equiv d) \begin{array}{|cccc|} \hline 1&1&1&1 \\ \hline 0&0&0&0 \\ \hline 1&0&1&0 \\ \hline 0&1&0&1 \\ \hline \end{array}; \quad (a \not\equiv d) \wedge (b \not\equiv c) \begin{array}{|cccc|} \hline 1&1&0&0 \\ \hline 0&0&1&1 \\ \hline 1&0&1&0 \\ \hline 0&1&0&1 \\ \hline \end{array}.$$

Note that only the last one remains creative.

The question addressed now is "Could an axiomatic view of analogical proportions offer a kind of intrinsic justification that only the 6 patterns obeying (1)–(2) are acceptable?"

3   Analogy and Its Lattice of Boolean Models

Analogy, viewed as a quaternary relation R, is supposed to obey 3 axioms (e.g., [7,10]):

1. $\forall a, b, R(a, b, a, b)$ (reflexivity);
2. $\forall a, b, c, d, R(a, b, c, d) \rightarrow R(c, d, a, b)$ (symmetry);
3. $\forall a, b, c, d, R(a, b, c, d) \rightarrow R(a, c, b, d)$ (central permutation).

These axioms are clearly inspired by numerical proportions. From them, some basic properties can be deduced by proper applications of symmetry and central permutation:

- $\forall a, b, R(a, a, b, b)$ (identity);
- $\forall a, b, c, d, R(a, b, c, d) \rightarrow R(b, a, d, c)$ (inside pair reversing);
- $\forall a, b, c, d, R(a, b, c, d) \rightarrow R(d, b, c, a)$ (extreme permutation).

In fact, another (less standard) axiom expected from a natural analogy is:

$$\forall a, b, R(a, a, b, x) \implies x = b \text{ (unicity)}$$

All these properties fit with our intuition of what may be an analogical proportion. In this paper, we focus on $\mathbb{B} = \{0, 1\}$ as interpretation domain. In that case, R should be interpreted as a subset of $\mathbb{B}^4$: removing the emptyset leaves $2^{16} - 1$ candidate models. It is straightforward to get a basic model. By applying reflexivity, we see that $0101, 1010$ should belong to the relation and $0000, 1111$ as well since we may have $a = b$, and central permutation then leads to add 0011 and 1100. Thus, we get the model $\Omega_0 = \{0000, 1111, 0101, 1010, 0011, 1100\}$, which is

---

[1] There are 3 companion approximations that involve the two additional patterns of A_K:

$$(a \equiv d) \wedge (b \equiv c) \begin{array}{|cccc|} \hline 1&1&1&1 \\ \hline 0&0&0&0 \\ \hline 1&0&0&1 \\ \hline 0&1&1&0 \\ \hline \end{array}; \quad (a \not\equiv b) \wedge (c \not\equiv d) \begin{array}{|cccc|} \hline 1&0&0&1 \\ \hline 0&1&1&0 \\ \hline 1&0&1&0 \\ \hline 0&1&0&1 \\ \hline \end{array}; \quad (a \not\equiv c) \wedge (b \not\equiv d) \begin{array}{|cccc|} \hline 1&1&0&0 \\ \hline 0&0&1&1 \\ \hline 1&0&0&1 \\ \hline 0&1&1&0 \\ \hline \end{array}.$$

stable under symmetry. Ω_0 is the *smallest* model for analogical proportion over $\mathbb{B}^2$. However, one may ask about other models, and we can show the following:

Property 1. *There are exactly 8 models of analogy (satisfying the 3 first axioms) over $\mathbb{B}$. There are exactly 2 models of analogy (satisfying the 3 first axioms plus unicity).*

Proof. Any model should include Ω_0. Let us note that a bigger model should necessarily have an even cardinality due to the following facts:

- To be bigger than Ω_0, it should contain a string s containing both 0 and 1.
- Thanks to symmetry or central permutation axioms, it should contain the symmetric *cdab* of $s = abcd$ and the central permutation *acbd* of s: necessarily, one of these 2 strings is different from s (otherwise, we get $a = b = c = d$).

So we have to look for models of cardinality 8, 10, 12, 14 and 16. Obviously $\mathbb{B}^4$ of cardinality 16 is a model, the biggest one. Due to the axioms, we have to add to Ω_0 subsets of $\mathbb{B}^4$ that are stable w.r.t. symmetry and central permutation. We have exactly:

- one subset with 2 elements: $S_2 = \{1001, 0110\}$
- two subsets with 4 elements: (i) $S_3 = \{1110, 1101, 1011, 0111\}$; (ii) $S_4 = \{0001, 0010, 0100, 1000\}$.

Since every model has to be built by adding to Ω_0 one of the previous subsets, we get the following models for analogy in $\mathbb{B}$:

(1) 1 model with 6 elements: Ω_0 (the smallest one)
(2) 1 model with 8 elements: $Kl = \Omega_0 \cup S_2 = \{0000, 1111, 0101, 1010, 0011, 1100, 0110, 1001\}$
    As previously explained, this model is due to Klein [6].
(3) 2 model with 10 elements:
    - $M_3 = \Omega_0 \cup S_3 = \{0000, 1111, 0101, 1010, 0011, 1100, 1110, 1101, 1011, 0111\}$,
    - $M_4 = \Omega_0 \cup S_4 = \{0000, 1111, 0101, 1010, 0011, 1100, 0001, 0010, 0100, 1000\}$
(4) 2 models with 12 elements:
    - $M_5 = M_3 \cup S_2 = \{0000, 1111, 0101, 1010, 0011, 1100, 1110, 1101, 1011, 0111, 0110, 1001\}$,
    - $M_6 = M_4 \cup S_2 = \{0000, 1111, 0101, 1010, 0011, 1100, 0001, 0010, 0100, 1000, 0110, 1001\}$,
(5) 1 model with 14 elements:
    - $M_7 = M_3 \cup S_4 = M_4 \cup S_3 = \Omega_0 \cup S_3 \cup S_4 = \{0000, 1111, 0101, 1010, 0011, 1100, 1110, 1101, 1011, 0111, 0100, 1000, 0110, 1001\}$,
(6) 1 model with exactly 16 elements: $\Omega = \Omega_0 \cup S_2 \cup S_3 \cup S_4 = \mathbb{B}$.

Finally Ω_0 and Kl satisfy unicity but $M3$ (containing 1100 and 1101) and $M4$ (containing 0000 and 0001) do not satisfy. This achieves the proof. $\qquad\square$

---

[2] Note that lower approximations of analogical proportions miss at least one axiom.

The set of models is a lattice with bottom element Ω_0 and top element $\mathbb{B}$, see Fig. 1. As can be seen, 8 models fit with the axioms in the Boolean case, including the 6-patterns model Ω_0 and the 8-patterns model Kl due to Klein. However, it is natural to privilege the smallest model, the minimal one that just accounts for the axioms and nothing more.

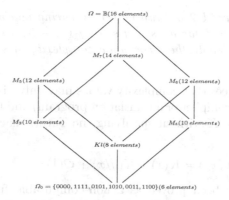

$\Omega = \mathbb{B}(16\ elements)$

$M_7(14\ elements)$

$M_5(12\ elements)$ $M_6(12\ elements)$

$M_3(10\ elements)$ $M_4(10\ elements)$

$Kl(8\ elements)$

$\Omega_0 = \{0000, 1111, 0101, 1010, 0011, 1100\}(6\ elements)$

Fig. 1. The lattice of Boolean models of analogy

We now investigate if another justification in favor of the minimal model Ω_0 can be obtained by minimizing an expression reflecting the information content of an analogical proportion in terms of Kolmogorov complexity. We now review the fundamentals of Kolmogorov complexity theory, also known as Algorithmic Complexity Theory.

4 Kolmogorov Complexity: A Brief Review

Kolmogorov complexity is not a new concept and the theory has been designed many years ago: see for instance [9] for an in depth study. This theory has not to be confused with Shannon information theory [16] despite the fact that they share some links.

The Starting Point. We need the help of a universal Turing machine denoted U. Then p denotes a program running on U. Two situations can happen: (i) either p does not stop for the input x, or (ii) p stops for the input x and outputs a finite string y. In that case, we write $U(p, x) = y$. The Kolmogorov complexity [9] of y w.r.t. x is then defined as:

$$K_U(y/x) = min\{|p|, U(p, x) = y\}.$$

$K_U(y/x)$ is the size of the shortest program able to reconstruct y with the help of x. The Kolmogorov complexity [9] of y is just obtained with the empty string ϵ:

$$K_U(y) = min\{|p|, U(p, \epsilon) = y\}.$$

Given a string s, $K_U(s)$ is an integer which, in some sense, is a measure of the information content of s: instead of sending s to somebody, we can send p from which s can be recovered as soon as this somebody has the machine U. K_U enjoys a lot of properties among which a kind of universality: this complexity is independent of the underlying Turing machine as we have the invariance theorem [9]:

Theorem 1. *If $U1$ and $U2$ are two universal Turing machines, there exists a constant c_{U1U2} such that for all string s: $|K_{U1}(s) - K_{U2}(s)| < c_{U1U2}$, where $K_{U1}(s)$ and $K_{U2}(s)$ denote the algorithmic complexity of s w.r.t. $U1$ and $U2$ respectively.*

This theorem guarantees that complexity values may only diverge by a constant c (e.g. the length of a compiler or a translation program) and for huge complexity strings, we can denote K without specifying the Turing machine U. It can also be shown that [9]:

Theorem 2. $\forall x, y, K(xy) = K(x) + K(y/x) + \mathcal{O}(1)$.

Unfortunately K has been proved as a non-computable function [9]. But in fact, K or an upper bound of K can be estimated in diverse ways that we investigate now.

Complexity Estimation. The first well known option available to estimate K is via lossless compression algorithm. For instance $bzip$ approximates better than $gzip$, and the PAQ family is still better than $bzip2$. Due to the invariance theorem, when the size of s is huge, using compression will provide a relatively stable approximation as the constant c in the theorem can be considered as negligible. It is obviously not the case when the size of s becomes small. When s is short, compression is not a valid option. On another side, the constant c can prevent for providing stable approximations of $K(s)$. Luckily, the works of [3,4,17] give means of providing sensible values for the complexity of short strings (i.e. less than 10bits). This job has been done by the Algorithmic Nature Group (https://algorithmicnature.org/). They have developed a tool OACC (http://www.complexitycalculator.com/) allowing to estimate the complexity of short strings. The authors derived their approach from a theorem from Levin [4,8] establishing the exact connection between $m(s)$ and $K(s)$, where $m(s)$ is a semi-measure known as the Universal Distribution defined as follows [18]: $m(s) = \Sigma_{p:U(p,\epsilon)=s} 2^{-|p|}$.

Theorem 3. *There exists a constant c depending only of the underlying Turing machine such that: $\forall s, |-log_2(m(s)) - K(s)| < c$.*

Rewriting the formula as $K(s) = -log_2(m(s)) + \mathcal{O}(1)$, shows that estimating K could also be done via estimating $m(s)$. Estimating $m(s)$ becomes realistic when s is short as we have to estimate the probability for s to be the output of a short program. Considering simple Turing machines as described in [17], over a Boolean alphabet $\{0, 1\}$ and a finite number n of states $\{1, ..., n\}$ plus a special Halt state denoted 0, there are exactly $(4n+2)^{2n}$ such Turing machines. Using clever optimizations [17], running these machines for $n = 4$ and $n = 5$ becomes realistic and provides an estimation of $m(s)$ and ultimately of $K(s)$. In the following, we denote $K'(s)$ this OACC estimation of $K(s)$.

Short Chains Complexity Estimation. Some properties are expected from a complexity calculator machinery to be in accordance with a cognitive process:

1. There is no way to distinguish strings of length 1 and it is absolutely clear that $K(0) = K(1)$ should hold whatever the considered universal Turing machine.
2. An important point is to be able to distinguish the 4 strings of length 2: $00, 11, 10, 01$ and we expect the following properties: $K(00) = K(11) < K(01) = K(10)$;
3. In terms of n bits strings, we expect $0\ldots0$ and $1\ldots1$ to be the simplest ones and to have the same complexity.

Observing the tables in [4], it appears that the properties above are satisfied, namely:

- Whatever the number of states of a 2-symbols Turing machine, $K'(0) = K'(1)$.
- Whatever the number of states of a 2-symbols Turing machine, $K'(00) = K'(11) = a$, $K'(01) = K'(10) = b$ and $a < b$.
- Whatever the number of states of a 2-symbols Turing machine, and for strings of length less than or equal to 10 (short strings) $K'(0\ldots0) = K'(1\ldots1) = a$ and a is the minimum value among the set of values.

Then the estimation of K via K' coming from the OACC estimator is a suitable candidate for our purpose. But before going further, we have first to check that OACC validate the above conditions. As we can check by examining Table 2 and column 4 of the final table in Sect. 6, these basic cognitive evidences are confirmed with the OACC tool. So we can start from OACC to check the properties required to validate the analogical hypothesis that we propose in the next section.

Table 2. Complexity of 1 bit and 2 bits chains with OACC

x	$K(x)$
0	3.5473880692782100
1	3.5473880692782100

$x_1 x_2$	$K(x_1 x_2)$
00	$5.4141012345247104 = a$
01	$5.4141040197301500 = b$
10	$5.4141040197301500 = b$
11	$5.4141012345247104 = a$

5 An Algorithmic Complexity View of Analogical Proportions

As described in our introduction, several attempts have been done to formalize analogy or analogical reasoning with mitigate success. In this paper, as it has been the case in the works of [2], we adopt a machine learning viewpoint. Our aim is to integrate analogical reasoning in the global landscape of predicting values from observable examples.

When stated in a machine learning perspective, the problem of analogical inference is as follows: for a given x_3, predict x_4 such that the *target* pair (x_3, x_4) is in the *same* relation that another given *source* pair (x_1, x_2) considered as an example. The pair (x_3, x_4) is the target pair which is partially known. In the case of classification where the 2nd element in a pair is the label, it amounts to predict the label of x_3 having only one classified example (x_1, x_2) at hand.

A functional view amounts to considering a hidden function f such that $x_2 = f(x_1)$ and we have to guess $x_4 = f(x_3)$. This functional view is the one developed in [2]: the problem of analogical inference strictly fits with a regression problem but with only one example. Ruling out any statistical models, this approach needs a brand new formalization that the authors extract from algorithmic complexity theory. Instead of trying to find regularities among a large set of observations (statistical approach), they consider the very meaning of each of the 3 observables $x_1, x_2 = f(x_1)$ and x_3. We start from this philosophy, but we depart from it as below:

- We focus on the Boolean case where the 3 objects under consideration are Boolean vectors. So we do not have to care about the change between the source domain representation and the target domain representation: these 2 domains are identical. The cost of this representation change is null in terms of algorithmic complexity.
- To be in line with the machine learning minimal assumption that there exists some unknown probability distribution P from which the data are drawn, we do not consider that x_2 is a (hidden) function of x_1. We just have a probability of observing x_2 having already observed x_1 which is more general than associating a fixed x_2 with every given x_1. It could be the case that for another x_2' we still have $x_1 : x_2' :: x_3 : x_4$.

As a consequence, we start from the following intuitions:

1. For $x_1 : x_2 :: x_3 : x_4$ to be accepted as a valid analogy, it is clear that the way we go from x_1 to x_2 should not be very different from the way we go from x_3 to x_4 (but it has not to be a functional link). We suggest to measure this expected proximity with the difference $|K(x_2/x_1) - K(x_4/x_3)|$. Considering $K(x_2/x_1)$ as the *difficulty* to build x_2 from x_1, the previous expression $|K(x_2/x_1) - K(x_4/x_3)|$, when small, tells us that it is not more difficult to build x_4 from x_3 than to build x_2 from x_1, and vice versa. This is what we call the atomic view of analogy. But this is obviously not enough.

2. In fact, the previous formula does not tell anything about the link between the pair (x_1, x_2) and the pair (x_3, x_4). For $x_1 : x_2 :: x_3 : x_4$ to be accepted as a valid analogy, the *difficulty* to apprehend the string $x_1 x_2$ from the string $x_3 x_4$ should be close to the difficulty to apprehend $x_3 x_4$ from $x_1 x_2$. We suggest to measure this expected proximity with the difference $|K(x_1 x_2 / x_3 x_4) - K(x_3 x_4 / x_1 x_2)|$. This difference is obviously symmetric and is linked to the symmetry of an analogy.

3. Above all, the global picture has to be "simple" i.e. telling that $x_1 : x_2 :: x_3 : x_4$ is a valid analogy should not be too disturbing, at least from a cognitive viewpoint. This means that the occurrence of the string $x_1 x_2 x_3 x_4$ in this order should be highly plausible. We suggest to measure this plausibility with $K(x_1 x_2 x_3 x_4)$ which is the size of the shortest program producing the binary string $x_1 x_2 x_3 x_4$ from a universal Turing machine.

Following the ideas of [2], we use the sum as aggregator operator and denote $k(x_1 x_2 x_3 x_4)$ the following formula measuring, in some sense, the quality of an analogy:

$$|K(x_2/x_1) - K(x_4/x_3)| + |K(x_1 x_2 / x_3 x_4) - K(x_3 x / x_1 x_2)| + K(x_1 x_2 x_3 x_4)$$

This leads us to postulate that the "best" x_4 we are looking for to build a valid analogy $x_1 : x_2 :: x_3 : x_4$ is the one minimizing this expression. So, we have: $x_4 = argmin_u k(x_1 x_2 x_3 u)$. Let us see if we can, at least from an empirical viewpoint, validate this model.

6   Validation in the Boolean Setting

As we are not in a position to prove something at this stage, let us just investigate now the empirical evidence for our formula. One point to start with is to check if this formula holds in the very basic Boolean case. Considering x_1, x_2, x_3, x_4 as Boolean values, we have to check how the 6 cases of valid analogical proportions actually behave w.r.t. the formula $k(x_1 x_2 x_3 x_4)$. Thus, we have to estimate formula $k(x_1 x_2 x_3 x_4)$ for every $x_1 x_2 x_3 x_4 \in \mathbb{B}^4$. The point is that our strings are very short: only 4 bits. So, as explained in Sect. 4, we have to rely on OACC instead of a compression estimation.

The Less Complex Analogical Chains. On top of that, we have to consider, not only pure Kolmogorov complexity K but also complexity w.r.t. a given string as in $K(x_3 x_4 / x_1 x_2)$. Generally, it is quite clear that $K(xy) \leq K(x) + K(y/x)$: roughly speaking, we can build a program whose output is xy by concatenating a program whose output is x to a program taking x as input and providing y as output. It is more difficult to get a more precise bound. Thanks to Theorem 2: $K(xy) = K(x) + K(y/x) + \mathcal{O}(1)$, which shows that we can approximate $K(y/x)$ with $K(xy) - K(x)$. As we now have all the tools needed to approximate formula k, it remains to use OACC to compute the estimation. The following table reports the results of this computation:

Using OACC http://www.complexitycalculator.com/

$$k(abcd) = |K(b/a) - K(d/c)| + |K(cd/ab) - K(ab/cd)| + K(abcd)$$

abcd	A K(b/a)	B K(d/c)	K(abcd)	C k(cd/ab)	D K(ab/cd)	|A-B|+|C-D| +K(abcd)
0000	1.8667131652	1.8667131652	11.2174683967	5.8033671621	5.8033671621	11.2174683967
1111	1.8667131652	1.8667131652	11.2174683967	5.8033671621	5.8033671621	11.2174683967
0101	1.8667159505	1.8667159505	11.7002793293	6.2861753096	6.2861753096	11.7002793293
1010	1.8667159505	1.8667159505	11.7002793293	6.2861753096	6.2861753096	11.7002793293
0001	1.8667131652	1.8667159505	11.5731249872	6.1590237527	6.3435937193	11.757697739
1110	1.8667131652	1.8667159505	11.5731249872	6.1590237527	6.3435937193	11.757697739
0111	1.8667159505	1.8667131652	11.5731249872	6.1590209675	6.3435965045	11.7577033094
1000	1.8667159505	1.8667131652	11.5731249872	6.1590209675	6.3435965045	11.7577033094
0011	1.8667131652	1.8667131652	11.8099819092	6.3958806747	6.3958806747	11.8099819092
1100	1.8667131652	1.8667131652	11.8099819092	6.3958806747	6.3958806747	11.8099819092
0100	1.8667159505	1.8667131652	11.757697739	6.3435937193	6.1590237527	11.9422704908
1011	1.8667159505	1.8667131652	11.757697739	6.3435937193	6.1590237527	11.9422704908
0010	1.8667131652	1.8667159505	11.757697739	6.3435965045	6.1590209675	11.9422760613
1101	1.8667131652	1.8667159505	11.757697739	6.3435965045	6.1590209675	11.9422760613
1001	1.8667159505	1.8667159505	12.0548412692	6.6407372495	6.6407372495	12.0548412692
0110	1.8667159505	1.8667159505	12.0548412692	6.6407372495	6.6407372495	12.0548412692

As can be seen for the 6 patterns of the model Ω_0 of analogical proportion, the unique solution of equation $a : b :: c : x$ always corresponds to a string $abcx$ that minimizes expression k wrt the other option $abc\overline{x}$ (where $\overline{x} = \neg x$), e.g. $k(1111) < k(1110)$. Besides 0101 is simpler than 0110 despite the fact that in the second case there is also an underlying function such that $x_2 = f(x_1)$ and $x_4 = f(x_3)$: the negation. Note that 0110 and 1001 exhibit the highest complexity as estimated by OACC. It eliminates Kl. As there is no known convergence result regarding K and that we cannot estimate the constant in the formula $K(s) = -log_2(m(s)) + \mathcal{O}(1)$, these experiences should only be considered as adding a bit of credibility to the smallest model.

7   Conclusion

We have given a complete description of the Boolean models of analogy. To choose the most relevant one among the possible 8 models beyond the minimality argument, we have proposed a complexity-based definition for Boolean analogical proportion. Using a set of calculations with OACC, the tool developed by the Algorithmic Nature Group (https://algorithmicnature.org/), we have checked that the truth table of the Boolean analogy fits with the fact that the corresponding combinations minimize the given complexity formula. It remains to consider the formula in a more general setting than the Boolean one. This would in particular allow to establish a link between transfer learning and Kolmogorov complexity. Another point would be of interest: to be able to solve the minimization problem associated to the formula. Doing so would be to solve the analogical equation $a : b :: c : x$. This might be the basis of a constructive process.

References

1. Bennett, C.H., Gács, P., Li, M., Vitányi, P., Zurek, W.H.: Information distance (2010). CoRR abs/1006.3520
2. Cornuéjols, A.: Analogy as minimization of description length. In: Nakhaeizadeh, G., Taylor, C. (eds.) Machine Learning and Statistics: The Interface, pp. 321–336. Wiley, Chichester (1996)
3. Delahaye, J.P., Zenil, H.: On the Kolmogorov-Chaitin complexity for short sequences (2007). CoRR abs/0704.1043
4. Delahaye, J.P., Zenil, H.: Numerical evaluation of algorithmic complexity for short strings: a glance into the innermost structure of randomness. Appl. Math. Comput. **219**(1), 63–77 (2012)
5. Goel, S., Bush, S.F.: Kolmogorov complexity estimates for detection of viruses in biologically inspired security systems: a comparison with traditional approaches. Complexity **9**(2), 54–73 (2003)
6. Klein, S.: Whorf transforms and a computer model for propositional/appositional reasoning. In: Proceedings of the Applied Mathematics colloquium, University of Bielefeld, West Germany (1977)
7. Lepage, Y.: Analogy and formal languages. Electr. Notes Theor. Comp. Sci. **53**, 180–191 (2002). Moss, L.S., Oehrle, R.T. (eds.) Proceedings of the Joint Meeting of the 6th Conference on Formal Grammar and the 7th Conference on Mathematics of Language
8. Levin, L.: Laws of information conservation (non-growth) and aspects of the foundation of probability theory. Probl. Inf. Transm. **10**, 206–210 (1974)
9. Li, M., Vitanyi, P.: An Introduction to Kolmogorov Complexity and Its Applications, 3rd edn. Springer, New York (2008)
10. Miclet, L., Delhay, A.: Relation d'analogie et distance sur un alphabet défini par des traits. Technical Report 1632, IRISA, July 2004
11. Miclet, L., Delhay, A.: Analogical dissimilarity: definition, algorithms and first experiments in machine learning. Technical Report RR-5694, INRIA, July 2005
12. Miclet, L., Prade, H.: Logical definition of analogical proportion and its fuzzy extensions. In: Annual Meeting of the North American Fuzzy Information Processing Society (NAFIPS), New-York, pp. 1–6. IEEE (2008)
13. Miclet, L., Prade, H.: Handling analogical proportions in classical logic and fuzzy logics settings. In: Sossai, C., Chemello, G. (eds.) ECSQARU 2009. LNCS (LNAI), vol. 5590, pp. 638–650. Springer, Heidelberg (2009). doi:10.1007/978-3-642-02906-6_55
14. Prade, H., Richard, G.: From analogical proportion to logical proportions. Logica Universalis **7**(4), 441–505 (2013)
15. Prade, H., Richard, G.: Homogenous and heterogeneous logical proportions. IfCoLog J. Logics Appl. **1**(1), 1–51 (2014)
16. Shannon, C.E.: A mathematical theory of communication. Bell Syst. Tech. J. **27**(3), 379–423 (1948)
17. Soler-Toscano, F., Zenil, H., Delahaye, J.P., Gauvrit, N.: Correspondence and independence of numerical evaluations of algorithmic information measures. Computability **2**, 125–140 (2013)
18. Solomonoff, R.J.: A formal theory of inductive inference. Part i and ii. Inf. Control **7**(1), 1–22 and 224–254 (1964)
19. Stroppa, N., Yvon, F.: Du quatrième de proportion comme principe inductif: une proposition et son application à l'apprentissage de la morphologie. Traitement Automatique des Langues **47**(2), 1–27 (2006)

Argumentation

Evaluation of Arguments
in Weighted Bipolar Graphs

Leila Amgoud$^{(\boxtimes)}$ and Jonathan Ben-Naim

IRIT – CNRS, 118, route de Narbonne, 31062 Toulouse, France
{amgoud,bennaim}@irit.fr

Abstract. The paper tackled the issue of arguments evaluation in
weighted bipolar argumentation graphs (i.e., graphs whose arguments
have basic strengths, and may be both supported and attacked). We
introduce axioms that an evaluation method (or semantics) could satisfy. Such axioms are very useful for judging and comparing semantics.
We then analyze existing semantics on the basis of our axioms, and finally
propose a new semantics for the class of acyclic graphs.

1 Introduction

Argumentation is a form of common-sense reasoning consisting of the justification of claims by arguments. The latter have generally basic strengths, and
may be attacked and/or supported by other arguments, leading to the so-called
bipolar argumentation graphs (BAGs). Several methods, called *semantics*, were
proposed in the literature for the evaluation of arguments in such settings. They
can be partitioned into two main families: *extension semantics* [1–5], and *gradual
semantics* [6–8]. The former extend Dung's semantics [9] for accounting for supports, and look for acceptable sets of arguments (called extensions). The latter
focus on the evaluation of individual arguments.

This paper extends our previous works on axiomatic foundations of semantics for unipolar graphs (support graphs [10] and attack graphs [11]). It defines
axioms (i.e. properties) that a semantics should satisfy in a bipolar setting.
Such axioms are very useful for judging and understanding the underpinnings
of semantics, and also for comparing semantics of the same family, and those of
different families. Some of the proposed axioms are simple combinations of those
proposed in [10,11]. Others are new and show how support and attack should be
aggregated. The second contribution of the paper consists of analyzing existing
semantics against the axioms. The main conclusion is that extension semantics do not harness the potential of support relation. Indeed, when the attack
relation is empty, the existing semantics declare all (supported, non-supported)
arguments of a graph as equally accepted. Gradual semantics take into account
supporters in this particular case, however they violate some key axioms. The
third contribution of the paper is the definition of a novel gradual semantics for
the sub-class of acyclic bipolar graphs. We show that it satisfies all the proposed
axioms. Furthermore, it avoids the *big jump* problem that impedes the relevance
of existing gradual semantics for practical applications, like dialogue.

© Springer International Publishing AG 2017
A. Antonucci et al. (Eds.): ECSQARU 2017, LNAI 10369, pp. 25–35, 2017.
DOI: 10.1007/978-3-319-61581-3_3

The paper is structured as follows: Sect. 2 introduces basic notions, Sect. 3 presents our list of axioms as well as some properties, Sect. 4 analyses existing semantics, and Sect. 5 introduces our new semantics and discusses its properties.

2   Main Concepts

This section introduces the main concepts of the paper. Let us begin with weightings:

Definition 1 (Weighting). *A weighting on a set X is a function from X to $[0, 1]$.*

Next, we introduce weighted bipolar argumentation graphs (BAGs).

Definition 2 (BAG). *A BAG is a quadruple $\mathbf{A} = \langle \mathcal{A}, w, \mathcal{R}, \mathcal{S} \rangle$, where $\mathcal{A}$ is a finite set of arguments, w a weighting on $\mathcal{A}$, $\mathcal{R} \subseteq \mathcal{A} \times \mathcal{A}$, and $\mathcal{S} \subseteq \mathcal{A} \times \mathcal{A}$.*

Given two arguments a and b, $a\mathcal{R}b$ (resp. $a\mathcal{S}b$) means a *attacks* (resp. *supports*) b, and $w(a)$ is the *intrinsic strength* of a. The latter may be the certainty degree of the argument's premises, trustworthiness of the argument's source, ...

We turn to the core concept of the paper. A semantics is a function transforming any weighted bipolar argumentation graph into a weighting on the set of arguments. The weight of an argument given by a semantics represents its *overall strength* or *acceptability degree*. It is obtained from the aggregation of its intrinsic strength and the overall strengths of its attackers and supporters. Arguments that get value 1 are *extremely strong* whilst those that get value 0 are *worthless*.

Definition 3 (Semantics). *A semantics is a function $\mathbf{S}$ transforming any BAG $\mathbf{A} = \langle \mathcal{A}, w, \mathcal{R}, \mathcal{S} \rangle$ into a weighting f on $\mathcal{A}$. Let $a \in \mathcal{A}$, we denote by $\mathrm{Deg}_{\mathbf{A}}^{\mathbf{S}}(a)$ the acceptability degree of a, i.e., $\mathrm{Deg}_{\mathbf{A}}^{\mathbf{S}}(a) = f(a)$.*

Let us recall the notion of *isomorphism* between graphs.

Definition 4 (Isomorphism). *Let $\mathbf{A} = \langle \mathcal{A}, w, \mathcal{R}, \mathcal{S} \rangle$ and $\mathbf{A}' = \langle \mathcal{A}', w', \mathcal{R}', \mathcal{S}' \rangle$ be two BAGs. An isomorphism from $\mathbf{A}$ to $\mathbf{A}'$ is a bijective function f from $\mathcal{A}$ to $\mathcal{A}'$ such that: (i) $\forall a \in \mathcal{A}$, $w(a) = w'(f(a))$, (ii) $\forall a, b \in \mathcal{A}$, $a\mathcal{R}b$ iff $f(a)\mathcal{R}'f(b)$, (iii) $\forall a, b \in \mathcal{A}$, $a\mathcal{S}b$ iff $f(a)\mathcal{S}'f(b)$.*

Notations: Let $\mathbf{A} = \langle \mathcal{A}, w, \mathcal{R}, \mathcal{S} \rangle$ be a BAG and $a \in \mathcal{A}$. We denote by $\mathrm{Att}_{\mathbf{A}}(a)$ the set of all attackers of a in $\mathbf{A}$ (i.e., $\mathrm{Att}_{\mathbf{A}}(a) = \{b \in \mathcal{A} \mid b\mathcal{R}a\}$), and by $\mathrm{sAtt}_{\mathbf{A}}(a)$ the set of all *significant attackers* of a, i.e., attackers x of a such that $\mathrm{Deg}_{\mathbf{A}}^{\mathbf{S}}(x) \neq 0$. Similarly, we denote by $\mathrm{Supp}_{\mathbf{A}}(a)$ the set of all supporters of a (i.e., $\mathrm{Supp}_{\mathbf{A}}(a) = \{b \in \mathcal{A} \mid b\mathcal{S}a\}$) and by $\mathrm{sSupp}_{\mathbf{A}}(a)$ the *significant supporters* of a, i.e., supporters x such that $\mathrm{Deg}_{\mathbf{A}}^{\mathbf{S}}(x) \neq 0$.

Let $\mathbf{A}' = \langle \mathcal{A}', w', \mathcal{R}', \mathcal{S}' \rangle$ be another BAG such that $\mathcal{A} \cap \mathcal{A}' = \emptyset$. We denote by $\mathbf{A} \oplus \mathbf{A}'$ the BAG $\langle \mathcal{A}'', w'', \mathcal{R}'', \mathcal{S}'' \rangle$ such that $\mathcal{A}'' = \mathcal{A} \cup \mathcal{A}'$, $\mathcal{R}'' = \mathcal{R} \cup \mathcal{R}'$, $\mathcal{S}'' = \mathcal{S} \cup \mathcal{S}'$, and $\forall x \in \mathcal{A}''$, the following holds: $w''(x) = w(x)$, if $x \in \mathcal{A}$; $w''(x) = w'(x)$, if $x \in \mathcal{A}'$.

3 Axioms for Acceptability Semantics

In what follows, we propose axioms that shed light on foundational principles behind semantics. In other words, properties that help us to better understand the underpinnings of semantics, and that facilitate their comparisons. The first nine axioms are simple *combinations* of axioms proposed for graphs with only one type of interactions (support in [10], attack in [11]). The three last axioms are new and show how the overall strengths of supporters and attackers of an argument should be aggregated.

The first very basic axiom, Anonymity, states that the degree of an argument is independent of its identity. It combines the two Anonymity axioms from [10,11].

Axiom 1 (Anonymity). *A semantics* S *satisfies* anonymity *iff, for any two BAGs* $\mathbf{A} = \langle \mathcal{A}, w, \mathcal{R}, \mathcal{S} \rangle$ *and* $\mathbf{A}' = \langle \mathcal{A}', w', \mathcal{R}', \mathcal{S}' \rangle$, *for any isomorphism* f *from* $\mathbf{A}$ *to* $\mathbf{A}'$, *the following property holds:* $\forall a \in \mathcal{A}, \mathrm{Deg}_{\mathbf{A}}^{\mathrm{S}}(a) = \mathrm{Deg}_{\mathbf{A}'}^{\mathrm{S}}(f(a))$.

Bi-variate independence axiom states the following: the acceptability degree of an argument a should be independent of any argument b that is not connected to it (i.e., there is no path from b to a, ignoring the direction of the edges). This axiom combines the two independence axioms from [10,11].

Axiom 2 (Bi-variate Independence). *A semantics* S *satisfies* bi-variate independence *iff, for any two BAGs* $\mathbf{A} = \langle \mathcal{A}, w, \mathcal{R}, \mathcal{S} \rangle$ *and* $\mathbf{A}' = \langle \mathcal{A}', w', \mathcal{R}', \mathcal{S}' \rangle$ *such that* $\mathcal{A} \cap \mathcal{A}' = \emptyset$, *the following property holds:* $\forall a \in \mathcal{A}, \mathrm{Deg}_{\mathbf{A}}^{\mathrm{S}}(a) = \mathrm{Deg}_{\mathbf{A} \oplus \mathbf{A}'}^{\mathrm{S}}(a)$.

Bi-variate directionality axiom combines Non-Dilution from [10] and Circumscription from [11]. It states that the overall strength of an argument should depend only on its incoming arrows, and thus not on the arguments it itself attacks or supports.

Axiom 3 (Bi-variate Directionality). *A semantics* S *satisfies* bi-variate directionality *iff, for any two BAGs* $\mathbf{A} = \langle \mathcal{A}, w, \mathcal{R}, \mathcal{S} \rangle$, $\mathbf{A}' = \langle \mathcal{A}', w', \mathcal{R}', \mathcal{S}' \rangle$ *such that* $\mathcal{A} = \mathcal{A}'$, $\mathcal{R} \subseteq \mathcal{R}'$, *and* $\mathcal{S} \subseteq \mathcal{S}'$, *the following holds: for all* $a, b, x \in \mathcal{A}$, *if* $\mathcal{R}' \cup \mathcal{S}' = \mathcal{R} \cup \mathcal{S} \cup \{(a,b)\}$ *and there is no path from* b *to* x, *then* $\mathrm{Deg}_{\mathbf{A}}^{\mathrm{S}}(x) = \mathrm{Deg}_{\mathbf{A}'}^{\mathrm{S}}(x)$. *Note that a path can mix attack and support relations, but the edges must always be directed from* b *to* x.

Bi-variate Equivalence axiom ensures that the overall strength of an argument depends *only* on the overall strengths of its direct attackers and supporters. It combines the two equivalence axioms from [10,11].

Axiom 4 (Bi-variate Equivalence). *A semantics* S *satisfies* bi-variate equivalence *iff, for any BAG* $\mathbf{A} = \langle \mathcal{A}, w, \mathcal{R}, \mathcal{S} \rangle$, *for all* $a, b \in \mathcal{A}$, *if:*

- $w(a) = w(b)$,
- *there exists a bijective function* f *from* $\mathrm{Att}_{\mathbf{A}}(a)$ *to* $\mathrm{Att}_{\mathbf{A}}(b)$ *such that* $\forall x \in \mathrm{Att}_{\mathbf{A}}(a)$, $\mathrm{Deg}_{\mathbf{A}}^{\mathrm{S}}(x) = \mathrm{Deg}_{\mathbf{A}}^{\mathrm{S}}(f(x))$, *and*

– *there exists a bijective function f' from $\mathrm{Supp}_A(a)$ to $\mathrm{Supp}_A(b)$ such that $\forall x \in$ $\mathrm{Supp}_A(a)$, $\mathrm{Deg}_A^S(x) = \mathrm{Deg}_A^S(f(x))$,*

then $\mathrm{Deg}_A^S(a) = \mathrm{Deg}_A^S(b)$.

Stability axiom combines Minimality [10] and Maximality [11] axioms. It says the following: if an argument is neither attacked nor supported, its overall strength should be equal to its intrinsic strength.

Axiom 5 (Stability). *A semantics* **S** *satisfies* stability *iff, for any BAG* **A** $=$ $\langle A, w, R, S \rangle$, *for any argument* $a \in A$, *if* $\mathrm{Att}_A(a) = \mathrm{Supp}_A(a) = \emptyset$, *then* $\mathrm{Deg}_A^S(a) = w(a)$.

Neutrality axiom generalizes Dummy axiom [10] and Neutrality one from [11]. It states that worthless attackers or supporters have no effect.

Axiom 6 (Neutrality). *A semantics* **S** *satisfies* neutrality *iff, for any BAG* **A** $= \langle A, w, R, S \rangle$, $\forall a, b, x \in A$, *if:*

– $w(a) = w(b)$,
– $\mathrm{Att}_A(a) \subseteq \mathrm{Att}_A(b)$,
– $\mathrm{Supp}_A(a) \subseteq \mathrm{Supp}_A(b)$,
– $\mathrm{Att}_A(b) \cup \mathrm{Supp}_A(b) = \mathrm{Att}_A(a) \cup \mathrm{Supp}_A(a) \cup \{x\}$, *and* $\mathrm{Deg}_A^S(x) = 0$,

then $\mathrm{Deg}_A^S(a) = \mathrm{Deg}_A^S(b)$.

Bi-variate Monotony states the following: if an argument a is equally or less attacked than an argument b, and equally or more supported than b, then a should be equally strong or stronger than b. This axiom generalizes 4 axioms from the literature (Monotony and Counting [10] for supports, and the same axioms from [11] for attacks).

Axiom 7 (Bi-variate Monotony). *A semantics* **S** *satisfies* bi-variate monotony *iff, for any BAG* **A** $= \langle A, w, R, S \rangle$, *for all* $a, b \in A$ *such that:*

– $w(a) = w(b) > 0$,
– $\mathrm{Att}_A(a) \subseteq \mathrm{Att}_A(b)$,
– $\mathrm{Supp}_A(b) \subseteq \mathrm{Supp}_A(a)$,

the following holds:

– $\mathrm{Deg}_A^S(a) \geq \mathrm{Deg}_A^S(b)$,   *(Monotony)*
– *if* $(\mathrm{Deg}_A^S(a) > 0$ *or* $\mathrm{Deg}_A^S(b) < 1)$ *and* $(\mathrm{sAtt}_A(a) \subset \mathrm{sAtt}_A(b)$, *or* $\mathrm{sSupp}_A(b) \subset$ $\mathrm{sSupp}_A(a))$, *then* $\mathrm{Deg}_A^S(a) > \mathrm{Deg}_A^S(b)$.   *(Strict Monotony)*

The next axiom combines the Reinforcement axioms of [10,11]. It states that any argument becomes stronger if the quality of its attackers is reduced and the quality of its supporters is increased.

Axiom 8 (Bi-variate Reinforcement). *A semantics* **S** *satisfies* bi-variate reinforcement *iff, for any BAG* **A** $= \langle A, w, R, S \rangle$, *for all* $C, C' \subseteq A$, *for all* $a, b \in A$, *for all* $x, x', y, y' \in A \backslash (C \cup C')$ *such that*

- $w(a) = w(b) > 0$,
- $\text{Deg}_A^S(x) \leq \text{Deg}_A^S(y)$,
- $\text{Deg}_A^S(x') \geq \text{Deg}_A^S(y')$,
- $\text{Att}_A(a) = C \cup \{x\}$,
- $\text{Att}_A(b) = C \cup \{y\}$,
- $\text{Supp}_A(a) = C' \cup \{x'\}$,
- $\text{Supp}_A(b) = C' \cup \{y'\}$,

the following holds:

- $\text{Deg}_A^S(a) \geq \text{Deg}_A^S(b)$, $\hspace{3cm}$ *(Reinforcement)*
- *if* $(\text{Deg}_A^S(a) > 0$ *or* $\text{Deg}_A^S(b) < 1)$ *and* $(\text{Deg}_A^S(x) < \text{Deg}_A^S(y)$, *or* $\text{Deg}_A^S(x') > \text{Deg}_A^S(y'))$, *then* $\text{Deg}_A^S(a) > \text{Deg}_A^S(b)$. $\hspace{1cm}$ **(Strict Reinforcement)**

Our next axiom combines Imperfection axiom from [10] with Resilience axiom from [11]. Imperfection states that an argument whose basic strength is less than 1 cannot be fully rehabilitated by supports. In other words, it cannot get an acceptability degree 1 due to supports. This axiom prevents irrational behaviors, like fully accepting fallacious arguments that are supported. Below, the argument A remains fallacious even if it is supported by B.

A: Tweety needs fuel, since it flies like planes.
B: Indeed, Tweety flies. It is a bird.

Resilience in [11] states that an argument whose basic strength is positive cannot be completely destroyed by attacks. Assume that B is attacked by the argument C below. Despite the attack, the argument B is still reasonable.

C: Tweety does not fly since it is a penguin

Axiom 9 (Resilience). *A semantics* **S** *satisfies* resilience *iff, for any BAG* **A** $= \langle \mathcal{A}, w, \mathcal{R}, \mathcal{S} \rangle$, *for all* $a \in \mathcal{A}$, *if* $0 < w(a) < 1$, *then* $0 < \text{Deg}(a) < 1$.

The next three axioms are new and answer the same question: how the overall strengths of attackers and supporters of an argument are aggregated? To answer this question, it is important to specify which of the two types of interactions is more important. In this paper, we consider both relations as equally important. Hence, Franklin axiom states that an attacker and a supporter of equal strength should counter-balance each other. Thus, neither attacks nor supports will have impact on the argument.

Axiom 10 (Franklin). *A semantics* **S** *satisfies* franklin *iff, for any BAG* **A** $= \langle \mathcal{A}, w, \mathcal{R}, \mathcal{S} \rangle$, *for all* $a, b, x, y \in \mathcal{A}$, *if*

- $w(b) = w(a)$,
- $\text{Deg}_A^S(x) = \text{Deg}_A^S(y)$
- $\text{Att}_A(a) = \text{Att}_A(b) \cup \{x\}$,
- $\text{Supp}_A(a) = \text{Supp}_A(b) \cup \{y\}$,

then $\text{Deg}_A^S(a) = \text{Deg}_A^S(b)$.

We show that attacks and supports of equal strengths eliminate each others.

Proposition 1. *Let* **S** *be a semantics that satisfies Bi-variate Independence, Bi-variate Directionality, Stability and Franklin. For any BAG* $\mathbf{A} = \langle \mathcal{A}, w, \mathcal{R}, \mathcal{S} \rangle$, *for all* $a \in \mathcal{A}$, *if there exists a bijective function* f *from* $\mathrm{Att}_\mathbf{A}(a)$ *to* $\mathrm{Supp}_\mathbf{A}(a)$ *such that* $\forall x \in \mathrm{Att}(a)$, $\mathrm{Deg}_\mathbf{A}^\mathbf{S}(x) = \mathrm{Deg}_\mathbf{A}^\mathbf{S}(f(x))$, *then* $\mathrm{Deg}_\mathbf{A}^\mathbf{S}(a) = w(a)$.

Weakening states that if attackers overcome supporters, the argument should loose weight. The idea is that supports are not sufficient for counter-balancing attacks. Please note that this does not means that supports will not have an impact on the overall strength of an argument. They may mitigate the global loss due to attacks.

Axiom 11 (Weakening). *A semantics* **S** *satisfies* weakening *iff, for any BAG* $\mathbf{A} = \langle \mathcal{A}, w, \mathcal{R}, \mathcal{S} \rangle$, *for all* $a \in \mathcal{A}$, *if* $w(a) > 0$ *and there exists an injective function* f *from* $\mathrm{Supp}_\mathbf{A}(a)$ *to* $\mathrm{Att}_\mathbf{A}(a)$ *such that:*

- $\forall x \in \mathrm{Supp}_\mathbf{A}(a)$, $\mathrm{Deg}(x) \leq \mathrm{Deg}(f(x))$; *and*
- $\mathrm{sAtt}_\mathbf{A}(a) \backslash \{f(x) \mid x \in \mathrm{Supp}_\mathbf{A}(a)\} \neq \emptyset$ *or* $\exists x \in \mathrm{Supp}_\mathbf{A}(a)$ *s.t* $\mathrm{Deg}(x) < \mathrm{Deg}(f(x))$,

then $\mathrm{Deg}(a) < w(a)$.

Strengthening states that if supporters overcome attackers, the argument should gain weight. Indeed, attacks are not sufficient for counter-balancing supports, however, they may mitigate the global gain due to supports.

Axiom 12 (Strengthening). *A semantics* **S** *satisfies* strengthening *iff, for any BAG* $\mathbf{A} = \langle \mathcal{A}, w, \mathcal{R}, \mathcal{S} \rangle$, *for all* $a \in \mathcal{A}$, *if* $w(a) < 1$ *and there exists an injective function* f *from* $\mathrm{Att}_\mathbf{A}(a)$ *to* $\mathrm{Supp}_\mathbf{A}(a)$ *such that:*

- $\forall x \in \mathrm{Att}_\mathbf{A}(a)$, $\mathrm{Deg}(x) \leq \mathrm{Deg}(f(x))$; *and*
- $\mathrm{sSupp}_\mathbf{A}(a) \backslash \{f(x) \mid x \in \mathrm{Att}_\mathbf{A}(a)\} \neq \emptyset$ *or* $\exists x \in \mathrm{Att}_\mathbf{A}(a)$ *s.t.* $\mathrm{Deg}(x) < \mathrm{Deg}(f(x))$,

then $\mathrm{Deg}(a) > w(a)$.

It is worth mentioning that weakening and strengthening generalize their corresponding axioms in [10,11]. Indeed, when the support relation is empty, bipolar version of weakening coincides with weakening axiom in [11]. However, it handles additional cases when supports exist. Similarly, when the attack relation is empty, the axiom coincides with strengthening axiom in [10].

Almost all axioms are independent, i.e., they do not follow from others. A notable exception is Bivariate Monotony which follows from five axioms.

Proposition 2. *If a semantics satisfies Bi-variate Independence, Bi-variate Directionality, Stability, Neutrality and Bi-variate Reinforcement, then it satisfies Bivariate Monotony.*

All axioms are compatible, i.e., they can be satisfied all together by a semantics.

Proposition 3. *All the axioms are compatible.*

4   Formal Analysis of Existing Semantics

There are several proposals in the literature for the evaluation of arguments in bipolar argumentation graphs. They can be partitioned into two families: *extension* semantics [1–5] and *gradual* semantics [6–8].

Extension semantics extend Dung's semantics [9] for accounting for supports between arguments. They take as input an argumentation graph $\langle \mathcal{A}, w, \mathcal{R}, \mathcal{S} \rangle$ whose arguments have all the *same* basic strength, and return sets of arguments, called extensions. From the extensions, a three-valued qualitative degree is assigned to every argument. Indeed, an argument is *accepted* if it belongs to all extensions, *undecided* (or credulously accepted) if it belongs to some but not all extensions, and *rejected* if it does not belong to any extension. When the support relation is empty, the semantics proposed in [1–5] coincide with Dung's ones. Thus, they violate the axioms that are violated by Dung's semantics (see [11] for a detailed analysis of Dung's semantics). For instance, stable semantics violates Independence, Equivalence, Stability, Resilience, and strict monotony. When the attack relation is empty, the approaches from [1,2,4] return a single extension, which contains all the arguments of the BAG at hand. Thus, all arguments are equally accepted. This shows that the support relation does not play any role, and a supported argument is as acceptable as a non-supported one. To say it differently, these approaches violate strengthening axiom which captures the role of supports. The approaches developed in [3,5] return a single extension when the attack relation is empty. This extension coincides with the set of arguments when there are no cycles in the BAG. Thus, they also violate strengthening and the support relation may not be fully exploited in the evaluation of arguments.

The second family of gradual semantics was introduced for the first time in [6]. In their paper, the authors presented some properties that such semantics should satisfy (like a particular case of strengthening). However, they did not define concrete semantics. To the best of our knowledge, the first gradual semantics is QuAD, introduced in [7], for evaluating arguments in *acyclic* graphs. This semantics assigns a numerical value to every argument on the basis of its intrinsic strength, and the overall strengths of its attackers and supporters. It evaluates separately the supporters and the attackers before aggregating them. Due to lack of space, we do not provide the formal definitions.

Proposition 4. *QuAD satisfies Anonymity, Bi-variate Independence, Bi-variate Directionality, Bi-variate Equivalence, Stability, Neutrality, Monotony, Reinforcement.*

QuAD violates Strict Monotony, Strict Reinforcement, Resilience, Franklin, Weakening, and Strengthening.

As a consequence of violating Weakening and Strengthening, QuAD may behave irrationally. Consider a BAG where $\mathcal{A} = \{a, b_1, b_2, b_3\}$, $w(b_1) = w(b_2) = 0.8$, $w(b_3) = 0.9$, $\mathcal{R} = \{(b_2, a), (b_3, a)\}$, and $\mathcal{S} = \{(b_1, a)\}$. Thus, a has an attacker and a supporter of equal strengths, and an additional attacker b_3. Note that if $w(a) = 0.2$, then $\text{Deg}_{\text{A}}^{\text{S}}(a) = 0.422$ meaning that the single supporter is privileged to the two attackers. However, if $w(a) = 0.7$, $\text{Deg}_{\text{A}}^{\text{S}}(a) = 0.477$ meaning that attacks are privileged to support. More generally, we can show that if $w(a) \geq 0.5$, then $\text{Deg}_{\text{A}}^{\text{S}}(a) < w(a)$, else $\text{Deg}_{\text{A}}^{\text{S}}(a) > w(a)$. Hence, choosing which of support and attack should take precedence depends on the intrinsic strength of an argument.

QuAD was recently extended to DF-QuAD in [8]. The new semantics focuses also on *acyclic graphs*. Unlike QuAD, it uses the same function for aggregating supporters and attackers separately. It satisfies Franklin axiom, thus it treats equally attacks and supports. It violates Strengthening and Weakening in presence of attackers/supporters of degree 1. However, the semantics avoids the irrational behavior of QuAD.

Proposition 5. *DF-QuAD satisfies Anonymity, Bi-variate Independence, Bi-variate Directionality, Bi-variate Equivalence, Stability, Neutrality, Monotony, Reinforcement, and Franklin. DF-QuAD violates Strict Monotony, Strict Reinforcement, Resilience, Weakening, and Strengthening.*

Both semantics (QuAD and DF-QuAD) suffer from a *big jump* problem. Let us illustrate the problem with the BAG depicted in Fig. 1. Note that the argument i has a very low basic strength ($w(i) = 0.1$). This argument is supported by the very strong argument j. According to QuAD and DF-QuAD, $\text{Deg}_{\text{A}}^{\text{S}}(i) = 0.991$. Thus, the value of i makes a big jump from 0.1 to 0.991. The argument i became even stronger than its supporter j. There are two issues with such jump: First, the gain is enormous and not reasonable. Assume that i is the argument "Tweety needs fuel, since it flies like planes". It is hard to accept i even when supported. The supporter may increase slightly the strength of the argument but does not correct the wrong premises of the argument. Second, such jump impedes the discrimination between different cases where $w(i) > 0.1$ since whatever the value of $w(i)$, the overall strength is almost 1.

5   Euler-Based Graded Semantics

As shown in the previous sections, no existing semantics satisfies all our 12 axioms together. The goal of the present section is to handle this issue. More precisely, we construct a new semantics satisfying all axioms, but at the cost of a certain degree of coverage. Indeed, we only consider a subclass of BAGs: acyclic non-maximal BAGs.

Definition 5 (BAG properties).   *A BAG* $\mathbf{A} = \langle \mathcal{A}, w, \mathcal{R}, \mathcal{S} \rangle$ *is acyclic iff the following holds: for any non-empty finite sequence* $\mathbf{a} = \langle a_1, a_2, \ldots, a_n \rangle$ *of elements of* $\mathcal{A}$, *if* $\forall i \in \{1, 2, \ldots, n-1\}$, $\langle a_i, a_{i+1} \rangle \in \mathcal{R} \cup \mathcal{S}$, *then* $\langle a_n, a_1 \rangle \notin \mathcal{R} \cup \mathcal{S}$. *Next,* $\mathbf{A}$ *is non-maximal iff* $\forall a \in \mathcal{A}$, $w(a) < 1$.

Without loss of generality, the basic strengths of arguments are less than 1. Note that few arguments are intrinsically perfect. The probability of false information, exceptions, etc., is rarely 0. In contrast, the loss of cyclic BAGs is important. But, we consider that the class of all acyclic non-maximal BAGs is expressive enough to deserve attention.

Definition 6 (Restricted semantics). *A restricted semantics is a function* **S** *transforming any acyclic non-maximal BAG* $\mathbf{A} = \langle \mathcal{A}, w, \mathcal{R}, \mathcal{S} \rangle$ *into a weighting on* $\mathcal{A}$.

All notations and axioms for semantics are straightforwardly adapted to restricted semantics. Before presenting our semantics, we need to introduce a relation between arguments based on the longest paths to reach them (mixing support and attack arrows).

Definition 7 (Well-founded relation). *Let* $\mathbf{A} = \langle \mathcal{A}, w, \mathcal{R}, \mathcal{S} \rangle$ *be an acyclic BAG and* $a \in \mathcal{A}$. *A path to* a *in* $\mathbf{A}$ *is a non-empty finite sequence* $\mathbf{a} = \langle a_1, a_2, \ldots, a_n \rangle$ *of elements of* $\mathcal{A}$ *such that* $a_n = a$ *and* $\forall i \in \{1, 2, \ldots, n-1\}$, $\langle a_i, a_{i+1} \rangle \in \mathcal{R} \cup \mathcal{S}$. *We denote by* $\mathtt{Rel}(\mathbf{A})$ *the well-founded binary relation* $\prec$ *on* $\mathcal{A}$ *such that* $\forall x, y \in \mathcal{A}$, $x \prec y$ *iff* $\max\{n \mid$ *there exists a path to* x *of length* $n\} < \max\{n \mid$ *there exists a path to* y *of length* $n\}$. *Since* $\mathbf{A}$ *is acyclic, those maximum lengths are well-defined, so is* $\mathtt{Rel}(\mathbf{A})$.

We are ready to define the *Euler-based restricted semantics*. The general idea is to take into account supporters and attackers in an exponent E of **e** (the Euler's number). More precisely, the stronger or more-numerous the supporters, the greater and more-likely-positive that exponent. Obviously, the inverse is true with the attackers. Then, the overall strength of an argument a is naturally defined as $w(a)\mathbf{e}^E$. Finally, we need certain tweakings (including a double polarity reversal) to make our function a restricted semantics in the first place, and to have it satisfy all our axioms. More formally:

Definition 8 (Euler-based restricted semantics). *We denote by* Ebs *the restricted semantics such that for any acyclic non-maximal BAG* $\mathbf{A} = \langle \mathcal{A}, w, \mathcal{R}, \mathcal{S} \rangle$, $\mathtt{Ebs}(\mathbf{A})$ *is the weighting* f *on* $\mathcal{A}$ *recursively defined with* $\mathtt{Rel}(\mathbf{A})$ *as follows:* $\forall a \in \mathcal{A}$,

$$f(a) = 1 - \frac{1 - w(a)^2}{1 + w(a)\mathbf{e}^E} \quad \text{where} \quad E = \sum_{x \in \mathtt{Supp}(a)} f(x) - \sum_{x \in \mathtt{Att}(a)} f(x).$$

As an immediate corollary, we have:

Corollary 1. *Let* $\mathbf{A} = \langle \mathcal{A}, w, \mathcal{R}, \mathcal{S} \rangle$ *be an acyclic non-maximal BAG and* $a \in \mathcal{A}$. *The following holds:*

$$\mathtt{Deg}_{\mathbf{A}}^{\mathtt{Ebs}}(a) = 1 - \frac{1 - w(a)^2}{1 + w(a)\mathbf{e}^E} \quad \text{where} \quad E = \sum_{x \in \mathtt{Supp}(a)} \mathtt{Deg}_{\mathbf{A}}^{\mathtt{Ebs}}(x) - \sum_{x \in \mathtt{Att}(a)} \mathtt{Deg}_{\mathbf{A}}^{\mathtt{Ebs}}(x).$$

Below is an example where most axioms are exemplified. Every circle contains [argument name]:[intrinsic strength] and below [**overall strength**].

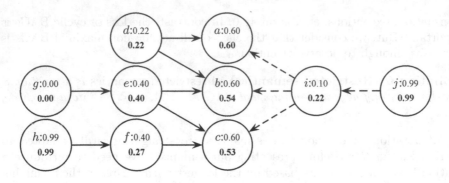

Fig. 1. Example of BAG

Example 1. The axiom neutrality can be checked with g and e, stability with e.g. d, bivariate monotony with a and b, bivariate reinforcement with b and c, Imperfection with i, Franklin with a, weakening with e.g. b, and strengthening with i.

Theorem 1. Ebs *satisfies all our 12 axioms.*

Note that being supported by an extremely strong argument does not cause a weak argument to become extremely strong as well, which shows that Ebs does not suffer from the big jump problem. Note that $\text{Deg}_{\mathbf{A}}^{\text{Ebs}}(i) = 0.22$ and thus the jump is not big. Note also that by satisfying Weakening and Strengthening, the semantics avoids the irrational behavior of QuAD.

6   Conclusion

The paper presented for the first time axioms that serve as guidelines for defining acceptability semantics in weighted bipolar settings. It also analyzed existing semantics with regard to the axioms. The results revealed that extension-based semantics like [1–5] fail to satisfy key properties. Furthermore, the role of support relation is a bit ambiguous since in case the attack relation of a BAG is empty, the argumentation graph has a single extension containing all the arguments. This means that supported and non-supported arguments are all equally acceptable. Gradual semantics defined in [7,8] satisfy more but not all the axioms. We proposed a novel semantics which satisfies all the 12 axioms. However, this semantics deals only with acyclic graphs. An urgent future work would be to prove whether the sequence of values it returns converges in case of arbitrary graphs. We also plan to investigate additional properties where attacks and supports do not have the same importance.

Acknowledgments. This work was supported by ANR-13-BS02-0004 and ANR-11-LABX-0040-CIMI.

References

1. Cayrol, C., Lagasquie-Schiex, M.C.: On the acceptability of arguments in bipolar argumentation frameworks. In: Godo, L. (ed.) ECSQARU 2005. LNCS, vol. 3571, pp. 378–389. Springer, Heidelberg (2005). doi:10.1007/11518655_33
2. Oren, N., Norman, T.: Semantics for evidence-based argumentation. In: Proceedings of COMMA, pp. 276–284 (2008)
3. Brewka, G., Woltran, S.: Abstract dialectical frameworks. In: Proceedings of KR (2010)
4. Boella, G., Gabbay, D.M., van der Torre, L., Villata, S.: Support in abstract argumentation. In: Proceedings of COMMA, pp. 111–122 (2010)
5. Nouioua, F., Risch, V.: Bipolar argumentation frameworks with specialized supports. In: International Conference on Tools with Artificial Intelligence, ICTAI 2010, pp. 215–218 (2010)
6. Cayrol, C., Lagasquie-Schiex, M.C.: Gradual valuation for bipolar argumentation frameworks. In: Godo, L. (ed.) ECSQARU 2005. LNCS, vol. 3571, pp. 366–377. Springer, Heidelberg (2005). doi:10.1007/11518655_32
7. Baroni, P., Romano, M., Toni, F., Aurisicchio, M., Bertanza, G.: Automatic evaluation of design alternatives with quantitative argumentation. Argum. Comput. 6(1), 24–49 (2015)
8. Rago, A., Toni, F., Aurisicchio, M., Baroni, P.: Discontinuity-free decision support with quantitative argumentation debates. In: Proceedings of KR, pp. 63–73 (2016)
9. Dung, P.M.: On the acceptability of arguments and its fundamental role in non-monotonic reasoning, logic programming and n-person games. Artif. Intell. 77, 321–357 (1995)
10. Amgoud, L., Ben-Naim, J.: Evaluation of arguments from support relations: axioms and semantics. In: Proceedings of IJCAI, pp. 900–906 (2016)
11. Amgoud, L., Ben-Naim, J.: Axiomatic foundations of acceptability semantics. In: Proceedings of KR, pp. 2–11 (2016)

Debate-Based Learning Game for Constructing Mathematical Proofs

Nadira Boudjani, Abdelkader Gouaich, and Souhila Kaci$^{(\boxtimes)}$

LIRMM, Montpellier, France
{boudjani,gouaich,kaci}@lirmm.fr

Abstract. Debate is a valuable and effective method of learning. It is an interactive process in which learners cooperate by exchanging arguments and counter-arguments to solve a common question. We propose a debate-based learning game for mathematics classroom to teach how to structure and build mathematical proofs. Dung's argumentation framework and its extensions are used as a means to extract acceptable arguments that form the proof. Moreover this allows instructors to provide continuous feedbacks to learners without information overload.

1 Introduction

Debate has been used as a learning method since antiquity notably by sophists such as Protagoras, Plato and Socrates. From that time, the principles of this method remained unchanged: a group of learners explore a common question by exchanging opinions, ideas in the form of arguments and counter-arguments. Such debates are based on a collaborative and progressive learning process. What has changed however are technologies that nowadays enable autonomous and ubiquitous learning.

The literature of educational science witnesses numerous examples of debate-based learning [14,20]. Furthermore, the use of debate in several areas such as medicine, natural sciences and humanities has been experimentally demonstrated as an effective learning tool [3,8]. In the fields of mathematics, [15] proposes a description of Lakatos's method by a dialogue game for collaborative mathematics. In short, we can synthesize main advantages of debate-based learning as follows: *(i)* it allows a group of heterogeneous learners with different backgrounds to collectively solve a common problem by using their own skills [1], *(ii)* it improves critical thinking skills, reasoning and communication within a group since each learner has to justify and defend her point of view and constructively criticize others [3,6], *(iii)* it increases motivation and involvement of learners by improving self-esteem and promoting social interaction [13], *(iv)* it makes explicit reasoning processes that led to conclusions. This provides instructors with information to identify misunderstandings and take actions to correct them.

The context. We are interested in this paper in learning how to structure and build mathematical proofs. Mathematical skills are increasingly becoming a central criterion in skills evaluation as they are an important selection criterion for

A. Antonucci et al. (Eds.): ECSQARU 2017, LNAI 10369, pp. 36–45, 2017.
DOI: 10.1007/978-3-319-61581-3_4

academic and professional applications. Therefore shortcomings in mathematics may be highly detrimental to students in both academic and professional opportunities. In the context of mathematical didactics several works have shown advantages of using debate in classes [5]. The object of the debate can be either to build a mathematical proof or to falsify a claim using sound mathematical deductions.

The problem and contribution. Despite the benefits witnessed in debates-based classes, we can mention some limitations of current approaches. The first limitation concerns learner's motivation and involvement. Although it has been demonstrated that learners show a high motivation in debate-based courses [8], it has also been acknowledged that this is influenced by instructor's animation abilities and the fact that these courses are formal and mandatory. In order to promote an autonomous and ubiquitous learning beyond institutional environment, we propose a *gamification* approach that considers the debate as a genuine game with its intrinsic motivation levers. The second limitation concerns assessment of debate outcomes and its effect on instructors' workload. In fact, assessment is fundamental in order to provide learners with continuous feedbacks. This task is often performed manually by the instructor who has to understand, evaluate and provide feedbacks based on the state of the debate. However, debate data are often overloaded by details of intermediate and erroneous stages of reasoning. While these stages are important to understand how learners have come to final conclusions, they do overload instructors with unnecessary details during assessment phases. As the basic ingredients in debates are arguments and counter-arguments, we use formal *argumentation* framework as a *means* to extract final conclusions from a debate before its assessment by instructors.

The remainder of this paper is structured as follows. In Sect. 2 we provide necessary background on argumentation. In Sect. 3 we introduce our system for debate-based learning through argumentation and games. In Sect. 4 we illustrate our system with an example on a group of learners. Lastly we conclude.

2   Background on Formal Argumentation Frameworks

Artificial intelligence witnesses a large amount of contributions in argumentation theory. In particular Dung's argumentation framework is a pioneer work in the topic [4].

Definition 1 (Dung's Framework). *An argumentation framework (AF) is a tuple $\langle \mathcal{A}, \text{Def} \rangle$, where $\mathcal{A}$ is a finite set of arguments and $\text{Def} \subseteq \mathcal{A} \times \mathcal{A}$ is a binary defeat[1] relation. Given $A, B \in \mathcal{A}$, A Def B stands for "A defeats B".*

The outcome of Dung's AF is sets of arguments, called *extensions*, that are robust against defeats [4]. We say that $\mathcal{T} \subseteq \mathcal{A}$ *defends* A if $\forall B \in \mathcal{A}$ s.t. B Def A, $\exists C \in \mathcal{T}$ such that C Def B. We say that $\mathcal{T} \subseteq \mathcal{A}$ is *conflict-free* if $\nexists A, B \in \mathcal{T}$

---

[1] Called *attack* in [4].

such that A Def B. A subset $\mathcal{T} \subseteq \mathcal{A}$ of arguments is an admissible extension iff it is conflict-free and it defends all elements in $\mathcal{T}$. Different acceptability semantics have been proposed. In particular $\mathcal{T} \subseteq \mathcal{A}$ is a complete extension of $\langle \mathcal{A}, Def \rangle$ iff it is admissible and contains all arguments it defends. $\mathcal{T}$ is a grounded extension $\langle \mathcal{A}, Def \rangle$ iff it is minimal (for set inclusion) complete extension. For other semantics, see [4].

Several authors have considered a new kind of interaction called the *support* relation [2,11,12]. An abstract bipolar argumentation framework is an extension of Dung's framework such that both defeat and support relations are considered.

Definition 2 (Bipolar argumentation framework). *An* abstract bipolar argumentation framework *(BAF) is a tuple* $\langle \mathcal{A}, \text{Def}, \text{Supp} \rangle$, *where* $\langle \mathcal{A}, \text{Def} \rangle$ *is Dung's AF and* $\text{Supp} \subseteq \mathcal{A} \times \mathcal{A}$ *is a binary support relation. For* $A, B \in \mathcal{A}$, A Supp B *means "A supports B".*

Defeat and support relations are combined to compute new defeat relations and recover Dung's framework from which acceptable extensions are computed [2,11,12].

In this paper we are interested in deductive reasoning, i.e. a conclusion is derived from a set of premises.

Definition 3 (Argument). *Let* Γ *be a set of formulas constructed from a given language* $\mathcal{L}$. *An argument over* Γ *is a pair* $A = \langle \Delta, \alpha \rangle$ *s.t. (i)* $\Delta \subseteq \Gamma$, *(ii)* $\Delta \nvdash_* \bot$, *(iii)* $\Delta \vdash_* \alpha$ *and, (iv) for all* $\Delta' \subset \Delta$, $\Delta' \nvdash_* \alpha$, *where* $\vdash_*$ *is the inference symbol.*

We say that $\langle \Delta, \alpha \rangle$ undercuts $\langle \Delta', \alpha' \rangle$ iff for some $\phi \in \Delta'$, α and ϕ are contradictory w.r.t. the language at hand. $\langle \Delta, \alpha \rangle$ rebuts $\langle \Delta', \alpha' \rangle$ iff α and α' are contradictory. Then, $\langle \Delta, \alpha \rangle$ defeats $\langle \Delta', \alpha' \rangle$ iff $\langle \Delta, \alpha \rangle$ rebuts/undercuts $\langle \Delta', \alpha' \rangle$.

3   A Debate-Based Learning Game

Our system builds on learners who exchange arguments for some purpose. The Oracle represents the instructor of the game. She opens the discussion by providing a first argument of the form $\langle P, C \rangle$, where P is a set of premises and C is a conclusion. Then learners are engaged in a debate in which they exchange arguments to construct a proof for C given P. A set of relations is provided by the Oracle to connect arguments. A learner may add an argument/relation or pass her turn. The discussion is closed when decided by the learners or stopped by the Oracle. The output of the debate is a debate graph which is composed of arguments and relations constructed during the debate. The graph is submitted to the Oracle for evaluation. We first discuss and give our design choices. Then we propose a game-based modeling of our learning system.

3.1 Design Choices

Different factors need to be considered for the game. Before we present our game-based modeling of the learning system let us expose these factors:

- *Argumentation mechanisms:* Three mechanisms have been distinguished to model the exchange of arguments between agents [17]: In a *Direct mechanism* every agent may propose a set of arguments at once. Then the process terminates. This mechanism is not appropriate in our setting because a learning process needs to be progressive. In a *synchronous mechanism* every agent may propose any set of arguments at the same time. The process is repeated until no agent wants to make more arguments. In a *dialectical mechanism* an order is assumed over agents to provide their arguments. Four variants (rigid, non rigid) × (single, multiple) can be obtained. A mechanism is rigid when an agent who passed her turn will no longer be allowed to propose arguments. The mechanism is not rigid if the agent is not discarded in such a situation. Also an agent may propose a single argument or multiple arguments when she takes her turn. As the purpose of the game is that *all* learners progressively collaborate to construct a proof *we use a dialectical, non rigid and single argument mechanism.*

- *Construction of the arguments & their validity:* Arguments may be provided by the Oracle or constructed by learners. In the second case, the Oracle provides a set of propositions upon which arguments will be constructed. *We choose the second option.* Now whatever arguments are constructed or provided we need to check their validity which refers to the satisfaction of conditions *(i)–(iv)* in Definition 3. Condition *(i)* will always be satisfied as arguments will be constructed from a set of propositions provided by the Oracle. An argument that does not satisfy condition *(ii)* should be removed. An argument that does not satisfy condition *(iv)* should be modified to make its set of premises minimal. Let us now consider condition *(iii)*. For example the argument $\langle \{p\}, q \rangle$ is not valid w.r.t. *(iii)* because q does not logically follow from p. This argument should be removed. On the other hand, the argument $\langle \{(2-\epsilon)(n+2) < 2n+1\}, 2-\epsilon < \frac{2n+1}{n+2} \rangle$ will be considered as valid although $2-\epsilon < \frac{2n+1}{n+2}$ does not logically follow from $(2-\epsilon)(n+2) < 2n+1$ because the set of premises of the argument is not complete. In fact we need an additional constraint, namely $n+2 > 0$. Such an argument can however be accepted in a debate. Then learners have to cooperate in order to defeat the argument or defend it by completing its set of premises. This is coherent with a debate-based learning process in which arguments can be both collaboratively and progressively constructed. We distinguish between two ways to deal with an argument that should be removed (e.g. $\langle \{p\}, q \rangle$ or condition *(ii)* not satisfied) or modified (condition *(iv)*). First notice that learners in the same group cooperate in order to solve a problem. Therefore they may be allowed to have a chat box by which they can discuss in order to convince a learner that her argument is not valid and should be removed/modified. We expect a cooperation from all learners. If not then the non valid argument will be refused by the Oracle (when the latter evaluates the debate graph at the end of the debate) which leads to the failure of the group to construct a correct proof. Another way to control the validity of an argument is to delegate this task to the Oracle in which case an argument is added to the debate graph only when it is valid. A penalty on the score of the group is applied each time a learner proposes a non valid argument. *We choose the first option.*

If constructed arguments do not comply with Definition 3 and not repaired/removed by learners then the group will fail to construct a correct proof.

- *Relations between arguments and their validity:*
  - *Defeat relation:* This relation is syntactically defined and should be in the background of learners. In contrast to existing works [16,17] in which the defeat relation is automatically stated by the system as soon as an agent proposes an argument we do believe that this relation should be stated by the learners themselves. This is a part of the learning process. Two ways to control the validity of a defeat relation are possible. Either we authorize defeats on defeats which can be captured by hierarchical argumentation frameworks [10], or the Oracle accepts a defeat relation in the debate graph only when it is valid. *We choose the second option in our game.*
  - *Support relation:* we do not control the support relation because the objective of the game is to construct a proof which is built using the support relation. So it is up to learners to discuss/agree on a given support relation without the intervention of the Oracle.
- *Termination of the game:* The game terminates when the group stops or the Oracle decides so. *We choose the first option in our game.* The debate graph is submitted to the Oracle for evaluation.
- *Score function:* A score which is a penalty degree is given to a group of learners when its debate graph is evaluated by the Oracle. We define a penalty degree as the number of irrelevant arguments present in the debate graph plus the number of non valid defeat relations refused by the Oracle during the debate.

3.2 Game-Based Modeling of the Learning System

Our game aims at creating a debate environment for learners so that they can exchange arguments to prove the Oracle's claim. Learners need to be maintained engaged in a game. For this purpose we use social levers of motivation: we divide learners into groups that will be made in competition. The Oracle initiates the debate by setting the question under the form of an argument. The group that wins the game is the one that manages to build a debate graph accepted by the Oracle with a minimal penalty. To construct such a graph, learners have to use their domain knowledge to build arguments and correctly set relationships between arguments. This section formalizes the game by describing states of the game and actions which are transitions among states.

3.2.1   State of the Game

To construct arguments we use a universe of discourse based on a given $\mathcal{L}$-language. We assume that this language is at least equipped with a conjunction operator. Given a set of arguments $\mathcal{A}$ and a set of relation labels L, a state of a debate is a sequence of relations indexed by natural numbers and labels: $S : L \times \mathbb{N} \to 2^{\mathcal{A} \times \mathcal{A}}$. $S(l, k)_{l \in L, \ k \in \mathbb{N}}$ represents the content of the relation l at the kth step. $\mathcal{S}$ denotes the set of all states.

3.2.2   Actions of the Game

Adding a relation. This action adds a new relationship between two (new or existing) arguments. This action takes as input the label of the relation and two arguments.

Definition 4. *Given a relation label $l \in L$ and a couple of arguments (A, B), the* add *action is a transition between states such that* $\forall S, S' \in \mathcal{S}, S \xrightarrow{add(l,A,B)} S'$ *iff*

$$
\exists k \in \mathbb{N}, \forall r \in L, \begin{cases} \forall m > k, S'(r, m) = S(r, m) = \emptyset \\ \forall n < k, S(r, n) = S'(r, n) \\ S(r, k) = \emptyset \\ S'(r, k) = S(r, k - 1) \ for \ r \neq l \\ S'(l, k) = S(l, k - 1) \cup \{(A, B)\} \end{cases}
$$

The *add* action makes a transition from S to S' if and only if the following conditions hold: *(i)* the game states contain only a finite number of known relation graphs. In other words, there exists an integer k for which all relation graphs of subsequent steps are empty, *(ii)* S and S' are equal until kth step. This means that S' copies S until the kth step, *(iii)* game state S does not contain any information about the relation l at step k, *(iv)* finally, game state S' at kth step is equal to the graph of l at the $(k-1)$th step to which the couple (A, B) is added.

Removing a relation. A player can remove a relation between arguments from the debate graph. This action takes as input the label of the relation and the couple to be removed. Specification of the remove action is similar to that of the add action, except that S' at kth step contains previous graph of l from which the pair (A, B) has been removed. That is $S'(l, k) = S(l, k - 1) \setminus \{(A, B)\}$.

Append action. The goal of this action is to make an argument's premises and conclusion more specific. It is an auxiliary action that is built as a composition of a remove and add actions. Before we define this action, let us introduce a connector $(\wedge)$ between arguments. Given $A = \langle P_1, C_1 \rangle$ and $B = \langle P_2, C_2 \rangle$, the notation $A \wedge B$ is an abbreviation for $\langle P_1 \cup P_2, C_1 \wedge C_2 \rangle^2$.

Definition 5. *The* append *relation between states S and S' is defined as:*

$$
\forall S, S' \in \mathcal{S}, S \xrightarrow{append(l,A,B,C)} S' \ iff \ \exists S'', S \xrightarrow{remove(l,A,B)} S'' \xrightarrow{add(l,A \wedge C, B)} S'.
$$

---

[2] Notice that $P_1 \cup P_2$ may be inconsistent but remind that the validity of arguments is decided by learners.

Submit action. This action ends the game and submits the final debate state to the Oracle for evaluation.

4   Example: Mathematical Proof

Let Q: "Prove that $\forall \epsilon > 0 \exists N \in \mathbb{N}$ such that $(n \geq N \Rightarrow 2 - \epsilon < \frac{2n+1}{n+2} < 2 + \epsilon)$" to be proved. We have the Oracle and 6 learners (players): $l_1, l_2, l_3, l_4, l_5, l_6$.

The Oracle provides a first argument A_0 corresponding to Q. We have $A_0 = \langle \{\epsilon > 0, \exists N \in \mathbb{N}, n \geq N\}, 2 - \epsilon < \frac{2n+1}{n+2} < 2 + \epsilon) \rangle$. The Oracle also provides a set of labels $L = \{defeat, support\}$ and a set of propositions $\mathcal{P}$:

$$\{n \in \mathbb{N}, \quad \frac{2n+1}{n+2} < 2, \quad 2 - \epsilon < \frac{2n+1}{n+2}, \quad \frac{2n+1}{n+2} < 2 + \epsilon, \quad \neg(\frac{2n+1}{n+2} < 2),$$
$$n \geq N, \quad \epsilon > 0, \quad N > \frac{3}{\epsilon} - 2, \quad n > \frac{3}{\epsilon} - 2,$$
$$\epsilon = -1, \quad n = 1, \quad \neg(2 - \epsilon < \frac{2n+1}{n+2}), n = -3, \quad \neg(N = [\frac{3}{\epsilon} - 2] + 1)$$
$$\neg(n = -3), \neg(\epsilon = -1), \exists N \in \mathbb{N}, \quad (2 - \epsilon)(n + 2) < 2n + 1\}.$$

The aim of the game is to demonstrate Q. Let $Group_1$ and $Group_2$ be two groups of learners: $G_1 = \{l_1, l_2, l_3\}$ and $G_2 = \{l_4, l_5, l_6\}$. Each group has to demonstrate the proof individually. Not only the groups have to argue to construct the proof but they also have to do that as fast as possible and with minimal penalty degree. Due to space limitation, we illustrate the game on $Group_1$ only. Initially, the debate graph contains only A_0. The discussion is dialectical: a random order is defined over the set of learners. Let l_1, l_2, l_3 be this order. Table 1 illustrates first states of the game.

Table 1. First game states during the game.

# state	Player	Argument	Action	State
0	$\emptyset$	A_0	$\emptyset$	$S_0 = \emptyset$
1	l_1	A_1	$add('support', A_1, A_0)$	$S_1('support', 1) = \{(A_1, A_0)\}$
2	l_2	A_2	$add('defeat', A_2, A_1)$	$S_2('support', 2) = \{(A_1, A_0)\}$
				$S_2('defeat', 2) = \{(A_2, A_1)\}$
3	l_3	A_3	$append('support', A_1, A_0, A_3)$	$S_3('support', 3) = \{(A_1 \wedge A_3, A_0)\}$
				$S_3('defeat', 3) = \{(A_2, A_1)\}$
4	l_1	$\emptyset$	$add('defeat', A_3, A_1)$ (not valid)	$S_4('support', 4) = \{(A_1 \wedge A_3, A_0)\}$
				$S_4('defeat', 4) = \{(A_2, A_1)\}$
...	...	...	...	...

The complete debate graph is given in Fig. 1, where

A_1: $\langle \{\epsilon > 0, \exists N \in \mathbb{N}, n \geq N\}, 2 - \epsilon < \frac{2n+1}{n+2}) \rangle$ A_2: $\langle \{\epsilon = -1, n = 1, \}, \neg(2 - \epsilon < \frac{2n+1}{n+2})) \rangle$

A_3: $\langle \{\epsilon > 0, \exists N \in \mathbb{N}, n \geq N\}, \frac{2n+1}{n+2} < 2 + \epsilon) \rangle$ A_4: $\langle \{\epsilon > 0, \frac{2n+1}{n+2} < 2\}, \frac{2n+1}{n+2} < 2 + \epsilon) \rangle$

A_5: $\langle \{\epsilon > 0\}, \epsilon > 0 \rangle$ A_6: $\langle \{n \in \mathbb{N}\}, \frac{2n+1}{n+2} < 2) \rangle$

A_7: $\langle \{n \in \mathbb{N}\}, n \in \mathbb{N} \rangle$ A_8: $\langle \{n \in \mathbb{N}, (2 - \epsilon)(n + 2) < 2n + 1\}, 2 - \epsilon < \frac{2n+1}{n+2}) \rangle$

A_9: $\langle \{\epsilon > 0, n > \frac{3}{\epsilon} - 2\}, (2 - \epsilon)(n + 2) < 2n + 1) \rangle$ A_{10}: $\langle \{n \geq N, N > \frac{3}{\epsilon} - 2\}, n > \frac{3}{\epsilon} - 2) \rangle$

A_{11}: $\langle \{N > \frac{3}{\epsilon} - 2\}, N > \frac{3}{\epsilon} - 2 \rangle$ A_{12}: $\langle \{n \geq N\}, n \geq N \rangle$

A_{13}: $\langle \{n = -3\}, \neg(\frac{2n+1}{n+2} < 2)) \rangle$ A_{14}: $\langle \{n \in \mathbb{N}\}, \neg(n = -3)) \rangle$

A_{15}: $\langle \{n \in \mathbb{N}, \epsilon > 0\}, \frac{2n+1}{n+2} < 2 + \epsilon \rangle$ A_{16}: $\langle \{\epsilon > 0\}, \neg(\epsilon = -1)) \rangle$

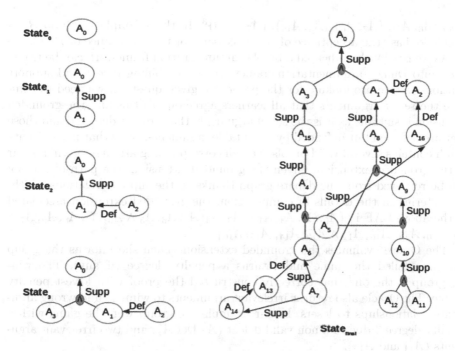

Fig. 1. Debate graph during the game.

In this example learners may add a symmetric defeat between $\langle\{\epsilon > 0\}, \epsilon > 0\rangle$ and $\langle\{\epsilon = -1\}, \epsilon = -1\rangle$. In this case learners need to discuss in a chat box and convince the learner who added the defeat relation from the latter to the former to remove her defeat because $\epsilon > 0$ appears in the premises of A_0 so it is a fact that overrides $\epsilon = -1$. We do not make any explicit distinction between propositions because distinguishing facts and giving them priority should be done by the learners. This is a part of the learning process. Assuming that the game ends when the learners agree to submit the debate graph to the Oracle. Thus $Group_1$ submits the graph of the $state_{final}$ using the action $submit$. Then the Oracle evaluates the debate graph by computing acceptable arguments. To this aim, she computes the argumentation framework corresponding to the debate graph in order to compute acceptable extensions. The debate graph is composed of nodes representing arguments and two types of relations, namely defeat and support relations. Therefore bipolar AF is suitable. We have $BAF = \langle\mathcal{A}, \mathrm{Def}, \mathrm{Supp}\rangle$, where $\mathcal{A} = \{A_0, A_1, A_1 \wedge A_3, A_2, A_3, A_4, A_5, A_6,$ $A_5 \wedge A_6, A_7, A_8, A_9, \ A_7 \wedge A_9, A_{10}, A_5 \wedge A_{10}, A_{11} \wedge A_{12}, \ A_{13}, A_{14}, A_{15}, A_{16}\}$, $\mathrm{Def} = \{(A_2, A_1), (A_{16}, A_2), (A_{13}, A_6), (A_{14}, A_{13})\}$ and $\mathrm{Supp} = \{(A_1 \wedge A_3, A_0),$ $(A_8, A_1), (A_9 \wedge A_7, A_8), (A_{10} \wedge A_5, A_9), (A_{11} \wedge A_{12}, A_{10}), (A_{15}, A_3), (A_4, A_{15}),$

$(A_5 \wedge A_6, A_4), (A_7, A_6), (A_5, A_{16}), (A_7, A_{14})\}^3$. In this example "$A$ Supp B" is interpreted as "the acceptance of A is necessary for the acceptance of B" [2,11].

As we previously indicated a bipolar argumentation framework can be translated into Dung's argumentation framework by combining defeat and support relations. As we are looking for the proof of a given question, we need to compute the set of arguments that all learners agree on. To this aim the grounded extension is suitable as it is the set of arguments that are not defeated and those that are defeated but defended by acceptable arguments. Note that if the debate graph contains several valid proofs then all corresponding arguments will appear in the grounded extension. Given the grounded extension, the path of a proof can be retrieved from the debate graph thanks to the support relations. Without entering in the details of computation, the grounded extension associated to the above BAF is $\{A_0, A_1, A_3, A_1 \wedge A_3, A_4, A_5, A_6, A_5 \wedge A_6, A_7, A_8, A_9, A_7 \wedge A_9, A_{10}, A_5 \wedge A_{10}, A_{11} \wedge A_{12}, A_{14}, A_{15}, A_{16}\}$.

The Oracle evaluates the grounded extension. Then she informs the group if it won/failed the game and returns a penalty degree (if any). The wining group is the one that correctly constructed the proof with lowest penalty degree. The Oracle also returns irrelevant arguments to winners and wrong arguments/relationships to losers. In our example the group won the game with a penalty degree 3 due to a non valid defeat (A_3 Def A_1) and two irrelevant arguments (A_{14} and A_{16}).

5   Conclusion and Ongoing Work

The present paper uses argumentation framework in the context of mathematical proofs. It offers a conceptual formal debate-based learning system whose advantages are twofold: *(i)* it offers a formal method to analyze and filter (generally huge amount of) information exchanged during the debate and computes valuable information (i.e., acceptable arguments) that serves to evaluate the debate, *(ii)* it also provides a game-modeling of the argumentation debate as a means to keep learners motivated.

Several works present argumentation as an abstract dialectical game [9,18]. However there are few genuine games based on the theoretical frameworks developed in AI. We can mention [19] which presents a simple graph of abstract arguments and players must select arguments in the graph to win the debate according to Dung's semantics [4]. The authors of [17] use structured argumentation framework in order to construct a syntactical path for the support relation. Our work does rely on both syntactical and semantical paths in the support relation in order to fit the mathematical domain. We do also give a greater importance to the output of the argumentation framework which serves to extract valuable information from the debate (which corresponds to the proof).

As perspective we intend to experimentally validate our learning system to assess both its learning value for learners and its acceptance by instructors. At

---

[3] A_{11} and A_{12} need not to be considered separately. In fact $A_{11} \wedge A_{12}$ is needed and sufficient.

the conceptual level we intend to consider preferences among learners in the argumentation framework [7]. In fact more experts learners need to be favored against less expert ones.

References

1. Amir, Y., Sharan, S., Ben-Ari, R., Desegregation, S.: Cross-Cultural Perspectives. Psychology Press, New York (1984)
2. Cayrol, C., Lagasquie-Schiex, M.C.: Bipolarity in argumentation graphs: Towards a better understanding. IJAR **54**(7), 876–899 (2013)
3. Davidson, N.: Enhancing Thinking through Cooperative Learning. Teachers College Press, New York (1992)
4. Dung, P.M.: On the acceptability of arguments and its fundamental role in non-monotonic reasoning, logic programming and n-person games. Artif. Intell. **77**, 321–357 (1995)
5. Durand-Guerrier, V., Boero, P., Douek, N., Epp, S., Tanguay, D.: Argumentation and proof in the mathematics classroom. In: Hanna, G., de Villiers, M. (eds.) Proof and Proving in Mathematics Education, pp. 349–367. Springer, Dordrecht (2012)
6. Johnson, D.W., Johnson, R.T.: Learning Together and Alone: Cooperative, Competitive, and Individualistic Learning. Allyn and Bacon, Boston (1999)
7. Kaci, S., van der Torre, L.: Preference-based argumentation: arguments supporting multiple values. IJAR **48**, 730–751 (2008)
8. Kennedy, R.R.: The power of in-class debates. Act. Learn. High Educ. **10**(3), 225–236 (2009)
9. Maudet, N., Moore, D.: Dialogue games as dialogue models for interacting with, and via, computers. Informal Logic **21**(3), 219–243 (2001)
10. Modgil, S.: Hierarchical argumentation. In: Fisher, M., Hoek, W., Konev, B., Lisitsa, A. (eds.) JELIA 2006. LNCS (LNAI), vol. 4160, pp. 319–332. Springer, Heidelberg (2006). doi:10.1007/11853886_27
11. Nouioua, F., Risch, V.: Bipolar argumentation frameworks with specialized supports. In: ICTAI 2010, pp. 215–218 (2010)
12. Oren, N., Norman, T.J.: Semantics for evidence-based argumentation. In: COMMA 2008, pp. 276–284 (2008)
13. Oros, A.L.: Let's debate: active learning encourages student participation and critical thinking. J. Polit. Sci. Educ. **3**(3), 293–311 (2007)
14. Park, C., Kier, C., Jugdev, K.: Debate as a teaching strategy in online education: a case study. Can. J. Learn. Technol. **37**(3), 17–20 (2011)
15. Pease, A., Budzynska, K., Lawrence, J., Reed, C.: Lakatos games for mathematical argument. In: COMMA 2014, pp. 59–66 (2014)
16. Rahwan, I., Larson, K.: Argumentation and game theory. In: Simari, G., Rahwan, I. (eds.) Argumentation in Artificial Intelligence, pp. 321–339. Springer, New York (2009)
17. Thimm, M., García, A.J.: Classification and strategical issues of argumentation games on structured argumentation frameworks. In: AAMAS 2010, pp. 1247–1254 (2010)
18. Weigand, E.: Argumentation: the mixed game. Argumentation **20**(1), 59–87 (2006)
19. Yuan, T., Svansson, V., Moore, D., Grierson, A.: A computer game for abstract argumentation. In: CMNA (IJCAI 2007 Workshop), pp. 62–68 (2007)
20. Zemplén, G.: History of Science and Argumentation in Science Education, pp. 129–140. SensePublishers, Rotterdam (2010)

Updating Probabilistic Epistemic States in Persuasion Dialogues

Anthony Hunter[1(✉)] and Nico Potyka[2]

[1] Department of Computer Science, University College London, London, UK
anthony.hunter@ucl.ac.uk
[2] Institute of Cognitive Science, University of Osnabrück, Osnabrück, Germany

Abstract. In persuasion dialogues, the ability of the persuader to model the persuadee allows the persuader to make better choices of move. The epistemic approach to probabilistic argumentation is a promising way of modelling the persuadee's belief in arguments, and proposals have been made for update methods that specify how these beliefs can be updated at each step of the dialogue. However, there is a need to better understand these proposals, and moreover, to gain insights into the space of possible update functions. So in this paper, we present a general framework for update functions in which we consider existing and novel update functions.

1 Introduction

The aim of persuasion is for the persuader to change the mind of the persuadee, and the provision of good arguments, and possibly counterarguments, is of central importance for this. Some recent developments in the field of computational persuasion have focused on the need to model the beliefs of the persuadee in order for the persuader to better select arguments to present to the persuadee. For instance, if the persuader wants to persuade the persuadee to give up smoking, and the persuader knows that the persuadee believes that if he gives up smoking, he will put on weight, then the persuader could start the dialogue by providing a counterargument to this, for example by saying that there is a local football team for ex-smokers who are looking for new players.

One approach to modelling the persuadee is to harness the epistemic approach to probabilistic argumentation [11]. In this, an argument graph (as defined by Dung [4]) is used to represent the arguments and attacks between them, and a probability distribution over the subsets of arguments is used to represent the uncertainty over which arguments are believed. The belief in an individual argument is then the sum of the belief in the subsets that contain this argument.

When a persuader starts a dialogue with a persuadee, the persuader identifies an appropriate probability distribution to represent what s/he thinks the persuadee believes. Then during the dialogue, the moves are made by the participants according to some protocol. After each move, the belief is updated using an update function (see Fig. 1). Some initial proposals for update functions have

© Springer International Publishing AG 2017
A. Antonucci et al. (Eds.): ECSQARU 2017, LNAI 10369, pp. 46–56, 2017.
DOI: 10.1007/978-3-319-61581-3_5

Fig. 1. Schematic representation of a dialogue $D = [m_1, \ldots, m_n]$ and user models P_i. Each user model P_i is obtained from P_{i-1} and move m_i using an update method.

been made (e.g. [10]) which seem intuitive and well-behaved, but there is a lack of a general understanding of what an update function is, of what the space of options are, and of how alternatives could be defined. The aim of this paper is to address these questions by proposing some basic properties for update functions, and then proposing a framework for update measures in which we show how existing and some useful novel update functions are situated.

2 Basics

We consider a finite argument graph G with arguments Args and attacks Attacks. For $A \in$ Args, we let $A^- = \{B \in$ Args $\mid (B, A) \in$ Attacks$\}$. Form denotes the set of propositional formulas over Args. That is, Form is the smallest set that contains Args and is closed under application of the usual logical connectives like $\neg$ and $\wedge$. An interpretation of Form is a subset $X \subseteq$ Args. X satisfies an atomic formula $A \in$ Args iff $A \in X$ and we write $X \models A$ in this case. The satisfaction relation is extended to complex formulas in the usual way. For instance, $X \models F_1 \wedge F_2$ iff $X \models F_1$ and $X \models F_2$. A probability distribution over Args is a function $P : 2^{\text{Args}} \to [0, 1]$ such that $\sum_{X \subseteq \text{Args}} P(X) = 1$. We let $\mathcal{P}$ denote the set of all probability distributions over Args. When speaking of topological properties of subsets of $\mathcal{P}$, we regard probability distributions as probability vectors and consider the usual topology on $\mathbb{R}^n$. Note that we can do so because 2^{Args} is finite (because Args is finite). For $F \in$ Form, we let $P(F) = P(\{X \subseteq \text{Args} \mid X \models F\})$. A complete conjunction over a subset $X \subseteq$ Args is a conjunction of the form $\bigwedge_{A \in X} L_A$, where either $L_A = A$ or $L_A = \neg A$. Let $\text{Conj}(X)$ denote the set of all complete conjunctions over X. In the following, we will make use of the fact that there is a 1-1 relationship between $\text{Conj}(\text{Args})$ and the interpretations 2^{Args}. More strictly speaking, a complete conjunction $\bigwedge_{A \in \text{Args}} L_A$ corresponds to the interpretation $\{A \in \text{Args} \mid L_A = A\}$ that contains all arguments that appear positive in the conjunction. Conversely, an interpretation $X \subseteq$ Args corresponds to the complete conjunction $\bigwedge_{A \in X} A \wedge \bigwedge_{A \in \text{Args} \setminus X} \neg A$.

Intuitively, a probability distribution over Args represents the epistemic state of an agent. Given an argument graph G, we want to impose certain constraints on probability distributions. We can consider some of the following rationality postulates for the epistemic state represented by P [11].

- **RAT:** P is *rational* iff for all $(A, B) \in$ Attacks, $P(A) > 0.5$ implies $P(B) \le 0.5$.

- **COH:** P is *coherent* iff for all $(A, B) \in$ Attacks, $P(A) \leq 1 - P(B)$.
- **SFOU:** P is *semi-founded* iff $A^- = \emptyset$ implies $P(A) \geq 0.5$.
- **FOU:** P is *founded* iff $A^- = \emptyset$ implies $P(A) = 1$.
- **SOPT:** P is *semi-optimistic* iff $A^- \neq \emptyset$ implies $P(A) \geq 1 - \sum_{B \in A^-} P(B)$.
- **OPT:** P is *optimistic* iff $P(A) \geq 1 - \sum_{B \in A^-} P(B)$.

For a subset $R \subseteq \{RAT, COH, SFOU, FOU, SOPT, OPT, JUS\}$ of rationality postulates, we write $P \models R$ iff P satisfies all constraints in R and for a subset $T \subseteq \mathcal{P}$, we write $T \models R$ iff $P \models R$ for all $P \in T$.

3   Properties of Update Functions

We can model the change of an agent's epistemic state in a dialogue by an update function [10]. Our goal here is to investigate the space of possible update functions systematically. Formally, we regard an update function as a function $U : \mathcal{P} \times \text{Form} \rightarrow 2^{\mathcal{P}}$ that takes a probability distribution and a formula and maps them to a set of probability distributions $U(P, F)$ that satisfy F in some way. In the following, we list several properties that might be interesting in this context. We start with a list of general properties.

- **Uniqueness:** $|U(P, F)| \leq 1$.
- **Completeness:** If $F \not\equiv \bot$ then $|U(P, F)| \geq 1$.
- **Tautology:** $U(P, \top) = \{P\}$.
- **Contradiction:** $U(P, \bot) = \emptyset$.
- **Representation Invariance:** If $F \equiv G$ then $U(P, F) = U(P, G)$.
- **Idempotence:** If $U(P, F) = \{P^*\}$ then $U(P^*, F) = \{P^*\}$.
- **Order Invariance:** $U(U(P, F_1), F_2) = U(U(P, F_2), F_1)$.

Uniqueness says that the solution of the update is always unique. Completeness says that a solution always exists when the new information is consistent. Tautology says that updating with a tautology should not change the epistemic state because we do not add any new information. Since our generated epistemic state should be consistent, Contradiction demands that updating with a contradictory formula should yield the empty set. Representation invariance says that semantically equivalent formulas should result in the same update. Idempotence says that if the update yields a unique solution, then updating again with the same information should not change the result. Order invariance says that the order in which we update does not affect the result.

Next, we consider some semantical properties. To begin with, we might want that updates take the structure of the argument graph into account. Therefore, we consider the following property for subsets $R \subseteq \{RAT, COH, SFOU, FOU, SOPT, OPT\}$ of rationality postulates:

- **R-Consistency:** If $P \models R$ then $U(P, F) \models R$.

In addition, the probability distributions in $U(P, F)$ should satisfy F in some way. We consider the following satisfaction conditions.

- **STRICT:** P satisfies F *strictly* iff $P(F) = 1$.
- **ϵ-WEAK:** P satisfies F *ϵ-weakly* iff $P(F) \geq 0.5 + \epsilon$ for $\epsilon \in (0, 0.5)$.

Remark 1. Note that strict satisfaction implies ϵ-weak satisfaction for all $\epsilon \in (0, 0.5)$.

For a satisfaction condition $S \in \{STRICT, \epsilon\text{-}WEAK\}$ and a formula $F \in \text{Form}$, we write $P \models_S F$ iff P satisfies F with respect to S and for a subset $T \subseteq \mathcal{P}$, we write $T \models_S F$ iff $P \models_S F$ for all $P \in T$. Analogous to rationality postulates, we consider the following property for $S \in \{STRICT, \epsilon\text{-}WEAK\}$:

- **S-Consistency:** $U(P, F) \models_S F$.

For a set of rationality postulates R and a satisfaction condition S, we define the set of *R-S-models* of $F \in \text{Form}$ by

$$\text{Mod}_{R,S}(F) = \{P \in \mathcal{P} \mid P \models R, P \models_S F\}$$

We call F *R-S-consistent* if $\text{Mod}_{R,S}(F) \neq \emptyset$ and *R-S-inconsistent* otherwise. If F is R-S-inconsistent, the condition of S-consistency becomes $\emptyset \models_S F$ and is trivially true. The following example illustrates an R-S-inconsistency.

Example 1. Consider an argument graph over A, B with Attacks $= \{(A, B)\}$. Let $R = \{RAT, FOU\}$. Then FOU implies $P(A) = 1$ for all $P \in \text{Mod}_{R,S}(\top)$ and therefore RAT implies $P(B) \leq 0.5$. Hence, $\text{Mod}_{R,\epsilon\text{-}WEAK}(B) = \emptyset$ for all $\epsilon > 0$.

Finally, we might want to update the epistemic state such that we minimally change the prior state. To this end, we can consider different change functions over $\mathcal{P}$. The first class of change measures that we consider measure the difference in probability mass that is assigned to interpretations.

- **Manhattan Distance:** $d_1(P, P^*) = \sum_{X \subseteq \text{Args}} |P(X) - P^*(X)|$.
- **Least Squares Distance:** $d_2(P, P^*) = \sum_{X \subseteq \text{Args}} (P(X) - P^*(X))^2$.
- **Maximum Distance:** $d_\infty(P, P^*) = \max_{X \subseteq \text{Args}} |P(X) - P^*(X)|$.
- **KL-divergence:** $d_{KL}(P^*, P) = \sum_{X \subseteq \text{Args}} P^*(X) \cdot \log \frac{P^*(X)}{P(X)}$.

Note that the KL-divergence is not a metric. In particular, it is asymmetric and we use the prior distribution P as the second argument. If we have $P^*(X) > 0 = P(X)$ for some $X \subseteq \text{Args}$, we let $d_{KL}(P^*, P) = \infty$ as usual.

When updating our belief with respect to a set of literals Φ, we might be interested only in the change with respect to atoms not appearing in Φ. The following two distance measures capture this intuition. Here, $X \subseteq \text{Args}$ denotes a set of arguments that is supposed to be updated.

- **Atomic Distance:** $d_{\text{At}}^X(P, P^*) = \sum_{B \in \text{Args} \setminus X} |P(B) - P^*(B)|$.
- **Joint Distance:** $d_{\text{Jo}}^X(P, P^*) = \sum_{C \in \text{Conj}(\text{Args} \setminus X)} |P(C) - P^*(C)|$.

Table 1. Illustration of different change measures.

A	B	P_0	P_1	P_2	P_3	P_4	d	d_1	d_2	d_∞	d_{KL}	$d_{At}^{\{A\}}$	$d_{Jo}^{\{A\}}$
0	0	0.3	0.4	0.5	0.5	0.3	$d(P_0, P1)$	0.3	0.03	0.1	0.16	0.1	0.1
0	1	0.3	0.2	0.2	0.4	0.3	$d(P_0, P2)$	0.4	0.06	0.2	0.09	0.1	0.2
1	0	0.3	0.2	0.2	0.1	0.2	$d(P_0, P3)$	0.6	0.1	0.2	∞	0	0
1	1	0.1	0.1	0.1	0	0.2	$d(P_0, P4)$	0.2	0.02	0.1	0.05	0.01	0.2

Both measures can be zero even though the distributions are unequal. This happens, when they have equal marginal probabilities on Args$\backslash X$ for the atomic distance measure and when they have equal marginal probabilities on Conj(Args$\backslash X$) for the joint distance measure. Hence, they are not metrics. However, they are pseudometrics as we explain in the full version[1]. We illustrate the different change measures in Fig. 1.

We consider the following minimality properties for each change measure d, set of rationality postulates R and satisfaction condition S:

– **R-S-d-minimality:** If $P^* \in U(P, F)$, then P^* minimizes the distance to P over $\mathrm{Mod}_{R,S}(F)$.

R-S-d-minimality demands that we update in such a way that we minimize the distance to the prior distribution among all probability distributions that satisfy the argument graph and the new information with respect to the chosen semantics (Table 1).

4 Refinement-Based Update Functions

In [10], several update functions have been proposed that are defined by means of the following refinement function. They are restricted in the sense that they are defined only for literals.

Definition 1. *Let $L \in$ Formulae(G) be a literal, let P be a probability distribution, and let $\lambda \in [0, 1]$. The* **refinement function** $H_\lambda : \mathcal{P} \times \{A, \neg A \mid A \in$ Args$\} \to \mathcal{P}$ *is defined by $H_\lambda(P, L) = P^*$ as follows where $X \subseteq$ Args*

$$P^*(X) = \begin{cases} P(X) + \lambda \cdot P(h_L(X)) & \text{if } X \models L \\ (1 - \lambda) \cdot P(X) & \text{if } X \models \neg L, \end{cases}$$

where $h_L(X) = X \backslash \{A\}$ if $L = A$ and $h_L(X) = X \cup \{A\}$ if $L = \neg A$ for some $A \in$ Args.

---

[1] Full version is at http://www0.cs.ucl.ac.uk/staff/a.hunter/papers/updatefunctionfull .pdf.

Table 2. Illustration of refinement-based updates for a graph with C attacks B and B attacks A. Note, by definition, $H_1(P, A) = U_{na}(P, A)$ and $H_1(P, B) = U_{na}(P, B)$.

A	B	C	P	$H_{0.75}(P, A)$	$H_1(P, A)$	$U_{na}(P, B)$	$U_{tr}(P, B)$	$U_{tr}(P, A)$	$U_{st}(P, B)$	$U_{st}(P, A)$
0	0	0	0.2	0.05	0	0	0	0	0	0.2
0	1	0	0.5	0.125	0	0.7	1	0	0.7	0.5
0	0	1	0	0	0	0	0	0	0	0
0	1	1	0.1	0.025	0	0.1	0	0	0.3	0.1
1	0	0	0	0.15	0.2	0	0	0.7	0	0
1	1	0	0	0.375	0.5	0	0	0	0	0
1	0	1	0.1	0.1	0.1	0	0	0.3	0	0.1
1	1	1	0.1	0.175	0.2	0.2	0	0	0	0.1

If we think of interpretations as bit vectors $(b_1, \ldots, b_n)$ where b_i is the truth state of the i-th argument, redistribution with respect to A_i can be explained as follows: for each bit vector $(b_1, \ldots, b_n)$, if $b_i = 1$, then move a fraction λ of the probability mass of $(b_1, \ldots, b_{i-1}, 0, b_{i+1}, \ldots, b_n)$ to $(b_1, \ldots, b_n)$. We illustrate this in Table 2.

Let us note that refinement functions are actually commutative in the sense that $H_{\lambda_2}(H_{\lambda_1}(P, L_1), L_2) = H_{\lambda_1}(H_{\lambda_2}(P, L_2), L_1)$, see [10], Proposition 8. Since the order in which we add literals is not important, refinement functions can also be applied to sets of literals Φ recursively, where we let $H_\lambda(P, \emptyset) = P$ and $H_\lambda(P, \Phi \cup \{L\}) = H_\lambda(H_\lambda(P, L), \Phi)$. As the following lemma explains, for $\lambda = 1$, updating with multiple literals comes down to shifting probability mass to the interpretations that satisfy the conjunction of these literals.

Lemma 1. Let $X = \{A_1, \ldots, A_k\} \subseteq \text{Args}$ and for $i = 1, \ldots, k$, let $L_i \in \{A_i, \neg A_i\}$. Let P be a probability distribution and let $H_1(P, \{L_1, \ldots, L_k\}) = P^*$. Then for all $C \in \text{Conj}(X)$ and $D \in \text{Conj}(\text{Args} \setminus X)$,

$$P^*(C \wedge D) = \begin{cases} P(C \wedge D) + \sum_{C' \in \text{Conj}(X) \setminus \{C\}} P(C' \wedge D) & \text{if } C = \bigwedge_{i=1}^k L_i \\ 0 & \text{else.} \end{cases}$$

We will now analyze some refinement-based update functions from [10] by means of the properties introduced in the previous section. Since the refinement-based update functions are only defined for atoms or literals, Tautology, Contradiction and Representation Invariance are not interesting here. However, it is reasonable to consider Idempotence and Order Invariance restricted to literals.

The naive update function shifts the probability mass from an interpretation X that violates L to the corresponding interpretation that is obtained from X by flipping the truth state of the argument in L.

Definition 2 ([10]). *The **naive update function*** $U_{na} : \mathcal{P} \times \{A, \neg A \mid A \in \text{Args}\} \rightarrow \mathcal{P}$ *is defined by* $U_{na}(P, L) = H_1(P, L)$.

U_{na} satisfies the following properties.

Proposition 1. U_{na} *satisfies Uniqueness, Completeness, Idempotence, Order Invariance and STRICT-Satisfaction.*

The naive update function is intended to model persuadees who believe any arguments that are posited in a dialogue. The function does not take the structure of the argument graph into account, and therefore can generally violate all rationality postulates that we introduced over argument graphs. However, given an update literal over the argument A, the naive update is guaranteed to be minimal with respect to $d_{J_O}^{\{A\}}$ - in fact, the change with respect to $d_{J_O}^{\{A\}}$ is 0 as we show in the full version of this paper.

The next two update functions maintain consistency with the argument graph by also considering arguments that are connected to the argument whose state we update. They are restricted to atomic arguments, however.

The trusting update refines the naive update by also shifting the probability mass from all interpretations that satisfy the attackers and attackees of the update argument.

Definition 3 ([10]). *The **trusting update function** $U_{\mathrm{tr}} : \mathcal{P} \times \mathsf{Args} \to \mathcal{P}$ is defined by $U_{\mathrm{tr}}(P, A) = H_1(P, \Phi)$, where $\Phi = \{A\} \cup \{\neg C \mid (A, C) \in \mathsf{Attacks}(G)$ or $(C, A) \in \mathsf{Attacks}(G)\}$.*

U_{tr} satisfies the following properties.

Proposition 2. U_{tr} *satisfies Uniqueness, Completeness, Idempotence, Order Invariance, STRICT-Satisfaction and R-Satisfaction for all $R \subseteq \{RAT, COH\}$.*

U_{tr} can violate the remaining R-Satisfaction properties, but it does guarantee that the joint distance to the prior distribution is 0. However, the joint distance is now not only defined with respect to the update argument, but also with all of its attackers and attackees as we show in the full version.

The strict update function conditionally updates the probability of an argument to 1. In order to maintain consistency with the argument graph, the update is only performed if no attackers of the argument are believed in the current epistemic state. If the update is performed, the belief in attacked arguments will additionally be set to 0.

Definition 4 ([10]). *The **strict update function** is a function $U_{\mathrm{st}} : \mathcal{P} \times \mathsf{Args} \to \mathcal{P}$. For $A \in \mathsf{Args}$, let $\Phi = \{A\} \cup \{\neg C \mid (A, C) \in \mathsf{Attacks}\}$ and let the constraint $C(P)$ be true iff for all $(B, A) \in \mathsf{Attacks}$, $P(B) \leq 0.5$. Then $U_{\mathrm{st}}(P, A) = P^*$ where*

$$P^* = \begin{cases} H_1(P, \Phi) & \textit{if } C(P) \\ P & \textit{else} \end{cases}$$

U_{st} satisfies the following properties.

Proposition 3. U_{st} *satisfies Uniqueness, Completeness, Idempotence and R-Satisfaction for all $R \subseteq \{RAT, COH, SFOU, FOU\}$.*

U_{st} does not satisfy Order Invariance, but it satisfies all semantical constraints except R-OPT and R-SOPT. U_{st} again guarantees joint distance 0, this time with respect to the update argument and all of its attackees. We refer again to the full version of this paper for more details and proofs.

In [10], $H_{0.75}$ is considered as an alternative to H_1 in the above definition, and this is used to model skeptical agents who do not entirely believe an argument when updating.

5 R-S-d Update Functions

We now consider another class of update functions. Whereas refinement-based update functions are based on the idea of shifting probability mass in a specific way, we will now consider a more declarative approach using tools from numerical optimization. R-S-d Update Functions are defined by minimizing some notion of distance subject to semantical constraints.

Definition 5. *Let $R \subseteq \{RAT, COH, SFOU, FOU, SOPT, OPT\}$, $S \in \{STRICT, \epsilon\text{-}WEAK\}$ and $d \in \{d_1, d_2, d_\infty, d_{At}^X, d_{Jo}^X\}$. An* **R-S-d Update Function** *$U_{R,S,d} : \mathcal{P} \times \text{Form} \to 2^{\mathcal{P}}$ is defined by*

$$U_{R,S,d}(P, F) = \arg \min_{P' \in \text{Mod}_{R,S}(F)} d(P, P').$$

Let us first note that most R-S-d update functions have some nice analytical properties (Table 3).

Lemma 2. *For each $R \subseteq \{COH, SFOU, FOU, SOPT, OPT\}$ (we left out RAT), $S \in \{STRICT, \epsilon\text{-}WEAK\}$ and $d \in \{d_1, d_2, d_\infty, d_m, d_{KL}, d_{At}^X, d_{Jo}^X\}$, computing $U_{R,S,d}(P, F)$ corresponds to a convex combination problem. In particular, the set $U_{R,S,d}(P, F)$ will be non-empty, convex and compact whenever $\text{Mod}_{R,S}(F) \neq \emptyset$.*

If R includes RAT, $U_{R,S,d}(P, F)$ will be non-empty and compact whenever $\text{Mod}_{R,S}(F) \neq \emptyset$.

We have the following general guarantees for R-S-d update functions.

Proposition 4. *For all $R \subseteq \{RAT, COH, SFOU, FOU, SOPT, OPT\}$, $S \in \{STRICT, \epsilon\text{-}WEAK\}$ and $d \in \{d_1, d_2, d_\infty, d_m, d_{At}^X, d_{Jo}^X\}$, $U_{R,S,d}$ satisfies Completeness (if the update argument is R-S-consistent), R-consistency, S-consistency and R-S-d-minimality.*

If we exclude RAT from R and $d \in \{d_2, d_{KL}\}$, $U_{R,S,d}$ also satisfies Uniqueness, Tautology, Contradiction, Representation Invariance and Idempotence.

We can give some stronger guarantees for some special cases, see the full paper for a detailed analysis.

Order Invariance can be violated for many combinations of semantical constraints and change measures. We give a simple example for the Euclidean distance without semantical constraints on the argument graph.

Table 3. Illustration of R-S-d updates with $R_1 = \{COH\}$, $R_2 = \{COH, SOPT\}$, $S = STRICT$ and $d = d_2$.

A	B	C	P	$U_{R_1,S,d}(P,A)$	$U_{R_1,S,d}(P,B)$	$U_{R_2,S,d}(P,A)$	$U_{R_2,S,d}(P,\top)$
0	0	0	0.2	0	0	0	0.17
0	1	0	0.5	0	1	0	0.49
0	0	1	0	0	0	0	0
0	1	1	0.1	0	0	0	0.07
1	0	0	0	0.45	0	0	0.02
1	1	0	0	0	0	0	0.5
1	0	1	0.1	0.55	0	1	0.09
1	1	1	0.1	0	0	0	0.12

Example 2. Consider an argument graph over A, B, let $R = \emptyset$, $S = STRICT$ and $d = d_2$. Let P be defined by $P(\{B\}) = 0.5$, $P(\{A, B\}) = 0.5$. Then $P_1 = U_{R,S,d}(U_{R,S,d}(P, A), B)$ is given by $P_1(\{B\}) = 0.125$, $P_1(\{A, B\}) = 0.875$, whereas $P_2 = U_{R,S,d}(U_{R,S,d}(P, B), A)$ is given by $P_2(\{A\}) = 0.25$, $P_2(\{A, B\}) = 0.75$.

What can we say about the relationship between refinement-based update functions and R-S-d update functions? We first note that R-S-d-update functions generalize the naive update function in the following sense.

Proposition 5. *Consider an arbitrary set of semantical constraints* $R \subseteq \{RAT, COH, SFOU, FOU, SOPT, OPT\}$, *a probability distribution* $P \in \mathcal{P}$ *and let* $L \in \{A, \neg A\}$ *be a literal for some* $A \in$ Args. *If there is a* $P^* \in$ $\mathrm{Mod}_{R,STRICT}(L)$ *such that* $d_{Jo}^{\{A\}}(P, P^*) = 0$ *then* $U_{R,STRICT,d_{Jo}^{\{A\}}}(P, L) = \{U_{\mathrm{na}}(P, L)\}$.

Remark 2. Note that if there is no $P^* \in \mathrm{Mod}_{R,STRICT}(L)$ such that $d_{Jo}^{\{A\}}(P, P^*) = 0$, then applying the Naive update function will violate some semantical constraint in R (because the probability distribution resulting from the naive update will have distance 0). Hence, $U_{R,STRICT,d_{Jo}^{\{A\}}}$ agrees with U_{na} whenever U_{na} is consistent with R. Otherwise, $U_{R,STRICT,d_{Jo}^{\{A\}}}$ will select probability distributions that are consistent with R and minimize the joint distance.

In particular, the Naive update function can be thought of as a special case of the following R-S-d-update function.

Corollary 1. $U_{\emptyset,STRICT,d_{Jo}^{\{A\}}}(P, L) = \{U_{\mathrm{na}}(P, L)\}$.

The trusting method can similarly be generalized by an R-S-d-update function.

Proposition 6. *Consider an arbitrary set of semantical constraints* $R \subseteq \{RAT, COH, SFOU, FOU, SOPT, OPT\}$, *a probability distribution* $P \in \mathcal{P}$ *and let* $L \in \{A, \neg A\}$ *be a literal for some* $A \in$ Args. *Let* $X' = \{C \mid (A, C) \in$ Attacks(G) *or* $(C, A) \in$ Attacks$(G)\}$ *and* $X = \{A\} \cup X'$. *If there is a* $P^* \in \text{Mod}_{R,STRICT}(L)$ *such that* $d_{J_o}^X(P, P^*) = 0$ *then* $U_{R,STRICT,d_{J_o}^X}(P, L \wedge \bigwedge_{C \in X'} \neg C) = \{U_{tr}(P, L)\}$.

Corollary 2. $U_{\emptyset,STRICT,d_{J_o}^X}(P, L \wedge \bigwedge_{C \in X'} \neg C) = \{U_{tr}(P, L)\}$.

We could get a similar result for the strict update using the joint distance over the update argument and its attackees. This would require a case differentiation analogous to the case differentiation that is used for the strict update.

6 Conclusions and Future Work

Most proposals for dialogical argumentation focus on protocols (*e.g.*, [2,5,14,15]) with strategies being under-developed. See [18] for a review of strategies in multi-agent argumentation. There are proposals for modelling the likelihood of the moves that an opposing agent might make (e.g. [6,7,16,17]). Note, however, that none of the above proposals consider the beliefs of the opposing agent. In [1], a planning system is used by the persuader to optimize choice of arguments based on belief in premises. However, there is no consideration of how the beliefs are updated during the dialogue.

The epistemic approach to probabilistic argumentation offers a formal framework for modelling a persuadee's beliefs in arguments. There are methods for updating beliefs during a dialogue [10], for efficient representation and reasoning with the pesuadee model [8], and for harnessing decision-theoretic decision rules for optimizing the choice of arguments based on the persuadee model [9]. Therefore, the framework for update functions presented in this paper clarifies and extends the space of update functions that we can harness in persuasion dialogues.

There are several interesting directions for future work. First, we can investigate different ways to deal with the problem of non-unique solutions. We might focus on some best solution or represent epistemic states by sets of probability distributions rather than by a single one. Second, we can deal with inconsistencies like in Example 1 in different ways. We might consider priorities over different semantical constraints [12] or select solutions that violate the constraints in a minimal way [3,13]. Third, we can try to include more expressive argumentation frameworks by introducing numerical constraints for other relations than attack relations.

Acknowledgements. This research was partly funded by EPSRC grant EP/N008294/1 for the Framework for Computational Persuasion project.

References

1. Black, E., Coles, A., Bernardini, S.: Automated planning of simple persuasion dialogues. In: Bulling, N., Torre, L., Villata, S., Jamroga, W., Vasconcelos, W. (eds.) CLIMA 2014. LNCS (LNAI), vol. 8624, pp. 87–104. Springer, Cham (2014). doi:10.1007/978-3-319-09764-0_6
2. Caminada, M., Podlaszewski, M.: Grounded semantics as persuasion dialogue. In: Proceedings of the International Conference on Computational Models of Argument (COMMA), pp. 478–485 (2012)
3. Daniel, L.: Paraconsistent probabilistic reasoning. Ph.D. thesis, L'École Nationale Supérieure des Mines de Paris (2009)
4. Dung, P.: On the acceptability of arguments and its fundamental role in non-monotonic reasoning, logic programming, and n-person games. Artif. Intell. **77**, 321–357 (1995)
5. Fan, X., Toni, F.: Assumption-based argumentation dialogues. In: Proceedings of IJCAI 2011, pp. 198–203 (2011)
6. Hadjinikolis, C., Siantos, Y., Modgil, S., Black, E., McBurney, P.: Opponent modelling in persuasion dialogues. In: Proceedings of IJCAI 2013, pp. 164–170 (2013)
7. Hadoux, E., Beynier, A., Maudet, N., Weng, P., Hunter, A.: Optimization of probabilistic argumentation with Markov decision models. In: Proceedings of IJCAI 2015, pp. 2004–2010 (2015)
8. Hadoux, E., Hunter, A.: Computationally viable handling of beliefs in arguments for persuasion. In: Proceedings of ICTAI 2016, pp. 319–326. IEEE Press (2016)
9. Hadoux, E., Hunter, A.: Strategic sequences of arguments for persuasion using decision trees. In: Proceedings of AAAI 2017, pp. 1128–1134. AAAI Press (2017)
10. Hunter, A.: Modelling the persuadee in asymmetric argumentation dialogues for persuasion. In: Proceedings of IJCAI 2015, pp. 3055–3061 (2015)
11. Hunter, A., Thimm, M.: On partial information and contradictions in probabilistic abstract argumentation. In: Principles of Knowledge Representation and Reasoning (KR 2016), pp. 53–62 (2016)
12. Potyka, N.: Reasoning over linear probabilistic knowledge bases with priorities. In: Beierle, C., Dekhtyar, A. (eds.) SUM 2015. LNCS (LNAI), vol. 9310, pp. 121–136. Springer, Cham (2015). doi:10.1007/978-3-319-23540-0_9
13. Potyka, N., Thimm, M.: Probabilistic reasoning with inconsistent beliefs using inconsistency measures. In: Proceedings of IJCAI 2015, pp. 3156–3163 (2015)
14. Prakken, H.: Coherence and flexibility in dialogue games for argumentation. J. Logic Comput. **15**(6), 1009–1040 (2005)
15. Prakken, H.: Formal sytems for persuasion dialogue. Knowl. Eng. Rev. **21**(2), 163–188 (2006)
16. Rienstra, T., Thimm, M., Oren, N.: Opponent models with uncertainty for strategic argumentation. In: Proceedings of IJCAI 2013, pp. 332–338 (2013)
17. Rosenfeld, A., Kraus, S.: Providing arguments in discussions on the basis of the prediction of human argumentative behavior. ACM Trans. Interact. Intell. Syst. **6**, 30:1–30:33 (2016)
18. Thimm, M.: Strategic argumentation in multi-agent systems. Kunstliche Intelligenz **28**, 159–168 (2014)

From Structured to Abstract Argumentation: Assumption-Based Acceptance via AF Reasoning

Tuomo Lehtonen[1], Johannes P. Wallner[2](✉), and Matti Järvisalo[1]

[1] HIIT, Department of Computer Science, University of Helsinki, Helsinki, Finland
[2] Institute of Information Systems, TU Wien, Vienna, Austria
wallner@dbai.tuwien.ac.at

Abstract. We study the applicability of abstract argumentation (AF) reasoners in efficiently answering acceptability queries over assumption-based argumentation (ABA) frameworks, one of the prevalent forms of structured argumentation. We provide a refined algorithm for translating ABA frameworks to AFs allowing the use of AF reasoning to answer ABA acceptability queries, covering credulous and skeptical acceptance problems over ABAs in a seamless way under several argumentation semantics. We empirically show that the approach is complementary with a state-of-the-art ABA reasoning system.

1 Introduction

Argumentation is today a vibrant area of modern AI research, providing formalisms for representing and reasoning about conflicting arguments, aiming at conflict resolution through detecting sets of non-conflicting arguments which together counter—or defend themselves against—all counterarguments.

Several argumentation formalisms have been proposed, with different desirable properties. Perhaps the simplest formalism for argumentation are abstract argumentation frameworks (AFs) [11]. AFs allow for representing conflicts—or attacks—between arguments as directed graphs, where nodes represent abstract arguments, and edges represent attacks. Several reasoning system implementations for AF reasoning exists today [5,6,17,18,24,25], especially for central AF reasoning problems such as credulous and skeptical acceptance of arguments under various AF semantics.

Another central formalism is structured argumentation [1–3,22,26] in which, in contrast to abstract argumentation, the internal structure of arguments is made explicit through derivations from more basic structures. One well-known approach to structured argumentation is assumption-based argumentation (ABA) [3,13,29]. In ABA arguments are represented compactly as graph-based derivations [7] from a given rule-based deductive system over sentences,

Work funded by Academy of Finland, grants 251170 COIN, 276412, and 284591; Research Funds of the University of Helsinki; and the Austrian Science Fund (FWF): I2854 and P30168.

A. Antonucci et al. (Eds.): ECSQARU 2017, LNAI 10369, pp. 57–68, 2017.
DOI: 10.1007/978-3-319-61581-3_6

starting from assumptions. A central approach to reasoning about acceptability of arguments over ABAs are so-called dispute derivations [7,12,14,20,21,28], implemented in various ABA reasoning systems [7–9,14,19–21,28]. The abagraph system [7] supporting credulous reasoning over ABAs under the admissible and grounded semantics represents the current state of the art.

While systems for reasoning over AFs and ABAs have been developed, the applicability of state-of-the-art abstract argumentation reasoners for reasoning about assumption-based argumentation frameworks has received little attention. To bridge this gap, we study the applicability of state-of-the-art abstract argumentation reasoners in efficiently answering acceptability queries over ABA frameworks. While theoretical work on mapping ABAs to AFs exists [4,14], here we concretely implement an approach to reasoning about acceptance of sentences in assumption-based argumentation via translating ABA frameworks into abstract argumentation frameworks, and thereafter using AF reasoning to decide acceptance of sentences. While it would be desirable to exactly compute a small, yet sufficient, set of AF arguments for a given ABA, we show that restricting argument construction to only those arguments satisfying a minimality condition in their supports, which we call relevant arguments, is computationally very demanding: we prove that counting the number of such relevant arguments is #P-complete. To overcome this obstacle, we propose an algorithm for overapproximating the set of relevant arguments for a given ABA framework. We implement the reasoning part by answer set programming (ASP) encodings specifically suited for the types of AFs the translation gives rise to. We show that a prototype implementation of the approach is complementary in terms of performance with the state-of-the-art abagraph system for credulous acceptance in ABA. Our approach is generic in that it covers both credulous and skeptical acceptance problems under several central argumentation semantics over ABAs in a seamless way. Proofs of the main theorems are available in the paper supplement online at https://cs.helsinki.fi/group/coreo/ecsqaru17.

2 Preliminaries

Assumption-Based Argumentation. We recall definitions related to assumption-based argumentation (ABA) [3,29], following [10]. We assume a deductive system $(\mathcal{L}, \mathcal{R})$ with $\mathcal{L}$ a formal language, i.e., a countable set of sentences, and $\mathcal{R}$ a set of inference rules over $\mathcal{L}$ with a rule $r \in \mathcal{R}$ having the form $a_0 \leftarrow a_1, \ldots, a_n$ with $a_i \in \mathcal{L}$. We denote the head of rule r by $head(r) = \{a_0\}$ and the (possibly empty) body of r by $body(r) = \{a_1, \ldots, a_n\}$. A sentence $a \in \mathcal{L}$ is derivable from a set $X \subseteq \mathcal{L}$ via rules $\mathcal{R}$, denoted by $X \vdash_\mathcal{R} a$, if there is a sequence of rules $(r_1, \ldots, r_n)$ s.t. $head(r_n) = a$ and for each rule r_i it holds that $r_i \in \mathcal{R}$ and each sentence in the body of r_i is derived from rules earlier in the sequence or in X, i.e., $body(r_i) \subseteq X \cup \bigcup_{j<i} head(r_j)$. The deductive closure for X w.r.t. rules $\mathcal{R}$ is given by $Th_\mathcal{R}(X) = \{a \mid X \vdash_\mathcal{R} a\}$.

An ABA framework is a tuple $(\mathcal{L}, \mathcal{R}, \mathcal{A}, ^-)$ with $(\mathcal{L}, \mathcal{R})$ a deductive system, a set of assumptions $\mathcal{A} \subseteq \mathcal{L}$, and a function $^-$ (contrary function) mapping assumptions $\mathcal{A}$ to sentences $\mathcal{L}$. We focus on flat ABA frameworks where assumptions cannot be derived. Let $D = (\mathcal{L}, \mathcal{R}, \mathcal{A}, {}^-)$ be an ABA framework. A set of assumptions $\Delta \subseteq \mathcal{A}$ attacks an assumption $b \in \mathcal{A}$ in the ABA framework D if the contrary of b is derivable from Δ in D, i.e., $\overline{b} \in Th_{\mathcal{R}}(\Delta)$. Further, Δ attacks a set of assumptions $\Delta' \subseteq \mathcal{A}$ in the ABA framework D if an assumption in Δ' is attacked by Δ, i.e., $Th_{\mathcal{R}}(\Delta) \cap \{\overline{a} | a \in \Delta'\} \neq \emptyset$.

Definition 1. *Let* $D = (\mathcal{L}, \mathcal{R}, \mathcal{A}, {}^-)$ *be an ABA framework. Further, let* $\Delta \subseteq \mathcal{A}$ *be a set of assumptions that does not attack itself in* D. *Set* Δ *is*

- admissible *in* D *if each set of assumptions* Δ' *that attacks* Δ *is attacked by* Δ;
- preferred *in* D *if* Δ *is admissible and there is no admissible set of assumptions* Δ' *in* D *with* $\Delta \subset \Delta'$; *and*
- stable *in* D *if each* $a \in \mathcal{A} \backslash \Delta$ *is attacked by* Δ.

We use the term σ-assumption-set to refer to an assumption set under a specific semantics $\sigma \in \{adm, stb, prf\}$.[1] Let $D = (\mathcal{L}, \mathcal{R}, \mathcal{A}, {}^-)$ be an ABA framework and σ a semantics. A sentence $s \in \mathcal{L}$ is credulously accepted in D under semantics σ if there is a σ-assumption-set Δ s.t. $s \in Th_{\mathcal{R}}(\Delta)$; and skeptically accepted in D under semantics σ if it holds that $s \in Th_{\mathcal{R}}(\Delta)$ for all σ-assumption-sets Δ.

Complexity of reasoning of (flat) ABA frameworks [10] is shown in Fig. 1.

	ABA		AF	
semantics	cred	skept	cred	skept
admissible	NP-c	P-c	NP-c	trivial
stable	NP-c	coNP-c	NP-c	coNP-c
preferred	NP-c	Π_2^p-c	NP-c	Π_2^p-c

Fig. 1. Complexity of reasoning.

Example 1. An ABA framework is shown in Fig. 2 (left) with $\mathcal{L} = \{a, b, c, d, e, f, g, h, i\}$, as well as the admissible, stable, and preferred assumption sets. Sentences g and h are credulously accepted under $\sigma \in \{adm, prf, stb\}$, since they can be derived from $\{a\}$ and $\{c\}$. Further, i is skeptically accepted under σ, since i is derivable from $\emptyset$.

Abstract Argumentation Frameworks. An abstract argumentation framework (AF) [11] is a pair $F = (A, R)$, where A is a finite non-empty set of arguments and $R \subseteq A \times A$ is the attack relation. The pair $(a, b) \in R$ indicates that a attacks b. A set $S \subseteq A$ attacks an argument b (in F) if there is an $a \in S$

rules $\mathcal{R}$		contr.	ass. sets σ	
$d \leftarrow a$	$g \leftarrow e$	$\overline{a} = h$	$\emptyset$	adm
$e \leftarrow a, b$	$d \leftarrow g$	$\overline{b} = e$	$\{a\}$	adm, prf, stb
$f \leftarrow c$	$h \leftarrow f$	$\overline{c} = d$	$\{c\}$	adm
$e \leftarrow d$	$i \leftarrow$		$\{b, c\}$	adm, prf, stb

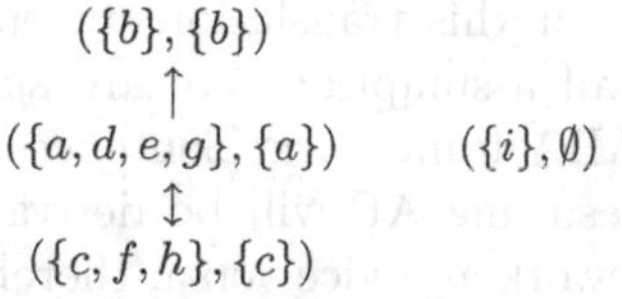

$(\{b\}, \{b\})$

$\updownarrow$

$(\{a, d, e, g\}, \{a\})$ $(\{i\}, \emptyset)$

$\updownarrow$

$(\{c, f, h\}, \{c\})$

Fig. 2. Example ABA with $\mathcal{A} = \{a, b, c\}$ (left) and the corresponding AF (right).

---

[1] We call, for reasons of uniformity and brevity, admissible sets a semantics; this is not meant to prescribe a particular logical stance to the frameworks.

s.t. $(a, b) \in R$. An argument $a \in A$ is *defended* (in F) by a set $S \subseteq A$ if, for each $b \in A$ such that $(b, a) \in R$, it holds that S attacks b.

AF semantics are defined through functions σ which assign to each AF $F = (A, R)$ a set $\sigma(F) \subseteq 2^A$ of extensions. We consider for σ the functions adm, stb, and prf.

Definition 2. *Let $F = (A, R)$ be an AF. A set $S \subseteq A$ is* conflict-free *(in F) if there are no $a, b \in S$ such that $(a, b) \in R$. We denote the collection of conflict-free sets of F by $cf(F)$. For a conflict-free set $S \in cf(F)$ it holds that*

- *$S \in adm(F)$ iff each $a \in S$ is defended by S;*
- *$S \in prf(F)$ iff $S \in adm(F)$ and $\nexists S' \in adm(F)$ with $S \subset S'$; and*
- *$S \in stb(F)$ iff each $a \in A \backslash S$ is attacked by S.*

We use "σ-extension" to denote an extension under a semantics σ. Let $F = (A, R)$ be an AF. An argument $a \in A$ is credulously accepted in F under σ if there is an $E \in \sigma(F)$ s.t. $a \in E$. An argument a is skeptically accepted in F under σ if a is contained in every $E \in \sigma(F)$. For complexity of AF reasoning [15] see Fig. 1.

3   From ABA to AF

The focus of this work is on studying the applicability of abstract argumentation reasoning tools for reasoning about acceptance of sentences in assumption-based argumentation frameworks. Given an ABA and a credulous/skeptical query as a sentence in the ABA, our approach to answer the query consists of the following two steps.

1. Translate the ABA framework into an AF in a way that the ABA query can be answered by applying AF reasoning principles on the resulting AF.
2. Adjust an AF reasoning system to answer the ABA query on the AF from step 1.

In this section we adapt translations of ABA frameworks [4,14] to AFs to suit our goal of computational feasibility. The idea of the approach is to view subsets of the assumptions, and sentences derived from these sets, as abstract arguments. The assumptions of such an argument are called support of the argument. A key point for this translation, to ensure correctness, is to construct arguments so that all assumption sets are sufficiently covered, not missing crucial parts of the ABA framework. Sentences contained in an argument in a σ-extension of the resulting AF will be derivable in a σ-assumption-set of the original ABA framework and vice versa, thereby aligning the corresponding reasoning tasks of ABA frameworks and AFs.

In order to make step 2 computationally feasible, care needs to be taken in order to ensure that the AF resulting from step 1 does not become restrictively large (in terms of the number of arguments) in order to enable reasoning on the AF. To this end, we consider constructing only those arguments, which we call *relevant arguments*, whose support is minimal, in the sense that there is a sentence derivable from the support, but the sentence is not derivable from any proper subset of the support. However, we will show that the complexity of computing (exactly) the set of relevant arguments is restrictive for practical purposes. Motivated by both the computational hardness result and the need for restriction of the number of arguments, we then, in the subsequent sections, propose an algorithm for over-approximating the set of relevant arguments of a given ABA, and detail an approach to step 2 via answer set programming.

Key to the translation of ABA frameworks to AFs are the arguments for the AF, which are viewed as pairs of a set of assumptions and sentences derived from the set of assumptions. With the aim of focusing on relevant arguments, we generalize and adapt the concept of support-minimality [7, Definition 4.11]. In [7] support-minimality is defined for arguments with a single claim (derivation for a single sentence).

Definition 3. *Let* $D = (\mathcal{L}, \mathcal{R}, \mathcal{A}, {}^-)$ *be an ABA framework. We define the set of sets of assumptions* $minsupp(D)$ *by* $\Delta \in minsupp(D)$ *iff* $\bigcup_{\Delta' \subset \Delta} Th_{\mathcal{R}}(\Delta') \subset Th_{\mathcal{R}}(\Delta)$.

In words, a set of assumptions Δ is a minimal support if there is a sentence derivable from Δ via $\mathcal{R}$ but not from any proper subset $\Delta' \subset \Delta$. Relevant arguments are defined as pairs of a set of sentences and a minimal support.

Definition 4. *Let* $D = (\mathcal{L}, \mathcal{R}, \mathcal{A}, {}^-)$ *be an ABA framework,* $L \subseteq \mathcal{L}$*, and* $\Delta \subseteq \mathcal{A}$*. A pair* (L, Δ) *is a relevant argument (for D) if the following two conditions hold: (i)* $\Delta \subset minsupp(D)$*; and (ii)* $L = Th_{\mathcal{R}}(\Delta) \backslash (\bigcup_{\Delta' \subset \Delta} Th_{\mathcal{R}}(\Delta'))$.

In words, a pair (L, Δ) is a relevant argument for a given ABA if Δ is in $minsupp(D)$ (first item), and L contains those sentences that are derivable from Δ but not any proper subset of Δ (second item).

Example 2. Consider the ABA framework from Example 1. The sets in $minsupp(D)$ are $\{a\}$, $\{b\}$, $\{c\}$, and $\emptyset$. The admissible assumption set $\{b, c\}$ is not in $minsupp(D)$ since all sentences derivable from $\{b, c\}$ are derivable from $\{b\}$ or $\{c\}$. For each set in $minsupp(D)$ there is a relevant argument, e.g., $(\{a, d, e, g\}, \{a\})$ is a relevant argument for the ABA framework and all sentences in $\{a, d, e, g\}$ can be derived from a, and all sentences derivable from $\{a\}$ but not $\emptyset$ are contained in the first component.

Definition 5. *Let* $D = (\mathcal{L}, \mathcal{R}, \mathcal{A}, {}^-)$ *be an ABA framework. An AF* $F = (A, R)$ *corresponds to the ABA D if the following two conditions hold. (i) A is the set of relevant arguments for D; and (ii)* $R = \{((L, \Delta), (L', \Delta')) | L \cap \{\overline{x} \mid x \in \Delta'\} \neq \emptyset\}$.

Briefly put, a corresponding AF for a given ABA framework contains the relevant argument for each set of assumptions in $minsupp(D)$, i.e., $|A| = |minsupp(D)|$, and attacks based on the supports and the derived sentences. In Fig. 2, the corresponding AF (right) for the ABA framework (left) is shown. In the following formal result, that follows the spirit of [14, Theorem 2.2] and

[4, Theorem 6], we show that we have a correspondence between the ABA framework and the corresponding AF in terms of the semantics, which allows for utilization of AF reasoners on the AF to answer ABA queries. We define $sentences(E) = \bigcup_{(L,\Delta)\in E} L$.

Theorem 1. *Let $D = (\mathcal{L}, \mathcal{R}, \mathcal{A},^-)$ be an ABA framework, $\Delta \subseteq \mathcal{A}$, and $\sigma \in \{adm, stb, prf\}$. For an AF $F = (A, R)$ that corresponds to D, and $E \subseteq A$, it holds that*

- *if Δ is a σ-assumption-set of D, then $E = \{(L, \Delta') \in A \mid \Delta' \subseteq \Delta\}$ is a σ-extension of F, and $Th_{\mathcal{R}}(\Delta) = sentences(E)$;*
- *if E is a σ-extension of F, then $\Delta = \bigcup_{(L,\Delta')\in E} \Delta'$ is a σ-assumption-set of D, and $Th_{\mathcal{R}}(\Delta) \supseteq sentences(E)$ for $\sigma = adm$, and $Th_{\mathcal{R}}(\Delta) = sentences(E)$ for $\sigma \in \{stb, prf\}$.*

Based on this formal correspondence, we can answer credulous (skeptical) acceptability queries in an ABA framework as specified in the next corollary.

Corollary 1. *Let $D = (\mathcal{L}, \mathcal{R}, \mathcal{A},^-)$ be an ABA framework, $l \in L$, $\sigma = \{adm, stb, prf\}$, $\sigma' = \{stb, prf\}$, and AF $F = (A, R)$ the corresponding AF for D. It holds that*

- *l is credulously accepted under σ in D iff there is a credulously accepted argument (L, Δ) under σ in F with $l \in L$;*
- *l is skeptically accepted under σ' in D iff for each σ'-extension E of F it holds that $l \in sentences(E)$.*

Skeptical acceptance under admissible semantics for ABA frameworks is polynomial-time decidable (Table 1), while our focus here is on the NP-hard acceptance problems. Omitting a relevant argument in a corresponding AF can directly lead to incorrect results w.r.t. acceptance queries of the original ABA framework. For instance, considering the corresponding AF shown in Example 2, removal of any of the relevant arguments of this AF would lead to missing sentences in the AF which are credulously accepted under, e.g., admissible semantics in the original ABA framework.

4   Computing Relevant Arguments

The authors of [7] conjecture that computing minimal supports may be computationally costly. We provide a formal result backing up this conjecture: we show that counting the number of minimal supports for a given ABA framework is intractable, in fact #P-complete under subtractive reductions [16] often used for showing hardness for counting complexity classes. (The prototypical #P-complete problem is that of counting satisfying assignments of a Boolean formula.)

Theorem 2. *For a given ABA framework, counting the number of minimal supports is #P-complete under subtractive reductions.*

To overcome this obstacle, we give an algorithm that overapproximates the set of relevant arguments. The algorithm traverses the rules backwards towards the assumptions. The underlying data structure operated on is a directed graph with vertices being both heads and bodies of rules in the ABA. There is a directed edge from a body to a head if there is a corresponding rule, and from a head to a body if the former is contained in the latter. We filter out non-derivable sentences. If the rules are acyclic, we can straightforwardly backward chain from the sinks to create all needed arguments. For the general (i.e. possibly cyclic) case, the presented algorithm also takes all heads of rules that are in non-trivial strongly connected components (SCCs), i.e., non-singleton SCCs, denoted by $SCC(D)$, as starting points. We store (partial) arguments with a set of sets of sentences, $Arg(X) = \{S_1, \ldots, S_n\}$ for a head or body X, indicating that X is derivable from any $S_1, \ldots, S_n$.

The main Algorithm 1 computes non-trivial SCCs, stores starting points in S, and recurses in the while loop (call by reference) with a picked sentence and the current derivation path P (for detecting cyclic derivations; initialized with $\emptyset$). After processing, the sentence is marked, indicating that all derivations have been exhausted. Algorithm 2, PROCESS-HEAD, marks the head s if not in a non-trivial

---

Algorithm 1. Argument Construction

Require: ABA $D = (\mathcal{L}, \mathcal{R}, \mathcal{A}, ^-)$
1: Compute $SCC(D)$ //non-trivial SCCs
2: $S = sinks(G) \cup (\bigcup SCC(D) \cap \mathcal{L})$
3: **while** $S \neq \emptyset$
4:   remove s from S
5:   PROCESS-HEAD$(s, \emptyset)$
6:   mark s visited

SCC and adds s to the derivation path P. In case s is an assumption (or $\top$ for sentences derived from $\emptyset$), we add a new argument $\{s\}$ for s. Otherwise call PROCESS-BODY(B, P) for each non-visited body B from which s can be derived, excluding P. Afterwards, we extend arguments for the bodies and add these arguments to $Arg(s)$.

Algorithm 3 takes care of bodies. We mark B if it is not a non-trivial SCC or when each element in the body is either marked or not in non-trivial SCC. To avoid cyclic derivations we check if an element in the body is contained in path P. If not, after adding body B to P, we call PROCESS-HEAD for each non-visited element. We collect all possible ways of deriving body B by taking all minimal combinations of arguments from which to directly derive each $s \in B$ (SUBDERIVATIONS). Arguments with the same support are then merged (there is at most one argument per set of assumptions) by calling MERGE-BY-SUP.

Each constructed argument contains only sentences derivable from its set of assumptions. For each $\Delta \in minsupp(D)$ an argument with Δ as its assumptions is constructed. Our algorithm approximates the set of relevant arguments in two senses. First, arguments with minimal support might contain more derived sentences, i.e., sentences also derivable from subsets of their support. Secondly, we might compute arguments with assumption sets not in $minsupp(D)$. Correctness of the overall approach is not affected by either approximation as long as the attacks are as specified in Definition 5.

Algorithm 2. PROCESS-HEAD(s, P)

1: **if** $s \notin \bigcup SCC(D)$ **then** mark s visited
2: $P = P \cup \{s\}$
3: **if** $s \in \mathcal{A} \cup \{\top\}$ **then**
4:     $Arg(s) = Arg(s) \cup \{\{s\}\}$
5: **else**
6:   **for each** $B \in \{body(r) | head(r) = s\}$
7:     **if** B not visited **then**
8:        PROCESS-BODY(B, P)
9:     $Arg(B) = \{A \cup \{s\} | A \in Arg(B)\}$
10:    $Arg(s) = Arg(s) \cup Arg(B)$
11: $P = P \backslash \{s\}$

Algorithm 3. PROCESS-BODY(B, P)

1: **if** $B \notin \bigcup SCC(D)$ **then** mark B visited
2: **if** each $s' \in B$ is marked or not in SCC
   then mark B visited
3: **if** $B \cap P \neq \emptyset$ **then** return
4: $P = P \cup \{B\}$
5: **for each** $s' \in B$
6:   **if** s' not visited **then**
7:      PROCESS-HEAD(s', P)
8: $Arg(B) = $ SUBDERIVATIONS(B)
9: MERGE-BY-SUPP$(Arg(B))$
10: $P = P \backslash \{B\}$

Special Cases. ABA acceptance can, in cases, be decided during the AF translation. Assume an ABA $D = (\mathcal{L}, \mathcal{R}, \mathcal{A}, {}^-)$ and a sentence $l \in \mathcal{L}$. For admissible and preferred semantics it holds that if $l \in Th_{\mathcal{R}}(\emptyset)$, then l is both credulously and skeptically accepted; and if $l \notin Th_{\mathcal{R}}(\mathcal{A})$ or each (L, Δ) with $l \in L$ is self-attacking, then l is neither credulously nor skeptically accepted. For stable semantics, it holds that if $l \in Th_{\mathcal{R}}(\emptyset)$, then l is skeptically accepted, and credulously accepted iff D has a stable assumption set. If $l \notin Th_{\mathcal{R}}(\mathcal{A})$ or each (L, Δ) with $l \in L$ is self-attacking, then l is not credulously accepted, and skeptically accepted iff D has no stable assumption set. In our approach, existence of stable assumption sets can be checked with an AF reasoner.

5   Reasoning About ABA Acceptance on AFs

For reasoning over the AFs (step 2) obtained from our ABA-to-AF translation (step 1) we encode the ABA acceptance problem over the AF obtained via step 1 using answer set programming (ASP). Interchangeably, one could apply essentially any of the e.g. SAT-based AF reasoning systems via similar minor modifications. We focus here on encodings for admissible and stable semantics; other central argumentation semantics can be encoded with relatively minor changes. Our encodings are similar to ones used in the ASP-based AF reasoning system ASPARTIX [18], except for one seemingly minor but essential difference: we represent the AF attack relation via its *complement*, using the predicate natt/2 (not attack) which is true for a pair (a, b) of nodes iff a *does not* attack b. This complement representation is vital as the edge relations of the AFs obtained via step 1 are typically very dense. This is in stark contrast to typical AF reasoning benchmark instances with relatively sparse attack relations [27].

Our encoding of the admissible semantics is

in(X) :- not out(X), arg(X).   :- in(X), in(Y), not natt(X, Y).

out(X) :- not in(X), arg(X).   defeated(X) :- in(Y), not natt(Y, X), arg(X).

out(X) :- arg(X), not natt(X, X).   :- in(X), arg(Y), not natt(Y, X), not defeated(Y).

Further minor changes to the original ASPARTIX encoding are that we include the rule on the lower left, and collapsed two rules to what is now here the last rule. For stable semantics, we simply replace the last rule by :- out(X), not defeated(X).

Implementing credulous ABA queries under NP-complete semantics such as admissible, preferred, and stable on the AF side is achieved by checking if there is an extension which includes an argument that contains the queried sentence during argument construction. This is implemented with the ASP constraint :- not in(a_1), ..., not in(a_n). for arguments a_i that contain queried sentence l, i.e., $a_i = (L, \Delta)$ with $l \in L$.

Implementing skeptical ABA queries for coNP-complete semantics such as stable is achieved by checking if there is a counterexample to the query, i.e., whether there is an extension of the AF that does not include any arguments containing the queried sentence. This is implemented by constraint :- in(a_i). for each argument a_i that contains the queried sentence, pruning the search space by partial instantiation. An alternative approach to skeptical acceptance would be to enumerate all AF extensions, and check whether each of them includes some argument that contains the queried sentence.

6 Experiments

For a first evaluation of the two-step approach to answering ABA queries via AF reasoning, we implemented a prototype translation (step 1) in Java 8. We compare our approach to the recently published state-of-the-art graph-based ABA reasoning system abagraph (http://www.doc.ic.ac.uk/~rac101/proarg/abagraph.html) implemented in Prolog, using SICStus Prolog 4.3.3. We used the "default" search strategy of abagraph. The experiments were run on 2.83-GHz Intel Xeon E5440 quad-core machines with 32-GB RAM under Linux using a 600-second timeout and a 16-GB memory limit per instance. As the running times of our approach, we report the combined time of the translation and the ASP solver Clingo 4.5.4 [23].

As benchmarks we use the 680 ABA frameworks provided by the authors of abagraph [7]. A benchmark instance consists of an ABA graph and a query on whether a given sentence in the ABA framework is credulously accepted under a specific semantics (recall that abagraph supports only credulous queries under admissible and grounded semantics). For each ABA framework, we used 10 queries per ABA. After filtering out 90 duplicate queries and trivial instances wrt the special cases outlined for admissible semantics in the previous section, we obtained 1466 final benchmark instances. The benchmarks are explicitly categorized wrt whether the rules of a framework give rise to cyclic dependencies, i.e., whether a framework is cyclic (804) or acyclic (662).

A comparison of abagraph and our approach is shown in Fig. 3 (left). Here we consider the credulous task of enumerating all admissible assumption sets containing the given query. The same task was used to evaluate abagraph in [7] and shown to outperform earlier state of the art. For enumeration in our approach, we used the built-in enumeration mode of Clingo. Figure 3 (left) shows that our approach and that of the dedicated abagraph approach are complementary in that there are instances on which each of the approaches is clearly better than the other. Figure 3 (right) corroborates this observation. The relative performance is essentially on-par on cyclic instances, while our approach is somewhat better on acyclic instances. To illustrate the generality of our approach, we also experimented on skeptical acceptance of sentences under stable semantics (a task not supported by abagraph). Our approach solved 6228 of the 6710 instances. The per-instance runtime was < 10 s on over 6000 instances. A majority of runtime was used in the AF translation on every instance, AF translation taking over 80% of the total runtime on approximately 95% of the solved instances. The ASP solving part was very efficient, finishing within 65 s on each instance.

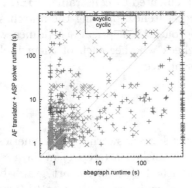

	Timeouts		Uniquely solved	
	abagraph	us	abagraph	us
acyclic	93	**56**	20	**57**
cyclic	**394**	402	**86**	78

Fig. 3. Left: running time comparison of abagraph and our approach on credulous reasoning under admissible semantics. Right: numbers of timeouts and uniquely solved instances.

7   Conclusions

We studied an approach to reasoning about acceptance in assumption-based argumentation via translating ABA frameworks into argumentation frameworks. We considered relevant ABA arguments as a sought after small yet sufficient set for reasoning about acceptance of ABA sentences on AFs. However, we showed that counting the number of relevant arguments is #P-complete, and hence proposed an algorithm for overapproximating the set of relevant arguments in order to translate ABAs to AFs, and ASP encodings specifically suited for the types of AFs obtained through the translation. Our prototype implementation yields complementary performance wrt the state-of-the-art dedicated ABA reasoning system abagraph. As a further benefit, our approach also allows for deciding skeptical acceptance in ABA, not supported by abagraph.

References

1. Besnard, P., García, A.J., Hunter, A., Modgil, S., Prakken, H., Simari, G.R., Toni, F.: Introduction to structured argumentation. Argum. Comput. 5(1), 1–4 (2014)
2. Besnard, P., Hunter, A.: Elements of Argumentation. The MIT Press, Cambridge (2008)
3. Bondarenko, A., Dung, P.M., Kowalski, R.A., Toni, F.: An abstract, argumentation-theoretic approach to default reasoning. Artif. Intell. 93, 63–101 (1997)
4. Caminada, M., Sá, S., Alcântara, J., Dvořák, W.: On the difference between assumption-based argumentation and abstract argumentation. In: Proceedings of BNAIC, pp. 25–32 (2013)
5. Cerutti, F., Dunne, P.E., Giacomin, M., Vallati, M.: Computing preferred extensions in abstract argumentation: a SAT-based approach. In: Black, E., Modgil, S., Oren, N. (eds.) TAFA 2013. LNCS, vol. 8306, pp. 176–193. Springer, Heidelberg (2014). doi:10.1007/978-3-642-54373-9_12
6. Cerutti, F., Giacomin, M., Vallati, M.: ArgSemSAT: solving argumentation problems using SAT. In: Proceedings of COMMA. FAIA, vol. 266, pp. 455–456. IOS Press (2014)
7. Craven, R., Toni, F.: Argument graphs and assumption-based argumentation. Artif. Intell. 233, 1–59 (2016)
8. Craven, R., Toni, F., Cadar, C., Hadad, A., Williams, M.: Efficient argumentation for medical decision-making. In: Proceedings of KR, pp. 598–602 (2012)
9. Craven, R., Toni, F., Williams, M.: Graph-based dispute derivations in assumption-based argumentation. In: Black, E., Modgil, S., Oren, N. (eds.) TAFA 2013. LNCS (LNAI), vol. 8306, pp. 46–62. Springer, Heidelberg (2014). doi:10.1007/978-3-642-54373-9_4
10. Dimopoulos, Y., Nebel, B., Toni, F.: On the computational complexity of assumption-based argumentation for default reasoning. Artif. Intell. 141(1/2), 57–78 (2002)
11. Dung, P.M.: On the acceptability of arguments and its fundamental role in non-monotonic reasoning, logic programming and n-person games. Artif. Intell. 77(2), 321–358 (1995)
12. Dung, P.M., Kowalski, R.A., Toni, F.: Dialectic proof procedures for assumption-based, admissible argumentation. Artif. Intell. 170(2), 114–159 (2006)
13. Dung, P.M., Kowalski, R.A., Toni, F.: Assumption-based argumentation. In: Rahwan, I., Simari, G.R. (eds.) Argumentation in Artificial Intelligence, pp. 25–44 (2009)
14. Dung, P.M., Mancarella, P., Toni, F.: Computing ideal sceptical argumentation. Artif. Intell. 171(10–15), 642–674 (2007)
15. Dunne, P.E., Wooldridge, M.: Complexity of abstract argumentation. In: Rahwan, I., Simari, G.R. (eds.) Argumentation in Artificial Intelligence, pp. 85–104 (2009)

16. Durand, A., Hermann, M., Kolaitis, P.G.: Subtractive reductions and complete problems for counting complexity classes. Theor. Comput. Sci. **340**(3), 496–513 (2005)
17. Dvořák, W., Järvisalo, M., Wallner, J.P., Woltran, S.: Complexity-sensitive decision procedures for abstract argumentation. Artif. Intell. **206**, 53–78 (2014)
18. Egly, U., Gaggl, S.A., Woltran, S.: Answer-set programming encodings for argumentation frameworks. Argum. Comput. **1**(2), 147–177 (2010)
19. Gaertner, D., Toni, F.: CaSAPI: a system for credulous and sceptical argumentation. In: Proceedings of NMR, pp. 80–95 (2007)
20. Gaertner, D., Toni, F.: Computing arguments and attacks in assumption-based argumentation. IEEE Intell. Syst. **22**(6), 24–33 (2007)
21. Gaertner, D., Toni, F.: Hybrid argumentation and its properties. In: Proceedings of COMMA. FAIA, vol. 172, pp. 183–195. IOS Press (2008)
22. García, A.J., Simari, G.R.: Defeasible logic programming: an argumentative approach. TPLP **4**(1–2), 95–138 (2004)
23. Gebser, M., Kaufmann, B., Kaminski, R., Ostrowski, M., Schaub, T., Schneider, M.T.: Potassco: the potsdam answer set solving collection. AI Comm. **24**(2), 107–124 (2011)
24. Nofal, S., Atkinson, K., Dunne, P.E.: Algorithms for decision problems in argument systems under preferred semantics. Artif. Intell. **207**, 23–51 (2014)
25. Nofal, S., Atkinson, K., Dunne, P.E.: Looking-ahead in backtracking algorithms for abstract argumentation. Int. J. Approx. Reasoning **78**, 265–282 (2016)
26. Prakken, H.: An abstract framework for argumentation with structured arguments. Argum. Comput. **1**(2), 93–124 (2010)
27. Thimm, M., Villata, S., Cerutti, F., Oren, N., Strass, H., Vallati, M.: Summary report of the first international competition on computational models of argumentation. AI Mag. **37**(1), 102 (2016)
28. Toni, F.: A generalised framework for dispute derivations in assumption-based argumentation. Artif. Intell. **195**, 1–43 (2013)
29. Toni, F.: A tutorial on assumption-based argumentation. Argum. Comput. **5**(1), 89–117 (2014)

On Relating Abstract and Structured Probabilistic Argumentation: A Case Study

Henry Prakken[1,2]([⊠])

[1] Department of Information and Computing Sciences,
Utrecht University, Utrecht, The Netherlands
h.prakken@uu.nl
[2] Faculty of Law, University of Groningen, Groningen, The Netherlands

Abstract. This paper investigates the relations between Timmer et al.'s proposal for explaining Bayesian networks with structured argumentation and abstract models of probabilistic argumentation. First some challenges are identified for incorporating probabilistic notions of argument strength in structured models of argumentation. Then it is investigated to what extent Timmer et al's approach meets these challenges and satisfies semantics and rationality conditions for probabilistic argumentation frameworks proposed in the literature. The results are used to draw conclusions about the strengths and limitations of both approaches.

1 Introduction and Motivation

There is a recent increase in interest in models of probabilistic argumentation. In argumentation theory, Hahn and others have advocated a probabilistic interpretation of argument schemes (e.g. [3]). A limitation of this work is that it does not deal with several crucial features of argumentation-based inference, such as attacks and combinations of arguments. Recent AI research on abstract models of probabilistic argumentation, e.g. [6,7], addresses the first limitation. However, since it says nothing about the structure of arguments and the nature of attack, the proposed models have so far been hard to interpret. For example, it is not easy to understand what the probability of an argument means, since in probability theory probabilities are assigned to the truth of statements or to outcomes of events, and an argument is in general neither a statement nor an event. What is required here is a precise account in terms of the structure of arguments and the nature of attack. The present paper aims to offer such an account.

In the literature two different uses of probability theory in argumentation can be seen, depending on whether the uncertainty is *in* or *about* the arguments. In the first use, probabilities are *intrinsic* to an argument in that they are used for capturing the strength of an argument given uncertainty concerning the truth of its premises or the reliability of its inferences. An example is default reasoning with probabilistic generalisations, as in *The large majority of Belgian people speak French, Mathieu is Belgian, therefore (presumably) Mathieu speaks French.* Clearly, if all premises of an argument are certain and it only makes deductive

© Springer International Publishing AG 2017
A. Antonucci et al. (Eds.): ECSQARU 2017, LNAI 10369, pp. 69–79, 2017.
DOI: 10.1007/978-3-319-61581-3_7

inferences, the argument should be given maximum probabilistic strength. [4] calls this use of probability the *epistemic* approach.

A second, *extrinsic* use of probability in argumentation (by [4] called the *constellations* approach) is for expressing uncertainty about whether arguments are accepted as existing by some arguing agent. [5] gives the example of a dialogue participant who utters an enthymeme and where the listener can imagine two reasonable premises that the speaker had in mind: the listener can then assign probabilities to these options, which translate into probabilities on which argument the speaker meant to construct. This uncertainty has nothing to do with the intrinsic strengths of the two candidate completed arguments: one might be stronger than the other while yet the other is more likely the argument that the speaker had in mind. Note that in this approach even deductive arguments from certain premises can have less than maximal strength.

This paper focuses on the first use of probability theory, unlike most recent work on probabilistic abstract argumentation, which instead focuses on the second use (cf. the overview in [6]). An exception is [4], who formally distinguishes and develops both approaches, followed-up in e.g. [6]. For its epistemic approach [4] instantiates probabilistic argumentation frameworks with classical argumentation and defines the strength of an argument as the probability of the conjunction of all its premises. However, while for classical argumentation (where all arguments are deductive) this makes sense, this is not the case for accounts where arguments make defeasible inferences from certain premises (as in the above example of default reasoning, where it is both certain that the large majority of Belgian people speaks French and that Mathieu is Belgian but where the conclusion does not deductively follow from them): here all arguments should according to [4] be given strength 1, which is clearly undesirable.

Accordingly, the problem studied in this paper is how to instantiate abstract probabilistic frameworks with an account of intrinsic probabilistic strength of structured arguments, where the premises of all arguments are certain but their inferences can be defeasible. The problem will be studied in the context of a simple instantiation of the $ASPIC^+$ framework [8]. In particular, [10]'s recent proposal will be studied to explain forensic Bayesian networks in terms of $ASPIC^+$-style structured argumentation frameworks ($SAFs$) with probabilistic argument strengths. Since $SAFs$ are an instance of [1]'s abstract argumentation frameworks (AFs), Timmer's probabilistic $SAFs$ are a suitable candidate for being related to abstract probabilistic frameworks.

This paper is organised as follows. Section 2 presents the formal preliminaries. Section 3 gives a conceptual analysis of the problem of defining probabilistic strengths of structured arguments. Section 4 summarises [10]'s structured model of probabilistic argumentation. Section 5 then formally investigates its relation with abstract probabilistic argumentation frameworks, while Sect. 6 draws some general conclusions on the relation between abstract and structured models of probabilistic argumentation.

2 Formal Preliminaries

An *abstract argumentation framework* (AF) is a pair $\langle \mathcal{A}, attack \rangle$, where $\mathcal{A}$ is a set arguments and $attack \subseteq \mathcal{A} \times \mathcal{A}$ is a binary relation. The theory of AFs addresses how sets of arguments (called *extensions*) can be identified which are internally coherent and defend themselves against attack. A key notion here is that of an argument being *acceptable with respect to*, or *defended by* a set of arguments: $A \in \mathcal{A}$ is defended by $S \subseteq \mathcal{A}$ if for all $B \in \mathcal{A}$: if B attacks A, then some $C \in S$ attacks B. Then relative to a given AF various types of extensions can be defined.

- E is *admissible* if E is conflict-free and defends all its members;
- E is a *preferred extension* if E is a $\subseteq$-maximal admissible set;
- E is a *stable extension* if E is admissible and attacks all arguments outside it;
- $E \subseteq \mathcal{A}$ is the *grounded extension* if E is the least fixpoint of operator F, where $F(S)$ returns all arguments defended by S.

Various proposals for extending abstract argumentation frameworks with probabilities exist. Here we focus on one of the simplest proposals, the one of [7] as adapted by [4]. A *probabilistic argumentation framework* $(PrAF)$ is a triple $\langle \mathcal{A}, attack, Pr \rangle$ where $\langle \mathcal{A}, attack \rangle$ is an abstract argumentation framework and $Pr : \mathcal{A} \mapsto [0, 1]$. Further notions concerning *PrAFs* will be discussed in Sect. 5 below.

$ASPIC^+$ [8] is a general framework for structured argumentation. It abstracts from the logical language $\mathcal{L}$ except that it assumes a binary contrariness relation defined over $\mathcal{L}$. In the present paper $\mathcal{L}$ will be a language of propositional or predicate-logic atoms. Arguments are constructed from a knowledge base expressed in $\mathcal{L}$ by chaining inference rules defined over $\mathcal{L}$ into graphs (which are trees if no premise is used more than once). For present purposes only certain (non-attackable) premises and defeasible (attackable) inference rules are needed. All this reduces to the following definitions:

Definition 1 (Argumentation System). *An argumentation system (AS) is a tuple* $AS = (\mathcal{L}, ^-, \mathcal{R})$ *where:*

- $\mathcal{L}$ *is a logical language consisting of propositional or predicate-logic atoms*
- $^- : \mathcal{L} \mapsto Pow(\mathcal{L})$ *is a contrariness function over* $\mathcal{L}$
- $\mathcal{R}$ *is a set of (defeasible) inference rules of the form* $\phi_1, \ldots, \phi_n \Rightarrow \phi$ *(where* ϕ, ϕ_i *are meta-variables ranging over wff in* $\mathcal{L}$*).*

Definition 2 (Knowledge Bases and Arguments). *An argument A on the basis of a* knowledge base $\mathcal{K} \subseteq \mathcal{L}$ *in an argumentation system AS is:*

1. φ *if* $\varphi \in \mathcal{K}$ *with:* $\mathtt{Prem}(A) = \{\varphi\}$; $\mathtt{Conc}(A) = \varphi$; $\mathtt{Sub}(A) = \{\varphi\}$;
2. $A_1, \ldots A_n \Rightarrow \psi$ *if* $A_1, \ldots, A_n$ *are arguments such that* $\mathtt{Conc}(A_1), \ldots,$ $\mathtt{Conc}(A_n) \Rightarrow \psi \in \mathcal{R}$ *with:* $\mathtt{Prem}(A) = \mathtt{Prem}(A_1) \cup \ldots \cup \mathtt{Prem}(A_n)$, $\mathtt{Conc}(A) = \psi$, $\mathtt{Sub}(A) = \mathtt{Sub}(A_1) \cup \ldots \cup \mathtt{Sub}(A_n) \cup \{A\}$;

An argument A is said to *attack* an argument B iff A *rebuts* B, where A rebuts B (on B') iff $\mathrm{Conc}(A) = \overline{\varphi}$ for some $B' \in \mathrm{Sub}(B)$ of the form $B_1'', \ldots, B_n'' \Rightarrow \varphi$.

The $ASPIC^+$ counterpart of an abstract argumentation framework is a structured argumentation framework.

Definition 3 *(Structured Argumentation Frameworks).* *Let AT be an argumentation theory $(AS, \mathcal{K})$. A structured argumentation framework (SAF) defined by AT, is a triple $\langle \mathcal{A}, \mathcal{C}, \preceq \rangle$ where $\mathcal{A}$ is the set of all finite arguments constructed from $\mathcal{K}$ in AS, $\preceq$ is an ordering on $\mathcal{A}$, and $(X, Y) \in \mathcal{C}$ iff X attacks Y.*

A relation of *defeat* is then defined as follows ($A \prec B$ is defined as usual as $A \preceq B$ and $B \not\preceq A$). A *defeats* B iff A rebuts B on B' and $A \not\prec B'$. Abstract argumentation frameworks are then generated from $SAFs$ by letting the attacks from an AF be the defeats from a SAF.

Definition 4 *(Argumentation Frameworks).* *An abstract argumentation framework (AF) corresponding to a $SAF = \langle \mathcal{A}, \mathcal{C}, \preceq \rangle$ (where $\mathcal{C}$ is $ASPIC^+$'s attack relation) is a pair $(\mathcal{A}, attack)$ such that attack is the defeat relation on $\mathcal{A}$ determined by SAF.*

3   Probabilistic Argument Strength: A Conceptual Analysis

We can now make the problem studied in this paper even more specific. The problem is: in the context of the just-described simple instantiation of $ASPIC^+$, how can a probabilistic notion of argument strength be defined such that for two arguments A and B we have that $A \preceq B$ just in case $strength(A) \leq strength(B)$? In other words: how can probabilistic argument strength be used to resolve attacks into defeats?

A first challenge here is that a higher internal strength does not necessarily make an argument dialectically stronger. Suppose 90% of the birds can fly, and 80% of penguins cannot fly. Should the argument *Tweety can fly since it is a penguin so it is a bird* be stronger than the argument *Tweety cannot fly since it is a penguin*? Of course not, since probability theory requires that all evidence is taken into account, and since penguins are a special kind of bird, we should (defeasibly) conclude that Tweety cannot fly. More formally, if we have $Pr(q|p) = x$ and $Pr(q|p \wedge r) = y$ and both p and r are given, then we should base our inference on $Pr(q|p \wedge r) = y$. For this reason, probability-based comparisons between arguments should, either explicitly or implicitly, involve a kind of specificity principle. For example, [2] do so in their notion of specificity defeat for probabilistic assumption-based argumentation. More generally this shows that the probabilistic strength of an argument cannot be calculated independent of its attackers.

This issue arises in a different way in case of attacks between arguments that do not have a specificity relation. Consider the following well-known example from nonmonotonic logic: *Quakers are usually pacifists, Republicans are usually not pacifists, Nixon was a quaker and a republican.* It is wrong to compare $Pr(P|Q)$ with $Pr(\neg P|R)$. What counts is $Pr(P|Q \wedge R)$ and in general the latter probability is independent of the former probabilities (although in special cases this may be different).

A third challenge arises from the step-by-step nature of arguments. Consider *I see smoke, my observations are usually correct, therefore (presumably) there is smoke. Where there is smoke, there usually is fire,- therefore (presumably) there is fire.* Can a recursive definition of argument strength be given where the strength of the entire argument depends on the strength of its subargument for *there is smoke* and the strength of its final step? This is not a trivial problem, since $Pr(fire|seesmoke)$ does in general not follow from $Pr(fire|smoke)$ and $Pr(smoke|seesmoke)$.

4   Explaining Bayesian Networks with Argumentation

In this section we summarise [10]'s method for explaining (forensic) Bayesian networks with argumentation. A Bayesian network is a graphical representation of a joint probability distribution. Formally, it is a pair (G, Pr) where G is a directed acyclic graph $(\mathbf{V}, \mathbf{E})$, with a finite set of variables $\mathbf{V}$ connected by edges $\mathbf{E}$ from $\mathbf{V} \times \mathbf{V}$, and Pr is a probability function which specifies for every variable V_i the probability distribution $Pr(V_i|parents(V_i))$ of its outcomes conditioned on its parents $parents(V_i)$ in the graph. [10] assume that all variables are boolean.

[10] first set the language $\mathcal{L}$ of the $ASPIC^+$ argumentation system to the set of all $V = v$ expressions where $V \in \mathbf{V}$ and v is a possible value of V. Then $\varphi \in \overline{\psi}$ iff φ and ψ assign different values to the same variable. [10] then derive an $ASPIC^+$ SAF from a BN plus a set of instantiated variables (the *evidence*) in terms of an intermediate structure called a *support graph*, which for a given variable of interest from $\mathbf{V}$ captures the potential reasoning paths through the BN. Entering evidence in a BN prunes all branches of the support graph that do not end in evidence. Arguments can be constructed from the support graph by making its non-premise nodes either true or false.

The basic idea is illustrated with Figs. 1 and 2, displaying an example from [10]. The variable of interest in the BN is whether the suspect committed the *Crime*. Evidence for this can be a *DNA_match* between the suspect's DNA and DNA found at the crime scene. Such a DNA match may also be explained by the existence of a *Twin* brother. The existence of a *Motive* makes the crime more likely. Evidence for a motive may be given in a psychological report (*Psych_report*). After the evidence *Psych_report = True* and *DNA_match = True* is entered into the BN, the chain in the support graph from *Twin* to *Crime* is pruned away. The arguments generated from this are all inference trees corresponding to the pruned graph or a subgraph with all non-evidence nodes instantiated in any possible way (i.e., with either *True* or *False*). Figure 2 displays the arguments when *Psych_report* and *Crime* are both true. Formally (with contrariness for convenience encoded with negation, even though $\neg$ is strictly speaking not part of $\mathcal{L}$) both these pieces of evidence are arguments and moreover:

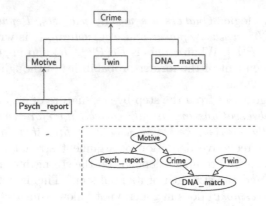

Fig. 1. A Bayesian network (below) and a support graph (above).

$$A_1 = Psych\_report \Rightarrow Motive \quad A_2 = A_1, DNA\_match \Rightarrow Crime$$

But, for instance, the following arguments can also be constructed:

$$B_1 = Psych\_report \Rightarrow \neg\ Motive \quad B_2 = B_1, DNA\_match \Rightarrow \neg\ Crime$$
$$C = A_1, DNA\_match \Rightarrow \neg\ Crime$$

Suppose that now *Twin* also becomes available as evidence. At first sight, one might expect that this gives rise to rebuttal of A_2 of the form $Twin \Rightarrow \neg\ Crime$. However, this is not how the method works. Instead, it captures all variables relevant to a conclusion in a single argument concerning that conclusion. So, A_2, B_2 and C are modified to:

$$A'_2 = A_1, DNA\_match, Twin \Rightarrow Crime \quad B'_2 = B_1, DNA\_match, Twin \Rightarrow \neg\ Crime$$
$$C' = A_1, DNA\_match, Twin \Rightarrow \neg\ Crime$$

So for every constructable argument there is a rebuttal with the same 'skeleton' but with some truth values of non-premises flipped, and there are no other (direct) rebuttals.

Space limitations prevent listing the formal definitions of the support graph and the induced SAF. Essentially, the knowledge base consists of the evidence while the set of defeasible rules corresponds to links in the support graph. For present purposes all that is relevant is that the following definition of argument strength, when used to resolve attacks into defeats, gives the induced SAF a number of 'good' properties, which means that [10]'s way to define probabilistic structured argumentation as an instantiation of Dung-style AFs makes sense (with some simplified notation compared to [10]):

$$strength(A) = Pr(\textsf{Conc}(A)|\mathcal{K})$$

(where $\mathcal{K}$ is the knowledge base of the AT induced by the BN-with-evidence). Thus the strength of an argument A equals the posterior probability of $\textsf{Conc}(A)$ in the BN-with-evidence inducing the SAF. This definition implies that the strengths of two directly rebutting arguments always adds up to 1, since $Pr(Q|P) = 1 - Pr(\neg Q|P)$.

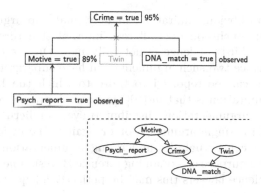

Fig. 2. Deriving arguments from the pruned support graph after entering evidence.

Figure 2 displays the strengths of the non-premise arguments in the figure, based on the probability tables in [10]. The strength of B_1 is $1 - strength(A_1) = 11\%$ and the strength of B_2 and C is $1 - strength(A_2) = 5\%$. In fact, to obtain the results listed below, for arguments for the variable of interest the definition of strength can be reduced to the equivalent definition $strength(A) = Pr(\texttt{Conc}(A)|\texttt{Prem}(A))$. However, for subarguments the inclusion of all of $\mathcal{K}$ is needed. For the reasons why see [10]. It should be noted that a strength of 1 of a non-premise argument does not mean that it should be modelled as applying strict rules, since $Pr(P|Q) = 1$ does not imply $Pr(P|Q \wedge R) = 1$. So even a non-premise argument of strength 1 is defeasible.

The 'good' properties (as proven by or easily following from [10]) are as follows.

1. The grounded extension equals the set of undefeated arguments.
2. The grounded extension satisfies subargument closure, direct and indirect consistency and (trivially) strict closure. For the definitions of these properties see [8].
3. If A is in the grounded extension, then A is the strongest argument for $\texttt{Conc}(A)$.
4. If A is in the grounded extension, then $strength(A) > 0.5$.

Interestingly, an argument can be stronger than some of its subarguments, An example is Fig. 2, where argument A_2 for *Crime* is stronger than its subargument A_2 for motive. This can happen since by combining arguments A_1 and *DNA_match*, argument A_2 aggregates the support given to its conclusion by two pieces of evidence. This reflects a general feature of probabilistic reasoning, namely, that the combination of pieces of evidence that each have weak probative force can have strong probative force.

Let us see how [10]'s approach deals with the challenges discussed in Sect. 3. The first two challenges are dealt with since, firstly, all arguments contain all variables from the BN that are relevant for their conclusions and, secondly, argument strength is defined relative to all (relevant) evidence. While this is good, it also has an obvious limitation, namely, that two arguments with the same premises and conclusion but different internal structure have the same strength. Thus the third challenge is not fully met. On the other hand, it is still partially met since two such arguments can have subarguments of different strengths, and this can be reported to those to which the BN is explained. Another possible limitation is that in [10] the reasons pro and con a conclusion are not, as usual in argumentation, distributed over conflicting arguments but are all contained in a single argument for or against the conclusion, which is not according to the conceptual idea underlying argumentation-based inference. Nevertheless, for the purpose of explaining forensic Bayesian networks to judges, prosecutors and defence lawyers this may be perfectly adequate.

5   Relating the Abstract and Structured Accounts

We now investigate whether the work of [10] can be seen as an instantiation of epistemically interpreted probabilistic abstract frameworks in the sense of [4]. The first step is obvious, namely, equating the probability of an argument in a $PrAF$ with the argument's strength according to [10]. However, the next step, instantiating the *attack* relation of $PrAFs$, is less obvious: should it be instantiated with $ASPIC^+$'s C relation of attack or with its defeat relation? Let us first assume that the probability of an argument is to be used to resolve attacks: can this be modelled at the abstract level or should this be modelled while taking the structure of arguments into account? [8] provide reasons for the latter approach, since the former approach cannot distinguish between direct and indirect attack relations and therefore runs the risk of applying the wrong probabilities to an attack. Consider an argument $A_3 = A_1, A_2 \Rightarrow \varphi$ and an argument B rebutting A_3 on A_1. [8] show that with a last-link ordering in terms of rule priorities it may be that $A_1 \prec B$ while $B \prec A_3$. Then resolving the attacks at the abstract level by saying that an argument A attacks (i.e., $ASPIC^+$-defeats) an argument B iff A rebuts B and $A \nprec B$ results in a grounded/preferred/stable/complete extension $\{A_2, A_3, B\}$, which is not closed under the subargument relation. This problem arises since the rebuttal of B on A_3 is incorrectly resolved with $B \prec A_3$ (so B does not attack A_3 in the AF) while it should be resolved with $A_1 \prec B$ (so B does attack A_3 in the AF), resulting in a grounded extension $\{A_2, B\}$. As shown by [8], the same problem can make conclusion sets of extensions violate consistency. Since, as noted above, in [10]'s approach an argument can also be stronger than some of its subarguments, all these problems also arise if the Pr function is used at the abstract level of $PrAFs$ to resolve attacks. This is an important lesson that can be learned from the present analysis.

  In [4], which instantiates $PrAFs$ with classical argumentation, the argument probabilities are not used in defining the attack relation between arguments, so (in terms of $ASPIC^+$), [4] instantiates the *attack* relation of $PrAFs$ with $ASPIC^+$'s C relation. Yet his approach does not necessarily suffer from the just-sketched problems, since [4] does not use the Pr functions to resolve the attacks in $PrAFs$. Instead, he defines the following notions. The *epistemic extension* of a $PrAF$ is $\{A \in \mathcal{A}|Pr(A) > 0.5\}$. A probability function is *rational* iff for every pair of arguments A and B such that A attacks B, if $Pr(A) > 0.5$ then $Pr(B) \leq 0.5$. A $PrAF$ is called rational if its Pr is rational. [4] then proves that for every rational $PrAF$ its epistemic extension is conflict-free with respect to the *attack* relation.

Yet the notion of an epistemic extension combined with the rationality constraint on Pr is not a good abstraction of [10]'s approach. To start with, the grounded extension in [10]'s approach does not always equal the epistemic extension. Consider a support graph $E \longrightarrow H_1 \longrightarrow H_2$ and consider the arguments $(E \Rightarrow H_1) \Rightarrow H_2$ and $(E \Rightarrow \neg H_1) \Rightarrow H_2$. Both have the same strength. Suppose their strength exceeds 0.5. Then they cannot be both in the grounded extension, since they have rebutting subarguments. Yet both are in the epistemic extension. This can happen since in [10]'s approach an argument can be stronger than some of its subarguments.

Next, [10]'s probability function is not guaranteed to be rational. Consider the same support graph, the above arguments and their respective subarguments $E \Rightarrow H_1$ and $E \Rightarrow H_2$, and suppose $strength((E \Rightarrow H_1) \Rightarrow H_2) = 0.6$ and $strength((E_1 \Rightarrow H_1)) = 0.4$ and $strength((E_1 \Rightarrow \neg H_1)) = 0.6$. The argument for $\neg H_1$ indirectly attacks the argument for H_2 but both have strength > 0.5. This can happen since the attack is indirect. In $PrAFs$ the distinction between direct and indirect attack cannot be modelled. However, if the rationality constraint is confined to direct attacks, then it holds, since the strengths of two directly rebutting arguments always add up to 1. This is a first indication of the importance of taking the structure of arguments into account.

Further indications follow from an analysis of the other "rationality conditions" on $PrAFs$ proposed in [6]. (Below for any $A \in \mathcal{A}$, $A^- = \{B|B \text{ attacks } A\}$).

COH Pr is *coherent* if for every $A, B \in \mathcal{A}$, if A attacks B then $Pr(A) \leq 1 - Pr(B)$.
INV Pr is *involutary* if for every $A, B \in \mathcal{A}$, if A attacks B then $Pr(A) = 1 - Pr(B)$.
SFOU Pr is *semi-founded* if $Pr(A) \geq 0.5$ for every unattacked $A \in \mathcal{A}$.
FOU Pr is *founded* if $Pr(A) = 1$ for every $A \in \mathcal{A}$ with $A^- = \emptyset$.
SOPT Pr is *semi-optimistic* if $Pr(A) \geq 1 - \Sigma_{B \in A^-} Pr(B)$ whenever $A^- \neq \emptyset$.
OPT Pr is *optimistic* if $Pr(A) \geq 1 - \Sigma_{B \in A^-} Pr(B)$ for every $A \in \mathcal{A}$.

We now investigate whether these properties hold for [10]'s approach, for both the $ASPIC^+$ $\mathcal{C}$ relation and its defeat relation. In doing so, we will use the support graph $E \longrightarrow H_1 \longrightarrow H_2$ and the various arguments it generates with evidence E, assuming that both $Pr(H_1|\mathcal{K}) > 0.5$ and $Pr(H_2|\mathcal{K}) > 0.5$. (For space limitations we omit a proof that a BN that generates such a support graph, arguments and strengths exists).

COH in general neither holds for $\mathcal{C}$ nor for defeat, since these relations can be indirect. For example, argument $(E \Rightarrow H_1) \Rightarrow H_2$ rebuts $(E \Rightarrow \neg H_1) \Rightarrow H_2$ on $E \Rightarrow \neg H_1$ but both arguments have strength > 0.5 (even though the latter's subargument for $\neg H_1$ has strength < 0.5). However, COH does hold when restricted to direct C or defeat relations, since for every pair of direct rebuttals their strengths add up to 1. All these observations also hold for INV.

SFOU holds in general for both C and defeat. For C, note that the only non-rebutted arguments are elements of K, which by definition have strength 1. For defeat, if B unsuccessfully directly rebuts A, then $strength(B) < strength(A)$ so since these strengths add up to 1, $strength(A) > 0.5$. Note further that every non-premise argument has at least one rebuttal, so every non-defeated argument has strength > 0.5.

FOU holds in general for C but not for defeat. For C FOU holds for the same reason as why SFOU holds. Our above example provides a counterexample for defeat if the strength of the argument for H_2 does not equal 1. This is also a counterexample for FOU restricted to direct defeats.

SOPT neither holds for C nor for defeat, and neither for the direct nor for the indirect relations. In our example, $(E \Rightarrow H_1) \Rightarrow H_2$ has two direct rebuttals, namely, $(E \Rightarrow H_1) \Rightarrow \neg H_2$ and $(E \Rightarrow \neg H_1) \Rightarrow H_2$. If the strength of the argument for H_2 is higher than 0.5 but below 0.75, then the strengths of its two rebuttals add up to above 0.5. This is also a counterexample to OPT.

The negative results do not indicate flaws of [10]'s approach, since they are due to two of its features which both are reasonable for probabilistic argumentation: the distinction between direct and indirect attack and the fact that an argument can be stronger than some of its subarguments. It can therefore be concluded that [6]'s set of rationality conditions cannot be seen as minimum conditions for the well-behavedness of $PrAFs$.

6   Conclusion

In this paper we have investigated to what extent [10]'s probabilistic version of $ASPIC^+$, proposed for explaining Bayesian networks, satisfies semantics and rationality conditions for probabilistic argumentation frameworks proposed in the literature. Some results were positive but other results were negative. The negative results do not seem to point at flaws of [10]'s approach but instead at limitations of current abstract models of probabilistic argumentation, in particular their failure to distinguish between direct and indirect relations of attack and defeat. One conclusion is that to make this distinction in a proper way, the structure of arguments and the nature of attack and defeat must be made explicit. Another conclusion is that not all rationality conditions for probabilistic models of argumentation proposed in the literature can be regarded as minimum requirements for the well-behavedness of these models.

This paper has also identified several challenges for attempts to use probabilistic argument strength for resolving attacks into defeats. These challenges arise from the difference between the global nature of Bayesian probabilistic reasoning (where all evidence has to be taken into account) and the local nature of argumentation (where particular conflicting arguments are compared). [10] found a way to meet these challenges but with some limitations. While [10]'s approach may suffice for its intended application of explaining forensic Bayesian networks, future research should study whether more general solutions are possible without these limitations.

Some related work was already discussed throughout this paper. In addition, [2] propose an extension of [1]'s abstract frameworks with probability and then extend assumption-based argumentation with the means to label literals in rules with probabilities. The abstract approach models more than what is of present concern, namely, some aspects of multi-agent argumentation, while the abstract and assumption-based parts are not formally related. In future research it would be interesting to investigate [2]'s probabilistic version of assumption-based argumentation in the same way as we did for [10]'s probabilistic version of $ASPIC^+$.

In this paper we studied the use of probability for two things: for resolving attacks into defeats within $ASPIC^+$ and for identifying epistemic extensions in the sense of [4]. In future research it would be interesting to investigate the use of probability to define graded notions of argument acceptability, as in e.g. [9]. We conjecture that here, too, it is important to take the structure of arguments and the nature of attack into account.

References

1. Dung, P.M.: On the acceptability of arguments and its fundamental role in non-monotonic reasoning, logic programming, and n-person games. Artif. Intell. **77**, 321–357 (1995)
2. Dung, P.M., Thang, P.M.: Towards (probabilistic) argumentation for jury-based dispute resolution. In: Baroni, P., Cerutti, F., Giacomin, M., Simari, G.R. (eds.) Computational Models of Argument. Proceedings of COMMA 2010, pp. 171–182. IOS Press, Amsterdam (2010)
3. Hahn, U., Hornikx, J.: A normative framework for argument quality: argumentation schemes with a Bayesian foundation. Synthese **193**, 1833–1873 (2016)
4. Hunter, A.: A probabilistic approach to modelling uncertain logical arguments. Int. J. Approx. Reason. **54**, 47–81 (2013)
5. Hunter, A.: Probabilistic qualification of attack in abstract argumentation. Int. J. Approx. Reason. **55**, 607–638 (2014)
6. Hunter, A., Thimm, M.: On partial information and contradictions in probabilistic abstract argumentation. In: Principles of Knowledge Representation and Reasoning: Proceedings of the Fifteenth International Conference (KR-16), pp. 53–62. AAAI Press (2016)
7. Li, H., Oren, N., Norman, T.: Probabilistic argumentation frameworks. In: Proceedings First Workshop on the Theory and Applications of Formal Argument, pp. 1–16 (2011)
8. Modgil, S., Prakken, H.: A general account of argumentation with preferences. Artif. Intell. **195**, 361–397 (2013)
9. Thimm, M.: A probabilistic semantics for abstract argumentation. In: Proceedings of the 20th European Conference on Artificial Intelligence (ECAI 2012), pp. 750–755 (2012)
10. Timmer, S., Meyer, J.-J.C., Prakken, H., Renooij, S., Verheij, B.: A two-phase method for extracting explanatory arguments from Bayesian networks. Int. J. Approx. Reason. **80**, 475–494 (2017)

Bayesian Networks

Structure-Based Categorisation of Bayesian Network Parameters

Janneke H. Bolt$^{(\boxtimes)}$ and Silja Renooij

Department of Information and Computing Sciences, Utrecht University,
P.O. Box 80.089, 3508 TB Utrecht, The Netherlands
{j.h.bolt,s.renooij}@uu.nl

Abstract. Bayesian networks typically require thousands of probability para-meters for their specification, many of which are bound to be inaccurate. Know-ledge of the direction of change in an output probability of a network occasioned by changes in one or more of its parameters, i.e. the qualitative effect of parameter changes, has been shown to be useful both for parameter tuning and in pre-processing for inference in credal networks. In this paper we identify classes of parameter for which the qualitative effect on a given output of interest can be identified based upon graphical considerations.

1 Introduction

A *Bayesian network* defines a unique joint probability distribution over a set of discrete random variables [8]. It combines an acyclic directed graph, representing the independencies among the variables, with a quantification of local discrete distributions. The individual probabilities of these local distributions are called the parameters of the network. A Bayesian network can be used to infer any probability from the represented distribution.

The effect of possible parameter inaccuracies on the output probabilities of a network can be studied with a sensitivity analysis. An output can be described as a fraction of two functions that are linear in any network parameter; the coefficients of the functions are determined by the non-varied parameters [6]. Depending on the coefficients, such so-called sensitivity functions are either monotone increasing or monotone decreasing functions in each parameter. Interestingly, as we showed in previous research, for some outputs and parameters, the sensitivity function is even always increasing (or decreasing) regardless of the specific values of the other network parameters [1]. That is, regardless of the specific quantification of a network, the coefficients of the sensitivity function for a certain parameter can be such that the gradient is always positive (or always negative). In such a case, the qualitative effect of a parameter change on an output probability can be predicted from properties of the network structure, without considering the values of the other network parameters.

Knowledge of the qualitative effect of parameter changes can be exploited for different purposes. Examples of applications can be found in pre-processing for inference in credal networks [1] and in multi-parameter tuning of Bayesian networks [2].

© Springer International Publishing AG 2017
A. Antonucci et al. (Eds.): ECSQARU 2017, LNAI 10369, pp. 83–92, 2017.
DOI: 10.1007/978-3-319-61581-3_8

In this paper we present a complete categorisation of a network's parameters with respect to their qualitative effect on some output, where we assume that the network is pruned before hand to a sub-network that is computationally relevant to the output. The paper extends the work in [1] in which only a partial categorisation of the network parameters was given. Compared to our previous results, the present results also enable a *meaningful* categorisation for a wider range of parameters.

2   Preliminaries

2.1   Bayesian Networks and Notation

A Bayesian network $\mathscr{B} = (G, \mathrm{Pr})$ represents a joint probability distribution Pr over a set of random variables $\mathbf{W}$ as a factorisation of conditional distributions [5]. The independences underlying this factorisation are read from the directed acyclic graph G by means of the well-known d-separation criterion. In this paper we use upper case W to denote a single random variable, writing lowercase $w \in W$ to indicate a value of W. For binary-valued W, we use w and $\overline{w}$ to denote its two possible value assignments. Boldfaced capitals are used to indicate sets of variables or sets of value assignments, the distinction will be clear from the context; boldface lower cases are used to indicate a joint value assignment to a set of variables.

Two value assignments are said to be compatible, denoted by $\sim$, if they agree on the values of the shared variables; otherwise they are said to be incompatible, denoted by $\not\sim$. We use $\mathbf{W}_{pa(V)} = \mathbf{W} \cap pa(V)$ to indicate the subset of $\mathbf{W}$ that is among the parents of V, and $\mathbf{W}_{\overline{pa}(V)} = \mathbf{W} \backslash \mathbf{W}_{pa(V)}$ to indicate its complement in $\mathbf{W}$; descendants of V are captured by $de(V)$. To conclude, $\langle \mathbf{T}, \mathbf{U} \,|\, \mathbf{V} \rangle_d$, $\mathbf{T}, \mathbf{U}, \mathbf{V} \subseteq \mathbf{W}$, denotes that all variables in $\mathbf{T}$ are d-separated from all variables in $\mathbf{U}$ given the variables in $\mathbf{V}$, where we assume that $\langle \mathbf{T}, \emptyset \,|\, \mathbf{V} \rangle_d = \mathrm{True}$.

A Bayesian network specifies for each variable $W \in \mathbf{W}$ exactly one local distribution $\mathrm{Pr}(W \,|\, \boldsymbol{\pi})$ over the values of W per value assignment $\boldsymbol{\pi}$ to its parents $pa(W)$ in G, such that

$$\mathrm{Pr}(\mathbf{w}) = \prod_{W \in \mathbf{W}} \mathrm{Pr}(w \,|\, \boldsymbol{\pi})\big|_{w\boldsymbol{\pi} \sim \mathbf{w}}$$

where the notation $|_{\mathrm{prop}}$ is used to indicate the properties the arguments in the preceding formula adhere to. The individual probabilities in the local distributions are termed the network's *parameters*. A Bayesian network allows for computing any probability over its variables $\mathbf{W}$. A typical query is $\mathrm{Pr}(\mathbf{h} \,|\, \mathbf{f})$, involving two disjoint subsets of $\mathbf{W}$, often referred to as hypothesis variables ($\mathbf{H}$) and evidence variables ($\mathbf{F}$); $\mathbf{W}$ can also include variables not involved in the output query of interest.

An example Bayesian network is shown in Fig. 1. For output $\mathrm{Pr}(ghk \,|\, def)$ we identify hypothesis variables $\mathbf{H} = \{G, H, K\}$ (double circles), evidence variables $\mathbf{F} = \{D, E, F\}$ (shaded), and remaining variables $\{R, S\}$. In addition to the graph, conditional probability tables (CPTs) need to be specified for each node.

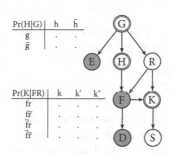

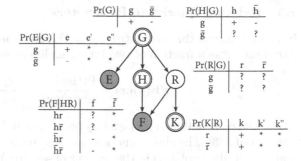

Fig. 1. An example Bayesian network (with just two of its CPTs).

Fig. 2. The example network from Fig. 1 after query dependent preprocessing given the output probability $\Pr(ghk|def)$.

2.2 Query Dependent Pre-processing

Prior to computing the result of a query, the Bayesian network can be pre-processed by removing parts of its specification that are easily identified as being irrelevant to the computations. Sets of nodes that can be removed based upon graphical considerations only are nodes d-separated from $\mathbf{H}$ given $\mathbf{F}$, irrelevant evidence nodes (effects blocked by other evidence), and barren nodes, that is, nodes in $\mathbf{W}\backslash(\mathbf{H}\cup\mathbf{F})$ which are leafs or have only barren descendants; the remaining nodes coincide with the so-called parameter sensitivity set [3,7]. In addition, evidence absorption can be applied, where the outgoing arcs of variables with evidence are removed and the CPTs of the former children are reduced by removing the parameters that are incompatible with the observed value(s) of their former parent(s) [4]. After evidence absorption, all variables with evidence correspond to leafs in the graph and the CPT parameters of their former children will all be compatible with the evidence.

From here on we consider Bayesian networks that are reduced to what we call its *query-dependent backbone* $\mathcal{B}_q$, using the above-mentioned pre-processing options[1]. $\mathcal{B}_q$, tailored to the original query $\Pr(\mathbf{h}|\mathbf{f})$, is assumed to be a Bayesian network over variables $\mathbf{V} \subseteq \mathbf{W}$ from which the now equivalent query $\Pr(\mathbf{h}|\mathbf{e})$ is computed for evidence variables $\mathbf{E} \subseteq \mathbf{F}$; the remaining variables $\mathbf{V}\backslash(\mathbf{H}\cup\mathbf{E})$ will be denoted by $\mathbf{R}$.

The backbone given output $\Pr(ghk|def)$ in our example network from Fig. 1 is depicted in Fig. 2. After evidence absorption, the arc from F to K and the last two rows of K's CPT are removed. The node S is removed since it is barren, and D is removed, since it is d-seperated from the variables in $\mathbf{H}$ given $\mathbf{F}$. In the backbone network we have the hypothesis variables $\mathbf{H} = \{G, H, K\}$, the evidence variables $\mathbf{E} = \{E, F\}$ and the remaining variable $\mathbf{R} = \{R\}$.

---

[1] Note that the parameters of the local distributions that are in $\mathcal{B}$ but not in $\mathcal{B}_q$ do not affect the output of interest in any way.

2.3 Relating Queries to Parameters

It is well-known that an output of a Bayesian network relates to a network
parameter x as a fraction of two functions linear in that parameter:

$$\Pr(\mathbf{h}|\mathbf{e})(x) = \frac{\Pr(\mathbf{he})(x)}{\Pr(\mathbf{e})(x)} = \frac{\tau_1 \cdot x + \tau_2}{\kappa_1 \cdot x + \kappa_2}$$

where the constants τ_1, τ_2, κ_1 and κ_2 are composed of network parameters independent of x [3]. The above function can be generalised to multiple parameters [6]
and is typically exploited in the context of sensitivity analysis, to determine how
a change in one or more parameters affects $\Pr(\mathbf{h}|\mathbf{e})$. We note that upon varying
a parameter x of a distribution, the other parameters of the same distribution
have to be co-varied to let the distribution sum to 1. If the distribution is associated with a binary variable, the co-varying parameter equals $1-x$. If a variable is
multi-valued, however, different co-variation schemes are possible [9]. Sensitivity
functions are monotonic functions in each parameter, and are either increasing or
decreasing functions in such a parameter. Here we consider increasing (decreasing) in a non-strict sense, that is, increasing (decreasing) includes non-decreasing
(non-increasing).

3 Categorisation of Parameters in a Backbone
   Network $\mathscr{B}_q$

We will discuss the parameters of the variables in $\mathbf{R}$, $\mathbf{H}$ and $\mathbf{E}$ of a backbone
network and categorise these parameters according to their qualitative effect on
$\Pr(\mathbf{h}|\mathbf{e})$ as summarised in Table 1. In the proofs of our propositions we repeatedly
use the definition of conditional probability and the factorisation defined by $\mathscr{B}_q$:

$$\Pr(\mathbf{h}|\mathbf{e}) = \frac{\Pr(\mathbf{he})}{\Pr(\mathbf{e})}$$

in which for the numerator $\Pr(\mathbf{he})$ we find

$$\Pr(\mathbf{he}) = \sum_{r \in \mathbf{R}} \Pr(\mathbf{rhe}) = \sum_{r \in \mathbf{R}} \prod_{V^* \in \mathbf{V}} \Pr(v^*|\boldsymbol{\pi}^*)\big|_{v^*\boldsymbol{\pi}^* \sim \mathbf{rhe}} \qquad (1)$$

and for the denominator $\Pr(\mathbf{e})$ we find that

$$\Pr(\mathbf{e}) = \sum_{r \in \mathbf{R}, h^* \in \mathbf{H}} \Pr(\mathbf{rh^*e}) = \sum_{r \in \mathbf{R}, h^* \in \mathbf{H}} \prod_{V^* \in \mathbf{V}} \Pr(v^*|\boldsymbol{\pi}^*)\big|_{v^*\boldsymbol{\pi}^* \sim \mathbf{rh^*e}} \qquad (2)$$

Parameters which are not present in Eqs. (1) and (2) cannot affect the output
directly and are categorised as '*'. The effect of all other parameters is investi-
gated by studying properties of their sensitivity functions $\Pr(\mathbf{h}|\mathbf{e})(x)$. Parame-
ters that are guaranteed to give a monotone increasing sensitivity function are
classified as '+'; parameters that are guaranteed to give monotone decreasing

Table 1. Categorisation of parameters $x = \Pr(v\,|\,\pi)$ of a Bayesian network $\mathscr{B}_q$ with respect to the output probability $\Pr(\mathbf{h}\,|\,\mathbf{e})$.

$V \in \mathbf{R}$	cat '?'				
$V \in \mathbf{H}$	$\mathbf{E} \cap de(V) = \emptyset$	$v \sim \mathbf{h}, \pi \sim \mathbf{h}$	cat '+'		
		$v \nsim \mathbf{h}$ or $\pi \nsim \mathbf{h}$	cat '*'		
	$\mathbf{E} \cap de(V) \neq \emptyset$	non-binary V	cat '?'		
		binary V	$\pi \nsim \mathbf{h}$	cat '?'	
			$\pi \sim \mathbf{h}$	$\neg\langle \mathbf{H}_{\overline{pa}(V)}, \mathbf{R}_{pa(V)} \,|\, \mathbf{E} \cup \mathbf{H}_{pa(V)} \cup V\rangle_d$	cat '?'
				$\langle \mathbf{H}_{\overline{pa}(V)}, \mathbf{R}_{pa(V)} \,|\, \mathbf{E} \cup \mathbf{H}_{pa(V)} \cup V\rangle_d$	$v \sim \mathbf{h}$ cat '+'
					$v \nsim \mathbf{h}$ cat '−'
$V \in \mathbf{E}$	$v \nsim \mathbf{e}$	cat '*'			
	$v \sim \mathbf{e}$	$\pi \nsim \mathbf{h}$	cat '−'		
		$\pi \sim \mathbf{h}$	$\neg\langle \mathbf{H}_{\overline{pa}(V)}, \mathbf{R}_{pa(V)} \,|\, \mathbf{E} \cup \mathbf{H}_{pa(V)}\rangle_d$	cat '?'	
			$\langle \mathbf{H}_{\overline{pa}(V)}, \mathbf{R}_{pa(V)} \,|\, \mathbf{E} \cup \mathbf{H}_{pa(V)}\rangle_d$	cat '+'	

sensitivity functions as '−'. Parameters for which the sign of the derivative of the sensitivity function depends on the actual quantification of the network will be categorised as '?'. Note that sensitivity functions for parameters of category '*' are not necessarily constant: variation of such a parameter may result in co-variation of a parameter from the same local distribution which *is* present in Eqs. (1) or (2). As such, parameters of category '*' may be indirectly affecting output $\Pr(\mathbf{h}\,|\,\mathbf{e})$, yet for the computation of $\Pr(\mathbf{h}\,|\,\mathbf{e})$ it suffices to know the values of parameters in the categories '+', '−' and '?'. In the relation between parameter changes and output changes these latter parameters are pivotal.

For the backbone network of our example in Fig. 2, the categories of its parameters are indicated in the CPTs.

4 Categorisation of the Parameters of Variables in R and H

4.1 Parameters of Variables in R

For a variable $R \in \mathbf{R}$ the qualitative effect of a change in one of its parameters x cannot be predicted without additional information: the sensitivity function $\Pr(\mathbf{h}\,|\,\mathbf{e})(x)$ can be either monotone increasing or decreasing. Therefore, these parameters are categorised as '?'.

Proof of the above claim, and of all further propositions concerning parameters in the category '?', is omitted due to space restrictions. All these proofs are based on demonstrating that additional knowledge—for example the specific network quantification—is required to determine whether the one-way sensitivity function is increasing or decreasing.

4.2 Parameters of Variables in H Without Descendants in E

The propositions in this section concern parameters $x = \Pr(v\,|\,\pi)$ of nodes $V \in \mathbf{H}$ without descendants in $\mathbf{E}$. The parameters of such a node which are fully compatible with $\mathbf{h}$ have a monotone increasing sensitivity function $\Pr(\mathbf{h}|\mathbf{e})(x)$ and therefore are classified as '+'. The parameters not fully compatible with $\mathbf{h}$ are not used in the computation of $\Pr(\mathbf{h}|\mathbf{e})$ and therefore are classified '*'.

Proposition 1. *Consider a query-dependent backbone Bayesian network $\mathscr{B}_q$ with probability of interest $\Pr(\mathbf{h}|\mathbf{e})$. Let $x = \Pr(v|\boldsymbol{\pi})$ be a parameter of a variable $V \in \mathbf{H}$ such that $de(V) \cap \mathbf{E} = \emptyset$. If both $v \sim \mathbf{h}$ and $\boldsymbol{\pi} \sim \mathbf{h}$, then $\Pr(\mathbf{h}|\mathbf{e})(x)$ is a monotone increasing function.*

Proof. Let $\mathbf{r}_\pi$ denote the configuration of $\mathbf{R}_{pa(V)}$ compatible with $\boldsymbol{\pi}$. In addition, let $\mathbf{h} = v\mathbf{h}_\pi\mathbf{h}_{\overline{\pi}}$, where $\mathbf{h}_\pi$ and $\mathbf{h}_{\overline{\pi}}$ are assignments, compatible with $\mathbf{h}$, to $\mathbf{H}_{pa(V)}$ and $\mathbf{H}_{\overline{pa}(V)}$, respectively. First consider the general form of $\Pr(\mathbf{he})$ given by Eq. (1). We observe that under the given conditions we can write:

$$\Pr(\mathbf{he})(x) = x \cdot \Pr(\mathbf{h}_{\overline{\pi}} | v\mathbf{h}_\pi \mathbf{er}_\pi) \cdot \frac{\Pr(v\mathbf{h}_\pi \mathbf{er}_\pi)}{\Pr(v|\boldsymbol{\pi})} + \sum_{\mathbf{r}^+ \in \mathbf{R}_{pa(V)}} \Pr(v\mathbf{h}_\pi \mathbf{h}_{\overline{\pi}} \mathbf{er}^+) \Big|_{\mathbf{r}^+ \neq \mathbf{r}_\pi}$$

where $\Pr(v\mathbf{h}_\pi \mathbf{er}_\pi)/\Pr(v|\boldsymbol{\pi})$ represents a sum of products of parameters no longer including $\Pr(v|\boldsymbol{\pi})$. This expression thus is of the form $x \cdot \tau_1 \cdot \tau_2 + \tau_3$, for non-negative constants τ_1, τ_2, τ_3.

For $\Pr(\mathbf{e})$, as given in Eq. (2), we observe that since V has no descendants in $\mathbf{E}$, this node in fact is barren with respect to $\Pr(\mathbf{e})$. As a result, none of V's parameters are relevant to the computation and $\Pr(\mathbf{e})(x)$ therefore equals a constant $\kappa_1 > 0$.

The sensitivity function for parameter x thus is of the form $\Pr(\mathbf{h}|\mathbf{e})(x) = (x \cdot \tau_1 \cdot \tau_2 + \tau_3)/\kappa_1$ with $\tau_1, \tau_2, \tau_3 \geq 0$ and $\kappa_1 > 0$. The first derivative of this function equals $(\tau_1 \cdot \tau_2)/\kappa_1$, which is always non-negative. $\square$

Proposition 2. *Let $\mathscr{B}_q$ and $\Pr(\mathbf{h}|\mathbf{e})$ be as before. Let $\Pr(v|\boldsymbol{\pi})$ be a parameter of a variable $V \in \mathbf{H}$ such that $de(V) \cap \mathbf{E} = \emptyset$. If $v \nsim \mathbf{h}$ or $\boldsymbol{\pi} \nsim \mathbf{h}$, then $\Pr(v|\boldsymbol{\pi})$ is not used in the computation of $\Pr(\mathbf{h}|\mathbf{e})$.*

Proof. We again consider $\Pr(\mathbf{he})$ as given by Eq. (1) and observe that a parameter $\Pr(v|\boldsymbol{\pi})$ with $v \nsim \mathbf{h}$ or $\boldsymbol{\pi} \nsim \mathbf{h}$ is not included in this expression. Moreover, as argued in the proof of Proposition 1, no parameter of V is used in computing $\Pr(\mathbf{e})$ from Eq. (2). $\Pr(v|\boldsymbol{\pi})$ is therefore not used in the computation of $\Pr(\mathbf{h}|\mathbf{e})$. $\square$

4.3   Parameters of Variables in $\mathbf{H}$ with Descendants in $\mathbf{E}$

For a parameter of a non-binary variable $V \in \mathbf{H}$ with at least one descendant in $\mathbf{E}$, we cannot predict without additional knowledge whether the sensitivity function $\Pr(\mathbf{h}|\mathbf{e})(x)$ is monotone increasing or monotone decreasing. These parameters therefore are classified as '?'. The same observation applies to a parameter of a binary variable $V \in \mathbf{H}$ with descendants in $\mathbf{E}$ for which $\boldsymbol{\pi} \nsim \mathbf{h}$ or for which $\mathbf{H}_{\overline{pa}(V)}$ and $\mathbf{R}_{pa(V)}$ are not d-separated given $\mathbf{H}_{pa(V)}$, V itself and the evidence.

If $V \in \mathbf{H}$ is binary, $\boldsymbol{\pi} \sim \mathbf{h}$ and $\mathbf{H}_{\overline{pa}(V)}$ and $\mathbf{R}_{pa(V)}$ are d-separated given $\mathbf{H}_{pa(V)}$, V itself and the evidence, then we do have sufficient knowledge to determine the qualitative effect of varying parameter $x = \Pr(v \mid \boldsymbol{\pi})$ of V. If $v \sim \mathbf{h}$ then $\Pr(\mathbf{h} \mid \mathbf{e})(x)$ is monotone increasing, and the parameter is classified as '+'. If $v \not\sim \mathbf{h}$ then $\Pr(\mathbf{h} \mid \mathbf{e})(x)$ is monotone decreasing, and the parameter is classified as '−' These observations are captured by Proposition 3. This proposition extends Proposition 2 in [1] by replacing the condition that $\mathbf{R}_{pa(V)} = \emptyset$ by the less strict d-separation condition mentioned above.

Proposition 3. *Let $\mathscr{B}_q$ and $\Pr(\mathbf{h} \mid \mathbf{e})$ be as before. Let $\Pr(v \mid \boldsymbol{\pi})$ with $\boldsymbol{\pi} \sim \mathbf{h}$ be a parameter of a binary variable $V \in \mathbf{H}$ and let $\langle \mathbf{H}_{\overline{pa}(V)}, \mathbf{R}_{pa(V)} \mid \mathbf{E} \cup \mathbf{H}_{pa(V)} \cup V \rangle_d$. If $v \sim \mathbf{h}$ then $\Pr(\mathbf{h} \mid \mathbf{e})(x)$ is a monotone increasing function; if $v \not\sim \mathbf{h}$ then $\Pr(\mathbf{h} \mid \mathbf{e})(x)$ is a monotone decreasing function.*

Proof. First consider the case where $v \sim \mathbf{h}$. Under the given conditions we have from the proof of Proposition 1 that $\Pr(\mathbf{he})$ takes on the form $\Pr(\mathbf{he})(x) = x \cdot \tau_1 \cdot \tau_2 + \tau_3$, for constants $\tau_1, \tau_2, \tau_3 \geq 0$.

For $\Pr(\mathbf{e})$ and binary V we note that Eq. (2) can be written as

$$\Pr(\mathbf{e})(x) = x \cdot \frac{\Pr(v\mathbf{h}_\pi \mathbf{r}_\pi \mathbf{e})}{\Pr(v \mid \pi)} + (1 - x) \cdot \frac{\Pr(\overline{v}\mathbf{h}_\pi \mathbf{r}_\pi \mathbf{e})}{\Pr(\overline{v} \mid \pi)} + \sum_{\mathbf{r}^+ \in \mathbf{R}_{pa(V)}} \Pr(v\mathbf{h}_\pi \mathbf{r}^+ \mathbf{e})\big|_{\mathbf{r}^+ \neq \mathbf{r}_\pi}$$

$$+ \sum_{\mathbf{r}^+ \in \mathbf{R}_{pa(V)}} \Pr(\overline{v}\mathbf{h}_\pi \mathbf{r}^+ \mathbf{e})\big|_{\mathbf{r}^+ \neq \mathbf{r}_\pi} + \sum_{\mathbf{h}^+ \in \mathbf{H}_{pa(V)}} \Pr(\mathbf{h}^+ \mathbf{e})\big|_{\mathbf{h}^+ \neq \mathbf{h}_\pi}$$

which takes on the following form: $\Pr(\mathbf{e})(x) = x \cdot \tau_2 + (1 - x) \cdot \kappa_2 + \kappa_1 + \kappa_3 + \kappa_4$, with constants $\kappa_i \geq 0$, $i = 1, \ldots, 4$.

The sign of the derivative of the sensitivity function is determined by the *numerator* $\tau_1 \cdot \tau_2 \cdot (\kappa_1 + \kappa_2 + \kappa_3 + \kappa_4) - \tau_3 \cdot (\tau_2 - \kappa_2)$ of $\Pr(\mathbf{h} \mid \mathbf{e})'(x)$. We observe that given $\langle \mathbf{H}_{\overline{pa}(V)}, \mathbf{R}_{pa(V)} \mid \mathbf{E} \cup \mathbf{H}_{pa(V)} \cup V \rangle_d$ we find that $\tau_1 \cdot \kappa_1 = \tau_3$ which guarantees the derivative to be non-negative. This implies that, for $v \sim \mathbf{h}$, $\Pr(\mathbf{h} \mid \mathbf{e})(x)$ is a monotone increasing function.

Now consider the case where $v \not\sim \mathbf{h}$. Since V is binary, this implies that $\overline{v} \sim \mathbf{h}$. The proof for this case follows by replacing, in the above formulas, every occurrence of v by $\overline{v}$ and, hence, every x with $1 - x$. As a result we find that in this case $\Pr(\mathbf{h} \mid \mathbf{e})(x)$ is a monotone decreasing function. □

5 Categorisation of the Parameters of the Variables in E

5.1 Parameters $\Pr(v \mid \boldsymbol{\pi})$ of a Variable $V \in \mathbf{E}$ with $v \not\sim \mathbf{e}$

Recall that after evidence absorption all parameters with $\boldsymbol{\pi} \not\sim \mathbf{e}$ are removed from the network. For a parameter $\Pr(v \mid \boldsymbol{\pi})$ of a variable in $V \in \mathbf{E}$, however, we may still find that $v \not\sim \mathbf{e}$; these parameters are in category '*'.

Proposition 4. *Let $\mathscr{B}_q$ and $\Pr(\mathbf{h}|\mathbf{e})$ be as before. Let $\Pr(v|\boldsymbol{\pi})$ be a parameter of a variable $V \in \mathbf{E}$. If $v \nsim \mathbf{e}$, then $\Pr(v|\boldsymbol{\pi})$ is not used in the computation of $\Pr(\mathbf{h}|\mathbf{e})$.*

Proof. This proposition is equivalent to Proposition 3 in [1], but stated for $\mathscr{B}_q$ rather than for $\mathscr{B}$. $\qquad\square$

5.2   Parameters $\Pr(v|\boldsymbol{\pi})$ of a Variable $V \in \mathbf{E}$ with $v \sim \mathbf{e}$ and $\boldsymbol{\pi} \nsim \mathbf{h}$

We now consider the parameters $\Pr(v|\boldsymbol{\pi})$ of $V \in \mathbf{E}$, with $v \sim \mathbf{e}$ and $\boldsymbol{\pi} \nsim \mathbf{h}$. The one-way sensitivity functions of such parameters are monotone decreasing. These parameters therefore are categorised as '$-$'.

Proposition 5. *Let $\mathscr{B}_q$ and $\Pr(\mathbf{h}|\mathbf{e})$ be as before. Let $x = \Pr(v|\boldsymbol{\pi})$ be a parameter of $V \in \mathbf{E}$ such that $v \sim \mathbf{e}$. If $\boldsymbol{\pi} \nsim \mathbf{h}$, then $\Pr(\mathbf{h}|\mathbf{e})(x)$ is a monotone decreasing function.*

Proof. This proposition is equivalent to Proposition 1 in [1], but stated for $\mathscr{B}_q$ rather than for $\mathscr{B}$. $\qquad\square$

5.3   Parameters $\Pr(v|\boldsymbol{\pi})$ of a Variable $V \in \mathbf{E}$ with $v \sim \mathbf{e}$ and $\boldsymbol{\pi} \sim \mathbf{h}$

We now consider parameters $\Pr(v|\boldsymbol{\pi})$ of $V \in \mathbf{E}$ with $v \sim \mathbf{e}$ and $\boldsymbol{\pi} \sim \mathbf{h}$. The one-way sensitivity functions of such parameters are monotone increasing under the condition that $\mathbf{H}_{\overline{pa}(V)}$ is d-separated from $\mathbf{R}_{pa(V)}$ given $\mathbf{H}_{pa(V)}$ and the evidence. Under this condition, these parameters therefore can be categorised as '$+$'. This proposition extends Proposition 1 in [1] by replacing the condition that $\mathbf{R}_{pa(V)} = \emptyset$ by the less strict d-separation condition mentioned above.

Proposition 6. *Let $\mathscr{B}_q$ and $\Pr(\mathbf{h}|\mathbf{e})$ be as before. Let $x = \Pr(v|\boldsymbol{\pi})$ with $v \sim \mathbf{e}$ and $\boldsymbol{\pi} \sim \mathbf{h}$ be a parameter of $V \in \mathbf{E}$ and let $\langle \mathbf{H}_{\overline{pa}(V)}, \mathbf{R}_{pa(V)} | \mathbf{H}_{pa(V)} \cup \mathbf{E} \rangle_d$. Then $\Pr(\mathbf{h}|\mathbf{e})(x)$ is a monotone increasing function.*

Proof. For $\Pr(\mathbf{he})$ we observe that Eq. (1) can be written as the expression in the proof of Proposition 1, but with v included in $\mathbf{e}$ rather than in $\mathbf{h}$. We therefore have that $\Pr(\mathbf{he})(x) = x \cdot \tau_1 \cdot \tau_2 + \tau_3$ for constants $\tau_1, \tau_2, \tau_3 \geq 0$.

For $\Pr(\mathbf{e})$, given by Eq. (2), we observe that we can write

$$\Pr(\mathbf{e})(x) = x \frac{\Pr(\mathbf{h}_\pi \mathbf{r}_\pi \mathbf{e})}{\Pr(v|\boldsymbol{\pi})} + \sum_{\mathbf{r}^+ \in \mathbf{R}_{pa(V)}} \Pr(\mathbf{h}_\pi \mathbf{e} \mathbf{r}^+)\big|_{\mathbf{r}^+ \neq \mathbf{r}_\pi} + \sum_{\mathbf{h}^+ \in \mathbf{H}_{pa(V)}} \Pr(\mathbf{h}^+ \mathbf{e})\big|_{\mathbf{h}^+ \neq \mathbf{h}_\pi}$$

which is of the form $x \cdot \tau_2 + \kappa_1 + \kappa_2$, for constants $\kappa_1, \kappa_2 \geq 0$.

We now find that the *numerator* of the first derivative of the sensitivity function equals $\tau_1 \cdot \tau_2 \cdot (\kappa_1 + \kappa_2) - \tau_2 \cdot \tau_3$. We observe that given $\langle \mathbf{H}_{\overline{pa}(V)}, \mathbf{R}_{pa(V)} | \mathbf{H}_{pa(V)} \cup \mathbf{E} \rangle_d$ we find that $\tau_1 \cdot \kappa_1 = \tau_3$ which guarantees the derivative to be non-negative. This implies that $\Pr(\mathbf{h}|\mathbf{e})(x)$ is a monotone increasing function. $\qquad\square$

In case the above mentioned d-separation property does not hold, we need additional information to predict whether the sensitivity function $\Pr(\mathbf{h}|\mathbf{e})(x)$ of a parameter x of a variable $V \in \mathbf{E}$ with $v \sim \mathbf{e}$ and $\boldsymbol{\pi} \sim \mathbf{h}$ is monotone increasing or monotone decreasing, without additional knowledge. If the property does not hold, therefore, these parameters are in category '?'.

6 Discussion

In this paper we presented fundamental results concerning the qualitative effect of parameter changes on the output probabilities of a Bayesian network. Based on the graph structure and the query at hand, we categorised all network parameters into one of four categories: parameters not included in the computation of the output, parameters with guaranteed monotone increasing sensitivity functions, parameters with guaranteed monotone decreasing sensitivity functions, and parameters of which the qualitative effect cannot be predicted without additional information. Previously we demonstrated that knowledge of the qualitative effects can be exploited in inference in credal networks [1] and in multiple-parameter tuning of Bayesian networks [2]. In our previous research only a partial categorisation of the parameters was given. Our present paper allocates a wider range of parameters into one of the meaningful categories '+', '−' and '*'. For future research we would like to further study properties of the additional information required to predict the qualitative effect of the parameters in category '?'.

Acknowledgements. This research was supported by the Netherlands Organisation for Scientific Research (NWO).

References

1. Bolt, J.H., De Bock, J., Renooij, S.: Exploiting Bayesian network sensitivity functions for inference in credal networks. In: Proceedings of the 22nd European Conference on Artificial Intelligence, vol. 285, pp. 646–654 (2016)
2. Bolt, J.H., van der Gaag, L.C.: Balanced sensitivity functions for tuning multi-dimensional Bayesian network classifiers. Int. J. Approx. Reason. **80c**, 361–376 (2017)
3. Coupé, V.M.H., van der Gaag, L.C.: Properties of sensitivity analysis of Bayesian belief networks. Ann. Math. Artif. Intell. **36**(4), 323–356 (2002)
4. van der Gaag, L.C.: On evidence absorption for belief networks. Int. J. Approx. Reason. **15**(3), 265–286 (1996)
5. Jensen, F.V., Nielsen, T.D.: Bayesian Networks and Decision Graphs, 2nd edn. Springer, New York (2007)
6. Kjærulff, U., van der Gaag, L.C.: Making sensitivity analysis computationally efficient. In: Boutilier, C., Goldszmidt, M. (eds.) Proceedings of the Sixteenth Conference on Uncertainty in Artificial Intelligence, pp. 317–325. Morgan Kaufmann Publishers, San Francisco (2000)

7. Meekes, M., Renooij, S., Gaag, L.C.: Relevance of evidence in Bayesian networks. In: Destercke, S., Denoeux, T. (eds.) ECSQARU 2015. LNCS, vol. 9161, pp. 366–375. Springer, Cham (2015). doi:10.1007/978-3-319-20807-7_33
8. Pearl, J.: Probabilistic Reasoning in Intelligent Systems: Networks of Plausible Inference. Morgan Kaufmann Publishers, Palo Alto (1988)
9. Renooij, S.: Co-variation for sensitivity analysis in Bayesian networks: properties, consequences and alternatives. Int. J. Approx. Reason. 55, 1022–1042 (2014)

The Descriptive Complexity of Bayesian Network Specifications

Fabio G. Cozman[1(✉)] and Denis D. Mauá[2]

[1] Escola Politécnica, Universidade de São Paulo, São Paulo, Brazil
fgcozman@usp.br
[2] Instituto de Matemática e Estatística, Universidade de São Paulo,
São Paulo, Brazil

Abstract. We adapt the theory of descriptive complexity to Bayesian networks, by investigating how expressive can be specifications based on predicates and quantifiers. We show that Bayesian network specifications that employ first-order quantification *capture* the complexity class PP; that is, any phenomenon that can be simulated with a polynomial time probabilistic Turing machine can be also modeled by such a network. We also show that, by allowing quantification over predicates, the resulting Bayesian network specifications *capture* the complexity class PP^NP, a result that does not seem to have equivalent in the literature.

1 Introduction

One can find a variety of "relational" Bayesian networks in the literature, where constructs from first-order logic are used to represent whole populations and repetitive patterns [7,15,16]. Such networks may be specified by diagrams or by text; typically they are viewed as templates that can be grounded into finite propositional Bayesian networks whenever needed.

It is only natural to ask what is the *expressivity* of Bayesian network specifications based on predicates and quantifiers. That is, what can and what cannot be modeled by these specifications, and at what computational costs. To address these questions, we are inspired by the well-known theory of *descriptive complexity* for logical languages [4,9]. It does not seem that the descriptive complexity of Bayesian networks has been investigated in previous work, and to do so we adapt existing insights and results. Because the topic is novel, most of this paper consists of building a framework in which to operate.

In fact, one should expect "relational" Bayesian network specifications to exhibit properties that cannot be matched by propositional networks — much as first-order logic goes beyond propositional logic. This sort of phenomenon has been already noted in connection with *lifted* inference algorithms [10,18], but has not been characterized in terms of descriptive complexity.

We define precisely what we mean by "Bayesian network specifications" in Sect. 2, and present some necessary background in Sect. 3. We then move to our main results in Sects. 4 and 5.

© Springer International Publishing AG 2017
A. Antonucci et al. (Eds.): ECSQARU 2017, LNAI 10369, pp. 93–103, 2017.
DOI: 10.1007/978-3-319-61581-3_9

We show that Bayesian network specifications that employ first-order quantification *capture* the complexity class PP. That is, a language is in PP *if and only if* its strings encode valid inferences in a Bayesian network specified with predicates and first-order quantifiers. Note that this is a much stronger statement than PP-completeness. And then we look at specifications that allow quantification over predicates, and show that such "second-order" Bayesian networks capture the complexity class PP^{NP}. It does not seem that previous results on descriptive theory have reached this latter complexity class.

Intuitively, these results can be interpreted as follows: suppose we have a (physical, social, economic) phenomenon that can be simulated by a probabilistic Turing machine in polynomial time: given an input, the machine will run for a number of steps that is polynomial in the length of the input, and the machine will stop with the same YES-NO answer as produced by the phenomenon. Our results show that the phenomenon can be modeled by a Bayesian network specification based on predicates and first-order quantification in the sense that, given the input as evidence, an inference with the network will produce the same result as the phenomenon. But what happens if the phenomenon is so complex that it requires even more computation to be simulated? For instance, what happens if the phenomenon requires a polynomial time probabilistic Turing machine with another nondeterministic Turing machine as oracle? This phenomenon cannot be modeled by a "relational" Bayesian network specification, unless widely accepted assumptions about complexity classes collapse. However, our results show that this phenomenon *can* be modeled by a Bayesian network specification that allows quantification over predicates.

We further discuss the intellectual significance of these results in the concluding Sect. 6.

2   Specifying Bayesian Networks with Logical Constructs

In the Introduction we loosely mentioned "relational" Bayesian networks. To make any progress, we must precisely define what is here allowed in specifying Bayesian networks. Our strategy, described in this section, is to adopt a proposal by Poole [14] to mix probabilistic assessments and logical equivalences [2].

2.1   Preliminaries

First, to recap: a *Bayesian network* is a pair consisting of a directed acyclic graph $\mathbb{G}$ whose nodes are random variables, and a joint probability distribution $\mathbb{P}$ over the variables in the graph, so that $\mathbb{G}$ and $\mathbb{P}$ satisfy the Markov condition (a random variable is independent of its nondescendants given its parents). The Markov condition induces a factorization of probabilities [11].

In this paper every random variable is binary with values 0 and 1, respectively signifying false and true.

We only consider textual specifications, mostly relying on formulas of function-free first-order logic with equality (denoted by FFFO). That is, most formulas we contemplate are well-formed formulas of first-order logic with equality but without functions, containing predicates from a finite relational vocabulary, negation ($\neg$), conjunction ($\wedge$), disjunction ($\vee$), implication ($\Rightarrow$), equivalence ($\Leftrightarrow$), existential quatification ($\exists$) and universal quantification ($\forall$). Later we discuss second-order quantification over predicates.

First-order theories are interpreted as usual [5], using *domains*, that are just sets, and *interpretations* that associate predicates with relations, and constants with elements; that is, an interpretation is a truth assignment for every grounding of every predicate. A pair domain/interpretation is a *structure*. We only deal with finite domains in this paper.

Throughout the paper it will be convenient to view each grounded predicate $r(\vec{a})$, for a fixed vocabulary/domain, as a random variable over interpretations (note: an overline arrow denotes a tuple). That is, given a domain $\mathcal{D}$, we understand $r(\vec{a})$ as a function over all possible interpretations of the vocabulary, so that $r(\vec{a})(\mathbb{I})$ yields 1 if $r(\vec{a})$ is true in interpretation $\mathbb{I}$, and 0 otherwise.

For instance, say we have two unary predicates r and s, and we are given a domain $\mathcal{D} = \{a, b\}$. Then we have groundings $\{r(a), r(b), s(a), s(b)\}$, and there are 2^4 possible interpretations. Each interpretation assigns true or false to $r(a)$, and similarly to every grounding. So $r(a)$ can be viewed as a random variable over the possible interpretations.

2.2   Relational Bayesian Network Specifications

A *relational Bayesian network specification*, abbreviated RELBN, is a directed acyclic graph where each node is a predicate (from a finite relational vocabulary), and where

1. each *root* node r is associated with a probabilistic assessment

$$\mathbb{P}\left(r(\vec{x}) = 1\right) = \alpha, \tag{1}$$

2. while each *non-root* node s is associated with a formula (called the *definition of s*)

$$s(\vec{x}) \Leftrightarrow \phi(\vec{x}), \tag{2}$$

where $\phi(\vec{x})$ is a formula in FFFO with free variables $\vec{x}$.

Given a domain, a RELBN can be grounded into a unique Bayesian network:

1. by producing every grounding of the predicates;
2. by associating with each grounding $r(\vec{a})$ of a root predicate the grounded assessment $\mathbb{P}\left(r(\vec{a}) = 1\right) = \alpha$;
3. by associating with each grounding $s(\vec{a})$ of a non-root predicate the grounded definition $s(\vec{a}) \Leftrightarrow \phi(\vec{a})$, and by replacing univeral/existential quantification by conjunction/disjunction over the domain;

4. finally, by drawing a graph where each node is a grounded predicate and where there is an edge into each grounded non-root predicate $s(\vec{a})$ from each grounding of a predicate that appears in the grounded definition of $s(\vec{a})$.

Consider, as an example, the following model of asymmetric friendship, where an individual is always a friend of herself, and where two individuals are friends if they are both fans (of a writer, say) or if there is some "other" reason for it:

$$\mathbb{P}(\mathsf{fan}(x)) = 0.2, \qquad \mathbb{P}\big(\mathsf{other}(x,y)\big) = 0.1,$$
$$\mathbb{P}\big(\mathsf{friends}(x,y)\big) \Leftrightarrow (x = y) \vee (\mathsf{fan}(x) \wedge \mathsf{fan}(y)) \vee \mathsf{other}(x,y), \qquad (3)$$

Suppose we have domain $\mathcal{D} = \{a,b,c\}$. Figure 1 depicts the Bayesian network generated by $\mathcal{D}$ and Expression (3).

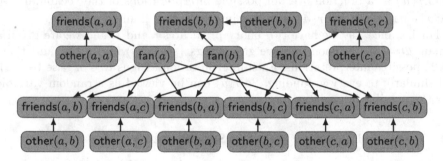

Fig. 1. The Bayesian network generated by Expression (3) and domain $\mathcal{D} = \{a,b,c\}$.

For a given RELBN τ and a domain $\mathcal{D}$, denote by $\mathbb{B}(\tau, \mathcal{D})$ the Bayesian network obtained by grouning τ with respect to $\mathcal{D}$. The set of all relational Bayesian network specifications is denoted by $\mathcal{B}(\mathsf{FFFO})$.

3   A Bit of Descriptive Complexity

We employ several concepts from finite model theory [4,8,12] and complexity theory [13]. We consider input strings in the alphabet $\{0,1\}$; that is, a *string* is a sequence of 0s and 1s. A *language* is a set of strings; a *complexity class* is a set of languages. A language is *decided* by a Turing machine if the machine accepts each string in the language, and rejects each string not in the language. The complexity class NP contains each language that can be decided by a nondeterministic Turing machine with a polynomial time bound. If a Turing machine is such that, whenever its transition function maps to a non-singleton set, the transition is selected with uniform probability within that set, then the Turing machine is a *probabilistic Turing machine*. The complexity class PP is the set of languages that are decided by a probabilistic Turing machine in polynomial time, with an error probability strictly less than 1/2 for all input strings. This complexity class can be equivalently defined as follows: a language is in PP if and only if there is a polynomial nondeterministic Turing machine such that a string is in the language if and only if more than half of the computation paths of the machine end in the accepting state when the string is the input.

If a formula $\phi(\vec{x})$ has free logical variables $\vec{x}$, then structure $\mathfrak{A}$ is a *model* of $\phi(\vec{a})$ iff $\phi(\vec{x})$ is true in structure $\mathfrak{A}$ when the logical variables $\vec{x}$ are replaced by elements $\vec{a}$ of the domain.

A formula ϕ in existential function-free second-order logic (denoted by ESO) is a formula of the form $\exists r_1 \ldots \exists r_m \phi'$, where ϕ' is a sentence of FFFO containing predicates $r_1, \ldots, r_m$. Such a sentence allows existential quantification over the predicates themselves. Note that again we have equality in the language (that is, the built-in predicate $=$ is always available). Here a structure $\mathfrak{A}$ is a pair domain/interpretation, but the interpretation does not touch predicates that are existentially quantified (that is, if ϕ contains predicates $r_1, \ldots, r_m$ and $s_1, \ldots, s_M$, but $r_1, \ldots, r_m$ are all existentially quantified, then a model for ϕ contains an intepretation for $s_1, \ldots, s_M$).

As an example, consider the following formula of ESO [8]:

$$\exists \text{partition} : \forall x : \forall y : \big(\text{edge}(x, y) \Rightarrow (\text{partition}(x) \Leftrightarrow \neg\text{partition}(y))\big). \qquad (4)$$

Here a domain can be viewed as a set of nodes, and an interpretation can be viewed as a set of edges; the formula is satisfied if and only if it is possible to partition the vertices into two subsets such that if a node is in one subset, it is not in the other (that is, the graph is bipartite).

There is an *isomorphism* between structures $\mathfrak{A}_1$ and $\mathfrak{A}_2$ when there is a bijective mapping g between the domains such that if $r(a_1, \ldots, a_k)$ is true in $\mathfrak{A}_1$, then $r(g(a_1), \ldots, g(a_k))$ is true in $\mathfrak{A}_2$, and moreover if $r(a_1, \ldots, a_k)$ is true in $\mathfrak{A}_2$, then $r(g^{-1}(a_1), \ldots, g^{-1}(a_k))$ is true in $\mathfrak{A}_1$ (where g^{-1} denotes the inverse of g). A set of structures is *isomorphism-closed* if whenever a structure is in the set, all structures that are isomorphic to it are also in the set.

We assume that every structure is given as a string, encoded as follows for a fixed vocabulary (encoding from Ref. [12, Sect. 6.1]). First, if the domain contains elements $a_1, \ldots, a_n$, then the string begins with n symbols 0 followed by 1. The vocabulary is fixed, so we take some order for the predicates, $r_1, \ldots, r_m$. We then append, in this order, the encoding of the interpretation of each predicate. Focus on predicate r_i of arity k. To encode it with respect to a domain, we need to order the elements of the domain, say $a_1 < a_2 < \cdots < a_n$. This total ordering is assumed for now to be always available; it will be important later to check that the ordering itself can be defined. In any case, with a total ordering we can enumerate lexicographically all k-tuples over the domain. Now suppose $\vec{a}_j$ is the jth tuple in this enumeration; then the $j$the bit of the encoding of r_i is 1 if $r(\vec{a}_j)$ is true in the given interpretation, and 0 otherwise. Thus the encoding is a string containing $n + 1 + \sum_{i=1}^{m} n^{\text{arity}(r_i)}$ symbols (either 0 or 1).

We can now state the celebrated theorem by Fagin on descriptive complexity:

Theorem 1. *Let S be an isomorphism-closed set of finite structures of some non-empty finite vocabulary. Then S is in* NP *if and only if S is the class of finite models of a sentence in* ESO.

The significance of Fagin's theorem is that it offers a definition of NP that is not tied to any computational model; rather, it is tied to the expressivity of the language that is used to specify problems. The surprising part of Fagin's theorem is that every language in NP can be exactly encoded by an ESO sentence.

4 $\mathcal{B}$(**FFFO**) Captures PP

Given a RELBN and a domain, an *evidence piece* $\mathbf{E}$ is a partial interpretation; that is, an evidence piece assigns a truth value to some groundings of predicates.

We encode a pair domain-evidence $(\mathcal{D}, \mathbf{E})$ with the same strategy used in the previous section to encode a structure; however, we must take into account the fact that a particular grounding of a predicate can be either assigned true or false or be left without assignment. So we use a pair of symbols in $\{0, 1\}$ to encode each grounding; we assume that 00 means false and 11 means true, while say 01 means lack of assignment.

Say there is an *isomorphism* between pairs $(\mathcal{D}_1, \mathbf{E}_1)$ and $(\mathcal{D}_2, \mathbf{E}_2)$ when there is a bijective mapping g between the domains such that if $r(a_1, \ldots, a_k)$ is true in $\mathbf{E}_1$, then $r(g(a_1), \ldots, g(a_k))$ is true in $\mathbf{E}_2$, and moreover if $r(a_1, \ldots, a_k)$ is true in $\mathbf{E}_2$, then $r(g^{-1}(a_1), \ldots, g^{-1}(a_k))$ is true in $\mathbf{E}_1$ (where again g^{-1} denotes the inverse of g). A set of pairs domain-evidence is *isomorphism-closed* if whenever a pair is in the set, all pairs that are isomorphic to it are also in the set.

Suppose a set of pairs domain-evidence is given with respect to a fixed vocabulary σ. Once encoded, these pairs form a language L that can for instance belong to NP or to PP. One can imagine building a relational Bayesian network specification τ on an extended vocabulary consisting of σ plus some additional predicates, so as to decide this language L of domain-evidence pairs. For a given input pair $(\mathcal{D}, \mathbf{E})$, the Bayesian network specification and the domain lead to a Bayesian network $\mathbb{B}(\tau, \mathcal{D})$; this network can be used to compute the probability of some groundings, and that probabiility in turn can be used to accept/reject the input. This is the sort of strategy we pursue.

The point is that we must determine some prescription by which, given a Bayesian network and an evidence piece, one can generate an actual decision so as to accept/reject the input pair domain/evidence. We adopt the following strategy. Assume that in the extended vocabulary of τ there are two sets of distinguished auxiliary predicates $A_1, \ldots, A_{m'}$ and $B_1, \ldots, B_{m''}$ that are not in σ. We can use the Bayesian network $\mathbb{B}(\tau, \mathcal{D})$ to compute the probability $\mathbb{P}(\mathbf{A}|\mathbf{B}, \mathbf{E})$ where $\mathbf{A}$ and $\mathbf{B}$ are interpretations of $A_1, \ldots, A_{m'}$ and $B_1, \ldots, B_{m''}$ respectively. And then we might accept/reject the input on the basis of $\mathbb{P}(\mathbf{A}|\mathbf{B}, \mathbf{E})$. However, we cannot specify particular intepretations $\mathbf{A}$ and $\mathbf{B}$ as the related predicates are not in the vocabulary σ. Thus the sensible strategy is to fix attention to some selected, fixed, pair of intepretations for these predicates; we simply take the interpretations that assign true to every grounding.

In short: use the Bayesian network $\mathbb{B}(\tau, \mathcal{D})$ to determine whether or not $\mathbb{P}(\mathbf{A}|\mathbf{B}, \mathbf{E}) > 1/2$, where $\mathbf{A}$ assigns true to every grounding of $A_1, \ldots, A_{m'}$, and $\mathbf{B}$ assigns true to every grounding of $B_1, \ldots, B_{m''}$. If this inequality is satisfied, the input pair is *accepted*; if not, the input pair is *rejected*.

We refer to $A_1, \ldots, A_{m'}$ as the *conditioned predicates*, and to $B_1, \ldots, B_{m''}$ as the *conditioning predicates*.

Here is the main result:

Theorem 2. *Let S be an isomorphism-closed set of pairs domain-evidence of some non-empty finite vocabulary, where all domains are finite. Then S is in* PP *if and only if S is the class of domain-evidence pairs that are accepted by a fixed* RELBN *with fixed conditioned and conditioning predicates.*

Proof. First, if S is a class of domain-query pairs that are accepted by a fixed RELBN, they can be decided by a polynomial time probability Turing machine. To see that, note that we can build a nondeterministic Turing machine that guesses the truth value of all groundings that do not appear in the query (that is, not in $\mathbf{A} \cup \mathbf{B} \cup \mathbf{E}$), and then verifies whether the resulting complete interpretation is a model of the RELBN. Recall that model checking of a fixed first-order sentence is in P [12].

To prove the other direction, we adapt the proof of Fagin's theorem as described by Grädel [8], along the same lines as the proof of Theorem 1 by Saluja et al. [17]. So, suppose that $\mathcal{L}$ is a language decided by some probabilistic Turing machine. Equivalently, there is a nondeterministic Turing machine that determines whether the majority of its computation paths accept an input, and accepts/rejects the input accordingly. By the mentioned proof of Fagin's theorem, there is a first-order sentence ϕ' with vocabulary consisting of the vocabulary of the input plus additional auxiliary predicates, such that each interpretation of this joint vocabulary is a model of the sentence if it is encodes a computation path of the Turing machine, as long as there is an available additional predicate that is guaranteed to be a linear order on the domain. Due to the lack of space, details of the construction are omitted; suffice to say that the same predicates X_q (one per state), Y_σ (one per symbol), and Z, employed by Grädel, are to be used here, with the same associated definitions. Denote by A the zero arity predicate with associated definition $A \Leftrightarrow \phi' \wedge \phi_E$, where ϕ_E is satisfied when an accepting state is reached. Suppose a linear order is indeed available; then by creating a RELBN where all groundings are associated with probability $1/2$, and where a non-root node is associated with the sentence in the proof of Fagin's theorem, we have that the probability of the query is larger than $1/2$ iff the majority of computation paths accept. The challenge is to encode a linear order. To do so, introduce a new predicate $<$ and the first-order sentence ϕ'' that forces $<$ to be a total order, and a zero arity predicate B that is associated with definition $B \Leftrightarrow \phi' \wedge \phi''$. Now an input domain-pair $(\mathcal{D}, \mathbf{E})$ is accepted by the majority of computation paths in the Turing machine if and only if we have $\mathbb{P}(A|B, \mathbf{E}) > 1/2$. Note that there are actually $n!$ linear orders that satisfy B, but for each one of these linear orders we have the same assignments for all other predicates, hence the ratio between accepting computations and all computations is as desired.   □

We might picture this as follows. There is always a Turing machine $\mathbb{TM}$ and a corresponding triplet (τ, A, B) such that for any pair $(\mathcal{D}, \mathbf{E})$, we have

$(\mathcal{D}, \mathbf{E})$ as input to $\mathbb{TM}$ with output given by $\mathbb{P}(\mathbb{TM}$ accepts $(\mathcal{D}, \mathbf{E})) > 1/2$,

if and only if

$(\mathcal{D}, \mathbf{E})$ as "input" to (τ, A, B) with "output" given by $\mathbb{P}_{\tau, \mathcal{D}}(A|B, \mathbf{E}) > 1/2$,

where $\mathbb{P}_{\tau, \mathcal{D}}(A|B, \mathbf{E})$ denotes probability with respect to $\mathbb{B}(\tau, \mathcal{D})$. (Of course, there is no even need to be restricted to zero-arity predicates A and B, as Theorem 2 allows for sets of predicates.)

Note that the same result could be proved if every evidence piece was taken to be a complete interpretation for the vocabulary σ. In that case we could directly speak of structures as inputs, and then the result would more closely mirror Fagin's theorem. However it is very appropriate, and entirely in line with practical use, to take the inputs to a Bayesian network as the groundings of a partially observed interpretation. Hence we have preferred to present our main result as stated in Theorem 2.

5   Moving to Second-Order

Suppose we have a phenomenon whose simulation requires computational powers that go beyond a polynomial time probabilistic Turing machine. There are in fact description languages whose inferences require exponential time probabilistic Turing machines [2,3], but we need not go to such extremes. We might for instance have a phenomenon that requires a polynomial time probabilistic Turing machine to use as oracle a polynomial time nondeterministic Turing machine. That is, we might have a phenomenon with requirements within $\mathsf{PP}^{\mathsf{NP}}$, a level above PP in the polynomial counting hierarchy [19]. Given current beliefs about complexity classes, Theorem 2 shows that such a phenomenon cannot be modeled with a RELBN. Can we find a "reasonable" specification language that allows one to model such a phenomenon?

Indeed we can, by putting together Fagin's theorem and Theorem 2. Consider a specification language as follows: we have a directed acyclic graph where nodes are predicates, and where as before root and non-root nodes are respectively associated with assessments $\mathbb{P}(r(\vec{x}) = 1) = \alpha$ and definitions $s(\vec{x}) \Leftrightarrow \phi(\vec{x})$, but the latter are now enlarged so that ϕ is a formula in ESO. A directed graph associated with such assessments and formulas is referred to as a *existential second-order Bayesian network specification*, abbreviated ESOBN.

For instance, consider the model of friendship presented in Expression (3), and suppose we add a variable that indicates whether the graph of friends can be partitioned, using Expression (4) as follows:

partitioned $\Leftrightarrow \exists$partition : $\forall x : \forall y : (\text{edge}(x, y) \Rightarrow (\text{partition}(x) \Leftrightarrow \neg\text{partition}(y)))$.

The presence of probabilities and second-order quantification gives us the desired modeling power:

Theorem 3. *Let S be an isomorphism-closed set of pairs domain-evidence of some non-empty finite vocabulary, where all domains are finite. Then S is in* $\mathsf{PP}^{\mathsf{NP}}$ *if and only if S is the class of domain-evidence pairs that are accepted by a fixed* ESOBN *with fixed conditioned and conditioning predicates.*

Proof. To prove that a class of domain-query pairs that are accepted by a fixed ESOBN can be decided within $\mathsf{PP}^{\mathsf{NP}}$, put together the argument in the first paragraph of the proof of Theorem 2 and the fact that an ESO sentence can be evaluated within NP by Theorem 1.

To prove the other direction, the corresponding part of the proof of Theorem 2 must be enlarged. We introduce again the same predicates X_q, Y_σ and Z for the base machine, together with the whole machinery in Grädel's proof of Fagin's theorem [8], and we also introduce predicates X_q^o, Y_σ^o, Z^o for the oracle machine (the superscript o refers to the "oracle"). Additionally, in the proof of Theorem 2 it is necessary to introduce logical variables that "mark the steps" of the Turing machine (these steps are ordered by the introduced linear order). Here we need two sets of such logical variables and associated machinery; one set "marks the steps" of the base machine, while the other "marks the steps" of the oracle machine. And of course a linear order must also be built as in the proof of Theorem 2. Again, due to lack of space the details of the construction are omitted. And again, an input domain-query is accepted by the majority of computation paths in the (base) Turing machine if and only if we have $\mathbb{P}(A|B, \mathbf{E}) > 1/2$, where A, B and $\mathbf{E}$ are as in the proof of Theorem 2. $\square$

6   Conclusion: A Finite Model Theory of Bayesian Networks?

We have introduced a theory of descriptive complexity for Bayesian networks, a topic that does not seem to have received due attention so far. Our results can be extended in a variety of directions, for instance to various fixpoint logics that are the basis of logic programming [4,12]. To summarize, we have shown that relational Bayesian network specifications capture PP, and we have indicated how we can go beyond PP in our modeling tools. Specifically, we added existential second-order quantification to capture the complexity class $\mathsf{PP}^{\mathsf{NP}}$.

These results can be better appreciated by taking a broader perspective. For several decades now, there has been significant study of models that arise from combinations of probability and logic [1,6]. However, by dealing with domains of arbitrary cardinality, and with logics that include too many constructs (for instance, functions) and that exclude valuable techniques (for instance, the modularity introduced by independence relations), these previous investigations arrive at results that are often too weak — for instance, almost always obtaining undecidability or very high computational complexity. By focusing on modular tools such as Bayesian networks, and by focusing on finite domains, we are able to obtain much sharper results, nailing down specific complexity classes such as PP and PP^{NP}. In fact, the purpose here is to initiate a "finite model theory of Bayesian network specifications" that can pin down the expressivity and complexity of Bayesian networks, not only when they are propositional objects, but particularly when they are specified using logical constructs.

Our results are also interesting from a point of view centered on complexity theory. There has been little work on capturing counting/probabilistic classes; the most significant previous results capture #P using counting [17]. We offer a more concrete modeling language that captures PP, and we move into the counting hierarchy, up to PP^{NP} — we are not aware of any similar result in the literature. Our results show that classes in the counting hierarchy can be tied to the expressivity of modeling tools, not to any particular computational model (much as Fagin's theorem does for NP).

Acknowledgements. The first author is partially supported by CNPq, grant 308433/2014-9. This paper was partially funded by FAPESP grant #2015/21880-4 (project Proverbs).

References

1. Abadi, M., Halpern, J.Y.: Decidability and expressiveness for first-order logics of probability. Inf. Comput. **112**(1), 1–36 (1994)
2. Cozman, F.G., Mauá, D.D.: Bayesian networks specified using propositional and relational constructs: Combined, data, and domain complexity. In: AAAI Conference on Artificial Intelligence (2015)
3. Cozman, F.G., Polastro, R.B.: Complexity analysis and variational inference for interpretation-based probabilistic description logics. In: Conference on Uncertainty in Artificial Intelligence, pp. 117–125 (2009)
4. Ebbinghaus, H.-D., Flum, J.: Finite Model Theory. Springer, Heidelberg (1995)
5. Enderton, H.B.: A Mathematical Introduction to Logic. Academic Press, Cambridge (1972)
6. Gaifman, H.: Concerning measures on first-order calculi. Isr. J. Math. **2**, 1–18 (1964)
7. Getoor, L., Taskar, B.: Introduction to Statistical Relational Learning. MIT Press, Cambridge (2007)
8. Grädel, E.: Finite model theory and descriptive complexity. In: Grädel, E. (ed.) Finite Model Theory and its Applications, pp. 125–229. Springer, Heidelberg (2007)
9. Grädel, E.E., Kolaitis, P.G., Libkin, L., Marx, M., Spencer, J., Vardi, M.Y., Venema, Y., Weinstein, S.: Finite Model Theory and its Applications. Springer, Heidelberg (2007)
10. Jaeger, M.: Lower complexity bounds for lifted inference. Theor. Pract. Logic Program. **15**(2), 246–264 (2014)
11. Koller, D., Friedman, N., Models, P.G.: Probabilistic Graphical Models: Principles and Techniques. MIT Press, Cambridge (2009)
12. Libkin, L.: Elements of Finite Model Theory. Springer, Heidelberg (2004)
13. Papadimitriou, C.H.: Computational Complexity. Addison-Wesley, Boston (1994)

14. Poole, D.: Probabilistic programming languages: independent choices and deterministic systems. In: Dechter, R., Geffner, H., Halpern, J.Y. (eds.) Heuristics, Probability and Causality - A Tribute to Judea Pearl, pp. 253–269. College Publications, London (2010)
15. De Raedt, L.: Logical and Relational Learning. Springer, Heidelberg (2008)
16. De Raedt, L., Frasconi, P., Kersting, K., Muggleton, S.: Probabilistic Inductive Logic Programming. Springer, Heidelberg (2010)
17. Saluja, S., Subrahmanyam, K.V.: Descriptive complexity of #P functions. J. Comput. Syst. Sci. **50**, 493–505 (1995)
18. Van den Broeck, G.: On the completeness of first-order knowledge compilation for lifted probabilistic inference. In: Neural Processing Information Systems, pp. 1386–1394 (2011)
19. Wagner, K.W.: The complexity of combinatorial problems with succinct input representation. Acta Informatica **23**, 325–356 (1986)

Exploiting Stability for Compact Representation of Independency Models

Linda C. van der Gaag[1](✉) and Stavros Lopatatzidis[2]

[1] Department of Information and Computing Sciences, Utrecht University,
Princetonplein 5, 3584 CC Utrecht, The Netherlands
L.C.vanderGaag@uu.nl
[2] SYSTeMS Research Group, Ghent University,
Technologiepark-Zwijnaarde 914, 9052 Zwijnaarde, Belgium
stavros.lopatatzidis@ugent.be

Abstract. The notion of stability in semi-graphoid independency models was introduced to describe the dynamics of (probabilistic) independency upon inference. We revisit the notion in view of establishing compact representations of semi-graphoid models in general. Algorithms for this purpose typically build upon dedicated operators for constructing new independency statements from a starting set of statements. In this paper, we formulate a generalised strong-contraction operator to supplement existing operators, and prove its soundness. We then embed the operator in a state-of-the-art algorithm and illustrate that the thus enhanced algorithm may establish more compact model representations.

1 Introduction

Pearl and his co-researchers were among the first to formalise qualitative properties of (probabilistic) independency in an axiomatic system [6]. Known as the semi-graphoid axioms, the axioms from this system are often looked upon as derivation rules for generating new independencies from a starting set of independency statements; any set of independencies that is closed under finite application of these rules is called a *semi-graphoid independency model*. Semi-graphoid models have been well studied from various different perspectives. Researchers have addressed the axiomatic system itself, focusing on the question whether the notion of independency relative to the class of discrete probability distributions allows a finite axiomatisation [7]. Also, the implication problem for semi-graphoid models has received attention, that is, the problem of verifying whether a new independency statement logically follows from a given set of statements by application of the semi-graphoid rules [4]. The problem on which we shall focus in this paper is the *representation problem*, that is, the problem of finding a small set of independency statements which fully describes a given model [1,2,8].

From a practical point of view, probabilistic independency models are the key to dealing with the computational complexity of problem-solving tasks involving uncertainty. Given their importance, different classes of independency model

© Springer International Publishing AG 2017
A. Antonucci et al. (Eds.): ECSQARU 2017, LNAI 10369, pp. 104–114, 2017.
DOI: 10.1007/978-3-319-61581-3_10

have been studied, among which are the graphoid models describing independency relative to the class of strictly positive probability distributions, and stable
independency models. This latter class of models derives its importance from its
role in describing the dynamics of (probabilistic) independency upon inference.
Any independency model is a complete description of all independencies among
the variables concerned and thereby provides for all possible contexts of given
information. At any time in an iterative problem-solving process, however, only
some of the independencies from the model apply: these are the independencies that pertain to the current context of information. As inference progresses,
learning new information causes the set of relevant independency statements to
change dynamically. Stability of an independency now means that the independency remains to hold as the context of given information grows.

Several researchers have studied the representation of semi-graphoid models
in general and developed algorithms for finding compact representations [1,2,8].
These algorithms construct such a representation in an iterative fashion by applying dedicated operators to a set of given independency statements. De Waal and
Van der Gaag were the first to exploit the notion of stability for studying model
representations [9]. They formulated an algorithm in which the stable and unstable parts of an independency model are addressed separately; upon doing so, they
exploited the strong-union axiom of stable independency. Niepert et $al.$ [5] have
since then formulated a second axiom for stable independency, called the strong-
contraction axiom. In this paper, we design a new operator to accommodate application of this axiom for deriving new statements and embed this operator in the
existing algorithm for finding compact representations of semi-graphoid models in
general. We will illustrate, by means of a small example, that the thus enhanced
algorithm may establish more compact representations.

The paper is organised as follows. Section 2 revisits semi-graphoid models
and introduces our notations. Section 3 addresses stable independency models
and their representation; more specifically, it defines our new operator for stable
independency. Section 4 embeds the operator in the state-of-the-art algorithm for
computing compact model representations. The paper ends with our concluding
observations in Sect. 5.

2   Independency Models Revisited

We briefly review semi-graphoid independency models and their representation
[6,8]. To this end, we consider a finite, non-empty set V of discrete random
variables. A $triplet$ over V is a statement of the form $\langle A, B \mid C \rangle$, where $A, B, C \subseteq$
V are pairwise disjoint subsets with $A, B \neq \varnothing$; we use $X = A \cup B \cup C$ to denote
the set of all variables involved in the triplet. A triplet $\langle A, B \mid C \rangle$ is taken to
state that the sets of variables A and B are independent given the conditioning
set C; relative to a discrete joint probability distribution Pr over V, the triplet
thus states that $\Pr(A, B \mid C) = \Pr(A \mid C) \cdot \Pr(B \mid C)$ for all values combinations
of A, B, C. The set of all triplets over V is denoted by $V^{(3)}$.

A set of triplets constitutes a $semi\text{-}graphoid$ $independency$ $model$ if it satisfies
the four properties stated in the following definition.

Definition 1. *A* semi-graphoid independency model *is a subset of triplets* $J \subseteq V^{(3)}$ *which satisfies the following properties:*

G1: *if* $\langle A, B \mid C \rangle \in J$, *then* $\langle B, A \mid C \rangle \in J$ (Symmetry)
G2: *if* $\langle A, B \mid C \rangle \in J$, *then* $\langle A, B' \mid C \rangle \in J$ *for any non-empty subset* $B' \subseteq B$ (Decomposition)
G3: *if* $\langle A, B_1 \cup B_2 \mid C \rangle \in J$ *with* $B_1 \cap B_2 = \varnothing$, *then* $\langle A, B_1 \mid C \cup B_2 \rangle \in J$ (Weak Union)
G4: *if* $\langle A, B \mid C \cup D \rangle \in J$ *and* $\langle A, C \mid D \rangle \in J$, *then* $\langle A, B \cup C \mid D \rangle \in J$ (Contraction)

The four properties stated above have been proven logically independent and taken to constitute an axiomatic system for the qualitative notion of independency [6]. The system is sound relative to the class of discrete probability distributions, yet not complete; in fact, it has been shown that the probabilistic notion of independence does not allow a finite axiomatisation [7]. The four properties of independence are often viewed as derivation rules for generating possibly new triplets from a given triplet set. Given a set $J \subseteq V^{(3)}$ and a designated triplet $\theta \in V^{(3)}$, we write $J \vdash^* \theta$ if the triplet θ can be derived from J by finite application of the semi-graphoid rules G1–G4. The *closure* of the starting set J then is the semi-graphoid independency model composed of J and all triplets θ that can be derived from J, that is, all triplets θ such that $J \vdash^* \theta$.

Semi-graphoid models typically being exponentially large in size, Studený proposed a more compact representation than mere enumeration of their element triplets [8]. The idea of this representation is to explicitly capture only a subset of triplets from a semi-graphoid model, called a *basis*, and let all other triplets be defined implicitly through the derivation rules. Underlying this representation was the notion of dominance [8], which was later enhanced by Baioletti *et al.* to the notion of *g-inclusion* [1].

Definition 2. *Let* $J \subseteq V^{(3)}$ *be a semi-graphoid independency model, and let* $\theta_i = \langle A_i, B_i \mid C_i \rangle$, *with* $X_i = A_i \cup B_i \cup C_i$, $i = 1, 2$, *be triplets in* J. *Then,* θ_1 *is g-included in* θ_2, *denoted as* $\theta_1 \sqsubseteq \theta_2$, *if the following two conditions hold:*

- $C_2 \subseteq C_1 \subseteq X_2$; *and,*
- $A_1 \subseteq A_2$ *and* $B_1 \subseteq B_2$, *or* $B_1 \subseteq A_2$ *and* $A_1 \subseteq B_2$.

A triplet $\theta \in J$ *is* g-maximal *in* J *if it is not g-included in any triplet* $\tau \in J$ *with* $\tau \neq \theta, \theta^T$.

The conditions for g-inclusion capture all possible ways in which the triplet θ_1 may be derived from θ_2 by means of the derivation rules G1–G3. Since g-included triplets can be derived from other ones, they need not be represented explicitly in a basis.

For application of the contraction rule G4 to pairs of triplets, Studený formulated a dedicated operator, called the *gc-operator*, which constructs from two triplets θ_1, θ_2, triplets θ_1', θ_2' by application of the rules G1–G3, to which the contraction rule can be applied to yield a possibly new triplet θ [8]. This operator is defined as follows.

Definition 3. *For all triplets* $\theta_i = \langle A_i, B_i \mid C_i \rangle \in V^{(3)}$ *with* $X_i = A_i \cup B_i \cup C_i, i = 1, 2,$ *such that*

- $A_1 \cap A_2 \neq \varnothing$;
- $C_1 \subseteq X_2$ *and* $C_2 \subseteq X_1$; *and*
- $(B_2 \backslash C_1) \cup (B_1 \cap X_2) \neq \varnothing$,

the gc-operator is defined through:

$$gc(\theta_1, \theta_2) = \langle A_1 \cap A_2, \ (B_2 \backslash C_1) \cup (B_1 \cap X_2) \mid C_1 \cup (A_1 \cap C_2) \rangle$$

For all pairs of triplets θ_1, θ_2 *for which the conditions stated above do not all hold,* $gc(\theta_1, \theta_2)$ *is undefined.*

Building upon the *gc*-operator, Baioletti *et al.* [1] formulated a generalised contraction rule which constructs from a pair of triplets θ_1, θ_2, a set of possibly new triplets by applying the *gc*-operator to θ_1, θ_2 and the transposes θ_i^T obtained from θ_i, $i = 1, 2$, by a single application of the symmetry rule G1. Their rule was later extended by Lopatatzidis and Van der Gaag [2] to the rule stated in the following lemma.

Lemma 1. *Let* $J \subseteq V^{(3)}$ *be a semi-graphoid independency model. Then,* J *satisfies the following property:*

G4†: *if* $\theta_1, \theta_2 \in J$, *then* $GC(\theta_1, \theta_2) \cup GC(\theta_2, \theta_1) \subseteq J$

where $GC(\theta_i, \theta_j) = \{gc(\theta_i, \theta_j), gc(\theta_i, \theta_j^T), gc(\theta_i^T, \theta_j), gc(\theta_i^T, \theta_j^T)\}$ *for all* $\theta_i, \theta_j \in J$.

We note that for any two triplets θ_1, θ_2, eight potentially new triplets may result from application of the G4$^+$ rule; for a proof of the lemma, we refer to [2].

For establishing compact representations of semi-graphoid models in general, an algorithm is available which builds on the G4$^+$ derivation rule. The algorithm constructs a basis, from a given starting set of triplets, for the independency model defined by this set [1,2,8]. To this end, the algorithm initialises the basis by the starting set and iteratively adapts the current basis until it no longer changes. In each iteration, the G4$^+$ rule is applied to any pair of triplets from the current basis and the results are added. After removing all g-included triplets from the resulting set, the next iteration is started.

3   Stable Independency Models

Of the family of semi-graphoid independency models, the subfamily of stable models is of special interest for studying the dynamics of independency upon inference. Informally, two sets of variables are *stably independent* given a specific conditioning set of variables, if they are independent given this set and remain to be so as this set grows [9]. We review the notion of *stable triplet* in an independency model.

Definition 4. *Let* $J \subseteq V^{(3)}$ *be a semi-graphoid independency model and let* $\theta = \langle A, B \mid C \rangle \in J$. *Then,*

- θ *is* stable *in* J *if* $\langle A, B \mid C' \rangle \in J$ *for all sets* C' *with* $C \subseteq C' \subseteq V \backslash (A \cup B)$ *and is called* saturated *in* J *if* $C = V \backslash (A \cup B)$;
- θ *is* unstable *in* J *if it is not stable in* J.

The set of all stable triplets in J *is called the* stable part *of* J, *and is denoted with* J_S; *the set of all unstable triplets in* J *constitutes its* unstable part, *denoted by* J_U.

The stable part of a semi-graphoid independency model satisfies the four semi-graphoid properties from Definition 1, and hence constitutes a semi-graphoid model by itself [9]; in fact, there exist semi-graphoid models composed of a stable part only, which have also been termed *ascending* [3]. The unstable part of an independency model not necessarily obeys the decomposition, weak union and contraction properties, and therefore need not constitute an independency model in general. In this section, we focus on stable models, that is, on independency models with an empty unstable part; in Sect. 4, we return to models which involve a non-empty unstable part as well.

A stable independency model has a highly regular structure, which is described by two properties in addition to the four semi-graphoid properties.

Lemma 2. *Let* $J \subseteq V^{(3)}$ *be a stable model. Then,* J *satisfies the following properties:*

S5: *if* $\langle A, B \mid C \rangle \in J$, *then* $\langle A, B \mid C \cup D \rangle \in J$ *for all sets* $D \subseteq V \backslash (A \cup B \cup C)$
  (Strong Union)
S6: *if* $\langle A, B \mid C \cup D \rangle, \langle A, B \mid C \cup E \rangle, \langle D, E \mid C \rangle \in J$, *then* $\langle A, B \mid C \rangle \in J$
  (Strong Contraction)

The strong-union property follows directly from the definition of stable independency [6,9]. The strong-contraction property was formulated by Niepert *et al.* who proved that the properties G1–G4 and S5–S6 constitute a sound and complete axiomatisation of stable independency relative to the class of discrete probability distributions [5]. The properties S5–S6 now are taken as derivation rules, for stable independencies only.

In the previous section we addressed the representation of semi-graphoid models in general, and reviewed the notion of g-inclusion which underlies representation by a basis. De Waal and Van der Gaag [9] observed that stable models can be represented more concisely, since by including a stable triplet $\langle A, B \mid C \rangle$ in a basis, all triplets $\langle A, B \mid C' \rangle$ with $C \subseteq C' \subseteq V \backslash (A \cup B)$ can be derived by means of the S5 rule and hence be left implicit. To further explore this observation, we define the notion of *stable g-inclusion*.

Definition 5. *Let* $J \subseteq V^{(3)}$ *be a stable independency model, and let* $\theta_i = \langle A_i, B_i \mid C_i \rangle$, $i = 1, 2$, *be triplets in* J. *Then,* θ_1 *is stably g-included in* θ_2, *denoted* $\theta_1 \sqsubseteq_S \theta_2$, *if the following two conditions hold:*

- $C_2 \subseteq C_1$; and,
- $A_1 \subseteq A_2$ and $B_1 \subseteq B_2$, or $B_1 \subseteq A_2$ and $A_1 \subseteq B_2$.

A triplet $\theta \in J$ is stably g-maximal in J if it is not stably g-included in any triplet $\tau \in J$ with $\tau \neq \theta, \theta^T$.

Like the general notion of g-inclusion, stable g-inclusion pertains to a single triplet and the triplets that can be derived from it by means of the rules G1–G3. The difference is in the condition $C_1 \subseteq X_2$: while the context X_2 is required for g-inclusion in general, this restriction is no longer necessary for stable triplets, since any triplet with a conditioning part expanding beyond X_2 is still included in the model under study through the S5 derivation rule. The notion of stable g-inclusion thus covers application not just of the rules G1–G3, but of the strong-union rule as well. For application of the contraction rule to stable independencies, De Waal and Van der Gaag introduced a dedicated operator, analogous to the gc-operator. This gc_S-operator constructs from two stable triplets θ_1, θ_2, stably g-included triplets θ_1', θ_2' by means of the G1–G3 and S5 rules, to which the contraction rule can be applied to give a possibly new triplet. It is defined as follows.

Definition 6. For all triplets $\theta_i = \langle A_i, B_i \mid C_i \rangle \in V^{(3)}$, $i = 1, 2$, such that

- $A_1 \cap A_2 \neq \varnothing$; and
- $(B_2 \backslash C_1) \cup (B_1 \backslash B_2) \neq \varnothing$,

the gc_S-operator is defined through:

$$gc_S(\theta_1, \theta_2) = \langle A_1 \cap A_2, (B_2 \backslash C_1) \cup (B_1 \backslash B_2)) \mid C_1 \cup (C_2 \backslash B_1) \rangle$$

For all pairs of triplets θ_1, θ_2 for which the conditions stated above do not all hold, $gc_S(\theta_1, \theta_2)$ is undefined.

Building upon the gc_S-operator, we now formulate a generalised contraction rule for stable models, similar to the G4$^+$ rule for semi-graphoid models in general.

Lemma 3. Let $J \subseteq V^{(3)}$ be a stable model. Then, J satisfies the following property:

G4^S: if $\theta_1, \theta_2 \in J$, then $GC_S(\theta_1, \theta_2) \cup GC_S(\theta_2, \theta_1) \subseteq J$

with $GC_S(\theta_i, \theta_j) = \{gc_S(\theta_i, \theta_j), gc_S(\theta_i, \theta_j^T), gc_S(\theta_i^T, \theta_j), gc_S(\theta_i^T, \theta_j^T)\}$ for all $\theta_i, \theta_j \in J$.

We note that for any two triplets θ_1, θ_2, eight new ones may result from applying the G4^S rule. Proof of the lemma follows directly from properties of the gc_S-operator [9].

The gc_S-operator takes the strong-union rule into consideration in addition to the derivation rules G1–G3, for constructing triplets to which the contraction rule can be applied. The generalised contraction rule G4^S thus accommodates the rules G1–G4 and S5. It does not yet cover the strong-contraction rule, however. For accommodating this rule, we introduce a new operator, denoted as gsc_S. This operator constructs, from three triplets $\theta_1, \theta_2, \theta_3$, stably g-included triplets $\theta_1', \theta_2', \theta_3'$ by application of G1–G3 and S5, to which the strong-contraction rule can be applied. It is defined as follows.

Definition 7. *For all triplets $\theta_i = \langle A_i, B_i \mid C_i \rangle \in V^{(3)}$, $i = 1, 2, 3$, such that*

- $A_1 \cap A_2 \neq \varnothing$ and $B_1 \cap B_2 \neq \varnothing$;
- $\varnothing \neq C_1 \backslash (B_2 \cup C_2) \subseteq A_3$ and $\varnothing \neq C_2 \backslash (B_1 \cup C_1) \subseteq B_3$; and
- $C_3 \subseteq (C_1 \cap C_2) \cup (B_1 \cap C_2) \cup (B_2 \cap C_1)$,

the gsc_S-operator is defined through:

$$gsc_S(\theta_1, \theta_2, \theta_3) = \langle A_1 \cap A_2, \ B_1 \cap B_2 \mid (C_1 \cap C_2) \cup (B_1 \cap C_2) \cup (B_2 \cap C_1) \rangle$$

For all triplets $\theta_1, \theta_2, \theta_3 \in J$ for which the conditions stated above do not all hold, $gsc_S(\theta_1, \theta_2, \theta_3)$ is undefined.

We show that any stable model is closed under application of the gsc_S-operator.

Lemma 4. *Let $J \subseteq V^{(3)}$ be a stable independency model. For all triplets $\theta_i \in J$, $i = 1, 2, 3$, if $gsc_S(\theta_1, \theta_2, \theta_3)$ is defined, then $gsc_S(\theta_1, \theta_2, \theta_3) \in J$.*

Proof. We let $\theta_i = \langle A_i, B_i \mid C_i \rangle, i = 1, 2, 3$, and assume that $gsc_S(\theta_1, \theta_2, \theta_3)$ is defined. From the three triplets, we construct, by application of the symmetry, decomposition, and weak- and strong-union rules, the following stably g-included triplets:

$$\theta_1' = \langle A_1 \cap A_2, B_1 \cap B_2 \mid C_1 \cup (B_1 \cap C_2) \rangle = \langle A_1', B_1' \mid C_1' \rangle$$
$$\theta_2' = \langle A_1 \cap A_2, B_1 \cap B_2 \mid C_2 \cup (B_2 \cap C_1) \rangle = \langle A_2', B_2' \mid C_2' \rangle$$
$$\theta_3' = \langle C_1 \backslash (B_2 \cup C_2), \ C_2 \backslash (B_1 \cup C_1) \mid (C_1 \cap C_2) \cup (B_1 \cap C_2) \cup (B_2 \cap C_1) \rangle =$$
$$= \langle A_3', B_3' \mid C_3' \rangle$$

For the intersection of the two sets C_1' and C_2', we observe that $C_1' \cap C_2' = (C_1 \cap C_2) \cup (B_1 \cap C_2) \cup (B_2 \cap C_1) = C_3'$. We further have that $A_3' = C_1 \backslash (B_2 \cup C_2) \subseteq C_1 \cup (B_1 \cap C_2) = C_1'$; with $A_3' \cup C_3' = C_1'$, we thus find that $C_1' \backslash (B_2' \cup C_2') \subseteq A_3'$. Similarly, $C_2' \backslash (B_1' \cup C_1') \subseteq B_3'$. Since its conditions are all met, the strong-contraction rule can be applied to $\theta_1', \theta_2', \theta_3'$ to yield the following triplet:

$$\theta = \langle A_1 \cap A_2, B_1 \cap B_2 \mid (C_1 \cap C_2) \cup (B_1 \cap C_2) \cup (B_2 \cap C_1) \rangle = gsc_S(\theta_1, \theta_2, \theta_3)$$

Since only sound rules for stable independency were applied in the derivation, we conclude that $\theta \in J$, and hence that $gsc_S(\theta_1, \theta_2, \theta_3) \in J$. $\square$

Analogous to the property that the gc_S-operator can be used for constructing a basis for any stable model [9], we now prove a related property for our gsc_S-operator.

Lemma 5. *Let $D \subseteq V^{(3)}$ be a set of stable triplets such that for all $\theta_i \in D$, $i = 1, 2, 3$, the following property holds:*

if $gsc_S(\theta_1, \theta_2, \theta_3)$ is defined, then there is a $\theta \in D$ such that $gsc_S(\theta_1, \theta_2, \theta_3) \sqsubseteq_S \theta$

Then, the set $J = \{\theta' \mid there is a \theta \in D such that \theta' \sqsubseteq_S \theta\}$ is closed under the derivation rules G1–G3, S5 and S6.

Proof. From its definition, it is immediate that the set J is closed under the rules G1–G3 and S5. We now show that the set is also closed under the strong-contraction rule S6. To this end, we consider three triplets $\tau_i = \langle a_i, b_i \mid c_i \rangle$, $i = 1, 2, 3$, in J and suppose that the strong-contraction rule can be applied directly to these triplets, to give $\tau = \langle a, b \mid c \rangle$. By definition, there are triplets $\theta_i = \langle A_i, B_i \mid C_i \rangle, i = 1, 2, 3$, in D with $\tau_i \sqsubseteq_S \theta_i$, such that $\theta = gsc_S(\theta_1, \theta_2, \theta_3)$ is well defined. To show that $\tau \sqsubseteq_S \theta$, we have to show that $a \subseteq A$, $b \subseteq B$ and $C \subseteq c$. From $\tau_i \sqsubseteq_S \theta_i$, we have that $a_i \subseteq A_i, i = 1, 2$. Since the strong-contraction rule can be applied directly to τ_i, $i = 1, 2, 3$, we have that $a_1 = a_2$, from which it is readily seen that $a_1 \cap a_2 \subseteq A_1 \cap A_2$ and, hence, that $a \subseteq A$; similarly, $b \subseteq B$. From $\tau_j \sqsubseteq_S \theta_j$ we further have that $C_j \subseteq c_j$, $j = 1, 2$, from which it follows that $C_1 \cap C_2 \subseteq c_1 \cap c_2$. From the conditions for application of the gsc_S-operator to $\theta_1, \theta_2, \theta_3$, we have that $C_2 \backslash (B_1 \cup C_1) \subseteq B_3$, from which we find that $(B_1 \cap C_2) \cap B_3 = \varnothing$. Since $b_3 \subseteq B_3$, it follows that $(B_1 \cap C_2) \cap b_3 = \varnothing$. From $b_3 = c_2 \backslash c_1 \neq \varnothing$, we thus find that $B_1 \cap C_2 \subseteq c_1 \cap c_2$; similarly, $B_2 \cap C_1 \subseteq c_1 \cap c_2$. We conclude that

$$C = (C_1 \cap C_2) \cup (B_1 \cap C_2) \cup (B_2 \cap C_1) \subseteq c_1 \cap c_2 = c$$

which proves the property stated in the lemma. $\square$

Building upon the gsc_S-operator, we now formulate a generalised strong-contraction rule S6^S for stable independency, similar to the G4$^+$ and G4^S rules.

Lemma 6. *Let $J \subseteq V^{(3)}$ be a stable model. Then, J satisfies the following property:*

S6^S: *if $\theta_1, \theta_2, \theta_3 \in J$, then $\mathcal{GSC}_S(\theta_1, \theta_2, \theta_3) \subseteq J$*

where $\mathcal{GSC}_S(\theta_1, \theta_2, \theta_3) = \bigcup_{(i,j,k) = \pi(1,2,3)} GSC_S(\theta_i, \theta_j, \theta_k)$, with π taking a permutation of its arguments and $GSC_S(\theta_i, \theta_j, \theta_k) = \{gsc_S(\theta_x, \theta_y, \theta_z) \mid \theta_x \in \{\theta_i, \theta_i^T\}, \theta_y \in \{\theta_j, \theta_j^T\}, \theta_z \in \{\theta_k, \theta_k^T\}\}$ for all triplets $\theta_i, \theta_j, \theta_k$.

We note that for any three triplets $\theta_1, \theta_2, \theta_3$, in essence as many as 48 potentially new triplets may result from application of the S6^S derivation rule.

4   Exploiting Stability for Basis Computation

From Studený's original idea, a series of algorithms have originated for computing compact representations of semi-graphoid models [1,2,8]. As reviewed in

Sect. 2, these algorithms iteratively construct, from a starting triplet set, a basis for the model defined by this set; to this end, the algorithms apply in each iteration the G4+ rule to any pair of triplets from the current basis. De Waal and Van der Gaag introduced the notion of stability into this algorithmic framework [9], by using both the G4+ and G4^S derivation rules and keeping track of the stable and unstable parts of a model separately.

We now extend the existing algorithmic framework by incorporating the strong-contraction rule for stable independency. To this end, we consider a set D of stable triplets for which the following hold, for all $\theta_i \in D$, $i = 1, 2, 3$:

- if $gc_S(\theta_1, \theta_2)$ is defined, then there is a $\theta \in D$ such that $gc_S(\theta_1, \theta_2) \sqsubseteq_S \theta$; and
- if $gsc_S(\theta_1, \theta_2, \theta_3)$ is defined, then there is a $\theta \in D$ such that $gsc_S(\theta_1, \theta_2, \theta_3) \sqsubseteq_S \theta$.

From the properties proved for the gc_S-operator [9] and our Lemmas 4 and 5, we find that the set $J = \{\theta' \mid \text{there is a } \theta \in D \text{ such that } \theta' \sqsubseteq_S \theta\}$ is a stable independency model. From this finding, we conclude that for incorporating the strong-contraction rule, it suffices to establish the closure of the stable part of a model under both the gc_S- and gsc_S-operators. Figure 1 outlines the thus enhanced algorithm for basis construction.

The enhanced algorithm has the same outline as the algorithm by De Waal and Van der Gaag, yet includes recent advances [1, 2] and embeds our new results. The algorithm takes a pair (J_S, J_U) of stable and unstable starting sets, and

---

Algorithm for Computing a Basis for $J = (J_S, J_U)$

1: **function** *ComputeBasis* (J_S, J_U)
2:    $J_S^0 = N_S^0 \leftarrow J_S$
3:    $J_U^0 = N_U^0 \leftarrow J_U$
4:    $k \leftarrow 0$
5:    **repeat**
6:       $k \leftarrow k + 1$
7:       $N_S^k \leftarrow \bigcup_{\theta_1, \theta_2 \in J_S^{k-1} \cup N_S^{k-1}} G4^S(\theta_1, \theta_2) \cup \bigcup_{\theta_1, \theta_2, \theta_3 \in J_S^{k-1} \cup N_S^{k-1}} S6^S(\theta_1, \theta_2, \theta_3)$
8:       $N_U^k \leftarrow \bigcup_{\theta_1, \theta_2 \in J_U^{k-1} \cup N_U^{k-1}} G4^+(\theta_1, \theta_2)$
9:       **for all** $\theta_1 \in J_U^{k-1} \cup N_U^{k-1}, \theta_2 \in J_S^{k-1} \cup N_S^{k-1}$:
10:          **if** $C_2 \setminus X_1 = \varnothing$ **then** $N_U^k \leftarrow N_U^k \cup \bigcup_{\theta_2' \sqsubseteq_S \theta_2: \, C_1 \setminus X_2' = C_2' \setminus X_1 = \varnothing} G4^+(\theta_1, \theta_2')$
11:       **end for all**
12:       $(J_S^k, J_U^k) \leftarrow Reduce(J_S^{k-1}, J_U^{k-1}, N_S^k, N_U^k)$
13:    **until** $(J_S^k, J_U^k) = (J_S^{k-1}, J_U^{k-1})$
14:    **return** (J_S^k, J_U^k)
15: **end function**

Fig. 1. Our enhanced algorithm for computing a basis by exploiting stability.

updates these iteratively. Line 7 of the algorithm pertains to the stable part of the model under consideration and constructs a basis for this part by means of the $G4^S$ and $S6^S$ rules; Line 8 constructs new triplets for the model's unstable part. Lines 9–11 address the interaction between the two parts of the model; for further details of the computation steps involved, we refer to [9]. Line 12 constructs a new pair (J_S, J_U) for the next iteration, by removing all included triplets and identifying implicit stable ones. We note that Fig. 1 presents just an outline of the algorithm and upon implementation should be further optimised.

To study the effects of our new results on computed representations, we constructed, in Python, prototype implementations of the algorithm by De Waal and Van der Gaag and of our enhanced algorithm. With both algorithms, we implemented two simple rules for identifying implicit stable independencies. The first rule moves any saturated triplets from the unstable part J_U to the stable part J_S. Secondly, if triplets $\langle A, B \mid C \cup \{V_i\} \rangle$ for all $V_i \in V \backslash (A \cup B \cup C)$ can be derived from $J_S \cup J_U$, and at most one of these triplets is unstable, then the stable triplet $\langle A, B \mid C \rangle$ is added to J_S.

With the two algorithms alike, we ran some preliminary experiments, for different starting sets. The following simple example illustrates that our new algorithm may return a smaller basis than the algorithm by De Waal and Van der Gaag.

Example 1. We take the set of variables $V = \{1, \ldots, 5\}$ and use concatenation of elements as a shorthand for subsets of V. Table 1 states the stable and unstable starting sets, and summarises the results from the two algorithms. Our new algorithm constructs a smaller stable part as a result of a single application of the gsc_S-operator, yielding the stable triplet $\langle 3, 4 \mid \varnothing \rangle$. As the algorithm finds this triplet during the first iteration, it maintains smaller intermediate bases throughout the subsequent iterations. □

Our enhanced algorithm is not guaranteed to always yield a smaller basis than existing algorithms. Since the strong-contraction rule aims at constructing triplets with the smallest possible conditioning contexts, some triplets may thereby become uncovered and have to be added explicitly to the basis under construction. Even when a reduction in size is achieved for the representation

Table 1. An example starting set and the triplet sets of the bases returned by the two algorithms.

Starting sets		Existing algorithm		New algorithm	
Unstable	Stable	Unstable	Stable	Unstable	Stable
$\langle 1, 5 \mid 2 \rangle$	$\langle 2, 5 \mid 3 \rangle$	$\langle 1, 25 \mid \varnothing \rangle$	$\langle 3, 4 \mid 1 \rangle$	$\langle 1, 25 \mid \varnothing \rangle$	$\langle 4, 3 \mid \varnothing \rangle$
$\langle 12, 5 \mid \varnothing \rangle$	$\langle 2, 5 \mid 4 \rangle$	$\langle 5, 12 \mid \varnothing \rangle$	$\langle 3, 4 \mid 2 \rangle$	$\langle 5, 12 \mid \varnothing \rangle$	$\langle 2, 15 \mid \varnothing \rangle$
	$\langle 1, 2 \mid \varnothing \rangle$		$\langle 2, 15 \mid \varnothing \rangle$		
	$\langle 3, 4 \mid 1 \rangle$				
	$\langle 3, 4 \mid 2 \rangle$				

of a model's stable part, therefore, may the enhanced algorithm return a larger overall basis than existing algorithms.

5 Conclusions

Revisiting semi-graphoid independency models, we addressed the potential of exploiting the notion of stability for constructing compact representations. We formulated a new derivation rule to accommodate the strong-contraction property for stable independencies and embedded this rule into an existing algorithmic framework. We illustrated the potential of the new rule by means of a simple example. In the near future, we will perform more extensive experimentation to investigate the interaction between the stable and unstable parts of the intermediate bases computed for a semi-graphoid independency model. We will further address the question whether also other parts of such a model can be identified and exploited for the construction of compact representations.

References

1. Baioletti, M., Busanello, G., Vantaggi, B.: Conditional independence structure and its closure: inferential rules and algorithms. Int. J. Approx. Reason. **50**, 1097–1114 (2009)
2. Lopatatzidis, S., van der Gaag, L.C.: Concise representations and construction algorithms for semi-graphoid independency models. Int. J. Approx. Reason. **80**, 377–392 (2015)
3. Matúš, F.: Ascending and descending conditional independence relations. In: Proceedings of the Eleventh Prague Conference on Information Theory, Statistical Decision Functions and Random Processes: B, pp. 189–200 (1992)
4. Niepert, M., Van Gucht, D., Gyssens, M.: A lattice-theoretic approach. In: Parr, R., van der Gaag, L.C. (eds.) Proceedings of the Twenty-Fourth Conference on Uncertainty in Artificial Intelligence, pp. 435–443. AUAI Press, Arlington (2008)
5. Niepert, M., Van Gucht, D., Gyssens, M.: Logical and algorithmic properties of stable conditional independence. Int. J. Approx. Reason. **51**, 531–543 (2010)
6. Pearl, J.: Probabilistic Reasoning in Intelligent Systems: Networks of Plausible Inference. Morgan Kaufmann, Palo Alto (1988)
7. Studený, M.: Conditional independence relations have no finite complete characterization. In: Kubík, S., Vísek, J.Á. (eds.) Information Theory, Statistical Decision Functions and Random Processes, pp. 377–396. Kluwer, Amsterdam (1992)
8. Studený, M.: Complexity of structural models. In: Proceedings of the Joint Session of the 6th Prague Conference on Asymptotic Statistics and the 13th Prague Conference on Information Theory, Statistical Decision Functions and Random Processes, vol. 2, pp. 521–528 (1998)
9. de Waal, P., van der Gaag, L.C.: Stable independence and complexity of representation. In: Chickering, M., Halpern, J. (eds.) Proceedings of the Twentieth Conference on Uncertainty in Artificial Intelligence, pp. 112–119. AUAI Press, Arlington (2004)

Parameter Learning Algorithms for Continuous Model Improvement Using Operational Data

Anders L. Madsen[1,2(✉)], Nicolaj Søndberg Jeppesen[1], Frank Jensen[1],
Mohamed S. Sayed[3], Ulrich Moser[4], Luis Neto[5], Joao Reis[5], and Niels Lohse[3]

[1] HUGIN EXPERT A/S, Aalborg, Denmark
anders@hugin.com
[2] Department of Computer Science, Aalborg University, Aalborg, Denmark
[3] Loughborough University, Loughborough, UK
[4] IEF-Werner GmbH, Furtwangen, Germany
[5] Instituto de Sistemas e. Robotica Associacao, Porto, Portugal

Abstract. In this paper, we consider the application of object-oriented
Bayesian networks to failure diagnostics in manufacturing systems and
continuous model improvement based on operational data. The analysis
is based on an object-oriented Bayesian network developed for failure
diagnostics of a one-dimensional pick-and-place industrial robot developed by IEF-Werner GmbH. We consider four learning algorithms (batch
Expectation-Maximization (EM), incremental EM, Online EM and fractional updating) for parameter updating in the object-oriented Bayesian
network using a real operational dataset. Also, we evaluate the performance of the considered algorithms on a dataset generated from the
model to determine which algorithm is best suited for recovering the
underlying generating distribution. The object-oriented Bayesian network has been integrated into both the control software of the robot as
well as into a software architecture that supports diagnostic and prog-
nostic capabilities of devices in manufacturing systems. We evaluate the
time performance of the architecture to determine the feasibility of online learning from operational data using each of the four algorithms.

Keywords: Bayesian networks · Parameter update · Practical application

1 Introduction

The need for diagnostic and health monitoring capabilities in manufacturing
systems is becoming increasingly important as manufacturing organisations
continuously aim to reduce system downtime and unpredicted disturbances to
production. We have found that Bayesian networks (BNs) [3,6,17] and their
extension Object-Oriented Bayesian Networks (OOBNs) [8,13] are well-suited to
capture and represent uncertainty in root-cause analysis using both component-
level models and wider system-level models integrating component-level models.
The crucial need for diagnostic and health monitoring capabilities is accompa-
nied with the availability of increasing amounts of sensory data and decreasing

© Springer International Publishing AG 2017
A. Antonucci et al. (Eds.): ECSQARU 2017, LNAI 10369, pp. 115–124, 2017.
DOI: 10.1007/978-3-319-61581-3_11

costs of computation on the shop-floor level have opened new opportunities for component suppliers and system integrators to provide more competitive functionalities that go beyond traditional control and process monitoring capabilities.

In this paper, we consider the challenge of parameter learning for continuous model improvement using operational data. In particular, we investigate the use of four different approaches to improve the diagnostic performance of an OOBN using operational data. The four algorithms are the batch EM algorithm, incremental EM, Online EM and fractional updating. The investigation is performed using an OOBN for root-cause analysis of a pick-and-place industrial robot developed by IEF-Werner GmbH[1] (the Linear Axis shown in the center of Fig. 3). An initial OOBN for root-cause analysis has been developed based on expert knowledge [11]. The OOBN has been integrated into the control software of the component and is being deployed in a production line where efficient and effective root-cause analysis is required in case of failure. In order to improve the diagnostic performance of the OOBN different methods for continuous model update based on operational data are being investigated. This paper reports on the results of these investigations.

Inspired by the work of [18], a number of approaches are considered. Notice that our work differs from the work of [18] in three important ways: (1) we are considering parameter learning in OOBNs, (2) the objective is to improve diagnostic performance (not classification), and (3) while [18] compares three algorithms, we investigate four algorithms. We consider the EM algorithm [9] for parameter learning from a batch of data (referred to as batch EM). Using batch EM, the idea is to collect data in batches and learn parameters off-line, for instance, during maintenance hours as suggested by [18]. We use batch EM as a reference. Adaptive causal probabilistic networks and fractional updating are described in [16] who cites [21] while adaptive probabilistic networks are described in [1,19]. A similar gradient descent approach is described in [5]. [10] describes how the approach of [16] referred to as sequential learning has been implemented in the HUGIN tool. The online EM algorithm [2] is a stochastic gradient method that is faster than other gradient methods such as [19] which involves a difficult task of determining the step size between iterations.

2   Preliminaries and Notation

A BN $\mathcal{N} = (\mathcal{X}, G, \mathcal{P})$ consists of a directed, acyclic graph G specifying dependence and independence relations over a set of variables $\mathcal{X}$ and a set of conditional probability distributions (CPDs) $\mathcal{P}$ encoding the strengths of the dependence relations effectively combining elements of probability and graph theory. A BN is a representation of a joint probability distribution $P(\mathcal{X}) = P(X_1, \ldots, X_n) = \prod_{X_i \in \mathcal{X}} P(X_i | \pi_{X_i})$ where π_X are the parents of X in G. The CPD $P(X | \pi_X)$ consists of one probability distribution over the states of X for each configuration of π_X. We only consider discrete variables.

---

[1] http://www.ief-werner.de.

An OOBN is a BN augmented with network classes, class instances and an associated notion of interface and private variables [6,8,13]. A class instance is the instantiation of a network class representing a sub-network within another network class. The variables $\mathcal{X}(C)$ of network class C are divided into disjoint subsets of input $\mathcal{I}$, output $\mathcal{O}$ and hidden $\mathcal{H}$ variables such that $\mathcal{X}(C) = \mathcal{I} \cup \mathcal{O} \cup \mathcal{H}$ where the interface variables $\mathcal{I} \cup \mathcal{O}$ are used to link nested class instances, see Fig. 1. Inference in an OOBN is performed by creating a *run-time* instance of the model and doing inference in this model. A run-time instance of an OOBN is created by expanding it into a corresponding flat BN.

The Hellinger distance $D_H(P, Q)$ used to compare two probability distributions P and Q is defined as $D_H(P, Q) = \sqrt{\sum_i (\sqrt{p_i} - \sqrt{q_i})^2}$ [18] who cites [7]. It is similar to the Kullback-Leibler divergence, but defined for zero probabilities. To compare the results of parameter learning using two different algorithms on the same OOBN, the distance is computed as a sum of $D_H(P_i, Q_i)$ for $X_i \in \mathcal{X}$. This is similar to the approach taken by [18,22]. For each parent configuration π of each X in each network class C, $D_H(P_1(X|\pi), P_2(X|\pi))$ is computed where P_1 and P_2 are CPDs produced by the two learning algorithms. The values $D_H(P_1(X|\pi), P_2(X|\pi))$ are summed across parent configurations, variables and classes (ignoring bounded input nodes). In the weighted Hellinger distance $D_H^w(P_1(X|\pi), P_2(X|\pi))$, $D_H(P_1(X|\pi), P_2(X|\pi))$ is weighted by $P(\pi)$ in the reference model.

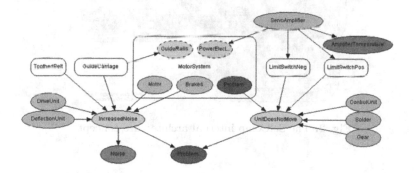

Fig. 1. The top level class of the Linear Axis Model.

3 The Linear Axis OOBN Model

The Linear Axis as a self-sustainable handling system that is designed to be a high performance machine with a demand to work 24 h/day seven days a week. Therefore, there is little or no time for maintenance and repair. This means that there is a need for system condition monitoring to prevent failures and for system failure diagnosis. The Linear Axis diagnosis model considered here is used for root-cause analysis under the assumption that a problem is observed and the five most likely root causes should be identified. Figure 1 shows the structure of the top-level class of the Linear Axis OOBN. In the figure, blue nodes denote possible root causes, orange nodes denote problem defining nodes, and green nodes denote possible observations such as sensor readings and operator feedback. The model has 35 variables, 27 failure states, 555 CPD entries, maximum CPD size of 128 and five class instances (two instances of the LimitSwitch class). The Linear Axis OOBN has been quantified using subject matter expert knowledge. We refer to this model as the knowledge driven model and its development is described in more detail in [11].

The diagnostic performance of the knowledge driven model has been assessed following the approach of [11]. The basic idea, is to iterate through the root causes where each root causes is instantiated to a failure state and all other root causes are instantiated to non-failure. For each such configuration, values for the observations are generated. The values for the observations are propagated in the model and the probabilities of the root causes recorded. This demonstrates how well the observations can distinguish the root causes.

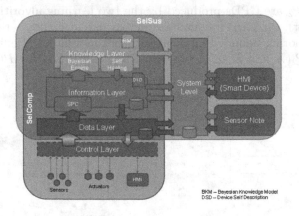

Fig. 2. The SelComp internal architecture concept.

4   The SelSus Architecture

The aim of the SelSus System Architecture is to provide an environment for highly effective, self-healing production resources and systems to maximize their performance over longer life times through highly targeted and timely repair, renovation and upgrading [20]. The architecture defines three levels of abstraction for its constituents: (1) Component Level, which relates directly to machines or its sub-components and is composed of smart sensory capabilities, methods for self-diagnostics and predictive maintenance. (2) Station Level, at this level the developments are constituted by previous capabilities plus human machine interfaces and tools to support the design and maintenance of the factory station. (3) Factory Level, previous levels capabilities are combined to create a semantic driven maintenance scheduling for large production factory plants.

The Linear Axis typically integrates a production cell, performing operations in collaboration with other machines (e.g., robotic arms and welding tools). To make operational and sensory data available to the SelSus System, the SelComp (SelSus Component) concept was designed. The SelComp (Fig. 2), is a self-aware entity that makes available to the SelSus system its internal state conditions, providing this way operational and structural knowledge. A SelComp also provides built-in models for state estimation based on sensor data which enables pro-active and predictive maintenance. These components have the ability to collect data from sensors that are mounted physically in the same device or in a near location and fuse this data to extend its models capabilities. The Linear Axis OOBN has been encapsulated in the Machine SelComp for the Linear Axis [12,20], see Fig. 3, as it represents a field device, machine or its sub-components. The goal is to provide diagnostic capabilities at component-level supporting system-level diagnostics. A Sensor SelComp [14,15], on the other hand, is designed to provide essentially smart sensor data to the SelSus system and more often to Machine SelComps. The Sensor SelComp component has *plug&play* capabilities in terms of physical sensors, data models and algorithms. The Linear Axis OOBN can also be abstracted as a service to provide outputs to and subscribe to inputs from the SelSus System and other SelComps.

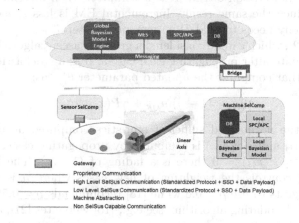

Fig. 3. The SelSus system architecture.

5   Parameter Learning Algorithms

Let $\mathcal{N} = (\mathcal{X}, G, \mathcal{P})$ be a BN with parameters $\Theta = (\theta_{ijk})$ such that $\theta_{ijk} = P(X_i = k \mid \pi_{X_i} = j)$ for each i, j, k. The task of parameter learning is to estimate the values of the parameters Θ given a dataset of cases $\mathcal{D} = \{c_1, \dots, c_N\}$. The cases of $\mathcal{D}$ are assumed independent and identically distributed (i.i.d.) with values missing at random or completely at random [4]. The generating probability distribution is assumed to be stationary. We first present the parameter learning algorithms for standard BNs followed by a description of how they are applied to OOBNs.

The EM algorithm [4,9] is well-suited for calculating maximum likelihood and maximum a posteriori (MAP) estimates in the case of missing data. The algorithm iterates the E-Step and M-Step until convergence. Given an initial value of the parameters Θ, the E-step computes the current expected sufficient statistics while the subsequent M-step maximizes the log-likelihood function $l(\Theta \mid D) = \sum_{i=1}^{N} \log P(c_i \mid \Theta)$. The E-step of the EM algorithm computes expected sufficient statistics for each family $fa(X_i)$ and parent configuration π_{ij} of each X_i under Θ as $n(Y) = \mathbb{E}_\Theta\{n(Y) \mid D\}$, where $n(\cdot)$ is counts for either π_{ij} or $X_i = k, \pi_{ij}$. The M-step computes new estimates of θ_{ijk}^* from the expected counts under θ_{ijk} and a virtual count $\theta_{ijk}\alpha_{ij}$ specified beforehand (MAP estimate):

$$\theta_{ijk}^* = \frac{n(X_i = k, \pi_{ij}) + \theta_{ijk}\alpha_{ij}}{n(\pi_{ij}) + \alpha_{ij}}, \tag{1}$$

where α_{ij} is the equivalent sample size (ESS) specified for π_{ij}.

The principle idea of the incremental EM algorithm, e.g. [18], is to divide the data $\mathcal{D}$ into disjoint subsets $\mathcal{D}_1, \ldots, \mathcal{D}_m$ and iteratively apply EM on $\mathcal{D}_i$. The estimates θ_{ijk} and α_{ij} produced by one iteration of EM are used as virtual counts in the next iteration of EM. If $\mathcal{D}$ is complete, then incremental EM and batch EM produce the same result. Incremental EM is less space demanding than EM as it only needs to hold $\mathcal{D}_i$ in memory at step i.

Online EM [2], which can be considered a gradient ascent algorithm, performs a parameter update after propagating each case c. It is a stochastic approximation algorithm that computes the updated parameter θ_{ijk}^* as:

$$\theta_{ijk}^* = (1 - \gamma)m_{ijk} + \gamma P(x_{ijk}|c), \tag{2}$$

where m_{ijk} is the normalized sufficient statistics computed as $m_{ijk} = \alpha_{ij} * p(x_{ik}|\pi_{ij})/\sum_j \alpha_{ij}$ and $p(x_{ijk}|c)$ is computed by propagating case c. Notice that even for π_{ij} with $P(\pi_{ij}) = 0$ there is a fading of $(1 - \gamma)$. The learning rate $\gamma = (1 + n)^{-\rho}$ controls the weighting of new cases where n is the iteration number. [2] suggests to use $\rho = 0.6$ while [18] recommends $\rho = 0.501$.

The fractional updating algorithm, see e.g., [16] who cites [21], also performs a parameter update after propagating each case c. In fractional updating the parameter θ_{ijk} is adjusted after propagating each case c as follows:

$$\theta_{ijk}^* = \frac{\alpha_{ijk} + P(x_{ijk}|c)}{\alpha_{ij} + P(\pi_{ij}|c)}. \tag{3}$$

Fading of past cases is controlled by a fading factor λ specified for each π_{ij} and a gradual fading is obtained by taking $P(\pi_{ij})$ into consideration. To improve performance, fractional updating is only performed for π_{ij} when $P(\pi_{ij}) > 0$ (an update would leave θ_{ijk}^* and α_{ij} unchanged when $P(\pi_{ij}) = 0$), see [10].

Both fractional updating and Online EM perform no update when $\alpha_{ij} = 0$. Also, fractional updating and Online EM are the least space demanding algorithms as they only need to hold the latest case in memory.

In the general case of OOBNs, we compute the average expected counts for the run-time instances of the node and increase the experience counts by the number of run-time instances. This applies to all four algorithms.

6   Empirical Evaluation

The empirical evaluation is organised into three different tests (1) we consider updating the parameters of the knowledge driven model where all distributions are made uniform using a dataset of 250,000 cases with 5% missing values generated completely at random from the knowledge driven model, (2) we consider updating the parameters of the knowledge driven model using an real operational dataset, and (3) we consider the time performance of updating the parameters in the knowledge driven model in a real setting. The evaluations are performed using different values of the parameters of the learning algorithms.

Figure 4 shows the results (1) where a random sample generated from the knowledge driven model is used to learn the parameters in the model with uniform distributions. Figure 4(left) shows the weighted Hellinger distance while (right) shows the time usage of each algorithm.

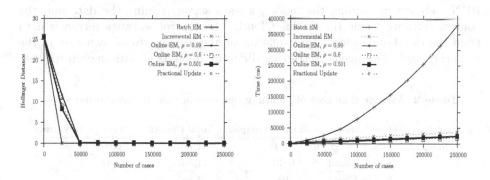

Fig. 4. Hellinger distance (left) and accumulated time is ms (right).

In the test, the values $\rho \in \{0.501, 0.6, 0.99\}$, $n_0 = 1$, $\sum_j \alpha_{ij} = 1$ for all i and $\lambda = 1$ are used. The value for $N = 0$ is the distance is between the uniform distribution and the knowledge driven model. The distances are quickly reduced in all cases and after a certain point no or little improvement is observed. D_H^w reduces the impact of large differences in distributions for π_{ij} with $\alpha_{ij} \ll 1$.

In (2) the operational dataset contains two sequences of 13,429 cases in total with six observed sensor readings represented in the model, i.e., there is 83% missing values due to the hidden variables alone. It contains both failure and non-failure cases. Table 1 shows the diagnostic performance of the five models considered where μ_{rank} refers to the average rank of the *true* root cause, i.e., the value 1 means perfect performance and 27 worst possible performance. In the test, the values $\rho = 0.99$, $\sum_j \alpha_{ij} = 13,429$ for all i and $n_0 = 13,429$ are used.

Table 1. The diagnostic performance of the five models considered.

Algorithm	Top-1	Top-5	μ_{rank}
Knowledge driven model	8	17	4.6
Batch EM	10	17	5.1
Online EM	9	17	4.5
Fractional update	10	21	3.4

For all algorithms, there is an increase in the number of true root causes identified as the cause with highest probability. For Top-5 there is a significant improvement using fractional updating. The value μ_{rank} is not improved for batch EM. This is due to three true root causes obtaining a significantly worse rank after learning (e.g., rank 2 before compared to rank 16 after learning).

Next (3), we report on a performance analysis of two levels of integration of the OOBN model into the SelSus architecture using Online EM and fractional updating for parameter learning. The first and most tight level of integration has been achieved by integrating the model directly into the component control software where data is read from file. The second configuration is to deploy a BBN web service holding the model, a data server holding the data and the control software inside the SelSus Cloud. The control software retrieves data from the cloud and requests propagation of and learning from each case in the data retrieved. We consider retrieving different amounts of data in each request.

Table 2. Average time cost of handling one case across the integration levels.

Algorithm	Configuration	Cases/Request	Total time (ms)	Average time (ms)
Online EM	Direct integration	1	1,730	0.067
	SelSus Cloud	1000	11,367	0.44
	SelSus Cloud	100	44,867	1.74
	SelSus Cloud	10	496,199	19.29
Fractional	Direct integration		1,533	0.067
updating	SelSus Cloud	1000	10,553	0.41
	SelSus Cloud	100	42,111	1.64
	SelSus Cloud	10	478,612	18.60

Table 2 shows the total and average time cost for each configuration. The analysis is performed using an operational data set of 25, 726 cases collected randomly. Here the focus is only on runtime and not the learning. As expected, there is a significant difference between direct integration and using a cloud service wrt. runtime. This means that the learning must be designed taking the data frequency into consideration.

7   Discussion

We have described the use of batch EM, incremental EM, Online EM and fractional updating on OOBNs for continuous model improvement using operational data. The objective was to improve the diagnostic performance of an OOBN for root-cause analysis by adjusting the model parameters using data.

The experimental results show that parameter learning for continuous model improvement using operation data is both feasible and will lead to better diagnostic performance. The Online EM algorithm is sensitive to the value of the ρ parameter and our results indicate that a value close to 1 is preferred. This is contrary to the [18] who suggests $\rho = 0.501$ and [2] who suggests $\rho \in [0.6; 0.9]$ and uses $\rho = 0.6$. The ρ-value controls the learning rate γ and higher γ-values (and lower ρ-values) means more emphasis on new cases (i.e., faster learning).

The results of the first experiment demonstrate that all four approaches quickly produce a model that has a low D_H^w relative to the knowledge driven models. There is no method that produce a significantly better result that the other algorithms. This is using a data set with 5% missing values. The running time of incremental EM with large batches and batch EM makes them infeasible in practice for this application.

For the Linear Axis OOBN with $||\mathcal{X}|| = 35$ only six are observed in the operational data. This data set has not been augmented with information on presence or absence of root causes nor any operator feedback. The data only includes sensor readings. Despite this fact, the algorithms all improve the diagnostic performance of the model compared to the initial knowledge driven model It is expected that enriching the operational data with information on absence or presence of root causes will improve the learning. This will reduce the bias of the data as much more non-failure than failure data must be expected. Fractional update enables the specification of different α_{ij} for different CPDs and parent configurations in this way controlling the impact of the expert assessed parameters whereas Online EM uses normalised sufficient statistics and uses ρ to control the learning rate.

When learning the parameters from operational data using a knowledge driven model as the starting point, a decision on the relative balance of the data and the expert elicited values must be made. In the experiments, we have defined an ESS equal to the size of the operational data. This decision is important when the data stream is, in principle, infinite in a real operational setting. In any case, it is important that the parameters are stable at least until sufficient data has been processed, which is dependent on the model complexity.

Acknowledgments. This work is part of the project "Health Monitoring and Life-Long Capability Management for SELf-SUStaining Manufacturing Systems (SelSus)" which is funded by the Commission of the European Communities under the 7th Framework Programme, Grant agreement no: 609382. We would like to thank Andres Masegosa for discussions on the Online EM algorithm and the reviewers for their insightful comments, which have helped to improve the paper.

References

1. Binder, J., Koller, D., Russell, S., Kanazawa, K.: Adaptive probabilistic networks with hidden variables. Mach. Learn. **29**(2), 213–244 (1997)
2. Cappe, O., Moulines, E.: Online EM algorithm for latent data models. J. Roy. Stat. Soc. Ser. B (Stat. Method.) **71**(3), 593–613 (2009)
3. Cowell, R.G., Dawid, A.P., Lauritzen, S.L., Spiegelhalter, D.J.: Probabilistic Networks and Expert Systems. Springer, New York (1999)
4. Dempster, A.P., Laird, N.M., Rubin, D.B.: Maximum likelihood from incomplete data via the EM algorithm. J. Roy. Stat. Soc. Ser. B **39**(1), 1–38 (1977)
5. Jensen, F.V.: Gradient descent training of Bayesian networks. In: Proceedings of the ECSQARU, pp. 190–200 (1999)
6. Kjærulff, U.B., Madsen, A.L.: Bayesian Networks and Influence Diagrams. A Guide to Construction and Analysis, 2nd edn. Springer, New York (2013)
7. Kokolakis, G., Nanopoulos, P.: Bayesian multivariate micro-aggregation under the Hellingers distance criterion. Res. Offic. Stat. **4**(1), 117–126 (2001)
8. Koller, D., Pfeffer, A.: Object-oriented bayesian networks. In: Proceedings of the UAI, pp. 302–313 (1997)
9. Lauritzen, S.L.: The EM algorithm for graphical association models with missing data. Comput. Stat. Anal. **19**, 191–201 (1995)
10. Madsen, A.L., Lang, M., Kjærulff, U.B., Jensen, F.: The Hugin tool for learning Bayesian networks. In: Proceedings of the ECSQARU, pp. 594–605 (2003)
11. Madsen, A.L., Søndberg-Jeppesen, N., Lohse, N., Sayed, M.: A methodology for developing local smart diagnostic models using expert knowledge. In: IEEE INDIN, pp. 1682–1687 (2015)
12. Madsen, A.L., Søndberg-Jeppesen, N., Sayed, M.S., Peschl, M., Lohse, N.: Applying object-oriented Bayesian networks for smart diagnosis and health monitoring at both component and factory level. Accepted for IEA/AIE 2017 (2017)
13. Neil, M., Fenton, N., Nielsen, L.M.: Building large-scale Bayesian networks. Knowl. Eng. Rev. **15**(3), 257–284 (2000)
14. Neto, L., Reis, J., Guimaraes, D., Concalves, G.: Sensor cloud: smartcomponent framework for reconfigurable diagnostics in intelligent manufacturing environments. In: IEEE INDIN, pp. 1706–1711 (2015)
15. Neto, L., Reis, J., Silva, R., Concalves, G.: Sensor SelComp, a smart component for the industrial sensor cloud of the future. In: IEEE ICIT, pp. 1256–1261 (2017)
16. Olesen, K.G., Lauritzen, S.L., Jensen, F.V.: aHUGIN: a system creating adaptive causal probabilistic networks. In: Proceedings of the UAI, pp. 223–229 (1992)
17. Pearl, J.: Probabilistic Reasoning in Intelligent Systems: Networks of Plausible Inference. Morgan Kaufmann Publishers, San Mateo (1988)
18. Ratnapinda, P., Druzdzel, M.J.: Learning discrete Bayesian network parameters from continuous data streams: what is the best strategy. J. Appl. Logic **13**, 628–642 (2015)
19. Russell, S., Binder, J., Koller, D., Kanazawa, K.: Local learning in probabilistic networks with hidden variables. In: Proceedings of IJCAI, pp. 1146–1152 (1995)
20. Sayed, M.S., Lohse, N., Søndberg-Jeppesen, N., Madsen, A.L.: SelSus: towards a reference architecture for diagnostics and predictive maintenance using smart manufacturing devices. In: IEEE INDIN, p. 6 (2015)
21. Titterington, D.M.: Updating a diagnostic system using unconfirmed cases. Appl. Stat. **25**, 238–247 (1976)
22. Zagorecki, A., Voortman, M., Druzdzel, M.J.: Decomposing local probability distributions in bayesian networks for improved inference and parameter learning. In: Proceedings of the FLAIRS, pp. 860–865 (2006)

Monotonicity in Bayesian Networks for Computerized Adaptive Testing

Martin Plajner[1,2][✉] and Jiří Vomlel[2]

[1] Faculty of Nuclear Sciences and Physical Engineering, Czech Technical University,
Prague, Trojanova 13, 120 00 Prague, Czech Republic
[2] Institute of Information Theory and Automation, Czech Academy of Sciences,
Pod Vodárenskou věží 4, 182 08 Prague 8, Czech Republic
{plajner,vomlel}@utia.cas.cz
http://staff.utia.cas.cz/plajner/
http://www.utia.cas.cz/vomlel/

Abstract. Artificial intelligence is present in many modern computer science applications. The question of effectively learning parameters of such models even with small data samples is still very active. It turns out that restricting conditional probabilities of a probabilistic model by monotonicity conditions might be useful in certain situations. Moreover, in some cases, the modeled reality requires these conditions to hold. In this article we focus on monotonicity conditions in Bayesian Network models. We present an algorithm for learning model parameters, which satisfy monotonicity conditions, based on gradient descent optimization. We test the proposed method on two data sets. One set is synthetic and the other is formed by real data collected for computerized adaptive testing. We compare obtained results with the isotonic regression EM method by Masegosa et al. which also learns BN model parameters satisfying monotonicity. A comparison is performed also with the standard unrestricted EM algorithm for BN learning. Obtained experimental results in our experiments clearly justify monotonicity restrictions. As a consequence of monotonicity requirements, resulting models better fit data.

Keywords: Computerized adaptive testing · Monotonicity · Isotonic regression EM · Gradient method · Parameters learning

1 Introduction

In our previous research Plajner and Vomlel (2015) we focused on Computerized Adaptive Testing (CAT) (Almond and Mislevy 1999; van der Linden and Glas 2000). We used artificial student models to select questions during the course of testing. We have shown that it is useful to include monotonicity conditions while learning parameters of these models (Plajner and Vomlel 2016b).

This work was supported by the Czech Science Foundation (project No. 16-12010S) and by the Grant Agency of the Czech Technical University in Prague, grant No. SGS17/198/OHK4/3T/14.

© Springer International Publishing AG 2017
A. Antonucci et al. (Eds.): ECSQARU 2017, LNAI 10369, pp. 125–134, 2017.
DOI: 10.1007/978-3-319-61581-3_12

Monotonicity conditions incorporate qualitative influences into a model. These influences restrict conditional probabilities in a specific way to avoid unwanted behavior. Some models we use for CAT include monotonicity naturally, but in this article we focus on a specific family of models, Bayesian Networks, which do not. Monotonicity in Bayesian Networks is discussed in literature for a long time. It is addressed, for example, by Wellman (1990), Druzdzel and Henrion (1993) and more recently by, e.g., Restificar and Dietterich (2013), Masegosa et al. (2016). Monotonicity restrictions are often motivated by reasonable demands from model users. In our case of CAT it means we want to make sure that students having certain skills will have a higher probability of answering questions depending on these skills correctly. Moreover, assuming monotonicity we can learn better models, especially when the data sample is small. In our work we have so far used monotonicity attained by logistic regression models of CPTs. This has proven useful but it is restrictive since it requires a prescribed CPT structure.

In this article we extends our results in the domain of Bayesian Networks. We present a gradient descent optimum search method for learning parameters of CPTs respecting monotonicity conditions. First, we establish our notation and monotonicity conditions in Sect. 2. Our method is derived in Sect. 3. We have implemented the method and performed tests. For testing we used two different data sets. First, we used a synthetic data set generated from a monotonic model (CPTs satisfying monotonicity) and second, we used real data set collected earlier. Experiments were performed on these data sets also with the isotonic regression EM (irem) method described by Masegosa et al. (2016) and the ordinary EM learning without monotonicity restrictions. In Sect. 4 of this paper we take a closer look at the experimental setup and present results of described tests. The last section brings an overview and a discussion of the obtained results.

2   BN Models and Monotonicity

2.1   Notation

In this article we use Bayesian Networks. Details about BNs can be found in, for example, Pearl (1988), Nielsen and Jensen (2007). We restrict ourselves to the following BN structure. Networks have two levels. In compliance with our previous articles, variables in the parent's level are addressed as skill variables S. The children level contains questions-answers variables X. Example network structures, which we also used for experiments, are shown in Figs. 1 and 2.

- We will use symbol $\boldsymbol{X}$ to denote the multivariable $(X_1, \ldots, X_n)$ taking states $\boldsymbol{x} = (x_1, \ldots, x_n)$. The total number of question variables is n, the set of all indexes of question variables is $\boldsymbol{N} = \{1, \ldots, n\}$. Question variables are binary and they are observable.
- We will use symbol $\boldsymbol{S}$ to denote the multivariable $(S_1, \ldots, S_m)$ taking states $\boldsymbol{s} = (s_1, \ldots, s_m)$. The set of all indexes of skill variables is $\boldsymbol{M} = \{1, \ldots, m\}$.

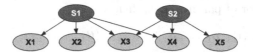

Fig. 1. Artificial model

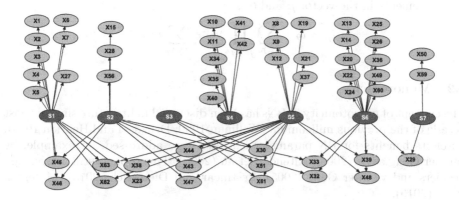

Fig. 2. CAT model network

Skill variables have variable number of states[1], the total number of states of a variable S_j is m_j and individual states are $s_{j,k}, k \in \{1, \ldots, m_j\}$. The variable $S^i = S^{pa(i)}$ stands for a multivariable same as S but containing only parent variables of the question X_i. Indexes of these variables are $M^i \subseteq M$. The set of all possible state configurations of S^i is $Val(S^i)$. Skill variables are all unobservable.

CPT parameters for a question variable X_i for all $i \in N, s^i \in Val(S^i)$ are

$$\theta_{i,s^i} = P(X_i = 0 | S^i = s^i), \ \boldsymbol{\theta}_i = (\theta_{i,s^i})_{s^i \in Val(S^i)}.$$

We will also use $\theta_{i,s} = \theta_{i,s^i}$ with the whole parent set S, where variables from $S \backslash S^i$ do not affect the value. Probabilities of a correct answer to a question X_i given state configuration s^i is $P(X = 1 | S^i = s^i) = 1 - \theta_{i,s^i}$ (binary questions).

Parameters of parent variables for $j \in M$ are

$$\rho_{j,s_j} = P(S_j = s_j), \ \boldsymbol{\rho}_j = (P(S_j = s_{j'})), \ j' \in \{1, \ldots, m_j\}.$$

Parameter vector $\boldsymbol{\rho}_j$ is constrained by a condition $\sum_{s_j=1}^{m_j} \rho_{j,s_j} = 1$. To remove this condition we reparametrize this vector to

$$\rho_{j,s_j} = \frac{exp(\mu_{j,s_j})}{\sum_{s_j'=1}^{m_i} exp(\mu_{j,s_j'})}.$$

---

[1] In our experiments we use parents with 3 states, but the following theory applies to any number of states.

The whole vector of parameters is then

$$\boldsymbol{\theta} = (\boldsymbol{\theta}_1, \ldots, \boldsymbol{\theta}_n, \boldsymbol{\rho}_1, \ldots, \boldsymbol{\rho}_m), \text{ or } \boldsymbol{\mu} = (\boldsymbol{\theta}_1, \ldots, \boldsymbol{\theta}_n, \boldsymbol{\mu}_1, \ldots, \boldsymbol{\mu}_m),$$

where the meaning of $\boldsymbol{\mu}_j$ is the same as $\boldsymbol{\rho}_j$ but in this case vectors contain reparametrized variables. The transition from $\boldsymbol{\mu}$ to $\boldsymbol{\theta}$ is simply done with the reparametrization above and will be used without further notice. The total number of elements in the vector $\boldsymbol{\mu}$ and $\boldsymbol{\theta}$ is

$$l_{\boldsymbol{\mu}} = l_{\boldsymbol{\theta}} = \sum_{i \in N} \prod_{j \in M^i} m_j + \sum_{l \in M} m_l.$$

2.2 Monotonicity

The concept of monotonicity in BNs has been discussed in literature since the last decade of the previous millennium (Wellman 1990; Druzdzel and Henrion 1993). Later its benefits for BN parameter learning were addressed, for example, by van der Gaag et al. (2004), Altendorf et al. (2005). This topic is still active, e.g., Feelders and van der Gaag (2005), Restificar and Dietterich (2013), Masegosa et al. (2016).

We will consider only variables with states from $\mathbb{N}_0$ with their natural ordering, i.e., the ordering of states of skill variable's S_j for $j \in \boldsymbol{M}$, is

$$s_{j,1} \prec \ldots \prec s_{j,m_j}.$$

For questions we use natural ordering of its states ($0 \prec 1$).

A variable S_j has monotone, resp. antitone, effect on its child if for all $k, l \in \{1, \ldots, m_j\}$:

$$s_{j,k} \preceq s_{j,l} \Rightarrow P(X_i = 1 | S_j = s_{j,k}, \boldsymbol{s}) \leq P(X_i = 1 | S_j = s_{j,l}, \boldsymbol{s}), \quad \text{resp.}$$
$$s_{j,k} \preceq s_{j,l} \Rightarrow P(X_i = 1 | S_j = s_{j,k}, \boldsymbol{s}) > P(X_i = 1 | S_j = s_{j,l}, \boldsymbol{s}).$$

where $\boldsymbol{s}$ is the configuration of other remaining parents of question i without S_j. For each question $X_i, i \in \boldsymbol{M}$ we denote by $\boldsymbol{S}^{i,+}$ the set of parents with a monotone effect and by $\boldsymbol{S}^{i,-}$ the set of parents with an antitone effect.

Next, we create a partial ordering $\preceq_i$ on all state configurations of parents $\boldsymbol{S}^i$ of the i-th question, where for all $\boldsymbol{s}^i, \boldsymbol{r}^i \in Val(\boldsymbol{S}^i)$:

$$\boldsymbol{s}^i \preceq_i \boldsymbol{r}^i \Leftrightarrow \left(s_j^i \preceq r_j^i, \ j \in \boldsymbol{S}^{i,+} \right) \text{ and } \left(r_j^i \preceq s_j^i, \ j \in \boldsymbol{S}^{i,-} \right).$$

The monotonicity condition then requires that the question probability of correct answer is higher for a higher order parent configuration, i.e., for all $\boldsymbol{s}^i, \boldsymbol{r}^i \in Val(\boldsymbol{S}^i)$:

$$\boldsymbol{s}^i \preceq_i \boldsymbol{r}^i \Rightarrow P(X_i = 1 | \boldsymbol{S}^i = \boldsymbol{s}^i) \leq P(X_i = 1 | \boldsymbol{S}^i = \boldsymbol{r}^i),$$
$$\boldsymbol{s}^i \preceq_i \boldsymbol{r}^i \Rightarrow P(Xi = 0 | \boldsymbol{S}^i = \boldsymbol{s}^i) \geq P(Xi = 0 | \boldsymbol{S}^i = \boldsymbol{r}^i) \Leftrightarrow \theta_{i,\boldsymbol{s}^i} \geq \theta_{i,\boldsymbol{r}^i}.$$

In our experimental part we consider only isotone effect of parents on their children. The difference with antitone effects is only in the partial ordering.

3   Parameter Gradient Search with Monotonicity

To learn parameter vector $\boldsymbol{\mu}$ we develop a method based on the gradient descent optimization. We follow the work of Altendorf et al. (2005) where they use a gradient descent method with exterior penalties to learn parameters. The main difference is that we consider models with hidden variables.

We denote by $\boldsymbol{D}$ the set of indexes of observations vectors. One vector $x^k, k \in \boldsymbol{D}$ corresponds to one student and an observation of i-th variable X_i is x_i^k. The number of occurrences of the k-th configuration vector in the data sample is d_k.

We use the model structure as described in Sect. 2, i.e., unobserved parent variables and observed binary children variables. With sets $\boldsymbol{I}_0^k$ and $\boldsymbol{I}_1^k$ of indexes of incorrectly and correctly answered questions, we create following products based on observations in the k-th vector:

$$p_0^k(\boldsymbol{\mu}, \boldsymbol{s}, k) = \prod_{i \in \boldsymbol{I}_0^k} \theta_{i,s}, \quad p_1^k(\boldsymbol{\mu}, \boldsymbol{s}, k) = \prod_{i \in \boldsymbol{I}_1^k} (1 - \theta_{i,s}), \quad p_\mu(\boldsymbol{\mu}, \boldsymbol{s}) = \prod_{j=1}^m exp(\mu_{j,s_j}).$$

We work with the log likelihood:

$$
\begin{aligned}
LL(\boldsymbol{\mu}) &= \sum_{k \in \boldsymbol{D}} d_k \cdot log \left(\sum_{\boldsymbol{s} \in Val(\boldsymbol{S})} \prod_{j=1}^m \frac{exp(\mu_{j,s_j})}{\sum_{s_j'=1}^{m_j} exp(\mu_{j,s_j'})} \cdot p_0^k(\boldsymbol{\mu}, \boldsymbol{s}, k) \cdot p_1^k(\boldsymbol{\mu}, \boldsymbol{s}, k) \right) \\
&= \sum_{k \in \boldsymbol{D}} d_k \cdot log \left(\sum_{\boldsymbol{s} \in Val(\boldsymbol{S})} p_\mu(\boldsymbol{\mu}, \boldsymbol{s}) \cdot p_0^k(\boldsymbol{\mu}, \boldsymbol{s}, k) \cdot p_1^k(\boldsymbol{\mu}, \boldsymbol{s}, k) \right) \\
&\quad - N \cdot \sum_{j=1}^m log \sum_{s_j'=1}^{m_j} exp(\mu_{j,s_j'}).
\end{aligned}
$$

The partial derivatives of $LL(\boldsymbol{\mu})$ with respect to θ_{i,s^i} for $i \in \boldsymbol{N}, s^i \in Val(\boldsymbol{S}^i)$ are

$$\frac{\delta LL(\boldsymbol{\mu})}{\delta \theta_{i,s^i}} = \sum_{k \in \boldsymbol{D}} d_k \cdot \frac{(-2x_i^k + 1) \cdot p_\mu(\boldsymbol{\mu}, \boldsymbol{s}^i) \cdot p_0^k(\boldsymbol{\mu}, \boldsymbol{s}^i, k) \cdot p_1^k(\boldsymbol{\mu}, \boldsymbol{s}^i, k)}{\theta_{i,s^i} \cdot \sum_{\boldsymbol{s} \in Val(\boldsymbol{S})} p_\mu(\boldsymbol{\mu}, \boldsymbol{s}) \cdot p_0^k(\boldsymbol{\mu}, \boldsymbol{s}, k) \cdot p_1^k(\boldsymbol{\mu}, \boldsymbol{s}, k)}.$$

and with respect to $\mu_{i,l}$ for $i \in \boldsymbol{M}, l \in \{1, \ldots, m_i\}$ are

$$
\begin{aligned}
\frac{\delta LL(\boldsymbol{\mu})}{\delta \mu_{i,l}} &= \sum_{k \in \boldsymbol{D}} d_k \cdot \frac{\sum_{\boldsymbol{s} \in Val(\boldsymbol{S})}^{s_i=l} p_\mu(\boldsymbol{\mu}, \boldsymbol{s}) \cdot p_0^k(\boldsymbol{\mu}, \boldsymbol{s}, k) \cdot p_1^k(\boldsymbol{\mu}, \boldsymbol{s}, k)}{\sum_{\boldsymbol{s} \in Val(\boldsymbol{S})} p_\mu(\boldsymbol{\mu}, \boldsymbol{s}) \cdot p_0^k(\boldsymbol{\mu}, \boldsymbol{s}, k) \cdot p_1^k(\boldsymbol{\mu}, \boldsymbol{s}, k)} \\
&\quad - N \cdot \frac{exp(\mu_{i,l})}{\sum_{l'=1}^{m_i} exp(\mu_{k,l'})}.
\end{aligned}
$$

3.1   Monotonicity Restriction

To ensure monotonicity we use a penalty function

$$p(\theta_{i,s^i}, \theta_{i,r^i}) = exp(c \cdot (\theta_{i,r^i} - \theta_{i,s^i}))$$

for the log likelihood:

$$LL'(\boldsymbol{\mu}, c) = LL(\boldsymbol{\mu}) - \sum_{i \in N} \sum_{\boldsymbol{s}^i \preceq_i \boldsymbol{r}^i} p(\theta_{i, \boldsymbol{s}^i}, \theta_{i, \boldsymbol{r}^i}),$$

where c is a constant determining the strength of the condition. Theoretically, this condition does not ensure monotonicity but, practically, selecting high values of c results in monotonic estimates. If the monotonicity is not violated, i.e. $\theta_{i,\boldsymbol{r}^i} < \theta_{i,\boldsymbol{s}^i}$ then the penalty value is close to zero. Otherwise, the penalty is raising exponentially fast with respect to $\theta_{i,\boldsymbol{r}^i} - \theta_{i,\boldsymbol{s}^i}$. In our experiments we have used the value of $c = 40$ but any value higher than 20 provided almost identical results.

Partial derivatives with respect to $\mu_{i,l}$ remain unchanged. Partial derivatives with respect to $\theta_{i,\boldsymbol{s}^i}$ are:

$$\frac{\delta LL'(\boldsymbol{\mu}, c)}{\delta \theta_{i,\boldsymbol{s}^i}} = \frac{\delta LL(\boldsymbol{\mu})}{\delta \theta_{i,\boldsymbol{s}^i}} + c \sum_{\boldsymbol{s}^i \preceq_i \boldsymbol{r}^i} p(\theta_{i,\boldsymbol{s}^i}, \theta_{i,\boldsymbol{r}^i}) - c \sum_{\boldsymbol{r}^i \preceq_i \boldsymbol{s}^i} p(\theta_{i,\boldsymbol{r}^i}, \theta_{i,\boldsymbol{s}^i})$$

Using the penalized log likelihood, $LL'(\boldsymbol{\mu}, c)$, and its gradient

$$\nabla(LL(\boldsymbol{\mu}, c)) = \left(\frac{\delta LL'(\boldsymbol{\mu}, c)}{\delta \theta_{i,\boldsymbol{s}^i}}, \frac{\delta LL(\boldsymbol{\mu})}{\delta \mu_{j,l}} \right),$$

for $i \in N, \boldsymbol{s}^i \in Val(\boldsymbol{S}^i)$, $j \in M, l \in \{1, \ldots, m_j\}$, we can apply the standard gradient method optimization to solve the problem. In order to ensure probability values of $\boldsymbol{\theta}_i, i \in N$ it is necessary to use a bounded optimization method.

4   Experiments

For testing we use two different Bayesian Network models. The first one is an artificial model and we use simulated data. The second model is one of the models we used for computerized adaptive testing and we work with real data (for details please refer to Plajner and Vomlel (2016a)). In both cases we learn model parameters from data. Parameters are learned with our gradient method, isotonic regression EM[2] and the standard unrestricted EM algorithm. The learned model quality is measured by the log likelihood of the whole data sample including the training subset. This is done in order to provide results comparable between different training set sizes.

---

[2] We have implemented the irem algorithm based on the article (Masegosa et al. 2016). We extended the method to work with parents with more states than 2 (the article considers only binary variables). Questions (children) remain binary which makes the extension easy.

4.1   Artificial Model

The first model is displayed in Fig. 1. This model was created to provide simulated data for testing. The structure of the model is similar to models we use in CAT modeling with two levels of variables. Parents S_1 and S_2 have 3 possible states and children $X_1, \ldots, X_5$ are binary. We have instantiated the model with random parameters vector θ^* satisfying monotonicity conditions. We drew a random sample of 100 000 cases from the model.

For parameters learning we use random subsets of size k of 10, 20, 50, 100, 1 000, 10 000, 50 000, and 100 000-(full data set) cases. For each size (except the last one) we use 10 different sets. Next, we prepared 15 initial parameter configurations for the fixed Bayesian Network structure (Fig. 1). These networks have starting parameters θ_i generated at random, but in such a way, that they satisfy monotonicity conditions. The assumption of monotonicity is part of our domain expert knowledge. Therefore we can use it to speed up the process and avoid local optima. Parameters of parent variables are uniform and initial vectors are the same for each method. In our experiment we learn network parameters for each initial parameter setup for each set in a particular set size (giving a total of 150 learned networks for one set size). The learned parameter vectors are $\theta_{i,j}$ for j-th subset of data.

The average log likelihood for the whole data sample

$$LL_A = \frac{\sum_{j=1}^{10} \sum_{i=1}^{15} LL(\theta_{i,j})}{150}$$

is shown in Fig. 3 for each set size. In case of this model we are also able to measure the distance of learned parameters from the generating parameters in addition to the log likelihood. First we calculate an average error for each learned model:

$$e_{i,j} = \frac{|\theta^* - \theta_{i,j}|}{l_\theta},$$

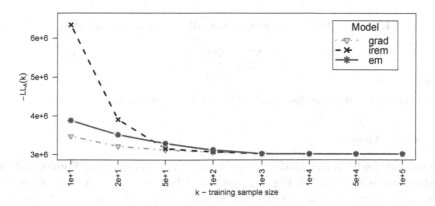

Fig. 3. Negative log likelihood for the whole sample and different training set sizes for the artificial model.

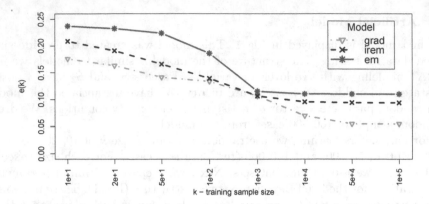

Fig. 4. Mean difference of parameters of learned and generating networks for different set sizes for the artificial model.

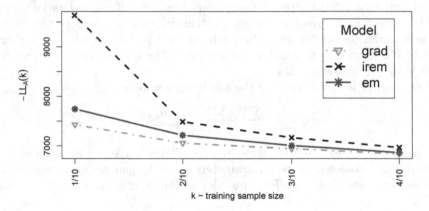

Fig. 5. Negative log likelihood for the whole sample and different training set sizes for the CAT model.

where $\|\ \|$ is L1 norm. Next we average over all results in one set size:

$$e = \frac{\sum_{j=1}^{10} \sum_{i=1}^{15} e_{i,j}}{150}.$$

Resulting values of e are displayed in Fig. 4 for each set size.

4.2   CAT Model

The second model is the model we used for CAT (Plajner and Vomlel 2016b). Its structure is displayed in Fig. 2. Parent variables $S_1, \ldots, S_7$ have 3 states and each one of them represents a particular student skill. Children nodes X_i are variables representing questions which are binary. Data associated with this model were collected from paper tests of mathematical skills of high school students.

In total the data sample has 281 cases. For more detailed overview of tests refer to Plajner and Vomlel (2016a). For learning we use random subsets of size of 1/10, 2/10, 3/10, and 4/10 of the whole sample. Similarly to the previous model, we drew 10 random sets for each size and initiated models by 15 different initial random monotonic starting parameters θ_i.

After learning we compute log likelihoods of the whole data set and we create averages for each set size $LL_A(k)$ as with the previous model. Resulting values are in Fig. 5. In this case we cannot compare learned parameters because the real parameters with real are unknown.

5   Conclusions

In this article we have presented a gradient based method for learning parameters of Bayesian Network under monotonicity restrictions. The method was described and then tested on two data sets. In Figs. 3 and 5 it is clearly visible that this method achieves the best results from three tested methods (especially for small training samples). The irem method has problems with small training samples and the log likehood in those cases is low. This is a consequence of the fact that it moves to monotonic solution from a poor EM estimate and in these cases ensuring monotonicity implies log likelihood degradation. We can also observe that for the training sets larger than 1000 data vectors the EM algorithm stabilizes in its parameter estimations. It means that at about $k = 1000$ the EM algorithm found the best model it can and increasing training size does not improve the result. Nevertheless, as we can observe in Fig. 4 parameters of learned networks are always closer to the generating parameters while considering monotonicity for both the irem and the gradient methods than for the standard EM.

These results verify usefulness of monotonicity for learning Bayesian Networks. A possible extension is to enlarge the theory of gradient based method to work with more general network structures.

References

Almond, R.G., Mislevy, R.J.: Graphical models and computerized adaptive testing. Appl. Psychol. Meas. **23**(3), 223–237 (1999)

Altendorf, E.E., Restificar, A.C., Dietterich, T.G.: Learning from sparse data by exploiting monotonicity constraints. In: Proceedings of the Twenty-First Conference on Uncertainty in Artificial Intelligence (UAI 2005) (2005)

Druzdzel, J., Henrion, M.: Efficient reasoning in qualitative probabilistic networks. In: Proceedings of the Eleventh National Conference on Artificial Intelligence, pp. 548–553. AAAI Press (1993)

Feelders, A.J., van der Gaag, L.: Learning Bayesian network parameters with prior knowledge about context-specific qualitative influences. In: Proceedings of the Twenty-First Conference on Uncertainty in Artificial Intelligence (UAI 2005) (2005)

Masegosa, A.R., Feelders, A.J., van der Gaag, L.: Learning from incomplete data in Bayesian networks with qualitative influences. Int. J. Approx. Reason. **69**, 18–34 (2016)

Nielsen, T.D., Jensen, F.V.: Bayesian Networks and Decision Graphs. Information Science and Statistics. Springer, New York (2007)

Pearl, J.: Probabilistic Reasoning in Intelligent Systems: Networks of Plausible Inference. Morgan Kaufmann Publishers Inc., San Francisco (1988)

Plajner, M., Vomlel, J.: Bayesian network models for adaptive testing. In: Proceedings of the Twelfth UAI Bayesian Modeling Applications Workshop, pp. 24–33. CEUR-WS.org, Amsterdam (2015)

Plajner, M., Vomlel, J.: Probabilistic models for computerized adaptive testing: experiments. Technical report, arXiv:1601.07929 (2016a)

Plajner, M., Vomlel, J.: Student skill models in adaptive testing. In: Proceedings of the Eighth International Conference on Probabilistic Graphical Models, pp. 403–414. JMLR.org (2016b)

Restificar, A.C., Dietterich, T.G.: Exploiting monotonicity via logistic regression in Bayesian network learning. Technical report, Oregon State University, Corvallis, OR (2013)

van der Gaag, L., Bodlaender, H.L., Feelders, A.J.: Monotonicity in Bayesian networks. In: 20th Conference on Uncertainty in Artificial Intelligence (UAI 2004), pp. 569–576 (2004)

van der Linden, W.J., Glas, C.A.W.: Computerized Adaptive Testing: Theory and Practice, vol. 13. Kluwer Academic Publishers, Dordrecht (2000)

Wellman, M.P.: Fundamental concepts of qualitative probabilistic networks. Artif. Intell. 44(3), 257–303 (1990)

Expert Opinion Extraction
from a Biomedical Database

Ahmed Samet[1(✉)], Thomas Guyet[1], Benjamin Negrevergne[3], Tien-Tuan Dao[2],
Tuan Nha Hoang[2], and Marie Christine Ho Ba Tho[2]

[1] Université Rennes 1/IRISA-UMR6074, Rennes, France
{ahmed.samet,thomas.guyet}@irisa.fr
[2] Sorbonne University, Université de technologie de Compiègne CNRS,
UMR 7338 Biomechanics and Bioengineering, Compiègne, France
{tien-tuan.dao,tuannha.hoang,mariechristinehoba.tho}@utc.fr
[3] LAMSADE, Université Paris-Dauphine, Paris, France
benjamin.negrevergne@dauphine.fr

Abstract. In this paper, we tackle the problem of extracting frequent opinions from uncertain databases. We introduce the foundation of an opinion mining approach with the definition of pattern and support measure. The support measure is derived from the commitment definition. A new algorithm called OpMiner that extracts the set of frequent opinions modelled as a mass functions is detailed. Finally, we apply our approach on a real-world biomedical database that stores opinions of experts to evaluate the reliability level of biomedical data. Performance analysis showed a better quality patterns for our proposed model in comparison with literature-based methods.

Keywords: Uncertain database · Data mining · Opinion · OpMiner

1 Introduction

Data uncertainty has challenged nearly all types of data mining tasks, creating a need for uncertain data mining. Uncertainty is inherent in data from many different domains, including social networks and cheminformatics [1]. The problem of pattern mining, or finding frequent patterns in data, has been extensively studied in deterministic databases [2] since its introduction by Aggrawal et al. [3] as well as in the field of uncertain databases [4]. The uncertain databases have brought more flexibility in data representation [5]. For instance, mass function of evidence theory are comparable to expert's opinion since it details answer to a question over a set of response elements. It also allows to model someone's degree of belief regarding an asked question. Therefore databases storing mass functions (commonly called evidential databases), are seen as a data support for expert opinions and imperfect data.

What classical approaches have in common is that they extract answers. They extract answer elements (fragment of the expert answer) as long they are

© Springer International Publishing AG 2017
A. Antonucci et al. (Eds.): ECSQARU 2017, LNAI 10369, pp. 135–145, 2017.
DOI: 10.1007/978-3-319-61581-3_13

redundant in the database. Therefore, the extracted information is limited and does not describe what experts have expressed. To illustrate this point, let us consider the example of several practitioners that have been asked to give their opinion regarding new treatments for a disease. We intend to extract knowledge from a set of experts' opinions asked about their evaluations of these treatments. Each practitioner gives his opinion regarding the efficiency of a treatment j among a set of evaluation possibilities $\{Good_j, Average_j, Bad_j\}$.

Table 1. Example of uncertain database.

Practitioner	Treatment 1	Treatment 2
P_1	$Bad_1^{0.3} \ Average_1^{0.7}$	$Good_2^1$
P_2	$\{Average_1 \cup Bad_1\}^1$	$Good_2^{0.5} \ Average_2^{0.5}$

The first practitioner hesitates between bad and average evaluation with a higher confidence to average. The second practitioner can not decide whether the treatment is average or bad. A classical pattern mining approach as [6] would extract answers as pattern. For instance, for a threshold of 0.7, $\{Treatment1 = Average_1\}$ is a frequent pattern[1]. Looking at Table 1, the pattern $\{Treatment1 = Average_1\}$ is a fraction of the opinion expressed by the practitioner P_1 and therefore the extracted information is not complete. Unfortunately, this type of output is generated by uncertain mining approaches [7–9]. An opinion pattern would be $Treatment_1 = Bad^{0.3} \ Average^{0.7}$ and is considered as frequent since it does not contradict with the opinion of P_2. In this work, we intend to shake this notion of answer pattern of uncertain databases and we aim to evaluate a pattern as a whole opinion.

Methodologically, we build the foundation of an opinion mining approach. We develop our mining approach on evidential databases. Evidential databases offer more knowledge representation with its simple formalism [10]. They bring more flexibility thanks to mass functions. In fact, it is possible to model all level of uncertainty from absolute certainty to total ignorance. From applicative point of view, we experiment our OpMiner algorithm on a real-world biomedical expert database. The results show the quality of retrieved patterns comparatively to classical ones. In addition, our algorithm shows interesting performances.

2   Preliminaries

The evidence theory or Dempster-Shafer theory [11,12] proposes a robust formalism for modeling uncertainty. The evidence theory is based on several fundamentals such as the Basic Belief Assignment (BBA). A BBA m is the mapping from elements of the power set 2^Θ onto [0, 1]:

---

[1] A pattern is called frequent if its computed support (i.e. frequency in the database) is higher than or equal to a fixed threshold set by an expert.

$$m : 2^{\Theta} \longrightarrow [0,1]$$

where Θ is the *frame of discernment*. It is the set of possible answers for a addressed problem and is composed of N exhaustive and exclusive hypotheses:

$$\Theta = \{H_1, H_2, ..., H_N\}.$$

A BBA m is constrained by:

$$\begin{cases} \sum_{A \subseteq \Theta} m(A) = 1 \\ m(\emptyset) = 0 \end{cases} . \tag{1}$$

Each subset X of 2^{Θ} fulfilling $m(X) > 0$ is called focal element. Constraining $m(\emptyset) = 0$ is the normalized form of a BBA and this corresponds to a closed-world assumption, while allowing $m(\emptyset) > 0$ corresponds to an open world assumption [13].

Dubois and Prade [14] have made three proposals to order BBAs. Let m_1 and m_2 be two BBA's on Θ. The statement that m_1 is at least as committed as m_2 is denoted $m_1 \sqsubseteq m_2$. Three types of ordering have been proposed:

- *pl-ordering* (plausibility ordering) if $Pl_1(A) \leq Pl_2(A)$ for all $A \subseteq \Theta$, we write $m_1 \sqsubseteq_{pl} m_2$,
- *q-ordering* (communality ordering) if $q_1(A) \leq q_2(A)$ for all $A \subseteq \Theta$, we write $m_1 \sqsubseteq_q m_2$,
- *s-ordering* (specialization ordering) if m_1 is a specialization of m_2, we write $m_1 \sqsubseteq_s m_2$,

In this paper, we develop our approach using the plausibility based commitment. The plausibility function $Pl(.)$ is defined as follows:

$$Pl(A) = \sum_{B \cap A \neq \emptyset} m(B). \tag{2}$$

Among all belief functions on Θ, the least committed belief function is the vacuous belief function (i.e. $m(\Theta) = 1$).

Finally, it is possible to store imperfect data modelled as BBAs into a database. This kind of database is commonly called evidential database. Formally, an *evidential database* is a triplet $\mathcal{EDB} = (\mathcal{A}_{\mathcal{EDB}}, \mathcal{O}, \mathcal{R}_{\mathcal{EDB}})$. $\mathcal{A}_{\mathcal{EDB}}$ is a set of attributes and $\mathcal{O}$ is a set of d transactions (i.e., rows). Each column A_j $(1 \leq j \leq n)$ has a domain Θ_j of discrete values. $\mathcal{R}_{\mathcal{EDB}}$ expresses the relationship between the ith transaction (i.e., row T_i) and the jth column (i.e., attribute A_j) by a normalized BBA $m_{ij} : 2^{\Theta_j} \to [0,1]$.

Example 1. We intend to extract knowledge from a set of experts' opinions asked about their evaluations of several treatment efficiencies for a disease. Each practitioner gives his opinion regarding a treatment j from a set of evaluation possibilities $\Theta_j = \{Good_j, Average_j, Bad_j\}$ (see Table 2).

3   Extraction Opinion Patterns over Evidential Databases

In the following subsection, we study the plausibility based commitment relation between two BBAs in the evidence theory.

Table 2. Example of evidential database

Practitioner	Treatment 1	Treatment 2
P_1	$m_{11}(Good_1) = 0.7$	$m_{12}(Good_2) = 0.4$
	$m_{11}(\Theta_1) = 0.3$	$m_{12}(Average_2) = 0.2$
		$m_{12}(\Theta_2) = 0.4$
P_2	$m_{21}(Good_1) = 0.6$	$m_{22}(Good_2) = 0.3$
	$m_{21}(\Theta_1) = 0.4$	$m_{22}(\Theta_2) = 0.7$

3.1   Plausibility Based Commitment Measure

Let us consider two BBAs m_1 and m_2 such as $m_1 \sqsubseteq_{pl} m_2$. We intend to develop a measure to estimate the commitment level of m_2 wrt m_1.

Definition 1. *Given the plausibility functions Pl_1 and Pl_2 of two BBAs m_1 and m_2, the plausibility $PL_{12}(.)$ expresses the difference between two plausibility functions and is computed as follows:*

$$PL_{12}(A) = Pl_1(A) - Pl_2(A). \tag{3}$$

Definition 2. *Assuming two BBAs m_1 and m_2. Assuming that $C(\cdot, \cdot)$ is a commitment measure between two BBAs. It is computed as follows,*

$$
C : 2^\Theta \times 2^\Theta \mapsto [0, 1]
$$
$$
(m_2, m_1) \rightarrow \begin{cases} 1 - ||PL_{21}|| = 1 - \sqrt{\sum_{A \subseteq \Theta} PL_{21}(A)^2} & \text{if } m_1 \sqsubseteq_{pl} m_2 \\ 0 & \text{Otherwise} \end{cases} \tag{4}
$$

Property 1. Assuming two BBAs m_1 and m_2 such as $m_2 \sqsubseteq_{pl} m_1$, Eq. 4 verifies the following properties:

- $C(m_2, m_1) \geq 0$ *(separation axiom)*;
- $C(m_2, m_1) = 1 \Leftrightarrow m_1 = m_2$ *(identity of indiscernible)*;
- $C(m_2, m_1) = C(m_1, m_2)$ *(symmetry)*;
- $C(m_2, m_3) \leq C(m_2, m_1) + C(m_1, m_3)$ *(triangle inequality)*.

3.2   Mining Opinions over Evidential Databases

In an evidential database, an *item* corresponds to a BBA. An *itemset* (so called *pattern*) corresponds to a conjunction of several BBAs having different domains $X = \{m_{ij} \in \mathcal{M}^\Theta\}$. We recall that i is the transaction id and j is the attribute id. $\mathcal{M}^\Theta$ denotes the set of all BBAs in $\mathcal{EDB}$.

Let us consider an evidential database $\mathcal{EDB}$ and the itemset X made of a set of BBAs. The frequency of appearance of an item $x = m_{i'j}$ in a transaction T_i can be computed as follows:

$$Sup_{T_i} : \mathcal{M}_i^{\Theta_j} \rightarrow [0, 1]$$

$$x \mapsto C(x, m_{ij}) \text{ where } m_{ij} \in \mathcal{M}_i^{\Theta_j}. \tag{5}$$

$\mathcal{M}_i^{\Theta_j}$ is the set of BBAs in the row T_i of the attribute j. As illustrated above, the Sup_{T_i} is a measure that computes whether x is in the row T_i. Even if the BBA is not in the studied row, we analyse if there is a BBA that generalizes it. Then, the support of an itemset X over the transaction T_i is computed as

$$Sup_{T_i}(X) = \prod_{x \in X} Sup_{T_i}(x). \tag{6}$$

Therefore, the support of m_{ij} over the database is computed as,

$$Sup_{\mathcal{EDB}}(X) = \frac{1}{d} \sum_{i=1}^{d} Sup_{T_i}(X). \tag{7}$$

Property 2. Assuming an itemset X, the measure of support fulfils the anti-monotony property, i.e.,

$$Sup_{\mathcal{EDB}}(X) \leq Sup_{\mathcal{EDB}}(X \cup m_{ij}). \tag{8}$$

Proof. Assuming an evidential database $\mathcal{EDB}$, let us consider two evidential itemsets X and $X \cup m_{ij}$. We aim at proving this relation $Sup_{\mathcal{EDB}}(X) \leq Sup_{\mathcal{EDB}}(X \cup m_{ij})$:

$$Sup_{T_i}(X \cup m_{ij}) = \prod_{m_{i'j'} \in X \cup m_{ij}} Sup_{T_i}(m_{i'j'})$$

$$Sup_{T_i}(X \cup m_{ij}) = \prod_{m_{i'j'} \in X} Sup_{T_i}(m_{i'j'}) \times Sup_{T_i}(m_{ij})$$

$$Sup_{T_i}(X \cup m_{ij}) \leq Sup_{T_i}(X) \text{ since } Sup_{T_i}(m_{ij}) \in [0, 1] \text{ then}$$

$$Sup_{\mathcal{EDB}}(X \cup m_{ij}) \leq Sup_{\mathcal{EDB}}(X).$$

Example 2. Assuming the evidential database given in Example 1. For a *minsup* $= 0.7$, the pattern $\{m_{11}, m_{12}\}$ have a support of $\frac{C(m_{11}, m_{11}) \times C(m_{12}, m_{12}) + C(m_{11}, m_{21}) \times C(m_{12}, m_{22})}{2} = 0.765$ and is then considered as frequent. Semantically, having a relatively good opinion on treatment 1 (i.e. m_{11}) and hesitant one regarding the treatment 2 (i.e. m_{12}) is redundant over 76.5% of asked practitioners. Moreover, comparatively to patterns of an evidential data mining algorithm, our output is more informative. In fact, a classical algorithm would provide the frequent pattern $\{Good_1, Good_2\}$ which contain less details than $\{m_{11}, m_{12}\}$.

In this section, we develop a new level-wise algorithm to mine opinions over evidential databases. OpMiner, shown in Algorithm 1 generates all BBAs of size one by favouring the most specific ones. Formally, for all $m_{ij}, m_{i'j} \in \mathcal{M}^\Theta$, we retain m_{ij} as long as $m_{ij} \sqsubseteq_{pl} m_{i'j}$. Thus, function $candidate\_gen$ reduces the set of frequent patterns to the set of the more specific ones. The other less specific BBAs are used to compute the support as described in Eq. 7. In addition, this selection aims at reducing time computing since candidate generation and support computing depends on the set of items (i.e. pattern with a single BBA). The patterns that have a support lower than the $minsup$ are pruned in line 5. The process stops until no candidate is left.

4   Experiments: Data Reliability Assessment Using Biomedical Expert Opinion

The investigation of the effects of muscles morphology and mechanics on motion, and the risks of injury, has been at the core of many studies, sometimes with conflicting results. Often different measurement methods have been used, making comparison of the results and drawing sound conclusions impossible [15]. In this section, we aim at studying the opinion of several experts on collected measurement data. To do so, we collected data by a systematic review process of 20 data sources (papers) from reliable search engines (PubMed and ScienceDirect). Data is described over 7 parameters regarding muscle morphology, mechanics and motion analysis. Four main questions were asked to experts about measuring technique (Q_1), experimental protocol (Q_2), number of samples (Q_3) and range of values (Q_4). An expert opinion database was built from an international panel of 20 contacted experts with different expertise (medical imaging, motion analysis). Five evaluation degrees were possible $\{Very\ high, High, Moderate, Low, Very\ low\}$. Each given degree was associated to a confidence value. In this study, as a first goal, we aim at finding frequent opinions in the database. Frequent opinions show correlation between opinions. The second goal is to evaluate the reliability of sources. To do so, the algorithm selects from the frequent set of patterns those that express a positive opinion regarding the same source.

Table 3 shows a small set of recorded answers from experts. For instance, row 1 details the opinions of expert 1 regarding the source S1 (data measures retrieved from a source). The expert expresses his opinion over 4 questions. The column $Conf_i$ shows the confidence of the expert regarding his given opinion for the question Q_i.

The evidential database is constructed by using the evaluation of the experts and their confidences. First, the evaluation of the expert is used to model a

Algorithm 1. OpMiner algorithm

Require: $\mathcal{EDB}, minsup, \mathcal{EDB}_{pl}, maxlen$

Ensure: $\mathcal{EIFF}$

1: $\mathcal{EIFF}, Items \leftarrow \emptyset, size \leftarrow 1$

2: $Items \leftarrow$ CANDIDATE_GEN($\mathcal{EDB}, \mathcal{EIFF}, Items$)

3: **While** ($candidate \neq \emptyset$ and $size \leq maxlen$)

4: **for all** $pat \in candidate$ **do**

5:      **if** SUPPORT($pat, minsup, \mathcal{EDB}_{pl}, Size\_\mathcal{EDB}$)$\geq minsup$ **then**

6:          $\mathcal{EIFF} \leftarrow \mathcal{EIFF} \cup pat$

7: $size \leftarrow size + 1$

8: $candidate \leftarrow$ CANDIDATE_GEN($\mathcal{EDB}, \mathcal{EIFF}, Items$)

9: **End While**

10: **function** SUPPORT($pat, minsup, \mathcal{EDB}_{pl}, d$)

11:      $Sup \leftarrow 0$

12:      **for** i=1 to d **do**

13:          **for all** $pl_{ij} \in \mathcal{M}_i$ **do**

14:              $pl \leftarrow mtopl(pat)\backslash\backslash$ computes the plausibility out of a BBA

15:              **if** $pl_{ij} \geq pl$ **then**

16:                  $Sup_{Trans} \leftarrow Sup_{Trans} \times 1 - ||pl_{ij} - pl||$

17:          $Sup \leftarrow Sup + Sup_{Trans}$

18:      **return** $\frac{Sup_I}{d}$

19: **function** CANDIDATE_GEN($\mathcal{EDB}, \mathcal{EIFF}, Items$)

20:      **if** $size(Items) = 0$ **then**

21:          **for all** $BBA \in \mathcal{EDB}$ **do**

22:              **while** $Items \neq \emptyset$ and $BBA \not\sqsubseteq_{pl} it$ **do**

23:                  **if** $Items = \emptyset$ **then**

24:                      $Add(BBA, Item)$

25:                  **else**

26:                      $Replace(BBA, it, Item)$

27:          **return** $Items$

28:      **else**

29:          **for all** $BBA \in \mathcal{EIFF}$ **do**

30:              **for all** $it \in Items$ **do**

31:                  **if** $!same\_attribute(it, BBA)$ **then**

32:                      $Cand \leftarrow Cand \cup \{BBA \cup it\}$

33:          **return** $Cand$

---

certain BBA[2]. Then, the confidence is used to integrate uncertainty into the BBA. To do so, the confidence is used as reliability measure and part of the mass initially given to the evaluation is then transferred to the ignorance mass. Formally, the discounting of a mass function m can be written as follows

$$\begin{cases} m^\alpha(B) = (1 - \alpha) \times m(B) & \forall B \subseteq \Theta \\ m^\alpha(\Theta) = (1 - \alpha) \times m(\Theta) + \alpha. \end{cases} \tag{9}$$

---

[2] A BBA is called a certain BBA when it has one focal element, which is a singleton. It is representative of perfect knowledge and the absolute certainty.

Table 3. Sample of the expert opinion data.

Expert	S1										Very high	Very high confidence
	Q_1	$Conf_1$	Q_2	$Conf_2$	Q_3	$Conf_3$	Q_4	$Conf_4$				
1	Hig	Hig	Hig	Hig	Mo	Hig	Hig	Mo			High	High confidence
2	Hig	Ver	Mo	Ver	Hig	Ver	Mo	Ver				
3	Hig	Hig	Hig	Hig	Hig	Hig	Hig	Hig			Moderate	Moderate confidence
4	Hig	Hig	Mo	Hig	Hig	Hig	Mo	Hig				
5	Lo	Ver	Lo	Ver	Mo	Ver	Mo	Ver			Low	Low confidence
6	Mo	Mo	Mo	Mo	Lo	Hig	Lo	Hig				
7	Mo	Ver	Mo	Ver	Hig	Ver	Mo	Ver			Very low	Very low confidence
8	Mo	Ver	Lo	Hig	Hig	Ver	Lo	Ver				
9	Mo	Ver	Mo	Hig	Hig	Ver	Mo	Hig				
10	Mo	Hig	Mo	Hig	Mo	Hig	Mo	Hig				
11	Ver	Ver	Ver	Ver	Ver	Ver	Ver	Ver				

α is the reliability factor and is in the set $\{0.8, 0.6, 0.4, 0.2, 0\}$. The higher α is the more mass is transferred to $m(\Theta)$.

In the following, we compare a classical evidential pattern mining approaches such as EDMA [16] and U-Apriori [6] with the output of OpMiner. To do so, we compare these three algorithms in terms of number of extracted patterns and computational time. Figure 1 illustrates the number of extracted patterns with regards to the threshold *minsup*. It is evident that the pattern mining approach EDMA finds the highest number of patterns for all fixed *minsup* comparatively to probabilistic approach approach and OpMiner. In fact, EDMA computes frequent patterns from a set of 28×2^5 items (i.e. sum of the size of all superset of attributes). Therefore, EDMA extracts more patterns than the probabilistic U-Apriori that mines from a set of 28×5 items (i.e. sum of the size of all frames of discernment). OpMiner is has a different approach since an item is a BBA and therefore the number of items is the number of BBAs in the database (i.e., 28×11). In addition, this number is reduced by selecting, at first, only the more committed BBAs. As a result OpMiner is more efficient than the two other approaches since it generates less candidates (see Fig. 2). OpMiner not only generates less frequent patterns but more informative ones since it regroups several information in a single item. Even if in our application, all treated BBAs are *simple*[3], OpMiner works perfectly on *normal* BBAs[4].

In order to test the quality of the patterns, we oppose the best pattern of EDMA relatively to the first four attributes shown in Table 4 to the best one provided by OpMiner. In fact, it is possible to select from the set of frequent patterns those having items of the four attributes. These patterns show the answer (opinion for OpMiner) that the majority of the experts have expressed. These patterns are representative of the quality of source $S1$ measures. As it is show in

---

[3] A BBA is said to be simple if it has at most two focal sets and, if it has two, Θ is one of those.

[4] A BBA is said to be normal if $\emptyset$ is not a focal set.

Frequent patterns   Time (s)

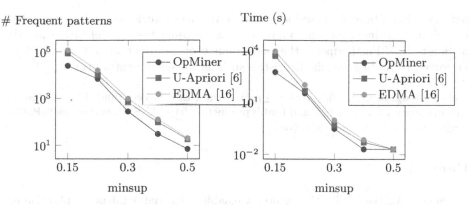

Fig. 1. Number of retrieved frequent patterns from the database.

Fig. 2. Number of retrieved valid association rules from the database.

Table 4, the construction of both patterns is not the same. EDMA's pattern is constructed from focal elements in contrary of OpMiner that contains BBAs. In addition, the interpretation of both patterns is different. EDMA's pattern shows a hesitation between *high* and *moderate* as an answer trend. Therefore, from this point, making an evaluation of source S1 is not straightforward. OpMiner pattern has a different meaning. It gives for each asked question the most shared opinion (i.e. BBA). It means that, for question 1, 2 and 4 the answer is *moderate* with *high* or *very high* confidence. For question 4, the trend is a high evaluation with a *very high* confidence. As a result, with an overall moderate evaluation of its measure, it is possible to conclude that source S1 is moderately reliable.

Table 4. EDMA's pattern vs. OpMiner's pattern

	EDMA S1 best pattern	OpMiner S1 best pattern
Pattern	{Q1 = Hig or Mod, Q2 = Hig or Mod, Q3 = Hig or Mod, Q4 = Hig or Mod}	$\{m_1(Mo_1) = 1, \begin{cases} m_2(Mo_2) = 0.8 \\ m_2(\Theta_2) = 0.2 \end{cases}$ $m_3(Hig_3) = 1, \begin{cases} m_4(Mo_4) = 0.8 \\ m_4(\Theta_4) = 0.2 \end{cases} \}$

5 Conclusion

In this paper, we introduced a new approach for mining opinion patterns from uncertain database. The uncertainty and the imprecision of the data are modelled with the evidence theory. The extraction is based on new anti-monotonic measures of support derived from the commitment relation. A mining algorithm OpMiner is then applied to retrieve frequent opinions patterns a from the database. The results on a real-world database shows more informative extracted patterns than literature-based approaches. In future work, we will be interested in refining the inclusion and support measure using the specialization matrix of Smets [17]. Furthermore, the performance of OpMiner algorithm could be improved by adding specific heuristics such as the decremental pruning [7].

Acknowledgements. This work is a part of the PEPS project funded by the French national agency for medicines and health products safety (ANSM), and of the SePaDec project funded by Region Bretagne.

References

1. Samet, A., Dao, T.T.: Mining over a reliable evidential database: application on amphiphilic chemical database. In: Proceeding of 14th International Conference on Machine Learning and Applications, Miami, Florida, pp. 1257–1262 (2015)
2. Aggarwal, C.C., Han, J. (eds.): Frequent Pattern Mining. Springer, Cham (2014)
3. Agrawal, R., Srikant, R.: Fast algorithm for mining association rules. In: Proceedings of International Conference on Very Large DataBases, VLDB, Santiago de Chile, Chile, pp. 487–499 (1994)
4. Aggarwal, C.C., Li, Y., Wang, J., Wang, J.: Frequent pattern mining with uncertain data. In: Proceedings of the 15th ACM SIGKDD International Conference on Knowledge Discovery and Data Mining, Paris, France, pp. 29–38 (2009)
5. Bell, D.A., Guan, J., Lee, S.K.: Generalized union and project operations for pooling uncertain and imprecise information. Data Knowl. Eng. **18**(2), 89–117 (1996)
6. Chui, C.K., Kao, B., Hung, E.: Mining frequent itemsets from uncertain data. In: Proceedings of the 11th Pacific-Asia Conference on Advances in Knowledge Discovery and Data Mining, Nanjing, China, pp. 47–58 (2007)
7. Aggarwal, C.C.: Managing and Mining Uncertain Data, vol. 3. Springer, New York (2010)
8. Hewawasam, K.R., Premaratne, K., Shyu, M.L.: Rule mining and classification in a situation assessment application: a belief-theoretic approach for handling data imperfections. IEEE Trans. Syst. Man Cybern. Part B **37**(6), 1446–1459 (2007)
9. Chen, Y., Weng, C.: Mining association rules from imprecise ordinal data. Fuzzy Set Syst. **159**(4), 460–474 (2008)
10. Bach Tobji, M.A., Ben Yaghlane, B., Mellouli, K.: Incremental maintenance of frequent itemsets in evidential databases. In: Proceedings of the 10th European Conference on Symbolic and Quantitative Approaches to Reasoning with Uncertainty, Verona, Italy, pp. 457–468 (2009)
11. Dempster, A.: Upper and lower probabilities induced by multivalued mapping. AMS **38**, 325–339 (1967)
12. Shafer, G.: A Mathematical Theory of Evidence. Princeton University Press, Princeton (1976)
13. Smets, P., Kennes, R.: The transferable belief model. Artif. Intell. **66**(2), 191–234 (1994)
14. Dubois, D., Prade, H.: The principle of minimum specificity as a basis for evidential reasoning. In: International Conference on Information Processing and Management of Uncertainty in Knowledge-Based Systems, Paris, France, pp. 75–84 (1986)

15. Hoang, T.N., Dao, T.T., Ho Ba Tho, M.C.: Clustering of children with cerebral palsy with prior biomechanical knowledge fused from multiple data sources. In: Proceedings of 5th International Symposium Integrated Uncertainty in Knowledge Modelling and Decision Making, Da Nang, Vietnam, pp. 359–370 (2016)

16. Samet, A., Lefèvre, E., Yahia, S.B.: Evidential data mining: precise support and confidence. J. Intell. Inf. Syst. **47**(1), 135–163 (2016)

17. Smets, P.: The application of the matrix calculus to belief functions. Int. J. Approximate Reasoning **31**(1–2), 1–30 (2002)

Solving Trajectory Optimization Problems by Influence Diagrams

Jiří Vomlel[1,2]($\boxtimes$) and Václav Kratochvíl[1,2]

[1] Institute of Information Theory and Automation, Czech Academy of Sciences,
Pod Vodárenskou Věží 4, 182 08 Prague 8, Czechia
`vomlel@utia.cas.cz`
[2] Faculty of Management, University of Economics,
Prague Jarošovská 1117/II, 377 01 Jindřichův Hradec, Czechia

Abstract. Influence diagrams are decision-theoretic extensions of Bayesian networks. In this paper we show how influence diagrams can be used to solve trajectory optimization problems. These problems are traditionally solved by methods of optimal control theory but influence diagrams offer an alternative that brings benefits over the traditional approaches. We describe how a trajectory optimization problem can be represented as an influence diagram. We illustrate our approach on two well-known trajectory optimization problems – the Brachistochrone Problem and the Goddard Problem. We present results of numerical experiments on these two problems, compare influence diagrams with optimal control methods, and discuss the benefits of influence diagrams.

Keywords: Influence diagrams · Probabilistic graphical models · Optimal control theory · Brachistochrone problem · Goddard problem

1 Introduction

Influence diagrams (IDs) were originally proposed by Howard and Matheson (1981). They extend Bayesian network models (Pearl 1988) by utility and decision nodes. They can be used to solve optimal decision problems. For a detailed introduction to influence diagrams, see, for example, Jensen (2001). In the regular influence diagrams proposed in 80's it was required that (a) a total ordering of decision nodes that specifies the order in which the decisions are made must be specified and (b) the total utility is the sum of utility values of all utility nodes in the influence diagram. In this paper we will see that both requirements are naturally satisfied for many trajectory optimization problems.

IDs have been applied to diverse decision problems. Kratochvíl and Vomlel (2016) applied IDs to the speed profile optimization problem. The experiments performed on a real problem – the speed control of a Formula 1 race car – revealed that IDs can provide a good solution of the problem very quickly and that this

---

This work was supported by the Czech Science Foundation through projects 16-12010S (V. Kratochvíl) and 17-08182S (J. Vomlel).

A. Antonucci et al. (Eds.): ECSQARU 2017, LNAI 10369, pp. 146–155, 2017.
DOI: 10.1007/978-3-319-61581-3_14

solution can be used as an initial solution for the methods of the optimal control theory, which significantly improves the convergence of these methods.

In this paper we will generalize the approach presented in (Kratochvíl and Vomlel 2016). In Sect. 2 we will describe how IDs can be used to solve trajectory optimization problems. We consider not only problems where the goal is to find the trajectory but also problems where we want to optimize certain criteria (e.g., the total time, the fuel consumption) for a given trajectory. We use the suggested approach to solve two well-known optimal control problems: the Brachistochrone problem in Sect. 3 and the Goddard problem in Sect. 4. We conclude the paper by a discussion in Sect. 5.

2   Influence Diagrams for Trajectory Optimization

In this section we will describe how one can use IDs to solve a trajectory optimization problem. Next we describe general guidelines for the construction of an ID for the trajectory optimization. We will illustrate this construction on two problems in the next sections.

1. Specify the state variables, the control variables, and the utility function.
2. Describe the system dynamics using a system of ordinary differential equations (ODEs). Often, the system dynamics is described with respect to time. In the trajectory optimization it is often more convenient or even necessary to rewrite the ODEs with respect to the trajectory.
3. Discretize the trajectory to short segments.
4. Find the analytical formula for the state transitions as a function of previous values of the state and control variables for one segment. If the analytical solution is not available use approximations by an appropriate method – candidates are, for example, the Euler method, a Runge-Kutta method, or an implicit method as the Gauss-Legendre method.
5. If necessary, discretize the state and control variables.
6. If states are discretized and the state transitions lead to states that are not in the set of state values then use the stochastic approximation of the state transition by a mixture of two nearest states whose probability is proportional to their closeness to the computed state value, so that the expected value (conditioned on previous state and control values) is equal to the computed state value, see (Kratochvíl and Vomlel 2016, Sect. 5.2).
7. Construct the ID with state variables as chance nodes, the control variables as decision nodes, and a utility node for each segment.
8. Specify the state transitions using conditional probability tables (CPTs).
9. Find and store the optimal policy for each state and control configuration and for each segment of the trajectory by solving the ID.
10. During the application in a real control problem the optimal policy for the actual observed values of state variables at each point of the trajectory is used. In practice, the controlled object often deviates from the optimal solution. The reason can be a measuring and control imprecision, a bias, unexpected interventions, etc. Therefore, it is very useful to have the optimal policy stored for all configurations of the parents of all decision nodes.

The additive utility requirement means that the utility function decomposes additively along the segments of the trajectory. Such utility functions are common in practice. This condition is satisfied, for example, by total time or by the total fuel consumption. The requirement of the total ordering of the decisions is also natural for trajectory optimization problems. The decisions at coordinates closer to the origin are taken before those more distant ones. Another natural total ordering is the ordering by time elapsed from the beginning.

3   Brachistochrone Problem

Formulated by Johan Bernoulli in 1696, the Brachistochrone Problem is: given two points find a curve connecting them such that a mass point moving along the curve under the gravity reaches the second point in minimum time. We will consider this problem as an optimal control problem (Bertsekas 2000, Example 3.4.2). The state variable is the vertical coordinate y and it is a function of the horizontal coordinate x. The variable u controls the derivative of y:

$$\frac{dy(x)}{dx} = u(x). \tag{1}$$

It is assumed that the initial speed at the origin is zero. Speed v is defined by the law of energy conservation – kinetic energy equals to the change of gravitational potential energy, which results in

$$v = \sqrt{-2 \cdot g \cdot y}. \tag{2}$$

For an infinitesimal segment of length dx with an infinitesimal change dy of the vertical position y we can write for the speed, which is the derivative of the position s with respect to time t:

$$v = \frac{ds}{dt} = \sqrt{\left(\frac{dy}{dt}\right)^2 + \left(\frac{dx}{dt}\right)^2} = \left(\sqrt{1 + \left(\frac{dy}{dx}\right)^2}\right)\frac{dx}{dt}. \tag{3}$$

By substituting (1) and (2) to (3) we get

$$dt = \frac{ds}{v} = \left(\frac{1}{\sqrt{-2 \cdot g \cdot y}} \cdot \sqrt{1 + u^2}\right) dx \stackrel{df}{=} q\,(y, u)\,dx. \tag{4}$$

The solution of the Brachistochrone problem is a function $y = f(x)$ that minimizes the total time T necessary to get from the point $(0,0)$ to the point (a, b), where $a > 0$ and $b < 0$.

3.1   Influence Diagram for the Brachistochrone Problem

Next, we will illustrate how an ID can be used to find an arbitrary precise solution of the problem. We will discretize the problem and use the following symbols:

the number of discrete intervals n, the distance discretization step $\Delta x = \dfrac{a}{n}$, the index of the interval $i = 0, 1, \ldots, n$, x-coordinate $x_i = i \cdot \Delta x$, y-coordinate y_i, the speed v_i, and time to get from x_{i-1} to x_i denoted as t_i. The control value u_i defines the vertical shift: $y_{i+1} = y_i + u_i$. In each segment we will assume that the path is a line segment[1], i.e. for $x \in [x_i, x_{i+1}]$ and for $y \in [y_i, y_{i+1}]$ it holds

$$y = \frac{u_i}{\Delta x} \cdot x + y_i. \tag{5}$$

By substituting (5) to (4) and solving the definite integral

$$\int_{x_i}^{x_{i+1}} q\left(y, u\right) dx = \int_{x_i}^{x_{i+1}} q\left(\frac{u_i}{\Delta x} \cdot x + y_i, u_i\right) dx \tag{6}$$

we get the formula for the time spent at the segment $[x_i, x_{i+1}]$:

$$t_{i+1} = \begin{cases} \dfrac{\Delta x}{\sqrt{-2 \cdot g \cdot y_i}} & \text{if } u_i = 0 \\ -\sqrt{\dfrac{2}{g}} \cdot \left(\dfrac{(\Delta x)^2 + u_i^2}{u_i}\right) \cdot \left(\sqrt{-y_i} - \sqrt{-u_i - y_i}\right) & \text{otherwise.} \end{cases} \tag{7}$$

The goal is to find a control strategy $u = (u_0, \ldots, u_{n-1})$, $u_i \in \mathbb{R}$, $i = 0, 1 \ldots, n-1$ so that we get from the initial point (x_0, y_0) to the terminal point (x_n, y_n) minimizing the total time $\sum_{i=1}^{n} t_i$ and satisfying the state constraints $y_i \leq y_0$ for $i = 1, \ldots, n$.

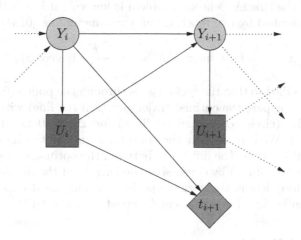

Fig. 1. One segment of the ID for the Brachistochrone Problem

---

[1] This is an approximation only, but the smaller distance discretization step the smaller the approximation error.

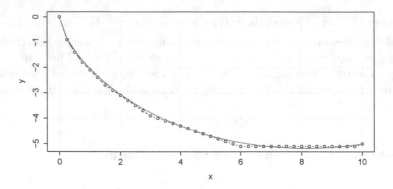

Fig. 2. Comparison of the optimal solution with the ID solution.

The structure of a segment of the ID for the discrete version of the Brachistochrone Problem is presented in Fig. 1. The state transition CPT is deterministic and defined as:

$$P(Y_{i+1} = y_{i+1} | U_i = u_i, Y_i = y_i) = \begin{cases} 1 & \text{if } y_{i+1} = y_i + u_i \\ 0 & \text{otherwise.} \end{cases} \tag{8}$$

The utility function for node t_{i+1} is defined by formula (7).

3.2   Experimental Results

The solution of the Brachistochrone Problem is known – it is a part of a cycloid, which can be specified by two functions of a parameter $\varphi \in [0, M]$, $M \leq 2\pi$:

$$x = \frac{K}{2}(\varphi + \sin \varphi) + L, \quad y = -\frac{K}{2}(1 - \cos \varphi). \tag{9}$$

K, L, M are specified so that the cycloid goes through the points $(0,0)$ and (a, b).

In Fig. 2 we compare the optimal trajectory (full red line) with the solution found by the ID (circles connected by lines) for $\Delta x = 0.25$, $\Delta y = 0.1$ and $(a, b) = (10, -5)$. We can see that the solution found by the ID approximates well the optimal solution. The difference between the optimal trajectory and the ID solution can be reduced by decreasing the lengths of the discretization steps Δx and Δy. More details about the experiments and the R code used for the experiments can be found in our research report (Vomlel 2017).

4   Goddard Problem

Formulated by Robert H. Goddard, see (Goddard 1919), the problem is to establish the optimal thrust profile for a rocket ascending vertically from the Earth's surface to achieve a given altitude with a given speed and pay load and with the minimum fuel expenditure. The aerodynamic drag and the gravitation vary with the altitude. We assume a bounded thrust. The problem has become a benchmark in the optimal control theory due to a characteristic singular arc behavior in connection with a relatively simple model structure.

In this paper we consider the normalized Goddard Problem. For the derivation of the normalized version and more details about the approximation methods see our research report (Vomlel and Kratochvíl 2017). We specify the Goddard Problem as an optimal control problem. The movement of the rocket is described by ordinary differential equations (ODEs). We describe the system dynamics with respect to the altitude h measured as the distance from the Earth's center. The rocket's mass m is composed from the pay load and the fuel, the latter is burnt during the rocket ascent. The speed is denoted by v.

The control variable u controls the engine thrust, which is the derivative of the rocket's mass m with respect to time t multiplied by the jet speed c, i.e.,

$$u = c \cdot \frac{dm}{dt} \;=\; c \cdot \frac{dm}{dh} \cdot \frac{dh}{dt} \;=\; \frac{dm}{dh} \cdot c \cdot v. \tag{10}$$

The derivatives of mass m and speed v with respect to h are defined using functions of g and f as it follows:

$$\frac{dm}{dh} = g(u, v) \stackrel{df}{=} \frac{u}{c \cdot v} \tag{11}$$

$$\frac{dv}{dh} = f(h, m, u, v) \stackrel{df}{=} -\frac{u}{m \cdot v} - \frac{v \cdot s \cdot c_D}{2 \cdot m} \cdot \rho_0 \cdot \exp\left(\beta \cdot (1 - h)\right) - \frac{1}{v \cdot h^2}, \tag{12}$$

where s is the cross-section area of the rocket, c_D is the drag constant, ρ_0 is the density of the air at the Earth's surface, and β is a dimensionless constant.

We will use model parameter values presented in (Tsiotras and Kelley 1991) and (Seywald and Cliff 1992). The aerodynamic data and the rocket's parameters originate from (Zlatskiy and Kiforenko 1983) and correspond roughly to the Soviet SA-2 surface-to-air missile, NATO code-named Guideline. The control will be restricted to $u \in [-3.5, 0]$. It is assumed that the rocket is initially at rest at the surface of the Earth and that its fuel mass is 40% of the rocket total mass. The nondimensionalized values of these constants and the initial and terminal values are:

$$\beta = 500, \; s \cdot \rho_0 = 12400, \; c_D = 0.05, \; c = 0.5,$$
$$h_0 = 1, \; h_T = 1.01, \; m_0 = 1, \; m_T \geq 0.6 \cdot m_0 = 0.6, \; v_0 = 0.$$

4.1   The Influence Diagram for the Goddard Problem

In each segment i of the ID there are (a) two state variables – a speed variable V_i and a mass variable M_i, (b) one decision variable U_i controlling the thrust of the rocket engine, (c) one utility node f_i representing the fuel consumption in the segment. The structure of one segment of the ID for the discrete version of the Goddard Problem is presented in Fig. 3.

We discretize the trajectory to segments of length Δh with a constant control. In each segment a solution of the system of two ODEs (11) and (12) is found by an

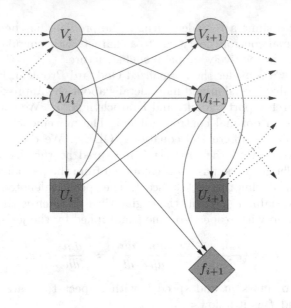

Fig. 3. One segment of the ID for the Goddard Problem

ODE approximation method. The solution provides values of the mass $m(h+\Delta h)$ and speed $v(h+\Delta h)$ at the end of the segment. ODE approximation methods can be used, e.g. the Euler, Runge–Kutta, and Gauss–Legendre methods. See (Vomlel and Kratochvíl 2017) for a derivation of these methods for the Goddard Problem. The computed mass and speed values will not lay in the discrete set of values of these variables. Therefore we will approximate the state transformations by non-deterministic CPTs $P(V_{i+1}|U_i, V_i, M_i)$ and $P(M_{i+1}|U_i, V_i, M_i)$ as it is described in (Kratochvíl and Vomlel 2016, Sect. 5.2).

4.2   Experimental Results

In Fig. 4 we compare the control, speed, and mass profiles of the optimal solution found by Bocop[2] (Team Commands, Inria Saclay 2016) with solutions found by IDs with different discretizations and different approximation methods. It is known (Miele 1963) that the optimal solution consists of three sub-arcs: (a) a maximum-thrust sub-arc, (b) a variable-thrust sub-arc, and (c) a coasting sub-arc, i.e., a sub-arc with the zero thrust.

    For the solutions found by IDs we use the following name schema v.u.m.M.h composed from the parameters used in the experiments:

– v ... the number of states of the speed variables,

---

[2] Bocop package implements a local optimization method. The optimal control problem is approximated by a nonlinear programming (NLP) problem using a time discretization. The NLP problem is solved by Ipopt, using sparse exact derivatives computed by ADOL-C.

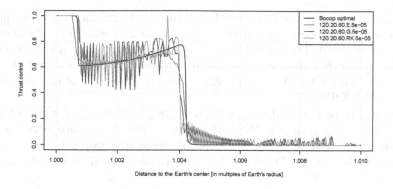

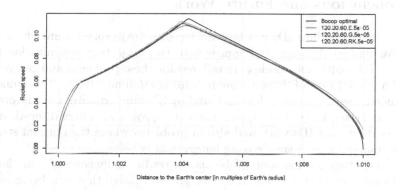

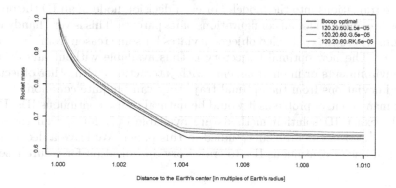

Fig. 4. Comparisons of the optimal solution with ID solutions.

- u ... the number of states of the control variables,
- m ... the number of states of the mass variables,
- M ... the discretization method for solving ODEs (E is the Euler method, G the Gauss–Legendre method, and RK the Runge–Kutta RK4 method), and
- h ... the length of the trajectory segment.

By looking at Fig. 4 we can conclude that the Euler method best approximates the optimal control and it suffers from smaller oscillations of the control than other methods. The control strategy found by Runge-Kutta and the Gauss-Legendre methods have larger oscillations. However, the speed and the mass profiles are similar for all methods and they are close to the optimal profiles found by BOCOP[3].

The quality of the solution is influenced by the number of states of speed, control, and mass variables. We had to find a proper balance between these parameters to avoid large oscillations. This issue deserves a further study to allow the application of IDs to problems where no optimal solution is known.

5   Conclusions and Future Work

We have described how IDs can be used to solve trajectory optimization problems. We applied the suggested approach to two trajectory optimization problems. The ID solution methods were tailored for these problems and can be considered a special case of dynamic programming (Bellman 1957). The numerical experiments reveal that the solutions found by IDs approximates well the optimal solution and the quality of the approximation improves with finer discretizations. It is important that IDs work well also in problems where the optimal strategy is more complex than a simple bang-bang strategy[4].

The trajectory optimization problems are traditionally solved by methods of the optimal control theory but IDs offer an alternative that can bring several benefits over the traditional approaches. IDs can incorporate uncertainty about the state transitions into the model. For each decision node in an ID the optimal decision is computed for all configurations of its parents. This is very handy in the situations where the controlled objects deviates for some reason from the optimal trajectory. The new optimal trajectory is thus available without any delay. In noisy environments or in environments with interactions with other objects the imposed deviations from the optimal trajectory can be quite common.

For many control problems it would be natural to use continuous IDs. Unfortunately, exact ID solution methods are available only for special cases that cannot be used in the problems studied in this paper. We leave a deeper study of applications of continuous IDs to trajectory optimization for future research.

---

[3] The initial mass of the rocket is the same for all methods but in the third plot of Fig. 4 we can see that the terminal mass slightly differs. The lowest fuel consumption is observed in case of RK method but we should conclude from that the RK method is optimal but rather that it has the largest approximation error.

[4] A bang-bang strategy is a strategy that consists of extreme values only, e.g. it consists of the full thrust and the zero thrust phases only. Bang-bang strategies are optimal solutions of a wide class of optimal control problems.

References

Bellman, R.: Dynamic Programming. Princeton University Press, Princeton (1957)

Bertsekas, D.P.: Dynamic Programming and Optimal Control, 2nd edn. Athena Scientific, Belmont (2000)

Goddard, R.H.: A method for reaching extreme altitudes, volume 71(2). Smithsonian Miscellaneous Collections (1919)

Howard, R.A., Matheson, J.E.: Influence diagrams. In: Howard, R.A., Matheson, J.E. (eds.) Readings on the Principles and Applications of Decision Analysis, vol. II, pp. 721–762. Strategic Decisions Group (1981)

Jensen, F.: Bayesian Networks and Decision Graphs. Springer, New York (2001)

Kratochvíl, V., Vomlel, J.: Influence diagrams for speed profile optimization. Int. J. Approximate Reasoning. (2016, in press). http://dx.doi.org/10.1016/j.ijar.2016.11.018

Miele, A.: A survey of the problem of optimizing flight paths of aircraft and missiles. In: Bellman, R. (ed.) Mathematical Optimization Techniques, pp. 3–32. University of California Press (1963)

Pearl, J.: Probabilistic Reasoning in Intelligent Systems: Networks of Plausible Inference. Morgan Kaufmann series in representation and reasoning. Morgan Kaufmann, Burlington (1988)

Seywald, H., Cliff, E.M.: Goddard problem in presence of a dynamic pressure limit. J. Guidance Control Dyn. **16**(4), 776–781 (1992)

Team Commands, Inria Saclay: BOCOP: an open source toolbox for optimal control (2016). http://bocop.org

Tsiotras, P., Kelley, H.J.: Drag-law effects in the goddard problem. Automatica **27**(3), 481–490 (1991)

Vomlel, J.: Solving the Brachistochrone Problem by an influence diagram. Technical report 1702.02032 (2017). http://arxiv.org/abs/1702.02032

Vomlel, J., Kratochvíl, V.: Solving the Goddard Problem by an influence diagram. Technical report 1703.06321 (2017). http://arxiv.org/abs/1703.06321

Zlatskiy, V.T., Kiforenko, B.N.: Computation of optimal trajectories with singular-control sections. Vychislitel'naia i Prikladnaia Matematika **49**, 101–108 (1983)

Belief Functions

Iterative Aggregation of Crowdsourced Tasks Within the Belief Function Theory

Lina Abassi[✉] and Imen Boukhris

LARODEC, Institut Supérieur de Gestion de Tunis,
Université de Tunis, Tunis, Tunisia
lina.abassi@gmail.com, imen.boukhris@hotmail.com

Abstract. With the growing of crowdsourcing services, gathering training data for supervised machine learning has become cheaper and faster than engaging experts. However, the quality of the crowd-generated labels remains an open issue. This is basically due to the wide ranging expertise levels of the participants in the labeling process. In this paper, we present an iterative approach of label aggregation based on the belief function theory that simultanously estimates labels, the reliability of participants and difficulty of each task. Our empirical evaluation demonstrate the efficiency of our method as it gives better quality labels.

Keywords: Aggregation · Crowd · Expectation-Maximization · Belief function theory · Expertise

1 Introduction

Recently, crowdsourcing has attracted an increasing interest as an effective and fast way for collecting training data for supervised machine learning methods. It emerged from the need to label a big amount of data at a low cost. Crowdsourcing systems such as the Amazon Mechanical Turk (AMT) provide a platform where requesters post tasks to be executed by human workers in exchange of few cents. Nevertheless, these advantages come at a cost of a lower results quality. In fact, workers are not always reliable and mistakes can occur even with ones who make real efforts.

As a result, several strategies arised in order to handle the label quality problem, among them, redundancy [11] which consists in assigning a task to more than one worker and then aggregating the results. In fact, the main objective of label aggregation is to find the unknown ground truth of a set of tasks. This objective is hardly reachable since workers are of different levels of expertise leading to high uncertainty in answers, likewise for tasks which difficulty can strongly influence how worker's expertise is determined.

In this paper, we provide a new solution to aggregate imperfect labels produced by multiple workers under the belief function theory (BFT). The BFT [1,2] is known to be a rich framework that deals effectively with imperfect information in addition of providing tools to both integrate sources confidence and

© Springer International Publishing AG 2017
A. Antonucci et al. (Eds.): ECSQARU 2017, LNAI 10369, pp. 159–168, 2017.
DOI: 10.1007/978-3-319-61581-3_15

combine information induced from them. The proposed method generates the true label, the worker reliability and the task's difficulty simultanously. These latter are all unknown a priori.

The remainder of the paper is structured as follows: We first give an overview the related work in Sect. 2 then we present the belief function theory fundamentals in Sect. 3. Section 4 introduces our proposed approach. We finally discuss experimental results in Sect. 5 and conclude with a summary and future work in Sect. 6.

2   Related Work

It has been shown that gathering multiple labels even if they are noisy [8] and then aggregate them according to the simplest manner such as majority decision (MD) is a better way than just having one label. Yet, a lot of work has attempted to improve even more the aggregation results since (MD) assumes that all workers have the same expertise. In [17], a benchmark is proposed comparing different methods and classifying them into non iterative and iterative approaches. The first class includes methods that benefit from some a priori ground truth also called gold data. The Expert Label Injected Crowd Estimation (ELICE) [10] uses gold data to estimate parameters such as worker's expertise and question difficulty to then integrate them in the aggregation process. As for the Honeypot (HP) [5], the gold data is used to filter workers that could not correctly label them and after that labels are aggregated according to (MD). In a previous work [15], the Belief Label Aggregation (BLA) estimates worker's expertise assuming that (MD) is the ground truth and then aggregates labels using BFT tools.

The second class of approaches covers EM-based methods that operate when no ground truth is available. The Expectation-Maximization (EM) algorithm used in the Dawid and Skene method (DS) [9] is an iterative technique that consists of two phases: The E step, and the M step. In the E step, answers are estimated given workers' error-rates. In the M step, the workers' expertise are re-estimated since the missing labels are known from the E step. Many methods resort to the EM algorithm such as the Supervised Learning from Multiple Experts (SLME) [7] that can be used only for binary labels and represents the worker's expertise by the sensitivity and specificity measures. Also the Generative model of Labels, Abilities and Difficulties (GLAD) [13] generates aside worker's expertise, another parameter which is the question difficulty. In [8], a bayesian approach is employed to estimate the error-rates and final labels. Another recent work, [16] suggests an EM-based method computing the aggregated label and worker's expertise and proposes to distinguish a positive expertise on positive labels and a negative expertise on negative labels and to incorporate them to the label aggregation step.

In this paper, we propose a model inspired by the EM method that iteratively estimates labels, worker's expertise and task difficulty. Unlike most related works, it benefits from the BFT power to cope with uncertainty generalizing the probabilistic and possibilistic frameworks and distinguishing between equiprobability and total ignorance. To improve quality results when aggregating labels, our idea is to adapt this theory to one of the most used optimization method namely the EM algorithm.

3   Belief Function Theory: Fundamentals

The theory of belief function [1,2] is among the most used theories for representing and reasoning with uncertainty. It is considered as a flexible and rich framework for dealing with imperfect information. Several interpretations have emerged including the Transferable Belief Model (TBM) [12].

3.1   Representation of Information

Suppose that Ω is a finite set of elementary, non empty and mutually exclusive events applied to a given problem, we call it the frame of discernment. All possible values that each subset of Ω can take is the power set of Ω denoted by 2^{Ω} defined as $2^{\Omega} = \{ E : E \subseteq \Omega \}$.

The impact of a piece of evidence on the whole subsets of Ω is represented by the basic belief assignement (bba). A bba is a function $m^{\Omega} : 2^{\Omega} \to [0, 1]$ that satisfies: $\sum_{E \subseteq \Omega} m^{\Omega}(E) = 1$.

Each subset E of Ω having a strictly positive mass $m^{\Omega}(E) > 0$ is referred as the focal element of the bba.

In order to express particular situations related to uncertainty, Shafer [1] proposed some special bbas:

- **Vacuous** bba defined by: $m(\Omega) = 1$ and $m(E) = 0$ for $E \neq \Omega$. It represents the total ignorance.
- **Categorical** bba has a unique focal element E.
- **Certain** bba is a categorical bbas except that its focal element is a singleton.
- **Simple support function** is a bba with at most one focal element other than Ω.

3.2   Discounting Information

It is possible to quantify the reliability of a source inducing degrees of support. Thus, when dealing with bba, degrees of reliability of its source has to be taken into account. Discounting [1] consists in weighting bba by a discount rate $\alpha \in [0, 1]$ with $(1 - \alpha)$ is the reliability of the source. The discounted mass function is given by:

$$\begin{cases} m^{\alpha}(E) = (1 - \alpha) \cdot m(E), & \forall E \subset \Omega, \\ m^{\alpha}(\Omega) = (1 - \alpha) \cdot m(\Omega) + \alpha. \end{cases} \tag{1}$$

- A source is fully reliable if $\alpha = 0$ accordingly $m^{\alpha}(E) = m(E)$
- A source is fully unreliable if $\alpha = 1$ leading to a vacuous bba: $m^{\alpha}(\Omega) = 1$

3.3 Combination of Information

There is a great number of combination rules proposed in the framework of belief function. They are intended to fuse a set of *bbas* into only one *bba* in order to simplify decision making. In what follows we present those related to our work. Let s_1 and s_2 be two distinct and cognitively independent reliable sources providing two different *bbas* m_1 and m_2 defined on the same frame of discernment Ω.

(1) Conjunctive rule of combination

This rule is introduced by Smets [6], a mass can be allocated to the empty set interpreted as the non exhaustivity of the frame of discernment. Therefore, the conjunctive rule of combination noted by ⓝ and defined as:

$$m_1 ⓝ m_2(E) = \sum_{F \cap G = E} m_1(F)m_2(G) \tag{2}$$

(2) Dempster's rule of combination

The rule was proposed by Dempster [2] but unlike the conjunctive rule, it generates a normalized *bba* (i.e. $m(\varnothing) = 0$), denoted by $\oplus$ and defined as:

$$m_1 \oplus m_2(G) = \begin{cases} \dfrac{m_1 ⓝ m_2(G)}{1 - m_1 ⓝ m_2(\varnothing)} & \text{if } E \neq \varnothing, \forall G \subseteq \Omega \\ 0 & \text{otherwise.} \end{cases} \tag{3}$$

(3) The Combination with Adapted Conflict

The Combination With Adapted Conflict (CWAC) rule [4] denoted by ⓗ is an adaptive weighting between the conjunctive and Dempster's rules. It actually behaves as the first rule if *bbas* are contradictory and as the second in the opposite case. It is defined as follows:

$$m_ⓗ(E) = (ⓗ m_i)(E) = D m_ⓝ(E) + (1 - D)m_\oplus(E) \tag{4}$$

with $\mathrm{D} = \max [d(m_i, m_j)]$ is the maximum distance between m_i and m_j. This distance is the Jousselme distance [3], a measure of dissimilarity between *bbas* that ensures the adaptation between the two combination rules. It is defined as follows:

$$d(m_1, m_2) = \sqrt{\frac{1}{2}(m_1 - m_2)^t \mathrm{D}(m_1 - m_2)}, \tag{5}$$

where D is the Jaccard index defined by:

$$\mathrm{D}(E, F) = \begin{cases} 0 & \text{if } E = F = \varnothing, \\ \dfrac{|E \cap F|}{|E \cup F|} & \forall E, F \in 2^\Omega. \end{cases} \tag{6}$$

3.4 Decision Process

To make a decision, Smets [12] proposes to convert beliefs to a probability function called the pignistic probability, denoted $BetP$. The pignistic transformation is presented as follows:

$$BetP(\omega_i) = \sum_{E \subseteq \Omega} \frac{|E \cap \omega_i|}{|E|} \cdot \frac{m(E)}{(1 - m(\varnothing))} \quad \forall \omega_i \in \Omega \tag{7}$$

4   I-BLA: Iterative Belief Label Aggregation

In this section, we describe our approach the Iterative Belief Label Aggregation. It simultaneously infers worker's expertise, task difficulty and aggregated labels. The calculation of aggregated labels and parameters will go through a sequence of EM iterations till convergence. It leads to the improvement of the estimation since labels and the parameters are mutually boosted at each iteration. Therefore, the aggregated label and parameters jointly depend on each other.

A set of X labelers gives answers on Y tasks. A labeler j receives a task i and contributes a label $l_{ij} \in \{0, 1, -1\}$. The (-1) label means that the worker skipped the task. L_i is the final aggregated label belonging to $(0, 1)$ for simplicity. Each worker's expertise is modelled by $(1 - \alpha_j)$ with $\alpha_j \in [0, 1]$ is his error-rate whereas a task difficulty is denoted by $\beta_i \in [0, 1]$.

Algorithm 1. I-BLA Algorithm

Input: All labels l_{ij} for i $\in [1, Y]$ and j $\in [1, X]$
Output: Estimation of α_j, β_i and the final aggregated label L_i
 1: Pre-processing: Labels representation by *bbas*
 2: Initialize worker's expertise with $\alpha_j = 0$ (considering all workers are equally reliable)
 3: **repeat**
 4:    E step: Compute the aggregated labels L_i for all tasks taking into account α_j
 5:    M step: Compute all α_j and all β_i
 6: **until** convergence
 7: **return** α_j, β_i and L_i for each worker and each task

---

We describe the Iterative Belief Label Aggregation in Algorithm 1. It is inspired by the EM algorithm and induces jointly the worker's expertise, the task difficulty and the final aggregated label. These parameters are mutually boosted. First, We initialize α_j to 0 assuming that all workers have equal expertise (like the majority decision method). Then, as a pre-processing step, *bbas* are generated for all labels. The algorithm iterates two steps until it reaches convergence: E step computes the aggregated labels integrating the workers' expertise and tasks difficulty, and the M step updates both the workers' expertise as measures of the aggreement between workers labels and aggregated label as well as the tasks difficulty according to the conflict degree induced when aggregating labels.

We detail these three steps in-depth in what follows:

4.1   Pre-processing

This pre-processing step consists in a *bba* generation where all labels are transformed into mass functions under the belief function theory. As a result, each label is changed into a *bba* m_{ij}^{Ω} with $\Omega = \{\omega_1, \ldots, \omega_n\}$. In our case, we are dealing with binary labelling therfore $\Omega = \{0, 1\}$.

In this work, three cases are possible:

– If $l_{ij} = 1$ then $m_{ij}(\{1\}) = 1$,
– If $l_{ij} = 0$ then $m_{ij}(\{0\}) = 1$,
– If $l_{ij} = -1$ then $m_{ij}(\Omega) = 1$ reflecting label ignorance.

4.2   E Step: Label Estimation

The Expectation (E) step consists in estimating all tasks true labels depending on two parameters namely worker's expertise and task difficulty. First, generated *bbas* are discounted by each worker's expertise $(1 - \alpha_j)$ (with α_j is the error-rate) with the discounting operation (Eq. 1) depending on task difficulty (β_i):

$$\begin{cases} \text{if } \beta_i = 1 \ and \ \alpha_j > 0.8 \ then \ \alpha_j = 1, \\ \text{if } \beta_i = 0 \ and \ \alpha_j < 0.2 \ then \ \alpha_j = 0, \\ \text{else } \alpha_j = \alpha_j. \end{cases} \qquad (8)$$

The first case takes away the worker's label (with a *bba* transformed into a vacuous *bba*) since that he has a low expertise $(\alpha_j > 0.8)$ and the task is hard $(\beta_i = 1)$ so we assume that he is most likely to give a wrong answer.

The second case reinforces the worker's label (keeping his initial certain *bba*) seen that he has a high expertise $(\alpha_j < 0.2)$ and the task is easy $(\beta_i = 0)$ so we assume that he is most likely to give a correct answer.

For the other cases, *bbas* are discounted and hence transformed into simple support functions.

We note that the error-rate (α_j) threshold values are fixed under the assumption that workers with more than 80% of bad answers are considered unskilled or even spammers, and those with less than 20% of bad answers must be experts or futur experts.

After the discounting step, all *bbas* are aggregated applying the combination with adapted conflict (CWAC) rule 4 in order to get one *bba* such as $\bigodot m_i = m_{i1}^{\alpha} \bigodot m_{i2}^{\alpha} \bigodot \ldots m_{ij}^{\alpha}$. This rule generates a conflict degree which is the mass on the empty set of the aggregated *bba* denoted by $c_i \in [0, 1]$.

When the number of mass functions are high, using the conjunctive rule leads to a conflict hitting very high values due to its absorbing power. So unlike the conjunctive rule, the CWAC induces a reasonable conflict degree that actually reflects how much the *bbas* are contradictory.

As a final step, decision about the possible label is made through the one that has the higher pignistic probability *(BetP)* (Eq. 7).

Example 1. Supposing that the aggregating results for a task is the following *bba*:

$$m_i(\varnothing) = 0.25, \ m_i(\{0\}) = 0.75$$

We obtain the following pignistic probability:
BetP$(\{0\}) = 1 \cdot (0.75/(1 - 0.25)) = 1$,
BetP$(\{1\}) = 0$
As a result, the decision is $L_i = 0$.

4.3 M Step: Parameters Estimation (α, β)

During the Maximization (M) step, the worker's expertise and task difficulty are estimated depending on the generated labels in the E step. In fact, the worker's expertise is induced from the comparison of the worker labels and the generated labels by calculating the error-rate α_j in $[0, 1]$ as follows:

$$\alpha_j = \frac{Number\_of\_incorrect\_labels}{Number\_of\_labeled\_tasks} \tag{9}$$

As for the second parameter, namely the task difficulty β in $\{0, 1\}$, it is induced from the conflict degree c_i. Indeed, according to [4] when c_i is above 0.8 then *bbas* are heteregenous and thus conflictual. Therefore, we conclude that:

– If $c_i > 0.8$ then the task i is considered hard ($\beta_i = 1$) as answers are very contradictory,
– Else the task i is considered easy ($\beta_i = 0$).

5 Experimentation

Datasets. Experiments are conducted on three real datasets namely the duchenne dataset [13], the event temporal ordering (Temp) dataset [14] and the recognizing textual entailment (RTE) dataset [14] covering various tasks of different difficulty, collected employing different numbers of workers. In Table 1, we present a description of these datasets.

Table 1. Datasets details

Dataset	Workers	Task	Labels	Proportion of labels($\neq (-1)$)
Duchenne	17	159	1221	0.45
Temp	76	462	4620	0.13
RTE	164	800	8000	0.06

5.1   Results Evaluation

In this section we validate our proposed method (I-BLA) by comparing it to a non-iterative and iterative baseline methods namely Majority Decision (MD) and the Dawid-Skene (DS) and the Belief Label Aggregation (BLA) which is also based on the belief function theory to aggregate labels.

The comparison is done considering accuracy which measures the proportions of correctly estimated labels.

Results of our (I-BLA), (MD), (DS) and (BLA) methods as functions of the number of workers are shown in Figs. 1, 2 and 3.

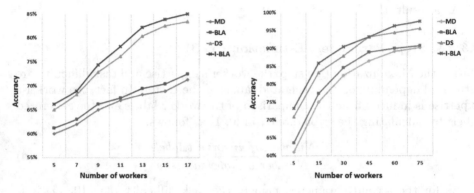

Fig. 1. Accuracies as function of workers' number for Duchenne

Fig. 2. Accuracies as function of workers' number for Temp

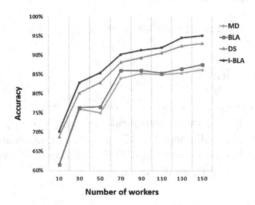

Fig. 3. Accuracies as function of workers' number for RTE

According to the plots, the accuracy of all techniques increases with the increase of the number of workers. In general, the iterative methods (I-BLA) and (DS) are the best performers when the number of workers is high. This is due to the fact that workers' labels are together justified through iterations. Yet, (I-BLA) is taking the lead and overpassing the (DS) and non-iterative techniques. Indeed, in Fig. 2, the leading method (I-BLA) accuracy goes from 72% when employing 5 labelers up to 97% at 75 labelers gaining about 2% performance over (DS) method.

In Table 2, the average accuracies of these figures are reported. For the three datasets, our method accomplish the best performance according to accuracy overpassing state-of-art methods. It records at best an improvement over 20% compared to (MD) and over 17% compared to (BLA) for the duchenne dataset. Moreover, it achieves at least an average of 2% compared to the (DS) method for all datasets.

Table 2. Accuracy of (MD), (BLA), (DS) and (I-BLA) for different datasets

Dataset	(MD)	(BLA)	(DS)	(I-BLA)
Duchenne	63%	66%	81%	83,4%
RTE	79%	81,3%	85,2%	88%
Temp	80,9%	82%	88%	90,1%

6 Conclusion and Future Works

In this paper, we proposed a new approach (I-BLA) for the aggregation of crowdsourced labels within the belief function theory. This approach is based on EM concept that iteratively computes the aggregated labels, worker's expertise and task difficulty all depending on each other. We demonstrated the efficiency of our method through extensive experimentation on three datasets.

References

1. Shafer, G.: A Mathematical Theory of Evidence, vol. 1. Princeton University Press, Princeton (1976)
2. Dempster, A.P.: Upper and lower probabilities induced by a multivalued mapping. Ann. Math. Stat. **38**, 325–339 (1967)
3. Jousselme, A.-L., Grenier, D., Bossé, É.: A new distance between two bodies of evidence. Inf. Fusion **2**, 91–101 (2001)
4. Lefèvre, E., Elouedi, Z.: How to preserve the conflict as an alarm in the combination of belief functions? Decis. Support Syst. **56**, 326–333 (2013)
5. Lee, K., Caverlee, J., Webb, S.: The social honeypot project: protecting online communities from spammers. In: International World Wide Web Conference, pp. 1139–1140 (2010)
6. Smets, P.: The combination of evidence in the transferable belief model. IEEE Trans. Pattern Anal. Mach. Intell. **12**(5), 447–458 (1990)

7. Raykar, V.C., Yu, S., Zhao, L.H., Jerebko, A., Florin, C., Valadez, G.H., Bogoni, L., Moy, L.: Supervised learning from multiple experts: whom to trust when everyone lies a bit. In: Proceedings of the 26th Annual International Conference on Machine Learning, pp. 889–896 (2009)
8. Raykar, V.C., Yu, S., Zhao, L.H., Valadez, G.H., Florin, C., Bogoni, L., Moy, L.: Learning from crowds. J. Mach. Learn. Res. **11**, 1297–1322 (2010)
9. Dawid, A.P., Skene, A.M.: Maximum likelihood estimation of observer error-rates using the EM algorithm. Appl. Stat. **28**, 20–28 (2010)
10. Khattak, F.K., Salleb, A.: Quality control of crowd labeling through expert evaluation. In: The Neural Information Processing Systems 2nd Workshop on Computational Social Science and the Wisdom of Crowds, pp. 27–29 (2011)
11. Sheng, V.S., Provost, F., Ipeirotis, P.G.: Get another label? improving data quality and data mining using multiple, noisy labelers. In: Proceedings of the 14th ACM SIGKDD International Conference on Knowledge Discovery and Data Mining, pp. 614–622 (2008)
12. Smets, P., Mamdani, A., Dubois, D., Prade, H.: Non Standard Logics for Automated Reasoning, pp. 253–286. Academic Press, London (1988)
13. Whitehill, J., Wu, T., Bergsma, J., Movellan, J.R., Ruvolo, P.L.: Whose vote should count more: optimal integration of labels from labelers of unknown expertise. In: Neural Information Processing Systems, pp. 2035–2043 (2009)
14. Snow, R., et al.: Cheap and fast but is it good? Evaluation non-expert annotations for natural language tasks. In: The Conference on Empirical Methods in Natural Languages Processing, pp. 254–263 (2008)
15. Abassi, L., Boukhris, I.: Crowd label aggregation under a belief function framework. In: Lehner, F., Fteimi, N. (eds.) KSEM 2016. LNCS, vol. 9983, pp. 185–196. Springer, Cham (2016). doi:10.1007/978-3-319-47650-6_15
16. Georgescu, M., Zhu, X.: Aggregation of crowdsourced labels based on worker history. In: Proceedings of the 4th International Conference on Web Intelligence, Mining and Semantics, pp. 1–11 (2014)
17. Quoc Viet Hung, N., Tam, N.T., Tran, L.N., Aberer, K.: An evaluation of aggregation techniques in crowdsourcing. In: Lin, X., Manolopoulos, Y., Srivastava, D., Huang, G. (eds.) WISE 2013. LNCS, vol. 8181, pp. 1–15. Springer, Heidelberg (2013). doi:10.1007/978-3-642-41154-0_1

A Clustering Approach for Collaborative Filtering Under the Belief Function Framework

Raoua Abdelkhalek$^{(\boxtimes)}$, Imen Boukhris, and Zied Elouedi

LARODEC, Institut Supérieur de Gestion de Tunis,
Université de Tunis, Tunis, Tunisia
abdelkhalek_raoua@live.fr, imen.boukhris@hotmail.com, zied.elouedi@gmx.fr

Abstract. Collaborative Filtering (CF) is one of the most successful approaches in Recommender Systems (RS). It exploits the ratings of similar users or similar items in order to predict the users' preferences. To do so, clustering CF approaches have been proposed to group items or users into different clusters. However, most of the existing approaches do not consider the impact of uncertainty involved during the clusters assignments. To tackle this issue, we propose in this paper a clustering approach for CF under the belief function theory. In our approach, we involve the Evidential C-Means to group the most similar items into different clusters and the predictions are then performed. Our approach tends to take into account the different memberships of the items clusters while maintaining a good scalability and recommendation performance. A comparative evaluation on a real world data set shows that the proposed method outperforms the previous evidential collaborative filtering.

Keywords: Collaborative filtering · Belief function theory · Clustering · Evidential C-Means

1 Introduction

During the last few years, Recommender Systems (RS) [1] have attracted considerable attention from several research communities and have reached a high level of popularity. The diversity of the information sources and the variety of domain applications gave birth to various recommendation approaches. According to the literature, CF is considered to be the most popular and the widely used approach in this area [1–3]. In order to provide recommendations, CF tends to predict the users' preferences based on the users or the items sharing similar ratings. To do so, this latter exploits the user-item matrix and computes the similarities between users (user-based [4]) or items (item-based [5]) in the system. Based on the computed similarities, the prediction process is then performed. CF has achieved widespread success in both academia and industry [2]. Despite its simplicity and efficiency, CF approach exhibits some limitations such as the scalability problems [6]. Actually, CF needs to search the whole user-item space in order to compute similarities. This computation increases with the number of items and users leading to poor scalability performance. To overcome the problem mentioned above, several recommendation approaches have

© Springer International Publishing AG 2017
A. Antonucci et al. (Eds.): ECSQARU 2017, LNAI 10369, pp. 169–178, 2017.
DOI: 10.1007/978-3-319-61581-3_16

been proposed using different model-based techniques such as Bayesian network [7], Singular Value Decomposition (SVD) [8] and clustering techniques [6,9,10]. The common point of these approaches is to forecast pre-trained models using an item-user matrix. For instance, in clustering CF approaches, items can be assigned to clusters based on their historical ratings and recommendations are performed accordingly. However, an item may potentially belong to more than only one cluster. This concept is referred to as soft clustering. This imprecision may impact the relationship between the items and therefore the final prediction. Indeed, we show in a previous work [11] the relevance of handling uncertainty in CF throughout the prediction process. In this paper, we treat uncertainty involved in the clustering CF approaches where we consider the cluster membership of each item to be uncertain. To this end, we opt for the belief function theory (BFT) [12–14] which offers a rich representation about all situations ranging from complete knowledge to complete ignorance. Several clustering methods have been proposed under this theory. For example, the belief K-modes (BKM) has been proposed by [15] to deal with uncertainty in the attribute values. On the other hand, the Evidential C-Means (ECM) [16] has been conceived to handle uncertainty for objects' assignment. Since we are in particular interested in assessing the uncertainty in items cluster membership, we involve the Evidential C-Means method which is based on the concept of credal partition. Taking advantage of the BFT in particular the ECM technique, we propose an evidential clustering CF. The new approach allows us to assign the items to soft clusters whilst handling challenges imposed from the CF framework.

This paper is organized as follows: Sect. 2 recalls the basic concepts of the belief function theory and the Evidential C-Means. Section 3 presents briefly some related works on clustering CF as well as CF under the belief function framework. Our proposed recommendation approach is presented in Sect. 4. Section 5 exposes the experimental results conducted on a real world data set. Finally, the paper is concluded and some future works are depicted in Sect. 6.

2   Clustering in a Belief Function Framework

The BFT [12–14] represents a flexible and rich framework for reasoning under uncertainty. In this section, we provide an overview about its basic concepts and we recall the Evidential C-Means [16] as a clustering method under an uncertain framework.

2.1   Belief Function Theory

In the BFT, a problem domain is represented by the frame of discernment Θ. The belief committed to each element of Θ is expressed by a basic belief assignment (bba) which is a mapping function $m : 2^{\Theta} \rightarrow [0, 1]$ such that: $\sum_{A \subseteq \Theta} m(A) = 1$

Each mass $m(A)$ quantifies the degree of belief exactly assigned to the event A of Θ. The subsets A of Θ such as $m(A) > 0$ are called focal elements.

To make decisions, beliefs can be represented by pignistic probabilities defined as:

$$BetP(A) = \sum_{B \subseteq \Theta} \frac{|A \cap B|}{|B|} \frac{m(B)}{(1 - m(\varnothing))} \quad for\ all\ A \in \Theta \qquad (1)$$

2.2 Evidential C-Means

The Evidential C-Means (ECM) [16] is a clustering technique based on the concept of credal partition. Given an object i, this method determines the mass m_{ij} representing partial knowledge regarding the cluster membership to any subset A_j of $\Theta = \{\omega_1, \omega_2, \ldots, \omega_n\}$ where n is the number of clusters. Every partition is represented by a center $v_k \in \mathbb{R}^p$ where p is the dimension of data. Each subset A_j of Θ is represented by the barycenter v_j of the centers v_k associated to the clusters composing A_j. The barycenter is computed as follows:

$$v_j = \frac{1}{c_j} \sum_{k=1}^{c} s_{kj} v_k \qquad (2)$$

where $c_j = |A_j|$ denotes the cardinal of A_j and s_{kj} is defined as follows:

$$s_{kj} = \begin{cases} 1 & if\ \omega_k \in A_j \\ 0 & otherwise \end{cases} \qquad (3)$$

The distance between an object i and any subset A_j of Θ is defined by:

$$d_{ij} = \|x_i - v_j\| \qquad (4)$$

Finally, the credal partition is determined by minimizing the following objective function:

$$J_{ECM} = \sum_{i=1}^{n} \sum_{\{j/A_j \neq \emptyset, A_j \subseteq \Theta\}} c_j^\alpha m_{ij}^\beta d_{ij}^2 + \sum_{i=1}^{n} \delta^2 m_{i\emptyset}^\beta \qquad (5)$$

α, β and δ are the input parameters such that $\alpha \geq 0$ is a weighting exponent for cardinality. $\beta > 1$ is a weighting exponent controlling the hardness of the partition and δ represents the distance between all instances and the empty set. More details about parameters and credal partition process can be found in [16].

3 Related Work on Collaborative Filtering

CF has shown a great applicability in a wide variety of domains [2]. The key idea is that if two users rated some items similarly or had similar behaviors in the past then, they would rate or act on other items similarly. CF approaches are divided into two categories namely, memory-based and model-based. Memory-based CF approaches exploit the whole user-item matrix to find similar users or items and generate recommendations accordingly. In contrast, model-based algorithms rely on the ratings matrix to infer a model which is then applied for predictions. The model building process can be performed using different methods. For example, Bayesian networks have been used in [7] for CF process. Clustering CF approaches that are based on a cluster model to reduce the time complexity have also been proposed [6,9,10,17]. In [6,17], authors have proposed a clustering approach for CF that classifies the users in different groups and neighborhood has been selected for each cluster. In [9], the users have been clustered from the views of both rating patterns and social trust relationships. Similarly, a CF approach has been implemented in [10] based on user's preferences clustering. All the clustering techniques mentioned above focus on user-based CF. In our work, we consider only item-based CF where items are clustered into groups rather than users. It is obvious that developing RSs that can quickly produce high quality recommendations have become more and more required in this area [6]. On the other hand, considering uncertainty during the recommendation process can be argued to be another important challenge in real-world problems [18]. The belief function theory [12–14] is among the most widely used ones for dealing with uncertainty. Recent studies have investigated the benefits of the adoption of such theory in RSs area. In fact, authors in [19] have represented the user's preferences through the BFT tools and integrate context information for predicting all unprovided ratings. Another approach developed in [20] relies on this theory to represent both user's preferences and community preferences extracted from social networks. The authors in [11] have proposed an evidential item-based CF where they considered the similar items as different pieces of evidence. They computed the similarities between the target item and the whole items in the system and the final prediction was an aggregation of the ratings corresponding to the similar items. However, a lot of heavy computations are needed in this case. This problem is referred to as the scalability problem which we tackle in our proposed recommendation approach.

4   Evidential Clustering Approach for CF

In this section, we represent our evidential CF method based on items clustering. Figure 1 gives the overall flow of the proposed recommendation approach.

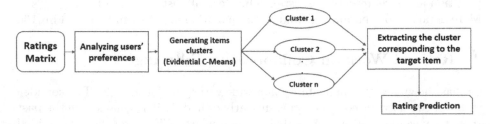

Fig. 1. A new clustering CF approach under the belief function theory

4.1 Items Clustering

Clustering is a crucial step in our approach since the predictions are then performed accordingly. The heart of this approach is to use the efficient soft clustering method, ECM [16] in order to provide a credal partition of the items. Hence, we allocate, for each item in the ratings matrix a mass of belief not only to single clusters, but also to any subsets of the frame of discernment Θ. Before performing the clustering process, we exploit the rating matrix and we randomly initialize the cluster centers commonly referred to as prototypes. Then, we compute the Euclidean distance between the items and the non empty subsets of Θ. We obtain the final credal partition when the objective function (Eq. 5) is minimized.

Example 1. *Let us consider the user-item matrix illustrated in Table 1.*

Table 1. User-item matrix

	Movie$_1$	Movie$_2$	Movie$_3$	Movie$_4$	Movie$_5$
User$_1$	3	?	4	1	2
User$_2$	4	4	2	?	?
User$_3$	3	2	4	3	2
User$_4$	?	1	5	2	3
User$_5$	5	2	0	2	5

Suppose that the number of clusters c = 3, the clustering process consists of providing a credal partition for the 5 movies. In other words, each movie in the system may belong to not only singleton clusters but also to disjunctions of clusters as represented in Table 2.

Table 2. The credal partition corresponding to the five movies

Movies	$\emptyset$	$\{C_1\}$	$\{C_2\}$	$\{C_1, C_2\}$	$\{C_3\}$	$\{C_1, C_3\}$	$\{C_2, C_3\}$	Θ
Movie$_1$	0.0025	0.9682	0.009	0.0078	0.0046	0.0043	0.0018	0.0017
Movie$_2$	0.0468	0.2946	0.2715	0.1106	0.1135	0.0731	0.0516	0.0382
Movie$_3$	0.0005	0.0010	0.0018	0.0004	0.9934	0.0009	0.0017	0.0004
Movie$_4$	0.0062	0.0212	0.8856	0.0174	0.0247	0.0107	0.0246	0.0097
Movie$_5$	0.0366	0.1484	0.4931	0.0909	0.0947	0.0479	0.0556	0.0327

4.2 Clusters Selection

In order to make a final decision about the cluster of the current item, we compute the pignistic probability $BetPi(C_k)$ (Eq. 1) induced by each *bba*. These values are interpreted as the degree of membership of the item i to cluster k. Finally, a hard partition can be easily obtained by assigning each object to the cluster with the highest pignistic probability.

Example 2. *Based on the credal partition derived in the first step, the bba's can be transformed into pignistic probablities in order to select the corresponding cluster having the highest value as shown in Table 3.*

Table 3. The pignistic probabilities corresponding to the five movies

Movies	C_1	C_2	C_3	Selected cluster
$Movie_1$	0.9773	0.0144	0.0083	C_1
$Movie_2$	0.4188	0.3833	0.1979	C_1
$Movie_3$	0.0017	0.0029	0.9953	C_3
$Movie_4$	0.0387	0.9155	0.0458	C_2
$Movie_5$	0.2374	0.5992	0.1633	C_2

4.3 Ratings Prediction

The selected clusters are used to obtain knowledge about the items that should be considered in the rating prediction. In order to perform the prediction task, only the items belonging to the same cluster as the target item are extracted. The predicted rating consists of the average of the ratings corresponding to the same clusters members. Given a target item, the prediction is performed as follows:

$$\widehat{R}_{u,i} = \frac{\sum_{j \in C_i(u)} R_{uj}}{|C_i(u)|} \tag{6}$$

where $C_i(u)$ is the set of items belonging to the cluster of the target item i and that have been rated by the user u. R_{uj} is the rating given by user u to item j. $|C_i(u)|$ is the number of items in cluster C_i which have been rated by user u.

Example 3. *For instance, to predict the rating $\widehat{R}_{1,2}$ given by $User_1$ to $Movie_2$, we simply average the ratings of the items belonging to the same cluster and that have been rated by $User_1$. In our case, only $Movie_1 \in C_1$. Then $\widehat{R}_{1,2} = \frac{3}{1} = 3$.*

5 Experimental Evaluation

In order to evaluate our proposal, we test our approach using a real world data set which is widely used in CF and publicly available on the MovieLens[1] website. It contains 100.000 ratings collected from 943 users in 1682 movies.

---

[1] http://movielens.org.

We conducted our experiments by following the experimental protocol suggested by [7]. The movies rated by the 943 users are ranked according to the number of the ratings given by the users. Rating matrix do not have enough data for accurate predictions, which is known as sparsity. The experimentation strategy consists on increasing progressively the number of the missing rates leading to different sparsity degrees. Hence, we obtain 10 different subsets containing a specific number of ratings provided by the 943 users for 20 different movies. For each subset, we randomly extract 20% of the available ratings as a testing data and the remaining 80% were considered as a training data.

5.1   Evaluation Metrics

We assume that involving an evidential clustering approach for CF may lead to a better performance over the predicted ratings as well the consuming time.

Prediction and Recommendation

In order to assess the prediction accuracy and to evaluate the quality of recommendations provided to the active user, we opt for two evaluation metrics commonly used in CF: the *Mean Absolute Error* (MAE) which belongs in this case to $[0, 4]$ and the precision belonging to $[0, 1]$ defined by:

$$MAE = \frac{\sum_{u,i} |\widehat{R}_{u,i} - R_{u,i}|}{\|\widehat{R}_{u,i}\|} \tag{7}$$

$$Precision = \frac{IR}{IR + UR} \tag{8}$$

where $R_{u,i}$ is the real rating for the user u on the item i and $\widehat{R}_{u,i}$ is the predicted value. $\|\widehat{R}_{u,i}\|$ is the total number of the predicted ratings over all the users. IR indicates that an interesting item has been correctly recommended while UR indicates that an uninteresting item has been incorrectly recommended. The lower the MAE values are, the more accurate the predictions are. Otherwise, the highest values of the precision indicate a better recommendation quality.

Scalability

We also investigated the performance of our approach in terms of scalablity. We recall that the purpose of scalability refers to the ability of a method to be run quickly by handling the evolution regarding the number of items and users.

5.2   Experimental Results

We performed various experiments over the 10 selected subsets by varying each time the number of clusters c. We used $c = 2$, $c = 3$, $c = 4$ and $c = 5$. For each subset, the results corresponding to the different number of clusters used in the experiments are then averaged. In other words, we compute the MAE and the precision measure for each value of c and we note the overall results. For all our experiments, we used $\alpha = 2$, $\beta = 2$ and $\delta^2 = 10$ as invoked in [16].

Unlike the evidential item-based CF (Evidential IB-CF) [11], the proposed evidential clustering item-based CF (Evidential Clustering IB-CF) relies on items clusters rather than the user-item matrix. Hence, we compare the two CF methods proposed under the BFT in order to evaluate the performance of our approach. Table 4 recapitulates the results of each evidential IB-CF considering different sparsity degrees.

Table 4. Comparison results in terms of MAE and precision

Evaluation metrics	Subsets	Sparsity degrees	Evidential IB-CF	Evidential clustering IB-CF
MAE	$Subset_1$	53%	0.751	0.749
Precision			0.79	0.792
MAE	$Subset_2$	56.83%	0.84	0.8
Precision			0.76	0.74
MAE	$Subset_3$	59.8%	0.761	0.747
Precision			0.77	0.785
MAE	$Subset_4$	62.7%	0.763	0.793
Precision			0.763	0.782
MAE	$Subset_5$	68.72%	0.831	0.845
Precision			0.735	0.752
MAE	$Subset_6$	72.5%	0.851	0.8
Precision			0.735	0.813
MAE	$Subset_7$	75%	0.744	0.733
Precision			0.78	0.805
MAE	$Subset_8$	80.8%	0.718	0.762
Precision			0.778	0.755
MAE	$Subset_9$	87.4%	0.840	0.873
Precision			0.707	0.73
MAE	$Subset_{10}$	95.9%	0.991	0.83
Precision			0.513	0.55
Overall MAE			0.809	**0.793**
Overall Precision			0,733	**0.75**

The proposed approach allows an improvement over the standard evidential item-based CF approach [11] by acquiring, in average the lowest error rates over the 10 subsets (0.793 compared to 0.809) as well as the highest overall precision (0.75 compared to 0.733). While the clustering CF proposed in [6] improves the scalability with a worse prediction quality compared to the traditional one, our evidential clustering CF outperforms the standard evidential CF in both cases.

Scalability Performance

We perform the scalability of our approach by varying the sparsity degree. We compare the results to the standard evidential CF as depicted in Fig. 2.

According to Fig. 2, the elapsed time corresponding to the clustering CF approach is substantially lower than the basic evidential CF. These results are

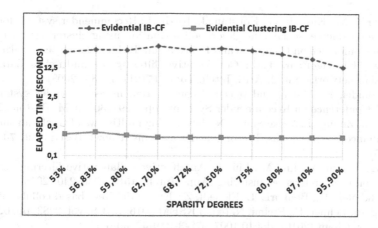

Fig. 2. Elapsed time of evidential clustering CF vs. evidential CF

explained by the fact that standard CF methods need to search the closest neighbors to the target item in the whole item space, which leads to huge computing amount.

6   Conclusion

In this paper, we have proposed a new clustering CF approach based on the Evidential C-Means method. Compared to a recent CF method under the belief function theory, elapsed time has been significantly improved, along with better prediction and recommendation performance. As future work, we intend to rely on the different $bba's$ corresponding to the different clusters rather than the most significant one.

References

1. Bobadilla, J., Ortega, F., Hernando, A., Gutiérrez, A.: Recommender systems survey. Knowl.-Based Syst. **46**, 109–132 (2013)
2. Park, Y., Park, S., Jung, W., Lee, S.G.: Reversed CF: A fast collaborative filtering algorithm using a k-nearest neighbor graph. Expert Syst. Appl. **42**(8), 4022–4028 (2015)
3. Su, X., Khoshgoftaar, T.M.: A survey of collaborative filtering techniques. Adv. Artif. Intell. **2009**, 1–19 (2009)
4. Zhao, Z.D., Shang, M.S.: User-based collaborative-filtering recommendation algorithms on hadoop. In: Third International Conference on Knowledge Discovery and Data Mining, pp. 478–481. IEEE, Phuket (2010)
5. Sarwar, B., Karypis, G., Konstan, J., Riedl, J.: Item-based collaborative filtering recommendation algorithms. In: International Conference on World Wide Web, pp. 285–295. ACM, Hong Kong (2001)

6. Sarwar, B.M., Karypis, G., Konstan, J., Riedl, J.: Recommender systems for large-scale e-commerce: scalable neighborhood formation using clustering. In: International Conference on Computer and Information Technology. IEEE, Dhaka (2002)

7. Su, X., Khoshgoftaar, T.M.: Collaborative filtering for multi-class data using bayesian networks. Int. J. Artif. Intell. Tools **17**(01), 71–85 (2008)

8. Symeonidis, P.: Matrix and tensor decomposition in recommender systems. In: ACM Conference on Recommender Systems, pp. 429–430. ACM, Boston (2016)

9. Guo, G., Zhang, J., Yorke-Smith, N.: Leveraging multiviews of trust and similarity to enhance clustering-based recommender systems. Knowl.-Based Syst. **74**, 14–27 (2015)

10. Zhang, J., Lin, Y., Lin, M., Liu, J.: An effective collaborative filtering algorithm based on user preference clustering. Appl. Intell. **45**(2), 230–240 (2016)

11. Abdelkhalek, R., Boukhris, I., Elouedi, Z.: Evidential item-based collaborative filtering. In: Lehner, F., Fteimi, N. (eds.) KSEM 2016. LNCS, vol. 9983, pp. 628–639. Springer, Cham (2016). doi:10.1007/978-3-319-47650-6_49

12. Dempster, A.P.: A generalization of bayesian inference. J. Roy. Stat. Soc. Series B (Methodological) **30**, 205–247 (1968)

13. Shafer, G.: A Mathematical Theory of Evidence, vol. 1. Princeton University Press, Princeton (1976)

14. Smets, P.: The transferable belief model for quantified belief representation. In: Smets, P. (ed.) Handbook of Defeasible Reasoning and Uncertainty Management Systems, vol. 1, pp. 267–301. Springer, Dordrecht (1998)

15. Hariz, S., Elouedi, Z., Mellouli, K.: Clustering approach using belief function theory. In: Euzenat, J., Domingue, J. (eds.) AIMSA 2006. LNCS, vol. 4183, pp. 162–171. Springer, Heidelberg (2006). doi:10.1007/11861461_18

16. Masson, M.H., Denoeux, T.: ECM: An evidential version of the fuzzy c-means algorithm. Pattern Recogn. **41**(4), 1384–1397 (2008)

17. Xue, G.R., Lin, C., Yang, Q., Xi, W., Zeng, H.J., Yu, Y., Chen, Z.: Scalable collaborative filtering using cluster-based smoothing. In: International ACM SIGIR Conference on Research and Development in Information Retrieval, pp. 114–121. ACM, Salvador (2005)

18. Nguyen, V.-D., Huynh, V.-N.: A community-based collaborative filtering system dealing with sparsity problem and data imperfections. In: Pham, D.-N., Park, S.-B. (eds.) PRICAI 2014. LNCS, vol. 8862, pp. 884–890. Springer, Cham (2014). doi:10.1007/978-3-319-13560-1_74

19. Nguyen, V.-D., Huynh, V.-N.: A reliably weighted collaborative filtering system. In: Destercke, S., Denoeux, T. (eds.) ECSQARU 2015. LNCS, vol. 9161, pp. 429–439. Springer, Cham (2015). doi:10.1007/978-3-319-20807-7_39

20. Nguyen, V.-D., Huynh, V.-N.: Integrating with social network to enhance recommender system based-on dempster-shafer theory. In: Nguyen, H.T.T., Snasel, V. (eds.) CSoNet 2016. LNCS, vol. 9795, pp. 170–181. Springer, Cham (2016). doi:10.1007/978-3-319-42345-6_15

A Generic Framework to Include Belief Functions in Preference Handling for Multi-criteria Decision

Sébastien Destercke(✉)

Sorbonne Université, UMR CNRS 7253 Heudiasyc,
Université de Technologie de Compiègne CS 60319, 60203 Compiègne Cedex, France
sebastien.destercke@hds.utc.fr

Abstract. Modelling the preferences of a decision maker about multi-criteria alternatives usually starts by collecting preference information, then used to fit a model issued from a set of hypothesis (weighted average, CP-net). This can lead to inconsistencies, due to inaccurate information provided by the decision maker or to a poor choice of hypothesis set. We propose to quantify and resolve such inconsistencies, by allowing the decision maker to express her/his certainty about the provided preferential information in the form of belief functions.

1 Introduction

Preference modelling and multi-criteria decision analysis (MCDA) are increasingly used in our everyday lives. Generally speaking, their goal is to help decision makers (DM) to model their preferences about multi-variate alternatives, to then formulate recommendations about unseen alternatives. Such recommendations can take various shapes, but three common problems can be differentiated [1]:

- **the choice problem**, where a (set of) best alternative is recommended to the DM;
- **the ranking problem**, where a ranking of alternatives is presented to the DM;
- **the sorting problem**, where each alternative is assigned to a sorted class.

In this paper, we will be interested in the two first problems, which are closely related since the choice problem roughly consists in presenting only those elements that would be ranked highest in the ranking problem.

One common task, in preference modelling as well as in MCDA, is to collect or elicit preferences of decision makers (DM). This elicitation process can take various forms, that may differ accordingly to the chosen model (Choquet Integral [6], CP-net [4],...). Anyway, in all cases, each piece of collected information then helps to better identify the preference model of the DM. A problem is then to ensure that the information provided by the DM are consistent with the chosen model. Ways to handle this problem is to identify model parameters minimising some error term [6], or to consider a probabilistic model [11].

© Springer International Publishing AG 2017
A. Antonucci et al. (Eds.): ECSQARU 2017, LNAI 10369, pp. 179–189, 2017.
DOI: 10.1007/978-3-319-61581-3_17

Such methods solve inconsistent assessments in principled ways, but most do not consider the initial information to be uncertain. Another problem within preference modelling problems is to choose an adequate family of models, expressive enough to capture the DM preferences, but sufficiently simple to be identified with a reasonable amount of information. While some works compare the expressiveness of different model families, few investigate how to choose a family among a set of possible ones.

In this paper, we propose to model uncertainty in preference information through belief functions, arguing that they can bring interesting answers to both issues (i.e., inconsistency handling and model choice). Indeed, belief functions are adequate models to model uncertainty about non-statistical quantities (in our case the preferences of a DM), and a lot of work have been devoted about how to combine such information and handle the resulting inconsistency. It is not the first work that tries to combine belief functions with MCDA and preference modelling, however existing works on these issues can be split into two main categories:

- those starting from a specific MCDA model and proposing an adaptation to embed belief functions within it [2];
- those starting from belief functions defined on the criteria and proposing preference models based on belief functions and evidence theory, possibly but not necessarily inspired from existing MCDA techniques [3].

The approach investigated and proposed in this paper differs from those in two ways:

- no a priori assumption is made about the kind of model used, as we do not start from an existing method to propose a corresponding extension. This means that the proposal can be applied to various methods;
- when selecting a particular model, we can retrieve the precise version of the model as a particular instance of our approach, meaning that we are consistent with it.

Section 2 describes our framework. We will use weighted average as a simple illustrative example, yet the described method applies in principle to any given set of models. Needed notions of evidence theory are introduced gradually. Section 3 then discusses how the framework of belief functions can be instrumental to deal with the problems we mentioned in this introduction: handling inconsistent assessments of the DM, and choosing a rich enough family of models.

2   The Basic Scheme

We assume that we want to describe preferences over alternatives X issued from a multivariate space $\mathcal{X} = \times_{i=1}^{C} \mathcal{X}^i$ of C criteria $\mathcal{X}^i$. For instance, $\mathcal{X}$ may be the space of hotels, applicants ... and a given criteria $\mathcal{X}^i$ may be the price, age, ... In the examples, we also assume that X^i is within $[0, 10]$, yet the presented scheme can be applied to criteria ranked on ordinal scales, or even on symbolic methods such as CP-net [4].

We will denote by $\mathbb{P}_{\mathcal{X}}$ the set of partial orders defined over $\mathcal{X}$. Recall that a strict partial order P is a binary relation over $\mathcal{X}^2$ that satisfies Irreflexivity (not $P(x, x)$ for any $x \in \mathcal{X}$), Transitivity ($P(x, y)$ and $P(y, z)$ implies $P(x, z)$ for any $(x, y, z) \in \mathcal{X}^3$) and Asymmetry (either $P(x, y)$ or $P(y, x)$, but not both) and where $P(x, y)$ can be read "x is preferred to y", also denoted $x \succ_P y$. When P concerns only a finite set $\mathcal{A} = \{a_1, \ldots, a_n\} \subseteq \mathcal{X}$ of alternatives, convenient ways to represent it are by its associated directed acyclic graph $\mathcal{G}_P = (V, E)$ with $V = \mathcal{A}$ and $(a_i, a_j) \in E$ iff $(a_i, a_j) \in P$, and by its incidence matrix whose elements denoted P_{ij} will be such that $P_{ij} = 1$ iff $(a_i, a_j) \in P$. Given a partial order P and a subset $\mathcal{A}$, we will denote by Max_P the set of its maximal elements, i.e., $Max_P = \{a \in \mathcal{A} : \nexists a' \in \mathcal{A} \text{ s.t. } a' \succ_P a\}$.

2.1 Elementary Information Item

Our approach is based on the following assumptions:

- the decision-maker (DM) provides items of preferential information $\mathcal{I}_i$ together with some certainty degree $\alpha_i \in [0, 1]$ ($\alpha_i = 1$ corresponds to a certain information). $\mathcal{I}_i$ can take various forms: comparison between alternatives of $\mathcal{X}$ ("I prefer menu A to menu B") or between criteria, direct information about the model, ...
- given a selected space $\mathcal{H}$ of possible models, each item $\mathcal{I}_i$ is translated into constraints inducing a subset H_i of possible models consistent with this information.
- Each model $h \in \mathcal{H}$ maps subsets of $\mathcal{X}$ to a partial order $P \in \mathbb{P}_{\mathcal{X}}$. A subset $H \subseteq \mathcal{H}$ maps subsets of $\mathcal{X}$ to the partial order $H(\mathcal{A}) = \cap_{h \in H} h(\mathcal{A})$ with $\mathcal{A} \subseteq \mathcal{X}$.

We model this information as a simple support mass function m_i over $\mathcal{H}$ defined as

$$m_i(H_i) = \alpha_i, \quad m_i(\mathcal{H}) = 1 - \alpha_i. \tag{1}$$

Mass functions are the basic building block of evidence theory. A mass function over space $\mathcal{H}$ is a non-negative mapping from subsets of $\mathcal{H}$ (possibly including the empty-set) to the unit interval summing up to one. That is, $m : \wp(\mathcal{H}) \to [0, 1]$ with $\sum m(E) = 1$ and $\wp(\mathcal{H})$ the power set of $\mathcal{H}$. The mass $m(\emptyset)$ is interpreted here as the amount of conflict in the information. A subset $H \subseteq \mathcal{H}$ such that $m(H) > 0$ is often called a *focal set*, and we will denote by $\mathcal{F} = \{H \subseteq \mathcal{H} : m(H) > 0\}$ the collection of focal sets of m.

Example 1. Consider three criteria $\mathcal{X}^1, \mathcal{X}^2, \mathcal{X}^3$ that are averages of student notes in Physics, Math, French (we will use P, M, F). $\mathcal{X}$ is then the set of students. We also assume that the chosen hypothesis space $\mathcal{H}$ are weighted averages: a model $h \in \mathcal{H}$ is then specified by a positive vector (w_1, w_2, w_3) where $\sum w_i = 1$. A student a_i is evaluated by $a_i = w_1 P + w_2 M + w_3 F$, and an alternative a_i is better than a_j if $a_i > a_j$.

Any subset of models can be summarized by a subset of the space $\mathcal{H} = \{(w_1, w_2) : w_1 + w_2 \leq 1\}$, since the last weight can be inferred from the two firsts. For instance, let us assume that the information item $\mathcal{I}$ is $(0, 8, 5) \succ (8, 4, 5)$, meaning that

$$0w_1 + 8w_2 + 5w_3 > 8w_1 + 4w_2 + 5w_3 \rightarrow w_2 > 2w_1$$

The resulting subspace H of models is then pictured in Fig. 1. The decision maker can then provide some assessment of how certain she/he is about this information by providing a value α. For instance, if the DM is certain to choose a student with grades $(0, 8, 5)$ over one with grades $(8, 4, 5)$, then α should be close to 1. Yet if the DM is quite uncertain about this choice, then α should be closer to 0.

2.2 Combining Elements of Information

In practice, the DM will deliver multiple items of information, that should be combined. If m_1 and m_2 are two mass functions over the space $\mathcal{H}$, then their conjunctive combination in evidence theory is defined as the mass

$$m_{1 \cap 2}(H) = \sum_{H_i \in \mathcal{F}_i, H_1 \cap H_2 = H} m_1(H_1) m_2(H_2), \tag{2}$$

which is applicable if we consider that the provided information items are distinct, a reasonable simplifying assumption in a preference learning setting where the DM usually does not answer a question by consciously thinking about the ones she/he already answered. If we have n masses $m_1, \ldots, m_n$ to combine, corresponding to n information items $\mathcal{I}_1, \ldots, \mathcal{I}_n$, we can iteratively apply Eq. (2), as it is commutative and associative. If each m_i has two focal elements (H_i and $\mathcal{H}$), then the number of focal elements of the combined mass double after each application of (2). This of course limits the number n we can consider, yet in frameworks where individual decision makers are asked about their preferences, this number is often small.

It may happen that the given preferential information items conflict, producing a non-null mass $m(\emptyset) > 0$, meaning that no models in $\mathcal{H}$ satisfies all preferential information items. In evidence theory, two main ways to deal with this situation exist:

Fig. 1. Information item subset

W1 Ignoring the fact that some conflicting information exists and normalise m into m'. There are many ways to do so [10], but the most commonly used consists in considering m' such that for any $H \in \mathcal{F} \setminus \emptyset$ we have $m'(H) = m(H)/1-m(\emptyset)$.

W2 Use the value of $m(\emptyset)$ as a trigger to resolve the conflicting situation rather than just relocating it. A typical solution is then to use alternative combination rules [8].

We discuss in Sect. 3 how $m(\emptyset)$ can be used in our context to select the relevant information or to select alternative hypothesis spaces.

Example 2. Consider again the setting of Example 1, The first information delivered, $H_1 = \{(w_1, w_2) \in \mathcal{H} : w_2 \geq 2w_1\}$ is that $(0, 8, 5) \succ (8, 4, 5)$ with a mild certainty, say $\alpha_1 = 0.6$. The second item of information provided by the DM is that for her/him, sciences are more important than language, which we interpret as the inequality

$$w_1 + w_2 \geq w_3 \rightarrow w_1 + w_2 \geq 0.5$$

obtained from the fact that $\sum w_i = 1$. The DM is pretty sure about it, resulting in $\alpha_2 = 0.9$ and $H_2 = \{(w_1, w_2) \in \mathcal{H} : w_2 + w_1 \geq 0.5\}$. The mass resulting from the application of (2) to m_1, m_2 is then

$$m(H_1) = 0.06, \ m(H_2) = 0.36, \ m(H_1 \cap H_2) = 0.54, \ m(\mathcal{H}) = 0.04.$$

2.3   Inferences: Choice and Ranking

When having a finite set $\mathcal{A} = \{a_1, \ldots, a_n\}$ of alternatives and a mass with k focal elements $H_1, \ldots, H_k$, two tasks in MCDA are to provide a recommendation to the DM, in the form of one alternative a^* or a subset A^*, and to provide a (partial) ranking of the alternatives in $\mathcal{A}$. We suggest some means to achieve both tasks.

Choice. When a partial order P is given over $\mathcal{A}$, a natural recommendation is to provide the set $A^* = Max_P$ of maximal items derived from P. Providing a choice in an evidential framework, based on the mass m, then requires to extend this notion. Assuming that the best representation of the DM preferences we could have is a partial order P^*, a simple way to do so is to measure the so-called belief and plausibility measures that a given subset $A \subseteq \mathcal{A}$ is a subset of the set of maximal elements, considering that the subset Max_{P_i} derived from the focal element H_i represents a superset of A^*. These two values are easy to compute, as under these assumptions we have

$$Pl(A \subseteq A^*) = \sum_{A \subseteq Max_{P_i}} m(H_i), \qquad (3)$$

$$Bel(A \subseteq A^*) = \sum_{A = 2^{Max_{P_i}} \setminus \emptyset} m(H_i) = \begin{cases} 0 \text{ if } |A| > 1, \\ \sum_{Max_{P_i} = \{a\}} m(H_i) \text{ if } A = \{a\}. \end{cases} \qquad (4)$$

The particular form of Bel is due to the fact that we have no information about which subset have to be necessarily contained in the set of maximal elements of the unknown partial order P^*. Some noteworthy properties of Eqs. (3)–(4) are the following:

- for an alternative $a \in \mathcal{A}$, $Pl(\{a\}) = 1$ iff $\{a\}$ is a maximal element of all possible partial orders (in particular, $m(\emptyset) = 0$).
- given $A \subseteq B \subseteq \mathcal{A}$, we can have $Pl(A \subseteq A^*) \geq Pl(B \subseteq A^*)$, meaning that it is sensible to look for the most plausible set of maximal elements, that may not be $\mathcal{A}$.

Example 3. Consider the four alternatives $\mathcal{A} = \{a_1, a_2, a_3, a_4\}$ presented in Table 1. We then consider the mass of four focal elements given in Example 2 with the renaming:

$$H_1 = H_1, \; H_2 = H_2, \; H_3 = H_1 \cap H_2, \; H_4 = \mathcal{H}$$

From these, we can for example deduce that $P_1 = \{(a_1, a_4), (a_2, a_3)\}$ using simple linear programming. That $(a_1, a_4) \in P_1$ comes from the fact that the difference between a_1 and a_4 evaluation is always positive in H_1, that is

$$\min_{(w_1, w_2, w_3) \in H_1} (4w_1 + 3w_2 + 9w_3) - (7w_1 + w_2 + 7w_3) > 0.$$

Similarly, we have $P_3 = \{(a_1, a_4), (a_2, a_1), (a_2, a_3), (a_3, a_4), (a_2, a_4)\}$ and $P_2 = P_4 = \{\}$, from which follows $Max_{P_1} = \{a_1, a_2\}$, $Max_{P_3} = \{a_2\}$, $Max_{P_2} = Max_{P_4} = \mathcal{A}$. Interestingly, this shows us that while information $\mathcal{I}_2$ leading to H_2 does not provide sufficient information to recommend any student in $\mathcal{A}$, combined with $\mathcal{I}_1$, it does improve our recommendation, as $|Max_{P_3}| = 1$.

Table 1. A set of alternatives

	P	M	F
a_1	4	3	9
a_2	5	9	6

	P	M	F
a_3	8	7	3
a_4	7	1	7

Table 2 gives the plausibilities and belief resulting from Eqs. (3)–(4) for subsets of one or two elements. Clearly, $\{a_2\}$ is the most plausible answer, as well as the most credible, and hence should be chosen as the predicted set of maximal elements.

Table 2. Plausibilities and belief on sets of one and two alternatives

	$\{a_1\}$	$\{a_2\}$	$\{a_3\}$	$\{a_4\}$	$\{a_1, a_2\}$	$\{a_1, a_3\}$	$\{a_1, a_4\}$	$\{a_2, a_3\}$	$\{a_2, a_4\}$	$\{a_3, a_4\}$
Pl	0.46	1	0.4	0.4	0.46	0.4	0.4	0.4	0.4	0.4
Bel	0	0.54	0	0	0	0	0	0	0	0

Ranking. The second task we consider is to provide a (possibly partial) ranking of the alternatives. Since each (non-empty) focal element can be associated to a partial order over $\mathcal{A}$, this problem is close to the one of aggregating partial orders [9]. Focusing on pairwise information, we can compute the plausibilities and belief that one alternative a_i is preferred to another a_j, as follows:

$$Pl(a_i \succ a_j) = \sum_{P_k, P_{k,ji} \neq 0} m(H_k), \quad Bel(a_i \succ a_j) = \sum_{P_k, P_{k,ij}=1} m(H_k), \quad (5)$$

where $P_{k,ij}$ is the (i,j) value of the incidence matrix of P_k. In practice, Pl comes down to sum all partial orders that have a linear extension with $a_i \succ a_j$, and Bel the partial orders whose all linear extensions have $a_i \succ a_j$. The result of this procedure can be seen as an interval-valued matrix R with $R_{i,j} = [Bel(a_i \succ a_j), Pl(a_i \succ a_j)]$. It can also be noted that, if $m(\emptyset) = 0$, we do have $Pl(a_i \succ a_j) = 1 - Bel(a_j \succ a_i)$. From this matrix, we then have many choices to build a predictive ranking: we can either use previous results about belief functions [7], or classical aggregation rules of pairwise scores to predict rankings [5]. For instance, a classical way is to compute, for each alternative a_i, the interval-valued score $[\underline{s}_i, \overline{s}_i] = \sum_{a_j \neq a_i} [Bel(a_i \succ a_j), Pl(a_i \succ a_j)]$ and then to consider the resulting partial order. This last approach is connected to optimizing the Spearman footrule, and has the advantage of being straightforward to apply.

Example 4. The matrix R and the scores $[\underline{s}_i, \overline{s}_i]$ resulting from Example 3 is

$$
\begin{array}{c}
\begin{array}{ccccc}
 & a_1 & a_2 & a_3 & a_4 & [\underline{s}_i, \overline{s}_i] \\
\end{array} \\
\begin{array}{c}
a_1 \\
a_2 \\
a_3 \\
a_4
\end{array}
\left(
\begin{array}{cccc}
0 & [0, 0.46] & [0, 1] & [0.6, 1] \\
[0.54, 1] & 0 & [0.6, 1] & [0.54, 1] \\
[0, 1] & [0, 0.4] & 0 & [0.54, 1] \\
[0, 0.4] & [0, 0.46] & [0, 0.46] & 0
\end{array}
\right)
\overset{\sum}{=}
\left(
\begin{array}{c}
[0.6, 2.46] \\
[1.68, 3] \\
[0.54, 2.4] \\
[0, 1.32]
\end{array}
\right)
\end{array}
$$

from which we get the final partial order $P^* = \{(a_2, a_4)\}$.

Note that, in practice, it could be tempting to first compute the set of maximal elements and to combine them, rather than combining the models then computing a plausible set of maximal elements, as the first solution is less constrained. However, this can only be done when a specific set $\mathcal{A}$ of interest is known.

3   Inconsistency as a Useful Information

So far, we have largely ignored the problem of dealing with inconsistent information, avoiding the issue of having a strictly positive $m(\emptyset)$. As mentioned in Sect. 2.2, this issue can be solved through the use of alternative combination rules, yet in the setting of preference learning, other treatments that we discuss in this section appear at least as equally interesting. These are, respectively, treatments selecting models of adequate complexity and selecting the "best" subset of consistent information. To illustrate our purpose, consider the following addition to the previous examples.

Example 5. Consider that in addition to previously provided information in Example 2, the DM now affirms us (with great certainty, $\alpha_3 = 0.9$) that the overall contribution of mathematics (X^2) should count for at least four tenth of the evaluation but not more than eight tenth. In practice, if $\mathcal{H}$ is the set of weighted means, this can be translated into $H_3 = \{(w_1, w_2) : 8/10 \geq w_2 \geq 4/10\}$. Figure 2 shows the situation, from which we get that H_1, H_2 and H_3 do not intersect, with $m(\emptyset) = 0.6 \cdot 0.9 \cdot 0.9 = 0.486$, a number high enough to trigger some warning.

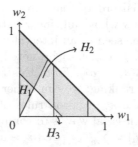

Fig. 2. Inconsistent information items

3.1   Model Selection

$m(\emptyset)$ can be high because the hypothesis space $\mathcal{H}$ is not complex enough to properly model a user preference. By considering more complex space $\mathcal{H}'$, we may decrease the value $m(\emptyset)$, as if $\mathcal{H} \subseteq \mathcal{H}'$, we will have that for any information $\mathcal{I}_i$, the corresponding sets of models will be such that $H_i \subseteq H_i'$ (as all models from $\mathcal{H}$ satisfying the constraints of $\mathcal{I}_i$ will also be in $\mathcal{H}'$), hence we may have $H_i \cap H_j = \emptyset$ but $H_i' \cap H_j' \neq \emptyset$.

Example 6. Consider again Example 5, where $\mathcal{H}'$ is the set of all 2-additive Choquet integrals. A 2-additive Choquet integral can be defined by a set of weights w_i and w_{ij}, $i \neq j$ where w_i and w_{ij} are the weights of groups of criteria $\{\mathcal{X}^i\}$ and $\{\mathcal{X}^i, \mathcal{X}^j\}$. The evaluation of alternatives for a 2-additive Choquet integral then simply reads

$$a_i = \sum_j w_j x_j + \sum_{j<k} w_{kj} \min(x_j, x_k).$$

For the evaluation function to respect the Pareto ordering, these weights must satisfy the following constraints

$$w_i \geq 0 \text{ for all } i,$$
$$w_{ij} + w_i + w_j \geq \max(w_i, w_j) \text{ for all pairs } i, j, \qquad (6)$$
$$\sum_i w_i + \sum_{ij} w_{ij} = 1.$$

Also, the contribution ϕ_i of a criterion i can be computed through the Shapley value

$$\phi_i = w_i + \frac{1}{2} \sum_{j \neq i} w_{ij}.$$

In the case of Example 5, this means that $\mathcal{H}$ corresponds to the set of vectors (w_i, w_{ij}) that satisfy the constraints given by Eq. (6). In this case, the information items H_1, H_2 provided in Example 1 and H_3 in Example 5 induce the following constraints:

$$H_1 = \{\mathbf{w} \in \mathcal{H}' : 4w_2 + w_{23} \geq 8w_1 + 4w_{12} + 5w_{13}\}$$

$$H_2 = \{\mathbf{w} \in \mathcal{H}' : \phi_1 + \phi_2 \geq \phi_3\} = \{\mathbf{w} \in \mathcal{H}' : w_1 + w_2 + w_{12} \geq w_{13}\}$$

$$H_3 = \{\mathbf{w} \in \mathcal{H}' : \frac{4}{10} \leq \phi_2 \leq \frac{8}{10}\} = \{\mathbf{w} \in \mathcal{H}' : \frac{4}{10} \leq w_2 + \frac{1}{2}w_{12} + \frac{1}{2}w_{13} \leq \frac{8}{10}\}$$

These constraints are not inconsistent, as for example the solution where $w_1 = 0.2, w_2 = 0.4, w_{23} = 0.4$ are the only non-null values is within H_1, H_2 and H_3. Among other things, this means that combining m_1, m_2, m_3 within the hypothesis space $\mathcal{H}'$ leads to $m(\emptyset) = 0$.

When considering a discrete nested sequence $\mathcal{H}^1 \subseteq \ldots \subseteq \mathcal{H}^K$ of hypothesis spaces, then a simple procedure to select a model is to iteratively increase its complexity is summarised in Algorithm 1, where H_j^i is the set of possible hypothesis induced by information $\mathcal{I}_j$ in space $\mathcal{H}^j$. It should be noted that the mass given to the empty set is guaranteed to decrease as the hypothesis spaces are nested. One could apply the same procedures to non-nested hypothesis spaces $\mathcal{H}^1, \ldots, \mathcal{H}^K$ (e.g., considering lexicographic orderings and weighted averages), yet in this case there would be no guaranteed relations between the conflicting mass induced by each hypothesis spaces.

Algorithm 1. Algorithm to select preference model

Input: Spaces $\mathcal{H}^1 \subseteq \ldots \subseteq \mathcal{H}^K$, Information $\mathcal{I}_1, \ldots, \mathcal{I}_F$, threshold τ, $i = 1$
Output: Selected hypothesis space $\mathcal{H}^*$
repeat
    foreach $j \in \{0, \ldots, m\}$ **do** Evaluate H_j^i;
    Combine $m_1^i, \ldots, m_F^i$ into m^i ;
    $i \leftarrow i + 1$
until $m^i(\emptyset) \leq \tau$ or $i = K + 1$;

3.2 Information Selection

If we assume that $\mathcal{H}$ is sufficiently rich to describe the DM preferences, then $m(\emptyset)$ results from the fact that the DM has provided some erroneous information. It then makes sense to discard those information items that are the most uncertain and introduce inconsistency in the result. In a short word, given a subset

$S \subseteq \{1, \ldots, n\}$, if we denote by m_S the mass obtained by combining the masses $\{m_i : i \in S\}$, then we can try to find the subset S such that $m_S(\emptyset) = 0$ and $Cer(S) = \sum_{i \notin S} \alpha_i$ is minimal with this property.

An easy but sub-optimal way to implement this strategy is to consider first the set $S^0 = \{1, \ldots, n\}$, and then to consider iteratively subsets by removing the set of sources having the lowest cumulated weight so far. In Example 5, this would come to consider first $S^1 = \{2, 3\}$ (with $Cer(S^1) = 0.6$), then either $S^2 = \{1, 3\}$ or $S^2 = \{1, 2\}$ (with $Cer(S^1) = 0.9$). From Fig. 2, we can see that for $S = \{2, 3\}$, we already have $m_S(\emptyset) = 0$, thus not needing to go any further. When n is small enough (often the case if MCDA), then such a naive search may remain affordable. Improving upon it then depends on the nature of the space $\mathcal{H}$. It seems also fair to assume that the DM makes his/her best to be consistent, and therefore the number of information items to remove from $S^0 = \{1, \ldots, n\}$ should be small in general.

One can also combine the two previously described approach, i.e., to first increase the model complexity if the conflict is important at first, and then to discard the most conflicting and uncertain information. There is a balance between the two: increasing complexity keeps all the gathered information but may lead to over-fitting and to computational problems, while letting go of some information reduces the computational burden, but also delivers more conservative conclusions.

4   Conclusion

In this paper, we have described a generic way to handle uncertain preference information within the belief function framework. In contrast with previous works, our proposal is not tailored to a specific method but can handle a great variety of preference models. It is also consistent with the considered preference model, in the sense that if enough fully reliable information is provided, we retrieve a precise preference model.

Our proposal is very general, and maybe more or less difficult to apply depending on the choice of $\mathcal{H}$. In the future, it would be interesting to study specific preference models and to propose efficient algorithmic procedures to perform the different calculi proposed in this paper. For instance, how do the computations look like where we consider numerical models? Indeed, all procedures described in this paper can be applied to numerical as well as to non-numerical models, but numerical models may offer specific computational advantages.

References

1. Benabbou, N., Perny, P., Viappiani, P.: Incremental elicitation of choquet capacities for multicriteria decision making. In: Proceedings of the Twenty-First European Conference on Artificial Intelligence, pp. 87–92. IOS Press (2014)
2. Beynon, M.J.: Understanding local ignorance and non-specificity within the DS/AHP method of multi-criteria decision making. Eur. J. Oper. Res. **163**(2), 403–417 (2005)

3. Boujelben, M.A., De Smet, Y., Frikha, A., Chabchoub, H.: A ranking model in uncertain, imprecise and multi-experts contexts: the application of evidence theory. Int. J. Approximate Reasoning **52**(8), 1171–1194 (2011)
4. Boutilier, C., Brafman, R.I., Domshlak, C., Hoos, H.H., Poole, D.: CP-nets: a tool for representing and reasoning with conditional ceteris paribus preference statements. J. Artif. Intell. Res. (JAIR) **21**, 135–191 (2004)
5. Destercke, S.: A pairwise label ranking method with imprecise scores and partial predictions. In: Machine Learning and Knowledge Discovery in Databases - European Conference, ECML PKDD 2013, Prague, Czech Republic, 23–27 September, Proceedings, Part II, pp. 112–127 (2013)
6. Grabisch, M., Kojadinovic, I., Meyer, P.: A review of methods for capacity identification in Choquet integral based multi-attribute utility theory: applications of the Kappalab R package. Eur. J. Oper. Res. **186**(2), 766–785 (2008)
7. Masson, M., Destercke, S., Denoeux, T.: Modelling and predicting partial orders from pairwise belief functions. Soft Comput. **20**(3), 939–950 (2016)
8. Pichon, F., Destercke, S., Burger, T.: A consistency-specificity trade-off to select source behavior in information fusion. IEEE Trans. Cybern. **45**(4), 598–609 (2015)
9. Rademaker, M., De Baets, B.: A threshold for majority in the context of aggregating partial order relations. In: 2010 IEEE International Conference on Fuzzy Systems (FUZZ), pp. 1–4. IEEE (2010)
10. Smets, P.: Analyzing the combination of conflicting belief functions. Inf. Fusion **8**, 387–412 (2006)
11. Viappiani, P., Boutilier, C.: Optimal bayesian recommendation sets and myopically optimal choice query sets. In: Advances in Neural Information Processing Systems, pp. 2352–2360 (2010)

A Recourse Approach for the Capacitated Vehicle Routing Problem with Evidential Demands

Nathalie Helal[1(✉)], Frédéric Pichon[1], Daniel Porumbel[2], David Mercier[1], and Éric Lefèvre[1]

[1] Univ. Artois, EA 3926, Laboratoire de Génie Informatique et d'Automatique de l'Artois (LGI2A), 62400 Béthune, France
nathalie.helal@ens.univ-artois.fr,
{frederic.pichon,david.mercier,eric.lefevre}@univ-artois.fr
[2] Conservatoire National des Arts et Métiers, EA 4629, Cedric, 75003 Paris, France
daniel.porumbel@cnam.fr

Abstract. The capacitated vehicle routing problem with stochastic demands can be modelled using either the chance-constrained approach or the recourse approach. In previous works, we extended the former approach to address the case where uncertainty on customer demands is represented by belief functions, that is where customers have so-called evidential demands. In this paper, we propose an extension of the recourse approach for this latter case. We also provide a technique that makes computations tractable for realistic situations. The feasibility of our approach is then shown by solving instances of this difficult problem using a metaheuristic algorithm.

Keywords: Vehicle routing problem · Stochastic Programming with Recourse · Belief function

1 Introduction

In the Capacitated Vehicle Routing Problem (CVRP), one aims at finding a set of routes of minimum cost, such that a fleet of vehicles initially located at a depot, collect goods from a set of customers with deterministic collect demands, while respecting the capacity restrictions of the vehicles. The CVRP with Stochastic Demands (CVRPSD) [14] is a modified version of this problem, where customers have stochastic demands such that, in general, the vehicle capacity limit has a non zero probability of being violated on any route. It is a stochastic integer linear program, which can be modelled by two main approaches: Chance Constrained Programming (CCP) and Stochastic Programming with Recourse (SPR) [1]. Modelling the CVRPSD via CCP consists in having constraints specifying that vehicle capacity limit on any route must not be violated with a high probability. While an SPR model for the CVRPSD allows so-called recourse actions to be performed along a route, such as returning to the depot

© Springer International Publishing AG 2017
A. Antonucci et al. (Eds.): ECSQARU 2017, LNAI 10369, pp. 190–200, 2017.
DOI: 10.1007/978-3-319-61581-3_18

to unload, in order to bring to feasibility a violated capacity limit. The cost of these actions is considered directly in the problem objective [14]. Specifically, the total expected travel cost is subject to minimisation, this cost covering the classical travel cost, *i.e.*, the cost of travel if no recourse action is performed, as well as the expected cost of the recourse actions. SPR models of the CVRPSD have a wider range of applications than CCP models, but they are generally more involved.

Recently [7], another variant of the CVRP was considered: the CVRP with Evidential Demands (CVRPED), where *evidential* means that uncertainty on customer demands is represented by belief functions [11]. Belief function theory is an alternative framework to probability theory for modelling uncertainty, and it can naturally account for uncertainty on customer demands in various situations, such as when pieces of information on customer demands are partially reliable. In [7], the CVRPED was modelled using an extension of the CCP approach used for CVRPSD, and subsequently solved using a metaheuristic, which is a classical means to tackle the CVRP, because it is NP-hard. In this paper, the CVRPED is modelled using an extension of the other main approach to modelling stochastic programs, that is by extending the SPR approach used for the CVRPSD, and then it is also solved using a metaheuristic algorithm.

Note that, to the best of our knowledge, this is the first time that an integer linear program involving uncertainty represented by belief functions is tackled using such a modelling approach. Indeed, besides [7], other works [8,9,12] handled optimisation problems involving uncertainty represented by belief functions in the case of *continuous* linear programs, which are usually much less difficult to solve than their discrete counterparts. In particular, Masri and Ben Abdelaziz [8] extended both CCP and SPR to model linear programs involving belief functions (so-called *belief* linear programs).

This paper is organised as follows. Necessary background on SPR modelling of CVRPSD and on belief function theory is recalled in Sect. 2. An extension of the recourse approach for the CVRPED is presented in Sect. 3. Experiments on CVRPED instances solved using a simulated annealing metaheuristic adapted from [6], are reported in Sect. 4. Section 5 concludes the paper.

2 Background

2.1 CVRPSD Modeled by SPR

In the CVRP, a fleet of m identical vehicles with a given capacity limit Q, initially located at a depot, must collect[1] goods from n customers, with $0 < d_i \leq Q$ the indivisible deterministic collect demand of client i, $i = 1, \ldots, n$. The objective is to find a set of m routes with minimum cost to serve all the customers such that (i) total customers demands on any route must not exceed Q; (ii) each route starts and ends at the depot; and (iii) each customer is serviced only once; we refer to [2] for a formal description of these constraints. Let R_k be the

---

[1] The problem can also presented in terms of delivery, rather than collection, of goods.

route associated to vehicle k and $c_{i,j}$ be the cost of traveling from customer i to customer j. The objective is thus to

$$\min \sum_{k=1}^{m} C(R_k),$$

where

$$C(R_k) = \sum_{i=0}^{n} \sum_{j=0}^{n} c_{i,j} w_{i,j,k}, \tag{1}$$

with $w_{i,j,k}$ a binary variable that equals 1 if vehicle k travels from i to j and serves them, and 0 if it does not.

In the CVRPSD, each client demand d_i, $i = 1, \ldots, n$, becomes a random variable, such that $P(d_i \leq Q) = 1$. As a consequence, a vehicle might not be able to load all of the actual customer demands on any given route having more than one customer. The SPR approach deals with this issue by permitting recourse actions, such as allowing vehicles to return to the depot to unload when they are full. These actions lead to extra costs for routes, which we call penalty costs, and it is generally possible to compute the expected penalty cost of a route induced by the stochastic demands. A general expression for SPR models of CVRPSD is then the following. The objective is to find a set of routes that

$$\min \sum_{k=1}^{m} C_{\mathrm{E}}(R_k),$$

where $C_{\mathrm{E}}(R_k)$ is the expected cost of R_k defined by

$$C_{\mathrm{E}}(R_k) = C(R_k) + C_{\mathrm{P}}(R_k),$$

with $C(R_k)$ the cost defined by (1) representing the cost of traveling along R_k if no recourse action is performed, and $C_{\mathrm{P}}(R_k)$ the expected penalty cost on R_k – $C_{\mathrm{P}}(R_k)$ may be defined in many different ways depending on the recourse policy used (see, e.g., [4,5]).

2.2   Belief Function Theory

Let us recall the concepts of belief function theory needed in this study. Let x be a variable taking its values in a domain X. In this theory, uncertain knowledge about x may be represented by a *Mass Function* (MF) defined as a mapping $m^X : 2^X \to [0,1]$ such that $m^X(\emptyset) = 0$ and $\sum_{A \subseteq X} m^X(A) = 1$. The quantity $m^X(A)$, for some $A \subseteq X$, represents the probability of knowing only that $x \in A$. Subsets $A \subseteq X$ such that $m^X(A) > 0$ are called focal sets. A MF whose focal sets are singletons, i.e., $m^X(A) > 0$ iff $|A| = 1$, corresponds to a probability mass function and is called a *Bayesian* MF. Furthermore, a variable x whose true value is known in the form of a MF will be called an evidential variable.

Finally, given a MF m^X and a function $h : X \to \mathbb{R}^+$, it is possible to compute its *upper* expected value $E^*(h, m^X)$ defined as [3]

$$E^*(h, m^X) = \sum_{A \subseteq X} m^X(A) \max_{x \in A} h(x).$$

3 Recourse Approach for the CVRPED

In this section, a recourse approach is proposed for the case where uncertainty on customer demands in the CVRP is represented by belief functions.

3.1 Formalisation

Assume customer demands d_i, $i = 1, \ldots, n$, are no longer deterministic or random, but evidential, *i.e.*, the actual demand of customer i is known with some uncertainty represented by a MF. In such case, one obtains a new problem called CVRPED. As shown in [7], this problem can be addressed via a constrained programming approach. However, similarly to what has been done for the case of belief linear programs [8], this problem may be also addressed using an extension of the other main approach to modelling stochastic programs, that is by extending the recourse approach of CVRPSD to CVRPED.

Specifically, we propose to extend the recourse approach, for the following policy and assumptions studied for the stochastic case in [4,5]. Each actual customer demand cannot exceed the vehicle capacity. In addition, when a vehicle arrives at a customer on its planned route, it is loaded with the actual customer demand up to its remaining capacity. If this remaining capacity is sufficient to pick-up the entire customer demand, then the vehicle continues its planned route. However, if it is not sufficient, *i.e.*, there is a failure, then the vehicle returns to the depot, is emptied, goes back to the client to pick-up the remaining customer demand and continues its originally planned route.

Consider a given route R containing N customers and, without lack of generality, that the i-th customer on R is customer i. According to the above setting, a failure cannot occur at the first customer on R. However, it can occur at any other customer on R, and there may even be failure at multiple customers on R (at worst, if the actual demand of each customer is equal to the capacity of the vehicle, failure occurs at each customer except the first one).

Formally, let us introduce a binary variable r_i that equals 1 if failure occurs at the i-th customer on R and 0 otherwise (by problem definition $r_1 = 0$). Then, the possible failure situations that may occur along R may be represented by the vectors $(r_2, r_3, \ldots, r_N) \in \{0, 1\}^{N-1}$. To simplify the exposition, we define the set $\Omega = \{\omega_1, \ldots \omega_{2^{N-1}}\}$ representing the possible failure situations along R, with failure situation $(r_2, r_3, \ldots, r_N)$ being in one-to-one correspondence with ω_j where $j = 1 + \sum_{i=2}^{N} r_i \times 2^{i-2}$. For instance, when R contains only $N = 3$ customers, we have $\Omega = \{\omega_1, \omega_2, \omega_3, \omega_4\}$, where ω_j, $j = 1, \ldots, 4$, mean that the vehicle needs to perform a round trip to the depot, respectively, "never", "when it reaches the second customer", "when it reaches the third customer", and "when it reaches both the second and third customers".

Furthermore, let $g : \Omega \rightarrow \mathbb{R}^+$ be a function representing the cost of each failure situation $\omega \in \Omega$. Since the penalty cost upon failure on customer i is $2c_{0,i}$ (a failure implies a return trip to the depot), the cost associated to failure ω_j is

$$g(\omega_j) = \sum_{i=2}^{N} r_i 2c_{0,i},$$

using the one-to-one correspondence $\omega_j \leftrightarrow (r_2, r_3, \ldots, r_N)$.

Let m^Ω be a MF representing uncertainty towards the actual failure situation occurring on R – as will be shown in the next section, evidential demands may induce such a MF.

Then, adopting a similar pessimistic attitude as in the recourse approach to belief linear programming [8], the upper expected penalty cost $C_P^*(R)$ of route R may be obtained as $C_P^*(R) = E^*(g, m^\Omega)$. Accordingly, the upper expected cost $C_E^*(R)$ of route R may be defined as

$$C_E^*(R) = C(R) + C_P^*(R),$$

with $C(R)$ the cost (1) of travelling along route R when no failure occurs.

The CVRPED under the above recourse policy, may then be modelled as the problem of finding a set of m routes optimising the following objective function

$$\min \sum_{k=1}^{m} C_E^*(R_k). \tag{2}$$

Since $C_E^*(R)$ is the upper, i.e., worst, expected cost of a route, we note that optimising (2) has some similarities with the protection against the worst case popular in robust optimisation [13].

The evaluation of the objective function (2) requires the computation for each route, of the MF m^Ω representing uncertainty on the actual failure situation occurring on the route. This is detailed in the next section.

3.2   Uncertainty on Recourses

We assume customer demands to be positive integers. Hence, evidential demands are defined on the finite set $\Theta = \{1, \ldots, Q\}$.

Consider again a route R containing N customers. In addition, let us first assume that MF m_i^Θ representing the evidential demand of the i-th client, $i = 1, \ldots, N$, on R is such that $\exists \theta_i \in \Theta$, $m_i^\Theta(\{\theta_i\}) = 1$, i.e., client demands are known without any uncertainty. Then, it is clear that the above recourse policy amounts to the following definition for the binary failure variables r_i:

$$r_i = \begin{cases} 1, & \text{if } q_{i-1} + \theta_i > Q, \\ 0, & \text{otherwise,} \end{cases} \quad \forall i \in \{2, \ldots, N\} \tag{3}$$

where q_j, $j = 1, \ldots, N$, denotes the load in the vehicle after serving the j-th customer such that $q_j = \theta_1$ for $j = 1$ and, for $j = 2, \ldots, N$,

$$q_j = \begin{cases} q_{j-1} + \theta_j - Q, & \text{if } q_{j-1} + \theta_j > Q, \\ q_{j-1} + \theta_j, & \text{otherwise.} \end{cases}$$

In other words, when it is known that the demand of the i-th customer is θ_i, $i = 1, \ldots, N$, then it can be deduced that the failure situation $\omega_j \leftrightarrow (r_2, r_3, \ldots, r_N)$, with r_i defined by (3), occurs. This can be encoded by a function $f : \Theta^N \to \Omega$, s.t. $f(\theta_1, \ldots, \theta_N) = \omega_j$, with ω_j the failure situation induced by demands θ_i. For example, suppose we have $N = 3$ customers on route R, with respective demands $\theta_1 = 3, \theta_2 = 3$ and $\theta_3 = 5$, and the vehicle capacity limit is $Q = 5$. In such case, failure situation $\omega_4 \leftrightarrow (r_2 = 1, r_3 = 1)$ occurs, hence $f(\theta_1, \theta_2, \theta_3) = \omega_4$.

Assume now that MF m_i^Θ, $i = 1, \ldots, N$, on R is such that $m_i^\Theta(A_i) = 1$, with $A_i \subseteq \Theta$, i.e., client demands are known imprecisely. In such case, it can only be inferred that the failure situation on R belongs to the subset $B \subseteq \Omega$ defined as (using a common abuse of notation for the image of a set)

$$B = f(A_1, \ldots, A_N) = \bigcup_{(\theta_1, \ldots, \theta_N) \in A_1 \times \cdots \times A_N} f(\theta_1, \ldots, \theta_N). \tag{4}$$

More generally, assume that MF m_i^Θ, $i = 1, \ldots, N$, have arbitrary numbers of focal sets and that the joint probability of knowing only that demands of customers $i = 1, \ldots, N$, belong, respectively, to $A_i \subseteq \Theta$, $i = 1, \ldots, N$, is equal to $\prod_{i=1}^{N} m_i^\Theta(A_i)$ (this latter equality is not necessary in our approach, but it simplifies the exposition and corresponds to the case considered in our experiments in Sect. 4). Then, uncertainty on the actual failure situation on R is represented by a MF m^Ω defined as

$$m^\Omega(B) = \sum_{f(A_1, \ldots, A_N) = B} \prod_{i=1}^{N} m_i^\Theta(A_i). \tag{5}$$

Computing m^Ω defined by (5) involves evaluating $f(A_1, \ldots, A_N)$ for all possible combinations of focal sets of MF m_i^Θ, $i = 1, \ldots, N$. Evaluating $f(A_1, \ldots, A_N)$ for some A_i, $i = 1, \ldots, N$, implies $|A_1| \times \cdots \times |A_N|$ (and thus at worst Q^N) times the evaluation of function f at some point $(\theta_1, \ldots, \theta_N) \in \Theta^N$. Hence, computing Eq. (5) is generally intractable. Nonetheless, in the particular and realistic case where the focal sets of MF m_i^Θ, $i = 1, \ldots, N$, are all intervals of positive integers (which will be the case in our experiments in Sect. 4), it becomes possible to compute $f(A_1, \ldots, A_N)$, and thus Eq. (5), with a much more manageable complexity. This is detailed in the next section.

We remark that if evidential demands of all customers are Bayesian, then we are actually dealing with a CVRPSD. In addition, m^Ω is in this case Bayesian on any given route R. Hence, the upper expected penalty cost $C_p^*(R)$ reduces to the classical (probabilistic) expected value of cost function g with respect to the probability mass function m^Ω, and thus our recourse modelling of the CVRPED clearly degenerates into the recourse modelling of the aforementioned CVRPSD.

Finally, we showed in [7] that the constrained programming modelling of CVRPED can be converted, in a particular case, into an equivalent CVRPSD modelled via constrained programming, by transforming each evidential demand represented by MF m_i into a stochastic demand represented by probability mass function p_i such that $p_i(\overline{A}) = m_i(A)$, $\forall A \subseteq \Theta$, with $\overline{A}$ the greatest value in A. It can be shown that under the recourse approach, this latter transformation cannot be used in general to convert a CVRPED into an equivalent CVRPSD.

3.3   Interval Demands

Let us consider a route R with N customers, such that the demand of customer i, $i = 1, \ldots, N$, is known in the form of an interval of positive integers, which we denote by $[\![\underline{A_i}; \overline{A_i}]\!]$, where $\underline{A_i} \geq 1$ and $\overline{A_i} \leq Q$. In this case, as explained above, the failure situation on R belongs to $f\left([\![\underline{A_1}; \overline{A_1}]\!], \ldots, [\![\underline{A_N}; \overline{A_N}]\!]\right) \subseteq \Omega$. Hereafter, we provide a method to efficiently compute $f\left([\![\underline{A_1}; \overline{A_1}]\!], \ldots, [\![\underline{A_N}; \overline{A_N}]\!]\right)$.

In a nutshell, this method consists in generating a rooted binary tree, which represents synthetically yet exhaustively what can possibly happen on R in terms of failure situations.

More precisely, this tree is based on the following remark. Suppose a vehicle traveling along R and all that is known about its load when it arrives at the i-th customer on R is that its load belongs to an interval $[\![\underline{q}; \overline{q}]\!]$. Let us denote by q_i its load after visiting the i-th customer. Then, there are three exclusive cases:

1. either $\overline{q} + \overline{A_i} \leq Q$, hence there will surely be no failure at that customer and all that is known is that $q_i \in [\![\underline{q}; \overline{q}]\!] + [\![\underline{A_i}; \overline{A_i}]\!]$;
2. or $\underline{q} + \underline{A_i} > Q$, hence there will surely be a failure at that customer and all that is known is that $q_i \in [\![\underline{q}; \overline{q}]\!] + [\![\underline{A_i}; \overline{A_i}]\!] - Q$;
3. or $\underline{q} + \underline{A_i} \leq Q < \overline{q} + \overline{A_i}$, hence it is not sure whether there will be or not a failure at that customer. However, we can be sure that if there is no failure at that customer, $i.e.$, the sum of the actual vehicle load and of the actual customer demand is lower or equal to Q, then it means that $q_i \in [\![\underline{q} + \underline{A_i}; Q]\!]$; and if there is a failure at that customer, then it means that $q_i \in [\![1; \overline{q} + \overline{A_i} - Q]\!]$.

By applying the above reasoning repeatedly, starting from the first customer and ending at the last customer, whilst accounting for and keeping track of all possibilities and their associated failures (or absence thereof) along the way, one obtains a binary tree. The tree levels are associated to the customers according to their order on R. The nodes at a level i represent the different possibilities in terms of imprecise knowledge about the vehicle load after the i-th customer, and they also store whether these imprecise pieces of knowledge about the load were obtained following a failure or an absence of failure at the i-th customer. The pseudo code of the complete tree induction procedure is provided in Algorithm 1, which is illustrated by Example 1.

Example 1. Let us illustrate Algorithm 1 on a route R where $Q = 10$ and containing 3 customers, with $[\![4; 8]\!]$, $[\![5; 7]\!]$ and $[\![7; 9]\!]$ the imprecise demands of the

---

Algorithm 1. Induction of Recourse Tree (RT)

Input: interval load $[\![q; \overline{q}]\!]$, Boolean failure variable r, next customer number i
Output: final tree $Tree$
1: create a root node containing interval load $[\![q; \overline{q}]\!]$ and Boolean failure r
2: **if** $i = N + 1$ **then**
3:     return $Tree = \{\text{root node}\}$
4: **else if** $\overline{q} + \overline{A_i} \leq Q$ **then**
5:     $Tree_L = RT([\![q; \overline{q}]\!] + [\![\underline{A_i}; \overline{A_i}]\!], 0, i + 1)$
6:     attach $Tree_L$ as left branch of $Tree$
7: **else if** $q + A_i > Q$ **then**
8:     $Tree_R = RT([\![q; \overline{q}]\!] + [\![\underline{A_i}; \overline{A_i}]\!] - Q, 1, i + 1)$
9:     attach $Tree_R$ as right branch of $Tree$
10: **else**
11:     $Tree_L = RT([\![q + A_i; Q]\!], 0, i + 1)$
12:     attach $Tree_L$ as left branch of $Tree$
13:     $Tree_R = RT([\![1; \overline{q} + \overline{A_i} - Q]\!], 1, i + 1)$
14:     attach $Tree_R$ as right branch of $Tree$
15: **end if**

---

first, second and third customers, respectively. Since the demand of the first customer is $[\![4; 8]\!]$, and there is no failure by definition at the first customer, and the customer following the first customer is the second customer, the tree is obtained with $RT([\![4; 8]\!], 0, 2)$ and is shown in Fig. 1.

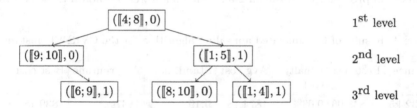

$([\![4; 8]\!], 0)$	1st level
$([\![9; 10]\!], 0)$ $([\![1; 5]\!], 1)$	2nd level
$([\![6; 9]\!], 1)$ $([\![8; 10]\!], 0)$ $([\![1; 4]\!], 1)$	3rd level

Fig. 1. Recourse tree constructed for Example 1

For a given branch of the tree, by concatenating in a vector the Boolean failure variable r_i at level i, $i = 2, \ldots, N$, we obtain the failure situation $\omega_j \leftrightarrow (r_2, r_3, \ldots, r_N)$. Hence, all the branches of the tree yield the subset $B \subseteq \Omega$. For instance, the rightmost branch of the tree in Fig. 1 yields the failure situation $(r_2 = 1, r_3 = 1) \leftrightarrow \omega_4$, the leftmost branch yields $(r_2 = 0, r_3 = 1) \leftrightarrow \omega_3$ and the remaining branch yields $(r_2 = 1, r_3 = 0) \leftrightarrow \omega_2$. The tree in this example yields thus the set $B = \{\omega_2, \omega_3, \omega_4\}$.

Proposition 1. *The set B built using the tree generated by Algorithm 1 verifies* $B = f\left([\![\underline{A_1}; \overline{A_1}]\!], \ldots, [\![\underline{A_N}; \overline{A_N}]\!]\right)$.

Worst-case complexity to obtain set B is $\mathcal{O}(2^{N-1})$ on a route R with N clients, which is the maximum number of leaf nodes in the tree.

4   Experiments

We used the CVRPED instances described in [7] and deriving from those of Augerat set A for the CVRP [10]. These CVRPED instances are obtained as follows. A customer deterministic demand d^{det} in the Augerat instances is transformed into an evidential demand with associated MF m^Θ defined by

$$m^\Theta(\{d^{det}\}) = \alpha, \quad m^\Theta([\lfloor d^{det} - \gamma \cdot d^{det} \rfloor; \lceil d^{det} + \gamma \cdot d^{det} \rceil]) = 1 - \alpha, \quad (6)$$

where $\alpha \in (0,1)$ and $\gamma \in [0,1]$. This transformation corresponds to assuming that the deterministic demand of each customer has been provided by a source, which is reliable with probability α and approximately (at $\pm\gamma * 100\%$) reliable with probability $1 - \alpha$. In addition, we assumed that these latter sources have independent probabilities of reliability.

Proposition 2. *For any α, the upper expected cost of an optimal set of routes for a CVRPED instance generated from a CVRP instance through transformation (6) and modelled via the recourse approach, is non decreasing in γ.*

Proposition 2 basically shows that the more a decision maker is uncertain (cautious) with respect to actual customer demands, *i.e.*, the greater γ is, the greater will be the (upper expected) cost of the optimal solution to his associated optimisation problem. Proposition 2 also yields a lower bound on the cost of the

Table 1. Results of the simulated annealing algorithm for the CVRPED instances

Instance	Best cost	Penalty cost	Avg cost	Stand. dev.	Avg runtime	Best cost $\gamma = 0$
A-n32-k5	843,06	0.03%	874,18	9,19	1837s	839,18
A-n33-k5	705,69	0.37%	724,11	8,39	2241s	697,12
A-n33-k6	773,55	0.75%	793,07	10,42	2271s	758,36
A-n34-k5	820,37	1.40%	837,04	9,19	2975s	812,16
A-n36-k5	884,51	0.34%	914,85	13,84	2715s	869,10
A-n37-k5	722,57	0%	753,51	12,86	2634s	720,85
A-n37-k6	1044,27	3.06%	1071,27	12,74	3111s	995,07
A-n38-k5	781,69	8.36%	816,67	18,44	4525s	748,64
A-n39-k5	890,88	1.57%	935,58	19	5068s	885,04
A-n39-k6	896,60	0.34%	916,91	16.11	3196s	884,09
A-n44-k6	1051,21	2.46%	1104,58	24,88	3922s	1019,07
A-n45-k6	1091,72	6.01%	1129,21	18,98	5444s	1006,90
A-n45-k7	1296,37	0.94%	1348,57	23,02	3237s	1246,14
A-n46-k7	1060,47	0.05%	1087,16	16	2865s	1045,93
A-n48-k7	1241,33	0.11%	1274,24	20,97	3119s	1227,79

optimal solution to any CVPRED instance built using (6): it is obtained by solving to optimality under the recourse approach the corresponding Augerat set A instance, since such instance corresponds to setting $\gamma = 0$ in (6).

In order to solve CVRPED instances under the recourse approach, we adapted a simulated annealing metaheuristic algorithm originally introduced for CVRP in [6]. However, we do not describe this adaptation here due to space limitation.

In our experiments, parameters α and γ of the CVRPED instances were set arbitrarily to 0.8 and 0.1, respectively. Each instance was solved 30 times and the best, average and standard deviation of costs are reported in Table 1. In addition, the contribution of the expected penalty costs to the overall costs of the best solutions is provided as percentages: as can be seen, it varies between 0% to 8%. Finally, the last column of Table 1 provides the best costs obtained with our metaheuristic when solving CVRPED instances with $\gamma = 0$ - these costs may be seen as an approximation of the lower bounds on the costs of the optimal solutions of the CVPRED instances generated through transformation (6).

5   Conclusions

Belief function theory was used to represent uncertainty on customer demands in the capacitated vehicle routing problem. We handled this problem by extending the recourse modelling approach of stochastic programming. In addition, we provided a technique that makes computations tractable in realistic cases. Instances of such cases were then solved using a simulated annealing algorithm. Future works include studying more elaborate recourse policies and improving the solving algorithm.

References

1. Birge, J.R., Louveaux, F.: Introduction to Stochastic Programming. Springer, New York (1997)
2. Bodin, L.D., Golden, B.L., Assad, A.A., Ball, M.O.: Routing and scheduling of vehicles and crews: the state of the art. Comput. Oper. Res. **10**(2), 63–212 (1983)
3. Denoeux, T.: Analysis of evidence-theoretic decision rules for pattern classification. Pattern Recogn. **30**(7), 1095–1107 (1997)
4. Dror, M., Laporte, G., Trudeau, P.: Vehicle routing with stochastic demands: properties and solution frameworks. Transport. Sci. **23**(3), 166–176 (1989)
5. Gauvin, C., Desaulniers, G., Gendreau, M.: A branch-cut-and-price algorithm for vehicule routing problem with stochastic demands. Comput. Oper. Res. **50**, 141–153 (2014)
6. Harmanani, H., Azar, D., Helal, N., Keirouz, W.: A simulated annealing algorithm for the capacitated vehicle routing problem. In: 26th International Conference on Computers and Their Applications, New Orleans, USA (2011)
7. Helal, N., Pichon, F., Porumbel, D., Mercier, D., Lefèvre, É.: The capacitated vehicle routing problem with evidential demands: a belief-constrained programming approach. In: Vejnarová, J., Kratochvíl, V. (eds.) BELIEF 2016. LNCS, vol. 9861, pp. 212–221. Springer, Cham (2016). doi:10.1007/978-3-319-45559-4_22

8. Masri, H., Ben Abdelaziz, F.: Belief linear programming. Int. J. Approx. Reason. **51**, 973–983 (2010)
9. Mourelatos, Z.P., Zhou, J.: A design optimization method using evidence theory. J. Mech. Design **128**, 901–908 (2006)
10. Vehicle Routing Data sets. http://www.coin-or.org/SYMPHONY/branchandcut/VRP/data/index.htm. Accessed 20 Mar 2016
11. Shafer, G.: A Mathematical Theory of Evidence. Princeton University Press, Princeton (1976)
12. Srivastava, R.K., Deb, K., Tulshyan, R.: An evolutionary algorithm based approach to design optimization using evidence theory. J. Mech. Design **135**(8), 081003-12 (2013)
13. Sungur, I., Ordónez, F., Dessouky, M.: A robust optimization approach for the capacitated vehicle routing problem with demand uncertainty. IIE Trans. **40**, 509–523 (2008)
14. Stewart Jr., W.R., Golden, B.L.: Stochastic vehicle routing: a comprehensive approach. Eur. J. Oper. Res. **14**(4), 371–385 (1983)

Evidential k-NN for Link Prediction

Sabrine Mallek[1,2(✉)], Imen Boukhris[1], Zied Elouedi[1], and Eric Lefevre[2]

[1] LARODEC, Institut Supérieur de Gestion de Tunis,
Université de Tunis, Tunis, Tunisia
sabrinemallek@yahoo.fr, imen.boukhris@hotmail.com, zied.elouedi@gmx.fr
[2] Univ. Artois, EA 3926, Laboratoire de Génie Informatique et d'Automatique
de l'Artois (LGI2A), 62400 Béthune, France
eric.lefevre@univ-artois.fr

Abstract. Social networks play a major role in today's society, they
have shaped the unfolding of social relationships. To analyze networks
dynamics, link prediction i.e., predicting potential new links between
actors, is concerned with inspecting networks topology evolution over
time. A key issue to be addressed is the imperfection of real world social
network data which are usually missing, noisy, or partially observed.
This uncertainty is perfectly handled under the general framework of the
belief function theory. Here, link prediction is addressed from a supervised learning perspective by extending the evidential k-nearest neighbors approach. Each nearest neighbor represents a source of information
concerning new links existence. Overall evidence is pooled via the belief
function theory fusion scheme. Experiments are conducted on real social
network data where performance is evaluated along with a comparative study. Experiment results confirm the effectiveness of the proposed
framework, especially when handling skewness in data.

Keywords: Link prediction · Social network · Belief function theory ·
Information fusion · Evidential k-nearest neighbor · Supervised learning

1 Introduction

Link prediction (LP) is an important task in social network analysis and graph
mining that plays a major role in the understanding of network evolution. It is
a powerful tool with a wide variety of applications such as prediction of protein-
protein interactions in bioinformatics [3], construction of recommendation systems for e-commerce [11], detection of criminals or terrorist cells for security
applications [24] or users aid to form new connections in social networks [15]. The
main goal is to accurately predict the existence of new links between unlinked
entities given a state of the network.

Supervised machine learning techniques have been intensively applied to
LP. Many classification models successfully addressed link prediction [4,9] (for
details, see [10]). Indeed, LP can be easily transformed into a two-class classifi-
cation problem. Given a social network graph $\mathcal{G}(\mathcal{V}, \mathcal{E})$ where $\mathcal{V}$ is the set nodes

© Springer International Publishing AG 2017
A. Antonucci et al. (Eds.): ECSQARU 2017, LNAI 10369, pp. 201–211, 2017.
DOI: 10.1007/978-3-319-61581-3_19

and $\mathcal{E}$ is the set of edges, one can partition the graph into two states by considering the edges that occurred at the time interval $[t_0, t_1]$ as the training set and those belonging to the time interval $[t_1, t_2]$ as the test set. To this end, LP is reformulated into a binary classification problem by assigning class labels to all pairs of nodes such that:

$$class(u, v | \mathcal{G}) = \begin{cases} 1 & \text{if } uv \in \mathcal{E} \\ 0 & \text{if } uv \notin \mathcal{E} \end{cases}$$

The most straightforward perception for LP is that similar nodes are likely to connect. That is, the challenge is how to evaluate this similarity accurately. From this point of view, we propose, in this paper, a framework that extends the k nearest neighbor (k-NN) classification approach to LP. We draw on the assumption that query links that are similar to links present in the network are likely to exist. Yet, the challenge is to evaluate the degree of support regarding the class membership of the new links. Actually, information given by the nearest neighbors cannot be considered completely trustworthy as a result of uncertainty and imperfection in the data. As pointed out in [2], real-world networks especially, the large scale ones, are characterized by shifting degrees of uncertainty. Besides, data from real world applications are inherently uncertain, they are frequently incomplete, noisy and sensitive to observation errors. In order to overcome data imperfection issues, k-NN extensions under uncertainty theories have been proposed. For example, fuzzy k-NN algorithm [14] is an extension of k-NN based on fuzzy set theory [30]. It applies a fuzzy editing that alters the membership of each training sample according to its k nearest neighbors. However, it does not allow to model and deal with imprecise or incomplete information effectively. Another example would be the evidential k-NN (EKNN) [7] based on the belief function theory (BFT) [6,25], a formal framework for reasoning under uncertainty that permits to manage and model imprecise information accurately.

Obviously, dealing with uncertainty is relatively correlated to the definition of fusion. It is important to determine the properties of objects and the relations among multiple ones. Yet, one has to quantify the uncertainty regarding certain characteristics of an object and the likelihood with which we can say that some elements are related. The BFT allows to carry out such fusion procedures. Furthermore, it enables to pool evidence while being cautious to the sources' reliability. The usage of the BFT to handle uncertainty in networks has been strongly recommended in the literature [5,13,29]. In that regard, we extend, in this paper, evidential k-NN proposed by [7] under the belief function framework [6,25] to address link prediction. The proposed framework combines topological properties and the intuition of the nearest neighbors approach. The similarity is evaluated using structural metrics as features. Each nearest neighbor is considered as a distinct item of evidence supporting the class membership of the query link. Finally, the overall evidence given by the k-nearest neighbors is fused using the belief function theory combination tools.

This paper is organized as follows: in the next two sections, we recall related work on link prediction and a brief background on the belief function theory.

In Sect. 4, our proposals for LP based on supervised learning under the belief function theory framework are presented. In Sect. 5, we report the conducted experiments to test the novel framework. Lastly, conclusions and possible future work are drawn in Sect. 6.

2 Related Work on Link Prediction

According to [17], LP approaches can be roughly classified into three groups: probabilistic models, maximum likelihood algorithms and similarity based methods [10]. The probabilistic models estimate the likelihood of links existence by building a joint probability distribution representing the graph and applying inference techniques. They are generally based on Markov Networks or Bayesian Networks [8]. On the other hand, maximum likelihood approaches concentrate on a given structure across the network (i.e. hierarchical structure, community structure, etc.) and try to fit the most likely structure through maximum likelihood algorithms. Finally, similarity-based algorithms compute similarity scores between the nodes based on some topological properties of the graph. These scores can be easily employed under a supervised learning. They are generic as they do not depend on the network domain and do not require an overall model. Additionally, each similarity score is independent from the others which allows to compute several ones separately and at the same time. They are the simplest of LP algorithms in terms of computational cost. Actually, LP applies to networks which are continuously evolving in size. In many cases, maximum likelihood and probabilistic models cannot even be checked due the large structure of the networks. On that point, similarity-based methods are more convenient since they do not only perform to large graphs but their performance is also impressive. We presented, in previous works [18–21], LP approaches inspired from similarity-based methods. However, the latter works are applicable to uncertain social networks i.e., edges are attached by uncertainty degrees regarding their existence. They operate merely using the BFT tools. Our proposed framework, in this paper, tackles LP under supervised learning. Furthermore, it applies to social networks without encapsulated uncertainty in their structure.

The similarity-based methods use topological information of the networks. This information is usually grouped into two types: local and global information. The first group of methods computes scores according to node-neighborhoods. An example would be the common neighbors of two nodes [23]. The intuition is that the more two nodes u and v share many common neighbors the more likely they tend to connect. This makes sense in many real world networks such as friendship networks, as two persons who have many mutual friends are very likely to become friends. In contrast, global information methods employ proximity in the network where two nodes are likely to connect if they are close in the network in distance terms. An example of such algorithms would be the shortest path between two nodes. Yet, these algorithms have higher computational complexity since they require all the topological information which is frequently not completely available. In this paper, we consider the node neighborhood based metrics as they are simple and not costly in computational terms.

For a node u, let $\tau(u)$ be the set if its neighbors in the network, called first level neighbors or direct neighbors. The second level neighbors of u, denoted $\tau(u)^2$, are the nodes connected to the direct neighbors of u. We recall here the most popular local similarity scores that proved their efficiency in many works from literature [15, 23, 31]:

- Common Neighbors (CN) [23] computes the common neighbors between a pair of nodes (u, v).
- Jaccard Coefficient (JC) [12] measures the ratio of the common neighbors of u and v and all their neighbors.
- Adamic Adar measure (AA) [1] weights all common neighbors of the pair (u, v) and penalizes the ones with high degrees.
- Resource Allocation (RA) [31] is inspired from the resource allocation process of networks. For an unlinked pair of nodes (u, v), each common neighbor plays the role of a transmitter of a single resource unit. As such, the similarity between u and v is the amount of resource v collected from u.
- Preferential Attachment (PA) [23] assumes that the probability that a new edge relate to u is proportional to $|\tau(u)|$. Thus, the score of uv is correlated to the number of neighbors of u and v.

Equations of the presented metrics are given in Table 1.

Table 1. Structural similarity measures based on local topological information between the pair of nodes (u, v) where $\tau(u)$ and $\tau(v)$ are respectively their sets of neighbors in the graph.

Common Neighbors (CN)	$|\tau(u) \cap \tau(v)|$
Adamic Adar (AA)	$\sum_{z \in (\tau(u) \cap \tau(v))} \frac{1}{log|\tau(z)|}$
Jaccard Coefficient (JC)	$\frac{|\tau(u) \cap \tau(v)|}{|\tau(u) \cup \tau(v)|}$
Resource Allocation (RA)	$\sum_{z \in (\tau(u) \cap \tau(v))} \frac{1}{|\tau(z)|}$
Preferential Attachment (PA)	$|\tau(u)| \cdot |\tau(v)|$

In this paper, local topological metrics are combined with EKNN to evaluate similarities for LP. A feature set is constructed using structural metrics to determine the nearest neighbors according to a distance measure. The assets of the belief function theory for information fusion are subsequently exploited to pool the information gathered from the nearest neighbors. We present, in the next section, some fundamental basic concepts of the BFT.

3   Background on the Belief Function Theory

In the belief function theory [6, 25], a problem is represented by a frame of discernment $\Omega = \{\omega_1, \omega_2, \ldots, \omega_n\}$, an exhaustive and finite set of mutually exclusive events. A basic belief assignment (bba), denoted by m, represents the knowledge committed to the elements of 2^Ω given a source of information. It is a mapping function $m : 2^\Omega \rightarrow [0,1]$, such that:

$$\sum_{A \subseteq \Omega} m(A) = 1 \tag{1}$$

An element A is called a focal element of the bba m if $m(A) > 0$. The belief committed to Ω represents the degree of ignorance. A state of total ignorance is defined by $m(\Omega) = 1$. When the bba has at most one focal element A different from Ω, it said to be a simple support function (ssf) and has the following form [26]:

$$\begin{cases} m(A) &= 1 - \omega \\ m(\Omega) &= \omega \end{cases} \tag{2}$$

for some $A \subset \Omega$ and $\omega \in [0,1]$.

Combining two basic assignments induced from two distinct sources of information over the same frame of discernment into one may be ensured using the conjunctive rule of combination denoted by $\textcircled{\cap}$. It is defined as [27]:

$$m_1 \textcircled{\cap} m_2(A) = \sum_{B,C \subseteq \Omega : B \cap C = A} m_1(B) \cdot m_2(C) \tag{3}$$

The combination rule permits to aggregate evidence by meaningfully outlining a corpus of data and making it simpler. Furthermore, it allows to fuse information induced from single and multiple sources.

4   Evidential k-nearest Neighbors for Link Prediction

The goal is to predict the existence of new links in a network graph $\mathcal{G}(\mathcal{V}, \mathcal{E})$ where $\mathcal{V}$ is the set of nodes and $\mathcal{E}$ is the set of edges. The set of classes is $\Omega = \{E, \neg E\}$, where E points up the existence of a link in $\mathcal{G}$ and $\neg E$ its absence. The class of each link in $\mathcal{E}$ is assumed to be known with certainty. The available information consists in a training set $\mathcal{T} = \{(e^1, \omega^1), \ldots, (e^{|\mathcal{E}|}, \omega^{|\mathcal{E}|})\}$ of single labeled links, where $e^i \in \mathcal{E}$, $i \in \{0, \ldots, |\mathcal{E}|\}$ and its corresponding class label is $\omega^i \in \Omega$.

Let e be a new link to be classified, where e may connect the pair of nodes (u, v). Let $\mathcal{L}^1$ be the set of links shared with the direct (1st level) neighbors of u and v in $\mathcal{G}$, where $|\mathcal{L}^1| = |\tau(u)| + |\tau(v)|$. Additionally, let $\mathcal{L}^2$ be the set of links unshared with the 2nd level neighbors of u and v in $\mathcal{G}$, where $|\mathcal{L}^2| = |\tau(u)^2| + |\tau(v)^2|$. The set $\mathcal{L}$ incorporates $\mathcal{L}^1$ and $\mathcal{L}^2$.

It is obvious that the class of the links in the set $\mathcal{L}^1$ is E. In contrast, the class of the links in $\mathcal{L}^2$ is $\neg E$ since it includes the links that are not shared between u and v and their respective 2nd level neighbors. As follows, the nearest neighbors of each unseen link are uncovered thought-out the neighborhood of its end points. Accordingly, instead of comparing each unseen link with all the possible edges in the network, which is computationally not feasible since there are $\frac{(|\mathcal{V}| \times (|\mathcal{V}|-1))}{2}$

possible links, it is compared to the neighboring ones. As such, the search space is reduced. Besides, the intuition of node neighborhood approaches is inherently employed as they are exactly intended to find similar nodes. In this context, the authors in [28], proposed a framework for LP using k-NN by considering the local similarity indexes to evaluate the similarity with the neighbors. The prediction of an edge uv is made by comparing the neighbors of u to v and vice versa. However, the proposed approach considers the nearest neighbors equally trustworthy. Besides, only one similarity index is considered in k-NN at a time.

In our proposed framework, the evidential k-NN classifier [7] operates in two stages. First, it computes the distances between a test link e and its neighborhood in $\mathcal{L}$ and retain the smallest k distances. Subsequently, the evidence given by the k nearest neighbors is combined to get an overview about the global belief regarding the existence of e. The steps are detailed in the following.

At first, one has to determine the k nearest neighbors. For that, we need to define a distance to evaluate the similarity between the edges in the test set and those in the train set. We propose to use the Euclidean distance $d(e, e^i)$ between the link e and its nearest neighbor $e^i \in \mathcal{L}$ by computing the similarities between their connecting nodes as follows:

$$d(e, e^i) = \sqrt{\sum_{j=1}^{n}(s_e^j - s_{e^i}^j)^2} \qquad (4)$$

where j is the index of a local similarity metric (e.g., CN, AA, JC, RA, PA), s_e and s_{e^i} are respectively its values for e and e^i and n is the number of local similarities considered.

Each link e^i in $\mathcal{L}$ represents a piece of evidence that increases our belief about e also belonging to ω^i. Yet, this information solely does not provide certain knowledge about the class of e. This situation is modeled in the BFT by simple support functions where only some part of the belief is committed to ω^i and the rest is affected to Ω. Therefore, we get the following bba:

$$\begin{cases} m_i(\{\omega^i\}) = \alpha\phi(d_i) \\ m_i(\Omega) = 1 - \alpha\phi(d_i). \end{cases} \qquad (5)$$

where $d_i = d(e, e^i)$, α is a parameter such that $0 < \alpha < 1$ and ϕ is a decreasing function. The closer e is to e^i according to the distance d, the more likely for e to have same class as e^i. In contrast, when e is far from e^i, in distance terms, then e^i would provide little information regarding the class of e. On that point, the function ϕ must verify $\phi(0) = 1$ and $lim_{d\to\infty}\phi(d) = 0$. Authors in [7] suggest to use the following decreasing function:

$$\phi(d_i) = e^{(-\gamma d_i^\beta)} \qquad (6)$$

where $\gamma > 0$ and $\beta \in \{1, 2, \dots\}$. β can be arbitrarily fixed to a small value (1 or 2).

As a result of considering each nearest neighbor in $\mathcal{L}$ as an independent source of evidence regarding the class of the e, we obtain k bba's that can be combined using the conjunctive rule of combination. Thus, a global bba m that synthesizes the belief regarding the existence of e is produced as follows:

$$m = m_1 \circledcirc \ldots \circledcirc m_k \tag{7}$$

Finally, decision about the membership of e to one of the classes in Ω is made by comparing $m(\{E\})$ and $m(\{\neg E\})$. If $m(\{E\}) > m(\{\neg E\})$ then e exists, it is absent otherwise.

5 Experiments

Experiments are conducted on a real social network component of 1 K nodes and 10 K edges of the Facebook dataset from [22]. Since network data with time information are not usually available, we must settle for the more drastic technique by randomly removing a partition of the edges from the network in order to use them as test set. That is, we remove a random 10% of the edges which we try to predict the existence along with randomly generated false links of the same size using the graph as a source composed by the remaining 90%. The results are obtained by averaging over 10 implementations with independently random divisions of testing set and training set. In order to reduce the computational time, a preprocessing phase is first conducted in which the local similarity scores of all the links from the train and test sets are computed. Evaluation is made according to accuracy which computes the number of correct predictions among all predictions and the precision which takes the fraction of predicted links that are relevant.

Required parameters are α, β, γ for the induced bba's and the number of nearest neighbors k. As discussed in [7], the parameters α and β do not have a great influence on the approach performance. Thus, as in [7], α is fixed to 0.95 and β to 1. We tested values of k ranging from 1 to 15. Tests for the optimization of the γ parameter allowed us to set it to the value of 0.12. A comparison with the standard k-NN method (KNN) is carried out, where the class of a link is predicted according to the majority classes of its k nearest neighbors. The results are reported in Fig. 1.

Figure 1 reports the results in terms of accuracy for different values of k. It can be seen that both methods have better performances as k increases. Indeed, as we boost the number of nearest neighbors we get more sources of evidence regarding the class membership of the links. Yet, as shown in Fig. 1, performance stops upon reaching a certain threshold. Rather, by increasing excessively the number of nearest neighbors, we get more distant ones (less similar). Consequently, the associated mass functions are close to the state of total ignorance. Thus, they have no impact in the combination and therefore in the prediction. The EKNN based framework has better classification performance than the standard KNN based LP method. It stands to reason that, EKNN performs better for some values of k as a results of taking implicitly the relevance of the information given by the sources into account unlike KNN as it considers all the nearest

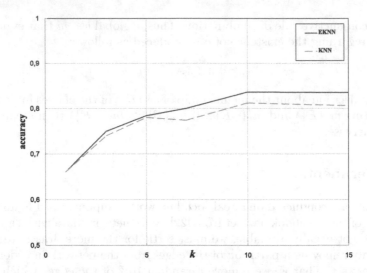

Fig. 1. Results according to the values of k

neighbors equally trustworthy. Although, the results are low for small values of k, it still gives acceptable results i.e., 75% accuracy for $k = 3$.

In a second stage, we conduct implementations by increasing the number of negative instances (non existing links) at each time to evaluate the behavior of our algorithms to class imbalance scenarios. Actually, LP is a very imbalanced class problem where the number of non existing links is much larger than the existing ones. The same parameters specifications are considered except that k is set to 15. Precision results are presented in Fig. 2 for different negative links number. Measuring precision is very important for evaluating LP since, in many cases, the main goal is to accurately predict the real existing edges. For example, in Facebook, it is more important to not miss actual friends and it does not really matter when unknown friends are recommended. As shown in Fig. 2, EKNN outperforms KNN for most values. Furthermore, the precision plot decreases as more non existing edges are predicted. However, the curve does not fall dramatically but rather slowly reaching 77% for 10 K false edges. These good performances are also obtained thanks to the advantages of supervised learning which permits to focus on class boundaries and balance data. As opposed to unsupervised methods which cannot address this imbalance well because they are agnostic to class distributions by nature [16]. We conjecture that our framework is capable of dealing with the class imbalance problem. Furthermore, it allows us to consider network topology and deal with uncertainty at the same time.

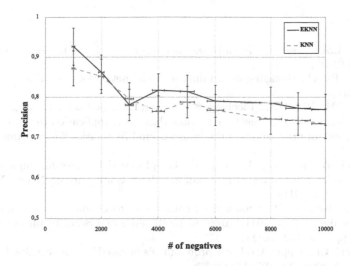

Fig. 2. Precision results for different negative links number

6   Conclusion

Local similarity measures naturally operate to detect similar nodes which make it simple to extend them to k-NN. In this paper, we propose a framework for link prediction that combines structural local topological properties and the evidential k-NN approach. Based on the direct and second level neighbors of two unlinked nodes u and v, local similarity indexes are computed and used as features to find their k-nearest neighbors. These latter are considered as items of evidence regarding the class membership of the link uv. Global evidence is pooled using the conjunctive rule of combination from the belief function theory to get an overall information abut new links existence. Tests on real world social network data proved the efficiency of the proposed framework. It is interesting to note that the novel framework handles skewness in social network data. It is capable of combating class imbalance that characterizes the link prediction task. Furthermore, it shows performance improvement over the baseline algorithm KNN which does not take into account uncertainty in the analysis.

A straightforward direction for future research is to take supplementary information into account such as node attributes. Obviously, nodes with similar attribute values are likely to share social relationships i.e., two authors who have the same affiliation and the same research fields. Besides, handling node attributes brings semantics to social connections. Therefore, it is an important source of information that may enhance the link prediction task.

References

1. Adamic, L.A., Adar, E.: Friends and neighbors on the web. Soc. Netw. **25**(3), 211–230 (2003)
2. Adar, E., Ré, C.: Managing uncertainty in social networks. Data Eng. Bull. **30**(2), 23–31 (2007)
3. Airoldi, E.M., Blei, D.M., Fienberg, S.E., Xing, E.P., Jaakkola, T.: Mixed membership stochastic block models for relational data with application to protein-protein interactions. In: Proceedings of the International Biometrics Society Annual Meeting (2006)
4. Cukierski, W., Hamner, B., Yang, B.: Graph-based features for supervised link prediction. In: Proceedings of International Joint Conference on Neural Networks, pp. 1237–1244 (2011)
5. Dahlin, J., Svenson, P.: A method for community detection in uncertain networks. In: Proceedings of the 2011 European Intelligence and Security Informatics Conference, pp. 155–162 (2011)
6. Dempster, A.P.: Upper and lower probabilities induced by a multivalued mapping. Ann. Math. Stat. **38**, 325–339 (1967)
7. Denoeux, T.: A k-nearest neighbor classification rule based on Dempster-Shafer theory. IEEE Trans. Syst. Man Cybern. **25**, 804–813 (1995)
8. Getoor, L., Taskar, B.: Introduction to Statistical Relational Learning. The MIT Press, Cambridge (2007)
9. Hasan, M.A., Chaoji, V., Salem, S., Zaki, M.J.: Link prediction using supervised learning. In: Proceedings of the 6th Workshop on Link Analysis, Counter Terrorism and Security (2006)
10. Hasan, M.A., Zaki, M.J.: A survey of link prediction in social networks. In: Aggarwal, C.C. (ed.) Social Network Data Analytics, pp. 243–275. Springer, Newyork (2011)
11. Huang, Z., Li, X., Chen, H.: Link prediction approach to collaborative filtering. In: Proceedings of the 5th ACMIEEE-CS Joint Conference on Digital Libraries, JCDL 2005, pp. 141–142. ACM (2005)
12. Jaccard, P.: Étude comparative de la distribution florale dans une portion des Alpes et des Jura. Bulletin de la Société Vaudoise des Sciences Naturelles **37**, 547–579 (1901)
13. Johansson, F., Svenson, P.: Constructing and analyzing uncertain social networks from unstructured textual data. In: Özyer, T., Erdem, Z., Rokne, J., Khoury, S. (eds.) Mining Social Networks and Security Informatics. Lecture Notes in Social Networks, pp. 41–61. Springer, Dordrecht (2014)
14. Keller, J.M., Gray, M.R., Givens, J.A.: A fuzzy k-nearest neighbor algorithm. IEEE Trans. Syst. Man Cybern. **SMC-15**(4), 580–585 (1985)
15. Liben-Nowell, D., Kleinberg, J.: The link prediction problem for social networks. J. Am. Soc. Inf. Sci. Technol. **58**(7), 1019–1031 (2007)
16. Lichtenwalter, R.N., Lussier, J.T., Chawla, N.V.: New perspectives and methods in link prediction. In: Proceedings of the 16th ACM SIGKDD International Conference on Knowledge Discovery and Data Mining, pp. 243–252 (2010)
17. Lu, L., Zhou, T.: Link prediction in complex networks: a survey. Phys. A **390**(6), 1150–1170 (2011)
18. Mallek, S., Boukhris, I., Elouedi, Z., Lefevre, E.: Evidential link prediction based on group information. In: Prasath, R., Vuppala, A.K., Kathirvalavakumar, T. (eds.) MIKE 2015. LNCS, vol. 9468, pp. 482–492. Springer, Cham (2015). doi:10.1007/978-3-319-26832-3_45

19. Mallek, S., Boukhris, I., Elouedi, Z., Lefevre, E.: The link prediction problem under a belief function framework. In: Proceedings of the IEEE 27th International Conference on the Tools with Artificial Intelligence, pp. 1013–1020 (2015)
20. Mallek, S., Boukhris, I., Elouedi, Z., Lefevre, E.: An evidential method for multi-relational link prediction in uncertain social networks. In: Proceedings of the 5th International Symposium on Integrated Uncertainty in Knowledge Modelling and Decision Making, pp. 280–292 (2016)
21. Mallek, S., Boukhris, I., Elouedi, Z., Lefevre, E.: Evidential missing link prediction in uncertain social networks. In: Proceedings of the 16th International Conference on Information Processing and Management of Uncertainty in Knowledge-Based Systems, pp. 274–285 (2016)
22. McAuley, J.J., Leskovec, J.: Learning to discover social circles in ego networks. In: Proceedings of the 26th Annual Conference on Neural Information Processing Systems 2012, pp. 548–556 (2012)
23. Newman, M.E.J.: Clustering and preferential attachment in growing networks. Phys. Rev. E **64**, 025102 (2001)
24. Rhodes, C.J., Jones, P.: Inferring missing links in partially observed social networks. JORS **60**(10), 1373–1383 (2009)
25. Shafer, G.R.: A Mathematical Theory of Evidence. Princeton University Press, Princeton (1976)
26. Smets, P.: The canonical decomposition of a weighted belief. In: Proceedings of the Fourteenth International Joint Conference on Artificial Intelligence, IJCAI 1995, vol. 14, pp. 1896–1901 (1995)
27. Smets, P.: Application of the transferable belief model to diagnostic problems. Int. J. Intell. Syst. **13**(2–3), 127–157 (1998)
28. Speegle, G., Bai, Y., Cho, Y.R.: Extending local similarity indexes with knn for link prediction. In: Proceedings of the 14th International Conference on Computational Science and Its Applications, ICCSA 2014, pp. 1–7 (2013)
29. Svenson, P.: Social network analysis of uncertain networks. In: Proceedings of the 2nd Skövde Workshop on Information Fusion Topics (2008)
30. Zadeh, L.: Fuzzy sets. Inf. Control **8**(3), 338–353 (1965)
31. Zhou, T., Lü, L., Zhang, Y.: Predicting missing links via local information. Eur. Phys. J. B-Condens. Matter Complex Syst. **71**(4), 623–630 (2009)

Ensemble Enhanced Evidential k-NN Classifier Through Random Subspaces

Asma Trabelsi[1,2(✉)], Zied Elouedi[1], and Eric Lefevre[2]

[1] Université de Tunis, Institut Supérieur de Gestion de Tunis, LARODEC,
Tunis, Tunisia
trabelsyasma@gmail.com, zied.elouedi@gmx.fr
[2] Univ. Artois, EA 3926, Laboratoire de Génie Informatique et d'Automatique de
l'Artois (LGI2A), 62400 Béthune, France
eric.lefevre@univ-artois.fr

Abstract. The process of combining an ensemble of classifiers has been deemed to be an efficient way for improving the performance of several classification problems. The Random Subspace Method, that consists of training a set of classifiers on different subsets of the feature space, has been shown to be effective in increasing the accuracy of classifiers, notably the nearest neighbor one. Since, in several real world domains, data can also be suffered from several aspects of uncertainty, including incompleteness and inconsistency, an Enhanced Evidential k-Nearest Neighbor classifier has been recently introduced to deal with the uncertainty pervading both the attribute values and the classifier outputs within the belief function framework. Thus, in this paper, we are based primarily on the Enhanced Evidential k-Nearest Neighbor classifier to construct an ensemble pattern classification system. More precisely, we adopt the Random Subspace Method in our context to build ensemble classifiers with imperfect data.

Keywords: Classifier ensemble · Random Subspace Method · Enhanced evidential k-NN · Belief function theory

1 Introduction

The core purpose of an ensemble classifier is to achieve a high accuracy for a given classification problem. The process of building an ensemble learning consists firstly of generating a set of base/weak classifiers from the training data and then perform actual classification by combining the output predictions of base classifiers. To gain a better accuracy, the basic classifiers should be diverse and independent [13]. Several ensemble classifier generation methods allow to achieve diversity among the base classifiers. Bagging [3] and Boosting [18] are widely used as ensemble methods but some authors have proven that these two techniques are not guaranteed to produce fully independent individual base classifiers [5]. Both theoretical and experimental researches conducted by the machine learning community have shown that the efficient method for achieving a good diversity consists of training the base classifiers on different feature subsets [4,24].

© Springer International Publishing AG 2017
A. Antonucci et al. (Eds.): ECSQARU 2017, LNAI 10369, pp. 212–221, 2017.
DOI: 10.1007/978-3-319-61581-3_20

This may be explained by the fact that a feature subset-based ensemble can reduce the correlation among the classifiers and also perform faster owing to the reduced size of input features [4,8,11]. The key problem of this kind of ensemble learning is how to yield attribute subsets with good predicting power. Several feature subsets techniques have been introduced till now where some of which are based on filter approaches [17], while others are relied on wrapper approaches [12]. Another more popular and effective tool is the Random Subspace Method (RSM) also called random subspacing [19] and has satisfactory yielded results particulary with the standard k-Nearest Neighbor classifier (k-NN) [2]. In this paper, we have to adapt the random subspace method in the real context of uncertain data. Precisely, we propose to design a classifier ensemble via random subspacing on the basis of the Enhanced Evidential k-NN (EEk-NN) classifier, which is proposed in [23], as a new technique for dealing with uncertain data represented within the belief function framework. The reminder of this paper is organized as follows: Sect. 2 is committed to highlighting the fundamental concepts of the belief function theory. In Sect. 3, we present the EEk-NN classifier that handles evidential databases. Section 4 is dedicated to describing our proposed ensemble classifier through random subspaces. Our experimentation on several synthetic databases is conducted in Sect. 5. Finally, the conclusion and our main future work directions are reported in Sect. 6.

2 Belief Function Theory: Background

The belief function theory, also referred to as evidence theory, is widely regarded as very effective and efficient basis for representing, managing and reasoning about uncertain knowledge. This section briefly reviews some important concepts underlying this theory.

2.1 Information Representation

Let $\Theta = \{\theta_1, \theta_2, \ldots, \theta_N\}$ denote the frame of discernment including a finite non empty set of N elementary hypotheses that are assumed to be exhaustive and mutually exhaustive. The power set of Θ, denoted by 2^Θ, is made up of all the subsets of Θ:

$$2^\Theta = \{\emptyset, \theta_1, \theta_2, \ldots, \theta_N, \ldots, \Theta\} \tag{1}$$

Expert's beliefs over the subsets of the frame of discernment Θ are represented by the so-called basic belief assignment (bba) denoted by m. It is carried out in the following manner:

$$\sum_{A \subseteq \Theta} m(A) = 1 \tag{2}$$

Each subset A of 2^Θ having fulfilled $m(A) > 0$ is called a focal element.

2.2   Combination Operators

For certain real world problems, we are clearly confronted with information issued from several sources. Therefore, a number of combination rules has been proposed and discussed for some past time. The conjunctive rule, introduced by Smets within the Transferable Belief Model (TBM) [21], is one of the best known ones. Given two information sources S_1 and S_2 with respectively m_1 and m_2 as bbas, the conjunctive rule, denoted by $\bigcirc\!\!\!\!\cap$, was established as follows:

$$m_1 \bigcirc\!\!\!\!\cap m_2(A) = \sum_{B \cap C = A} m_1(B)m_2(C), \quad \forall A \subseteq \Theta. \tag{3}$$

The belief completely associated to the empty set was recognised under the name of conflictual mass. A normalized version of the conjunctive rule has been proposed by Dempster [6] to retain the basic characteristics of the belief function theory. Indeed, it allows to manage the conflict while redistributing the conflictual mass over all focal elements. The Dempster rule is then set as follows:

$$m_1 \oplus m_2(A) = \frac{1}{1 - K} \sum_{B \cap C = A} m_1(B)m_2(C), \quad \forall A \subseteq \Theta \tag{4}$$

where the conflictual mass K caused by the combination of the two bbas m_1 and m_2 through the conjunctive rule, is given as follows:

$$K = \sum_{B \cap C = \emptyset} m_1(B)m_2(C) \tag{5}$$

2.3   Decision Making

The pignistic probability, denoted by $BetP$, has been proven to be an effective and efficient decision-making tool for selecting the most likely hypothesis relative to a given problem [20]. It consists of transforming beliefs into probability measures as follows:

$$BetP(A) = \sum_{B \cap A = \emptyset} \frac{|A \cap B|}{|B|} m(B), \quad \forall A \in \Theta \tag{6}$$

The hypothesis H_s that has to be chosen is the one with the highest pignistic probability:

$$H_s = argmax_A BetP(A), \quad \forall A \in \Theta \tag{7}$$

2.4   Dissimilarity Between bbas

In the research literature, several measures have been proposed to compute the degree of dissimilarity between two given bbas [10,16,22]. One of the earliest and best-known measures is the Jousselme distance. Formally, the Jousselme distance, for two given bbas m_1 and m_2, is defined by:

$$dist(m_1, m_2) = \sqrt{\frac{1}{2}(m_1 - m_2)^T D(m_1 - m_2)} \qquad (8)$$

where the Jaccard similarity measure D is set to:

$$D(X,Y) = \begin{cases} 1 & \text{if } X = Y = \emptyset \\ \dfrac{|X \cap Y|}{|X \cup Y|} & \forall X, Y \in 2^{\Theta} \end{cases} \qquad (9)$$

3   Nearest Neighbor Classifiers for Uncertain Data

Data uncertainty is regarded as one of the main issues of several real world applications that can affect experts' decisions. Two levels of uncertainty can be distinguished in the literature: the uncertainty that occurs in the attribute values and the one pervading the class labels. The process of constructing classifiers from totally uncertain data has not received the great attention till now. Drawing inspiration from the Evidential theoretic k-NN that incorporates classifier outputs uncertainties [7,9], we have proposed an EEk-NN classifier for handling not only the uncertainty associated with the classifier outputs but also that pervading the data, precisely the attribute values. Suppose we have to solve an M class classification problem. Let us denote by $X = \{x^i = (x_1^i, ..., x_n^i); L^i | i = 1, ..., N\}$ a collection on N n-dimensional training samples where each one is characterized by n uncertain attribute values x_j^i ($j \in \{1, ..., n\}$) represented within the belief function framework and a class label L^i demonstrating its membership to a specific class in $\Theta = \{\theta_1, ..., \theta_M\}$. Assume that $y = \{y_1, ..., y_n\}$ be a new query pattern to be classified on the basis of the training set X. The major idea underlying our proposed classifier is to compute the distance $d_{y,i}$ between the query pattern y and each instance $x^i \in X$ that corresponds to the sum of the absolute differences between the attribute values as follows:

$$d_{y,i} = \sum_{j=1}^{n} dist(x_j^i, y_j) \qquad (10)$$

Particulary, we have relied on the Jousselme distance measure $dist$ (see Eq. 8) for processing the uncertainty that characterizes the attribute values. It must be emphasised that $d_{y,i}$ can have values comprised within the range of from 0 to 1. A value of $d_{y,i}$ which is too small involves the situation that the instances y and x^i are described by the same class label L^i. In contrast, a high value of $d_{y,i}$ implies the situation of almost complete ignorance with regard to the class label of y. As a matter of fact, the uncertainty pervading the class label of the query pattern y can be modeled and represented within the belief function theory. Assume that the training instances are sorted in ascending order according to their distance from the test instance y, each training instance $x^i \in X$ provides an item of evidence denoted by $m^{(i)}(.|x^i)$ over Θ:

$$m^{(i)}(\{\theta_q\}|x^i) = \alpha\Phi_q(d_{y,i}) \tag{11}$$
$$m^{(i)}(\Theta|x^i) = 1 - \alpha\Phi_q(d_{y,i})$$
$$m^{(i)}(A|x^i) = 0, \forall A \in 2^\Theta \backslash \{\Theta, \theta_q\}$$

where the distance function $d_{y,i}$ should be calculated such as in Eq. 8, θ_q refers to the class label of x^i and α is a parameter satisfying $0 < \alpha < 1$. It has been proven that a value of α equal to 0.95 can lead to satisfactory or better outcomes [7]. The decreasing function Φ_q, checking $\Phi_q(0) = 1$ and $lim_{d\to\infty}\Phi_q(d) = 0$, will be given as follows:

$$\Phi_q(d) = exp(-\gamma_q d^2), \tag{12}$$

where γ_q displays a positive parameter of class θ_q. It can be optimized depending on the training samples. An exact method relied on a gradient search procedure can be used for small or medium data sets, while using a linearization approach for large data [25]. The best values of γ_q, for both exact and approximated methods, can be estimated by minimizing the mean squared classification error over the whole training set X of size N.

The final bba m^y regarding the class membership of the query pattern y can be obtained by merging the bbas issued from k nearest neighbors training instances of y through the Dempster rule of combination. The final bba will be defined as follows:

$$m^y = m^{(1)}(.|x^1) \oplus m^{(2)}(.|x^2) \oplus \ldots \oplus m^{(k)}(.|x^k) \tag{13}$$

The class label concerning the test pattern y, will be made by computing the pignistic probability $BetP$ of the bba m^y as shown in Eq. 6. The query pattern y is then assigned to the class label with the highest pignistic probability.

4   Ensemble Enhanced Evidential k-NN (Ensemble EEk-NN)

As already mentioned, the concept of diversity is regarded as a vital necessity for the ultimate success of ensemble classifier systems. Note however, that in this context, the RSM is a widely used technique addressed to ensure diversity between individual classifiers and has achieved satisfactory results notably for the ensembles of Evidential theoretic k-NNs [1]. Despite their relevance and success, such kind of ensemble systems cannot handle imperfect data, especially the uncertain ones. Get inspired from [1], in this paper, we propose a new ensemble system that fully benefits from the advantages of both RSM and EEk-NN. Our proposed ensemble classification system deals mainly with uncertain data where the uncertainty occurs precisely in the attribute values and is represented within the belief function framework. The suggested model is generally characterised by three main steps. Given a training data X, the first level concerns the generation of T feature subsets with size S from a uniform distribution over X. In the second level, the output label of each query pattern will be predicted through T EEk-NN classifiers that are trained with the different generated feature subsets. The final stage concerns the combination of the predictions yielded by the different classifiers. Let us remind that the output label of each individual classifier is expressed in terms of a mass function. The belief function theory has also been proven to be an efficient way for merging an ensemble of classifiers where each of which produces a belief function for each query instance. Different combination rules have been implemented within this framework and can be categorized according to the dependency between the merged sources. In this paper, we ultimately opted for the Dempster operator, which is the conventionally used rule within the belief function theory, for combining diverse classifiers.

Two substantial parameters need to be considered for our proposed framework:

- **The number of created classifiers:** A substantial key element when designing an ensemble classifiers is the number of individual classifiers used to get the final decision. There is no doubt that a huge number of classifiers may in the one hand increase the computational complexity and on the other hand decrease the comprehensibility. Several researches have been done to predefine a reasonable number of classifiers. The conclusion conducted following to the study of [15] shows that ensembles of 25 k-NN classifiers are sufficient for reducing the error rate and consequently for improving performance. For that very reason, in this paper, we set the number of combined EEk-NN classifiers to 25.
- **The size of feature subsets S:** The choice of the appropriate size of feature subsets is still being studied. Since a small subspace size can make the algorithm even faster, the chance to fall into missing informative features or also missing correlation between several features can ever be strong enough. To address that challenge, in this paper, we will randomly select the subspace size, relative to each individual EEk-NN classifier, in the range $[n/3; 2n/3]$, which means that at least one-third and at most two-thirds of the original feature set will be used to train each component classifier (i.e. the subspace size S varies from one classifier to another).

5 Experimentations

This section is devoted to studying the performance improvements of our Ensemble EEk-NN classifier in random subspaces compared with that in full feature space. Our comparative study will mainly be based on the percentage of correct classification (PCC) criterion. In what follows, we elaborate our experimentation settings (Sect. 5.1) and our experimentation results (Sect. 5.2).

5.1 Experimentation Settings

Since we are dealing specifically with uncertain knowledge, we have generated several synthetic databases while injecting a degree of uncertainty P, having values comprised within the range $[0, 1]$, to some well−known real data sets obtained from the UCI machine learning repository [14]. Table 1 provides a short description of the different tested databases where #Instances, #Attributes and #Classes denote, respectively, the number of instances, the number of attributes and the number of classes. Four uncertainty levels have been considered in this paper: certain case ($P = 0$), low uncertainty case ($0 < P < 0.4$), middle uncertainty case ($0.4 \leq P < 0.7$) and high uncertainty case ($0.7 \leq P \leq 1$).

Table 1. Description of databases

Databases	#Instances	#Attributes	#Classes
Voting Records	435	16	2
Heart	267	22	2
Monks	195	23	2
Lymphography	148	18	4
Audiology	226	69	24

Let D be a given database described by N instances x^i ($i \in \{1, \ldots, N\}$) and n attributes x^i_j ($j \in \{1, \ldots, n\}$). Let Θ_j be the frame of discernment associated to the attribute j. Suppose that $|\Theta_j|$ is the cardinality of Θ_j, every attribute value $v^i_{j,t}$ relative to an instance x^i such that $v^i_{j,t} \subseteq \Theta_j$ ($t \in \{1, \ldots, |\Theta_j|\}$) will be represented through the belief function framework as follows:

$$m^{\Theta_j}\{x^i\}(v^i_{j,t}) = 1 - P$$
$$m^{\Theta_j}\{x^i\}(\Theta_j) = P$$

(14)

5.2 Experimentation Results

To assesses our model performance, we have undertaken the 10-fold cross validation strategy. This technique splits randomly the treated data into ten equal sized parts where nine part is used as a training set and the remaining as testing sets. A major key issue in our proposed approach is related to the number of neighbors that may give satisfactory results, in our current experimentation tests, we evaluate five values of the nearest neighbors k which respectively correspond to 1, 3, 5, 7 and 9. The PCC results are given from Tables 2, 3, 4, 5 and 6.

According to the results given from Tables 2, 3, 4, 5 and 6, we can deduce that ensembles of the EEk-NN classifier through random subspacing has led to interesting results compared to the individual EEk-NN classifiers that are

Table 2. Results for Heart database (%)

	$k=1$		$k=3$		$k=5$		$k=7$		$k=9$	
	EEk-NN	Ensemble EEk-NN	EEk-NN	Ensemble EEk-NN	EEk-NN	Ensemble EEk-NN	EEk-NN	Ensemble EEk-NN	EEk-NN	Ensemble EEk-NN
No	61.15	**67.30**	63.84	**70.38**	67.30	**68.07**	70	**70.03**	71.15	**71.23**
Low	58.46	**68.84**	64.23	**66.15**	66.92	**69.23**	68.07	**68.07**	**79.03**	78.24
Middle	60	**69.23**	63.07	**65.38**	66.15	**67.69**	**69.61**	67.30	**68.07**	67.69
High	63.84	**68.46**	63.07	**65.76**	66.36	**66.53**	70.76	**71.13**	69.61	**70.03**

Table 3. Results for Vote Records database (%)

	$k=1$		$k=3$		$k=5$		$k=7$		$k=9$	
	EEk-NN	Ensemble EEk-NN	EEk-NN	Ensemble EEk-NN	EEk-NN	Ensemble EEk-NN	EEk-NN	Ensemble EEk-NN	EEk-NN	Ensemble EEk-NN
No	92.79	**92.05**	92.32	**92.65**	**93.02**	92.32	93.72	**94.01**	**93.72**	92.81
Low	92.09	**93.14**	93.02	**93.65**	92.55	**93.24**	93.25	**94.25**	93.25	**94.78**
Middle	91.62	**92.79**	91.39	**92.56**	91.39	**93.12**	91.86	**92.94**	92.32	**94.16**
High	84.18	**87.20**	87.67	**88.60**	88.60	**89.30**	**89.30**	86.97	89.76	**91.86**

Table 4. Results for Monks database (%)

	$k=1$		$k=3$		$k=5$		$k=7$		$k=9$	
	EEk-NN	Ensemble EEk-NN	EEk-NN	Ensemble EEk-NN	EEk-NN	Ensemble EEk-NN	EEk-NN	Ensemble EEk-NN	EEk-NN	Ensemble EEk-NN
No	72	**73.13**	59.81	**60.26**	60.54	**61.68**	70	69.03	79.81	**80.45**
Low	69.63	**71.01**	58.18	**59.49**	63.63	**94.16**	**70.90**	70.65	76.54	**77.88**
Middle	68.9	**69.85**	63.81	**64.23**	66.72	**68.9**	71.09	**72.84**	70.72	**72.13**
High	54.90	**56.14**	53.09	**53.68**	**52.54**	52.03	52.72	**53.26**	54.18	**55.36**

Table 5. Results for Audiology database (%)

	$k=1$		$k=3$		$k=5$		$k=7$		$k=9$	
	EEk-NN	Ensemble EEk-NN	EEk-NN	Ensemble EEk-NN	EEk-NN	Ensemble EEk-NN	EEk-NN	Ensemble EEk-NN	EEk-NN	Ensemble EEk-NN
No	63.18	**64.22**	60.45	**60.67**	52.72	**53.16**	50.45	**51.26**	44.54	**45.22**
Low	52.72	**52.98**	55.45	**55.67**	53.63	**53.87**	47.27	**47.56**	45.9	**46.81**
Middle	52.72	**53.24**	**48.18**	47.84	**44.54**	44.22	41.13	**42.76**	40.45	**41.68**
High	**15.45**	14.49	23.18	**24.01**	21.36	**22.45**	22.27	**23.46**	18.18	**18.96**

Table 6. Results for Lymphography database (%)

	$k=1$		$k=3$		$k=5$		$k=7$		$k=9$	
	EEk-NN	Ensemble EEk-NN	EEk-NN	Ensemble EEk-NN	EEk-NN	Ensemble EEk-NN	EEk-NN	Ensemble EEk-NN	EEk-NN	Ensemble EEk-NN
No	84.28	**84.51**	85	**85.07**	62.42	**63.45**	85.71	**86.25**	85	**85.42**
Low	80	**81.12**	85.71	**86.13**	**83.57**	82.56	**86.42**	81.17	85.71	**86.96**
Middle	82.14	**82.42**	**84.28**	83.96	86.42	**87.22**	84.28	**85.14**	82.14	**83.27**
High	58.57	**58.63**	61.42	**62.19**	59.28	**61.02**	58.57	**59.13**	65.71	**66.48**

learnt with the full feature space. In fact, the PCC yielded by an ensemble of classifiers is generally better than that yielded by an individual classifier for the most of cases. For instance, let us consider k equals 5, the PCC results yielded by the ensemble system on the Heart database with No, Low, Middle and High uncertainties are respectively equal to 67.30%, 66.92%, 66.15% and 66.36%. However, there are equal to 68.07%, 69.23%, 67.69% and 66.53% when using an individual system. This small difference may be explained by the existence of irrelevant and redundant features as a consequence of the random method.

6   Conclusion

In this paper, we have proposed an ensemble EEk-NN classifier through random subspaces with the aim of increasing the classification performance for a given classification problem. For assessing the performance of our proposed approach, we have carried out a comparative study between the ensemble EEk-NN classifier in random subspaces and that in full feature space when relied on the PCC assessment criterion. Although the RSM method can unfortunately increase the risk that irrelevant and redundant features may be part of the selected subsets, numerical results have shown that ensemble EEk-NN classifiers have contributed to somewhat more favorable PCC results for the different mentioned databases. To promote better and more effective classification performance, in our future studies and research projects, we look forward to solutions allowing to produce the best possible feature subsets.

References

1. Altınçay, H.: Ensembling evidential k-nearest neighbor classifiers through multi-modal perturbation. Appl. Soft Comput. **7**(3), 1072–1083 (2007)
2. Bay, S.D.: Combining nearest neighbor classifiers through multiple feature subsets. In: 15th International Conference on Machine Learning, vol. 98, pp. 37–45 (1998)
3. Breiman, L.: Bagging predictors. Mach. Learn. **24**(2), 123–140 (1996)
4. Bryll, R., Gutierrez-Osuna, R., Quek, F.: Attribute bagging: improving accuracy of classifier ensembles by using random feature subsets. Pattern Recogn. **36**(6), 1291–1302 (2003)
5. Cho, S.B., Won, H.-H.: Cancer classification using ensemble of neural networks with multiple significant gene subsets. Appl. Intell. **26**(3), 243–250 (2007)
6. Dempster, A.P.: Upper and lower probabilities induced by a multivalued mapping. Ann. Math. Stat. **38**, 325–339 (1967)
7. Denoeux, T.: A k-nearest neighbor classification rule based on Dempster-Shafer theory. IEEE Trans. Syst. Man Cybern. **25**(5), 804–813 (1995)
8. Günter, S., Bunke, H.: Feature selection algorithms for the generation of multiple classifier systems and their application to handwritten word recognition. Pattern Recogn. Lett. **25**(11), 1323–1336 (2004)
9. Jiao, L., Denœux, T., Pan, Q.: Evidential editing K-nearest neighbor classifier. In: Destercke, S., Denoeux, T. (eds.) ECSQARU 2015. LNCS, vol. 9161, pp. 461–471. Springer, Cham (2015). doi:10.1007/978-3-319-20807-7_42

10. Jousselme, A., Grenier, D., Bossé, E.: A new distance between two bodies of evidence. Inf. Fusion **2**(2), 91–101 (2001)
11. Kim, Y.: Toward a successful crm: variable selection, sampling, and ensemble. Decis. Support Syst. **41**(2), 542–553 (2006)
12. Kohavi, R., John, G.H.: Wrappers for feature subset selection. Artif. Intell. **97**(1–2), 273–324 (1997)
13. Kuncheva, L., Skurichina, M., Duin, R.P.: An experimental study on diversity for bagging and boosting with linear classifiers. Inf. Fusion **3**(4), 245–258 (2002)
14. Murphy, P., Aha, D.: UCI repository databases (1996). http://www.ics.uci.edu/mlear
15. Opitz, D., Maclin, R.: Popular ensemble methods: an empirical study. J. Artif. Intell. Res. **11**, 169–198 (1999)
16. Ristic, B., Smets, P.: The TBM global distance measure for the association of uncertain combat id declarations. Inf. Fusion **7**(3), 276–284 (2006)
17. Sánchez-Maroño, N., Alonso-Betanzos, A., Tombilla-Sanromán, M.: Filter methods for feature selection – a comparative study. In: Yin, H., Tino, P., Corchado, E., Byrne, W., Yao, X. (eds.) IDEAL 2007. LNCS, vol. 4881, pp. 178–187. Springer, Heidelberg (2007). doi:10.1007/978-3-540-77226-2_19
18. Schapire, R.E.: The boosting approach to machine learning: an overview. In: Denison, D.D., Hansen, M.H., Holmes, C.C., Mallick, B., Yu, B. (eds.) Nonlinear Estimation and Classification. LNS, vol. 171, pp. 149–171. Springer, New York (2003). doi:10.1007/978-0-387-21579-2_9
19. Skurichina, M., Duin, R.P.: Bagging, boosting and the random subspace method for linear classifiers. Pattern Anal. Appl. **5**(2), 121–135 (2002)
20. Smets, P.: Decision making in the TBM: the necessity of the pignistic transformation. Int. J. Approximate Reasoning **38**(2), 133 147 (2005)
21. Smets, P., Kennes, R.: The transferable belief model. Artif. Intell. **66**(2), 191–234 (1994)
22. Tessem, B.: Approximations for efficient computation in the theory of evidence. Artif. Intell. **61**(2), 315–329 (1993)
23. Trabelsi, A., Elouedi, Z., Lefevre, E.: A novel k-nn approach for data with uncertain attribute values. In: 30th International Conference on Industrial, Engineering and other Applications of Applied Intelligent Systems. Springer (2017, to appear)
24. Tumer, K., Ghosh, J.: Classifier combining: analytical results and implications. In: Proceedings of the National Conference on Artificial Intelligence, pp. 126–132 (1996)
25. Zouhal, L.M., Denoeux, T.: An evidence-theoretic k-nn rule with parameter optimization. IEEE Trans. Syst. Man Cybern. Part C (Appl. Rev.) **28**(2), 263–271 (1998)

Conditionals

Comparison of Inference Relations Defined over Different Sets of Ranking Functions

Christoph Beierle$^{(\boxtimes)}$ and Steven Kutsch

Department of Computer Science, University of Hagen, 58084 Hagen, Germany
christoph.beierle@fernuni-hagen.de

Abstract. Skeptical inference in the context of a conditional knowledge base $\mathcal{R}$ can be defined with respect to a set of models of $\mathcal{R}$. For the semantics of ranking functions that assign a degree of surprise to each possible world, we develop a method for comparing the inference relations induced by different sets of ranking functions. Using this method, we address the problem of ensuring the correctness of approximating c-inference for $\mathcal{R}$ by constraint satisfaction problems (CSPs) over finite domains. While in general, determining a sufficient upper bound for these CSPs is an open problem, for a sequence of simple knowledge bases investigated only experimentally before, we prove that using the number of conditionals in $\mathcal{R}$ as an upper bound correctly captures skeptical c-inference.

1 Introduction

For a knowledge base $\mathcal{R}$ containing conditionals of the form *If A then usually B*, various semantics have been proposed, e.g. [4,9]. Here, we will consider the approach of ranking functions (or *Ordinal Conditional Functions (OCF)* [10]), assigning a degree of surprise to each possible world. The models of $\mathcal{R}$ are then OCFs accepting all conditionals in $\mathcal{R}$, and every OCF model of $\mathcal{R}$ induces a nonmonotonic inference relation (e.g. [4,9,10]). For any set O of models of $\mathcal{R}$, skeptical inference with respect to O takes all elements of O into account. C-representations are particular ranking functions exibiting desirable inference properties [7], and c-inference is skeptical inference with respect to all c-representations of $\mathcal{R}$ [1].

The two main objectives of this paper are (1) to develop an approach for comparing the inference relations with respect to two different sets of OCFs O and O', and (2) to illustrate how this approach can be used for proving that in the context of c-representations [7], particular upper bounds in a finite domain constraint system are sufficient for correctly modeling skeptical c-inference [1] so that only a subset of all c-representations have to be taken into account.

For checking that the inference relations with respect to O and O' are identical, we introduce the notion of *merged order compatibility* and show that it suffices to check that their *inference cores* coincide if O and O' are merged order compatible. We demonstrate that there are knowledge bases $\mathcal{R}$ such that the set of all ranking modes of $\mathcal{R}$ is not merged order compatible, while at the

© Springer International Publishing AG 2017
A. Antonucci et al. (Eds.): ECSQARU 2017, LNAI 10369, pp. 225–235, 2017.
DOI: 10.1007/978-3-319-61581-3_21

same time the set of all c-representations of $\mathcal{R}$ is merged order compatible. We then investigate how this approach can be employed for c-representations [7] and skeptical c-inference [1]. For the sequence of knowledge bases $\mathcal{R}_n$ considered in [2] that contain only conditional facts of the form $(a|\top)$ we formally prove upper bounds that are sufficient for skeptical c-inference. This indicates that the concepts developed here may be helpful for addressing the open problem of determining upper bounds for general knowledge bases $\mathcal{R}$ that are sufficient for modelling skeptical c-inference for $\mathcal{R}$.

2 Background: Conditional Logic and OCFs

Let $\Sigma = \{v_1, ..., v_m\}$ be a propositional alphabet. A *literal* is the positive (v_i) or negated $(\overline{v_i})$ form of a propositional variable, $\dot{v}_i$ stands for either v_i or $\overline{v_i}$. From these we obtain the propositional language $\mathcal{L}$ as the set of formulas of Σ closed under negation $\neg$, conjunction $\wedge$, and disjunction $\vee$. For shorter formulas, we abbreviate conjunction by juxtaposition (i.e., AB stands for $A \wedge B$), and negation by overlining (i.e., $\overline{A}$ is equivalent to $\neg A$). Let Ω_Σ denote the set of possible worlds over $\mathcal{L}$; Ω_Σ will be taken here simply as the set of all propositional interpretations over $\mathcal{L}$ and can be identified with the set of all complete conjunctions over Σ; we will often just write Ω instead of Ω_Σ. For $\omega \in \Omega$, $\omega \models A$ means that the propositional formula $A \in \mathcal{L}$ holds in the possible world ω. For any propositional formula A let $\Omega_A = \{\omega \in \Omega \mid \omega \models A\}$ be the set of all possible worlds satisfying A.

A *conditional* $(B|A)$ with $A, B \in \mathcal{L}$ encodes the defeasible rule "if A then usually B" and is a trivalent logical entity with the evaluation [5,7]

$$[\![(B|A)]\!]_\omega = \begin{cases} true & \text{iff} \quad \omega \models AB \quad \text{(verification)} \\ false & \text{iff} \quad \omega \models A\overline{B} \quad \text{(falsification)} \\ undefined & \text{iff} \quad \omega \models \overline{A} \quad \text{(not applicable)} \end{cases}$$

An *Ordinal Conditional Function* (OCF, ranking function) [10] is a function $\kappa : \Omega \to \mathbb{N}_0 \cup \{\infty\}$ that assigns to each world $\omega \in \Omega$ an implausibility rank $\kappa(\omega)$: the higher $\kappa(\omega)$, the more surprising ω is. OCFs have to satisfy the normalization condition that there has to be a world that is maximally plausible, i.e., $\kappa^{-1}(0) \neq \emptyset$. The rank of a formula A is defined by $\kappa(A) = \min\{\kappa(\omega) \mid \omega \models A\}$. An OCF κ *accepts* a conditional $(B|A)$, denoted by $\kappa \models (B|A)$, iff the verification of the conditional is less surprising than its falsification, i.e., iff $\kappa(AB) < \kappa(A\overline{B})$. This can also be understood as a nonmonotonic inference relation between the premise A and the conclusion B: We say that A κ-*entails* B, written $A \vdash^\kappa B$, iff κ accepts the conditional $(B|A)$: $\kappa \models (B|A)$ iff $\kappa(AB) < \kappa(A\overline{B})$ iff $A \vdash^\kappa B$.

Note that κ-entailment is based on the total preorder on possible worlds induced by a ranking function κ as $A \vdash^\kappa B$ iff for all $\omega' \in \Omega_{A\overline{B}}$, there is a $\omega \in \Omega_{AB}$ such that $\kappa(\omega) < \kappa(\omega')$.

The acceptance relation is extended as usual to a set $\mathcal{R}$ of conditionals, called a *knowledge base*, by defining $\kappa \models \mathcal{R}$ iff $\kappa \models (B|A)$ for all $(B|A) \in \mathcal{R}$. This is synonymous to saying that κ is *admissible* with respect to $\mathcal{R}$ [6], or that κ is a *ranking model* of $\mathcal{R}$; the set of all ranking models of $\mathcal{R}$ is denoted by $Mod(\mathcal{R})$. $\mathcal{R}$ is *consistent* iff it has a ranking model.

3 Skeptical Inference and Merged Order Inference

While each OCF κ accepting $\mathcal{R}$ induces a nonmonotonic inference relation, also each set O of such ranking functions induces an inference relation determined by taking all elements of O into account.

Definition 1 (skeptical inference). *Let $\mathcal{R}$ be a knowledge base, $O \subseteq Mod(\mathcal{R})$, and $A, B \in \mathcal{L}$. Skeptical Inference over O in the context of $\mathcal{R}$, denoted by $\mathrel{\mathop{\vdash}\limits_{\mathcal{R}}^{O}}$, is defined by $A \mathrel{\mathop{\vdash}\limits_{\mathcal{R}}^{O}} B$ iff $A \mathrel{\vdash}^{\kappa} B$ for all $\kappa \in O$.*

Thus, $A \mathrel{\mathop{\vdash}\limits_{\mathcal{R}}^{O}} B$ holds if every $\kappa \in O$ accepts $(B|A)$. The skeptical inference relations defined over two different sets of OCFs may be identical. Instead of having to check the acceptance of all possible conditionals $(B|A)$ with respect to both sets of OCFs, we will investigate conditions under which it suffices to check only so-called base conditionals.

Definition 2 (base conditional). *A base conditional over the signature Σ is a conditional of the form $(\omega_1 | \omega_1 \vee \omega_2)$ with $\omega_1, \omega_2 \in \Omega_{\Sigma}$ and $\omega_1 \neq \omega_2$.*

Note that a base conditional $(\omega_1 | \omega_1 \vee \omega_2)$ is accepted by a ranking model κ, iff $\kappa(\omega_1) < \kappa(\omega_2)$. To characterize the behavior of an inference relation $\mathrel{\vdash}$ for these base conditionals, we define the *inference core* of an inference relation as the reduction of $\mathrel{\vdash}$ from pairs of formulas to pairs of possible worlds.

Definition 3 (inference core, $\lfloor \mathrel{\vdash} \rfloor$). *Let $\mathrel{\vdash}$ be an inference relation. The inference core of $\mathrel{\vdash}$, denoted by $\lfloor \mathrel{\vdash} \rfloor$, is the set of all pairs $(\omega_1, \omega_2) \in \Omega \times \Omega$ with $\omega_1 \neq \omega_2$, such that $\omega_1 \vee \omega_2 \mathrel{\vdash} \omega_1$, i.e., $\lfloor \mathrel{\vdash} \rfloor = \{(\omega_1, \omega_2) \mid \omega_1 \vee \omega_2 \mathrel{\vdash} \omega_1\}$.*

The notion of the inference core is based on an inference relation. The corresponding concept of a *merged order* is based solely on a set of ranking models:

Definition 4 (merged order, $<_O$). *Let O be a set of OCFs. The merged order $<_O$ is given by $<_O = \{(\omega_1, \omega_2) \mid \omega_1 \neq \omega_2, \kappa(\omega_1) < \kappa(\omega_2)$ for all $\kappa \in O\}$.*

Note that in general $<_O$ is a strict weak ordering, i.e. it is irreflexive, asymmetric, and transitive. The inference core of skeptical inference over a set of OCFs O coincides with the merged order induced by O:

Proposition 1 (inference core and merged order). *For any knowledge base $\mathcal{R}$ and any set $O \subseteq Mod(\mathcal{R})$ it holds that $\lfloor \mathrel{\mathop{\vdash}\limits_{\mathcal{R}}^{O}} \rfloor = <_O$.*

Proof

$$
\begin{aligned}
<_O &= \{(\omega_1, \omega_2) \in \Omega \times \Omega \mid \omega_1 \neq \omega_2, \kappa(\omega_1) < \kappa(\omega_2) \text{ for all } \kappa \in O\} \\
&= \{(\omega_1, \omega_2) \in \Omega \times \Omega \mid \omega_1 \neq \omega_2, \kappa \models (\omega_1 | \omega_1 \vee \omega_2) \text{ for all } \kappa \in O\} \\
&= \{(\omega_1, \omega_2) \in \Omega \times \Omega \mid \omega_1 \neq \omega_2, \omega_1 \vee \omega_2 \mathrel{\vdash}^{\kappa} \omega_1 \text{ for all } \kappa \in O\} \\
&= \lfloor \mathrel{\mathop{\vdash}\limits_{\mathcal{R}}^{O}} \rfloor \qquad\qquad\qquad\qquad\qquad\qquad\qquad\qquad\qquad\qquad\qquad \square
\end{aligned}
$$

We now define an inference relation with respect to $<_O$ in a similar way to inference with respect to the total pre-order on worlds induced by an OCF.

Definition 5 (inference relation induced by merged order, $\vdash_{\mathcal{R}}^{<_O}$). *Let $\mathcal{R}$ be a knowledge base, $O \subseteq Mod(\mathcal{R})$, and $A, B \in \mathcal{L}$. Then*

$$A \vdash_{\mathcal{R}}^{<_O} B \quad \textit{iff} \quad \textit{for all } \omega' \in \Omega_{A\overline{B}} \textit{ there is a } \omega \in \Omega_{AB} \textit{ such that } \omega <_O \omega'.$$

Proposition 2. *For any two sets of ranking models O and O' of $\mathcal{R}$ it holds that if $<_O = <_{O'}$ then $\vdash_{\mathcal{R}}^{<_O} = \vdash_{\mathcal{R}}^{<_{O'}}$.*

The inference relation induced by the merged order of a set of OCFs O approximates skeptical inference over O.

Proposition 3. *For any knowledge base $\mathcal{R}$ and $O \subseteq Mod(\mathcal{R})$ it holds that*

$$\vdash_{\mathcal{R}}^{<_O} \subseteq \vdash_{\mathcal{R}}^{O} \tag{1}$$

Proof.

$$
\begin{aligned}
A \vdash_{\mathcal{R}}^{<_O} B &\Leftrightarrow \forall \omega' \in \Omega_{A\overline{B}} \; \exists \omega \in \Omega_{AB} : \omega <_O \omega' \\
&\Leftrightarrow \forall \omega' \in \Omega_{A\overline{B}} \; \exists \omega \in \Omega_{AB} \; \forall \kappa \in O : \kappa(\omega) < \kappa(\omega') \\
&\Rightarrow \forall \kappa \in O : \min\{\kappa(\omega) \mid \omega \models AB\} < \min\{\kappa(\omega) \mid \omega \models A\overline{B}\} \\
&\Leftrightarrow \forall \kappa \in O : A \vdash^{\kappa} B \\
&\Leftrightarrow A \vdash_{\mathcal{R}}^{O} B \qquad\qquad\qquad\qquad\qquad\qquad\qquad \square
\end{aligned}
$$

While it is always the case that an inference over the merged order of a set O is also a skeptical inference over that set, the other direction of (1) does not hold in general.

Proposition 4. *There is a knowledge base $\mathcal{R}$ and a set $O \subseteq Mod(\mathcal{R})$ with*

$$\vdash_{\mathcal{R}}^{O} \not\subseteq \vdash_{\mathcal{R}}^{<_O}. \tag{2}$$

Proof. Consider $\mathcal{R} = \{(a|\top)\}$ over $\Sigma = \{a, b\}$. Let κ_1 and κ_2 be defined as:

$$
\kappa_1(\omega) = \begin{cases} 0 & \text{if } \omega = ab \\ 1 & \text{otherwise} \end{cases}
\qquad
\kappa_2(\omega) = \begin{cases} 0 & \text{if } \omega = a\overline{b} \\ 1 & \text{otherwise} \end{cases}
$$

Both κ_1 and κ_2 accept $\mathcal{R}$, but for $O = \{\kappa_1, \kappa_2\}$ it holds that $<_O = \emptyset$. Thus, since both OCFs accept $\mathcal{R}$ it holds that $\top \vdash_{\mathcal{R}}^{O} a$, but since $<_O$ is empty, $\top \not\vdash_{\mathcal{R}}^{<_O} a$. $\square$

Since (1) holds for all sets of OCF models, but the reverse direction does not hold in general, we introduce the notion of *merged order compatibility*, classifying the sets of OCFs for which the other direction of (1) holds.

Definition 6 (merged order compatible). *Let $\mathcal{R}$ be a knowledge base and $O \subseteq Mod(\mathcal{R})$. O is called* merged order compatible *iff $\vdash_{\mathcal{R}}^{O} \subseteq \vdash_{\mathcal{R}}^{<_O}$.*

Thus, for merged order comaptible O we immediately get:

Proposition 5. *If $O \subseteq Mod(\mathcal{R})$ is merged order compatible, then $\mathop{\vdash}\limits_{\mathcal{R}}^{<_O} = \mathop{\vdash}\limits_{\mathcal{R}}^{O}$.*

Since the merged order of a set of ranking models is equal to the inference core of the skeptical inference over that set of models, merged order compatibility ensures that equivalence of skeptical inference relations coincides with equivalence of inference cores.

Proposition 6. *For any two merged order compatible sets of ranking models O and O' of a knowledge base $\mathcal{R}$ it holds that:*

$$\lfloor \mathop{\vdash}\limits_{\mathcal{R}}^{O} \rfloor = \lfloor \mathop{\vdash}\limits_{\mathcal{R}}^{O'} \rfloor \text{ iff } \mathop{\vdash}\limits_{\mathcal{R}}^{O} = \mathop{\vdash}\limits_{\mathcal{R}}^{O'} \tag{3}$$

Proof. The direction from right to left trivially holds since base conditionals are a subset of all conditionals. For the other direction we have:

$$\lfloor \mathop{\vdash}\limits_{\mathcal{R}}^{O} \rfloor = \lfloor \mathop{\vdash}\limits_{\mathcal{R}}^{O'} \rfloor \Rightarrow <_O = <_{O'} \qquad \text{(Proposition 1)}$$
$$\Rightarrow \mathop{\vdash}\limits_{\mathcal{R}}^{<_O} = \mathop{\vdash}\limits_{\mathcal{R}}^{<_{O'}} \qquad \text{(Proposition 2)}$$
$$\Rightarrow \mathop{\vdash}\limits_{\mathcal{R}}^{O} = \mathop{\vdash}\limits_{\mathcal{R}}^{O'} \qquad \text{(Proposition 5)} \qquad \square$$

Note that according to Proposition 6, merged order compatibility provides a sufficient condition for reducing the question of skeptical inference equivalence to the equality of the inference cores.

4   C-Inference and Merged Order Compatibility

We will now illustrate merged order compatibility for a special kind of ranking models. C-Representations are special ranking models of a knowledge base $\mathcal{R}$, obtained by assigning individual impacts to the conditionals in $\mathcal{R}$. The rank of a possible world is then defined as the sum of impacts of falsified conditionals.

Definition 7 (c-representation [7,8]). *A c-representation of a knowledge base $\mathcal{R}$ is a ranking function κ constructed from integer impacts $\eta_i \in \mathbb{N}_0$ assigned to each conditional $(B_i|A_i)$ such that κ accepts $\mathcal{R}$ and is given by:*

$$\kappa(\omega) = \sum_{\substack{1 \leqslant i \leqslant n \\ \omega \models A_i \overline{B_i}}} \eta_i \tag{4}$$

Every c-representation exibits desirable inference properties, and two c-representations induce the same inference relation if they induce the same total preorder on worlds. In [1], a modeling of c-representations as solutions of a constraint satisfaction problem $CR(\mathcal{R})$ is given and shown to be correct and complete with respect to the set of all c-representations of $\mathcal{R}$. Recently, it has been suggested to take inferential equivalence of c-representations into account and to sharpen $CR(\mathcal{R})$ by introducing an upper bound for the impact values η_i.

Definition 8 ($CR^u(\mathcal{R})$) [3]). *Let $\mathcal{R} = \{(B_1|A_1), \ldots, (B_n|A_n)\}$ and $u \in \mathbb{N}$. The finite domain constraint satisfaction problem $CR^u(\mathcal{R})$ on the constraint variables $\{\eta_1, \ldots, \eta_n\}$ ranging over $\mathbb{N}$ is given by the conjunction of the constraints, for all $i \in \{1, \ldots, n\}$:*

$$\eta_i \geqslant 0 \tag{5}$$

$$\eta_i > \min_{\substack{\omega \models A_i B_i \\ \omega \models A_j \overline{B_j}}} \sum_{\substack{j \neq i \\ \omega \models A_j \overline{B_j}}} \eta_j - \min_{\substack{\omega \models A_i \overline{B_i} \\ \omega \models A_j \overline{B_j}}} \sum_{\substack{j \neq i \\ \omega \models A_j \overline{B_j}}} \eta_j \tag{6}$$

$$\eta_i \leqslant u \tag{7}$$

A solution of $CR^u(\mathcal{R})$ is an n-tuple $(\eta_1, \ldots, \eta_n)$ of natural numbers, its set of solutions is denoted by $Sol(CR^u(\mathcal{R}))$. For $\vec{\eta} \in Sol(CR^u(\mathcal{R}))$ and κ as in Eq. (4), κ is the *OCF induced by* $\vec{\eta}$, denoted by $\kappa_{\vec{\eta}}$, and the set of all induced OCFs is denoted by $\mathcal{O}(CR^u(\mathcal{R})) = \{\kappa_{\vec{\eta}} \mid \vec{\eta} \in Sol(CR^u(\mathcal{R}))\}$. The constraint satisfaction problem $CR(\mathcal{R})$, given in [1], is obtained by removing the constraints (7) from $CR^u(\mathcal{R})$.

C-inference is skeptical inference over the set of all c-representations.

Definition 9 (c-inference, $\vdash^c_\mathcal{R}$ [1]). *Let $\mathcal{R}$ be a knowledge base and let A, B be formulas. B is a (skeptical) c-inference from A in the context of $\mathcal{R}$, denoted by $A \vdash^c_\mathcal{R} B$, iff $A \vdash^\kappa B$ holds for all c-representations κ for $\mathcal{R}$.*

We will now illustrate c-representations, c-inference, and how the inference over the merged order of the set of all c-representations accepting a knowledge base can coincide with c-inference in the context of that knowledge base.

Example 1 ($\mathcal{R}_{lw}$). Consider $\Sigma_{lw} = \{l, w\}$ and $\mathcal{R}_{lw} = \{r_1, r_2, r_3\}$ with

$r_1 = (\overline{w}|l)$ *"land vehicles are usually not watercrafts"*
$r_2 = (l \vee \overline{w}|\top)$ *"usually, something is a land vehicle or not a watercraft"*
$r_3 = (w|\overline{l})$ *"things that are not land vehicles, are usually watercrafts"*

representing some default knowledge about vehicles in a country like Germany.

Using the verification and falsification behavior of the four possible worlds reveals that $\vec{\eta}_1, \ldots, \vec{\eta}_5$ as given in Table 1 are solutions to $CR(\mathcal{R}_{lw})$. Furthermore, there are no other solutions of $CR(\mathcal{R}_{lw})$ inducing an ordering on worlds that is different from every of the orderings induced by $\kappa_{\vec{\eta}_1}, \ldots, \kappa_{\vec{\eta}_5}$; for example, the solution $\vec{\eta}_6 = (3, 2, 3)$ induces the same ordering on worlds as $\kappa_{\vec{\eta}_5}$ and thus allows for exactly the same inferences.

Therefore, the merged order for $O = \{\kappa_{\vec{\eta}_1}, \ldots, \kappa_{\vec{\eta}_5}\}$, given in the lower right corner of Table 1, coincides with the merged order over all c-representations of $\mathcal{R}_{lw}$. Checking all pairs of formulas over Σ_{lw} shows that for $\mathcal{R}_{lw}$ there is no difference between merged order inference over O and skeptical c-inference.

The following example illustrates an interesting difference between the set of all ranking models of a knowledge base and the set of its c-representations and shows that there are knowledge bases $\mathcal{R}$ such that the former set is not merged order compatible while the latter set is merged order compatible.

Table 1. Verification (v), falsification (f), impacts (η_i), solution vectors $\vec{\eta}_i$, induced OCFs $\kappa_{\vec{\eta}_i}$, and merged order of $\{\kappa_{\vec{\eta}_1}, \ldots, \kappa_{\vec{\eta}_5}\}$ for $CR(\mathcal{R}_{lw})$ in Example 1.

ω	r_1: $(\overline{w}\lvert l)$	r_2: $(l \vee \overline{w}\lvert\top)$	r_3: $(w\lvert\overline{l})$	impact on ω	$\kappa_{\vec{\eta}_1}(\omega)$	$\kappa_{\vec{\eta}_2}(\omega)$	$\kappa_{\vec{\eta}_3}(\omega)$	$\kappa_{\vec{\eta}_4}(\omega)$	$\kappa_{\vec{\eta}_5}(\omega)$
$l\,w$	f	v	$-$	η_1	1	3	1	2	2
$l\,\overline{w}$	v	v	$-$	0	0	0	0	0	0
$\overline{l}\,w$	$-$	f	v	η_2	2	1	1	1	1
$\overline{l}\,\overline{w}$	$-$	v	f	η_3	3	2	2	3	2

impacts:	η_1	η_2	η_3	merged order:
$\vec{\eta}_1$	1	2	3	
$\vec{\eta}_2$	3	1	2	
$\vec{\eta}_3$	1	1	2	
$\vec{\eta}_4$	2	1	3	
$\vec{\eta}_5$	2	1	2	

merged order:

$$
\begin{array}{ccc}
 & \overline{l}\,\overline{w} & \\
 \nearrow & & \\
\overline{l}\,w & \big\downarrow & l\,w \\
 \nwarrow & & \nearrow \\
 & l\,\overline{w} &
\end{array}
$$

Example 2. Consider the knowledge base $\mathcal{R}$ and Σ from the proof of Proposition 4, and let $P = Mod(\mathcal{R})$ and let O be the set of all c-representations accepting $\mathcal{R}$. For both P and O, a can be inferred skeptically from $\top$ in the context of $\mathcal{R}$, i.e. $\top \vdash_{\mathcal{R}}^{P} a$ and $\top \vdash_{\mathcal{R}}^{O} a$. The two ranking functions κ_1 and κ_2 used in the proof of Proposition 4 both accept $\mathcal{R}$ and are thus elements of P. Since there are no two distinct worlds ω and ω' with $\kappa_1(\omega) < \kappa_1(\omega')$ and $\kappa_2(\omega) < \kappa_2(\omega')$, the merged order $<_P$ is empty, and therefore $\top \not\vdash_{\mathcal{R}}^{<_P} a$. On the other hand, for every c-representation κ accepting $\mathcal{R}$ it holds that $\kappa(ab) = \kappa(a\overline{b})$ and $\kappa(\overline{a}b) = \kappa(\overline{a}\overline{b})$ and $\kappa(ab) < \kappa(\overline{a}b)$. Thus, $<_O = \{(ab, \overline{a}b), (ab, \overline{a}\overline{b}), (a\overline{b}, \overline{a}b), (a\overline{b}, \overline{a}\overline{b})\}$ and hence $\top \vdash_{\mathcal{R}}^{<_O} a$. In fact, the set O of all c-representations accepting $\mathcal{R}$ is merged order compatible, while $\top \vdash_{\mathcal{R}}^{P} a$ and $\top \not\vdash_{\mathcal{R}}^{<_P} a$ shows that the set P of all ranking models of $\mathcal{R}$ is not merged order compatible.

For studying the exact relationship between $CR(\mathcal{R})$ and $CR^u(\mathcal{R})$, the concept of a *sufficient* $CR^u(\mathcal{R})$ was introduced in [3] to capture the idea that only a finite number of c-representations is needed for modeling c-inference.

Definition 10 (sufficient). *Let $\mathcal{R}$ be a knowledge base and let $u \in \mathbb{N}$. Then $CR^u(\mathcal{R})$ is called* sufficient *(for skeptical inference) iff for all formulas A, B we have*

$$A \vdash_{\mathcal{R}}^{c} B \quad \textit{iff} \quad A \vdash_{\mathcal{R}}^{\mathcal{O}(CR^u(\mathcal{R}))} B.$$

If $CR^u(\mathcal{R})$ is sufficient, we will also call u sufficient for $\mathcal{R}$.

In terms of the classical skeptical inference relation over a set of ranking models given in Definition 1 this means that $CR^u(\mathcal{R})$ is sufficient iff

$$A \vdash_{\mathcal{R}}^{\mathcal{O}(CR(\mathcal{R}))} B \quad \textit{iff} \quad A \vdash_{\mathcal{R}}^{\mathcal{O}(CR^u(\mathcal{R}))} B. \tag{8}$$

For various $\mathcal{R}$ and u, we will now use merged order compatibility for proving (8).

5   Proving Sufficient Upper Bounds

In this section, we continue the investigation from [2] and use the concepts from the previous section to formally prove an experimental result from [2].

Definition 11 ($\Sigma_n, \mathcal{R}_n$). *For* $n \geqslant 1$ *and* $\Sigma_n = \{a_1, \ldots, a_n\}$, $\mathcal{R}_n = \{(a_1|\top), \ldots, (a_n|\top)\}$ *is called the* knowledge base of n conditional facts.

Note that from the constraints in $CR(\mathcal{R}_n)$ and $CR^u(\mathcal{R}_n)$ it follows that for all impacts in c-representations accepting $\mathcal{R}_n$ it holds that $\eta_i \geqslant 1$. In the rest of this section, we investigate how the concepts of merged order compatibility and inference cores can be used to prove that for $\mathcal{R}_n$ the CSP $\mathcal{O}(CR^{n-1}(\mathcal{R}_n))$ is indeed sufficient. In [2] this was solely illustrated by means of some examples.

Because the structure of knowledge bases $\mathcal{R}_n$ is very simple, the rank of a world ω over Σ_n assigned by a c-representation depends on the set of falsified atoms in ω in a very predictable way.

Definition 12 ($f(\omega), <_f$). *For* $\omega \in \Omega_{\Sigma_n}$, $f(\omega) = \{i \mid \omega \models \overline{a_i}, i \in \{1, \ldots, n\}\}$ *is the set of indices of the negated literals in* ω. *The ordering* $<_f$ *on* Ω_{Σ_n} *is defined such that for two worlds* $\omega, \omega' \in \Omega_{\Sigma_n}$, $\omega <_f \omega'$ *iff* $f(\omega) \subsetneqq f(\omega')$.

As the ordering $<_O$ on worlds, also $<_f$ induces an inference relation.

Definition 13 ($\vdash\!\!\sim_{\mathcal{R}_n}^{<_f}$). *For* $n > 1$ *and formulas* $A, B \in \mathcal{L}_{\Sigma_n}$

$$A \vdash\!\!\sim_{\mathcal{R}_n}^{<_f} B \quad \text{iff} \quad \text{for every } \omega' \in \Omega_{A\overline{B}}, \text{ there is a } \omega \in \Omega_{AB} \text{ such that } \omega <_f \omega'.$$

The following proposition generalizes a proposition from [2] regarding the ranking of worlds ω and ω' incomparable in $<_f$.

Proposition 7. *Let* $n > 1$, $\omega' \in \Omega_{\Sigma_n}$ *and* $\Omega_V = \{\omega_1, \ldots, \omega_m\} \subseteq \Omega_{\Sigma_n}$. *If for all* $i \in \{1, \ldots, m\}$, $f(\omega') \not\subseteq f(\omega_i)$ *and* $f(\omega_i) \not\subseteq f(\omega')$, *then there exists a c-representation* κ *accepting* $\mathcal{R}_n$ *such that for all* $i \in \{1, \ldots, m\}$, $\kappa(\omega') \leqslant \kappa(\omega_i)$.

Proof. Let I be $I = (\bigcup_{i=1}^m f(\omega_i)) \setminus f(\omega')$. Note that because of the precondition $f(\omega') \not\subseteq f(\omega_i)$ and $f(\omega_i) \not\subseteq f(\omega')$, it holds that $I \neq \emptyset$. Let $\vec{\eta} = (\eta_1, \ldots, \eta_n)$ with

$$\eta_i = \begin{cases} 1 & i \notin I \\ n-1 & i \in I \end{cases}$$

Since for every $i \in f(\omega')$ the impact vector $\vec{\eta}$ assigns 1 to the corresponding conditional $(a_i|\top) \in \mathcal{R}_n$ and because we know that $\omega' \neq \overline{a_1} \ldots \overline{a_n}$, we get $\kappa_{\vec{\eta}}(\omega') = |f(\omega')| \leqslant n-1$. Because I is not empty, for every $i \in \{1, \ldots, m\}$, there is some $k \in f(\omega_i)$ such that $\eta_k = n-1$. Thus, we get $\kappa_{\vec{\eta}}(\omega_i) \geqslant n-1$. Therefore, it holds that $\kappa_{\vec{\eta}}(\omega') \leqslant \kappa_{\vec{\eta}}(\omega_i)$ for every $i \in \{1, \ldots, m\}$. $\square$

We now use Proposition 7 to show that the inference relation $\vdash\!\!\sim_{\mathcal{R}_n}^{<_f}$ defined over the ordering on worlds $<_f$ is equal to the skeptical inference over all c-representations accepting $\mathcal{R}_n$.

Proposition 8. *For $n > 1$ and $O = \mathcal{O}(CR(\mathcal{R}_n))$, $\mathrel|\joinrel\sim_{\mathcal{R}_n}^{<_f} = \mathrel|\joinrel\sim_{\mathcal{R}_n}^{O}$.*

Proof. Let A and B be arbitrary formulas from $\mathcal{L}_{\Sigma_n}$. If $A \mathrel|\joinrel\sim_{\mathcal{R}_n}^{<_f} B$ then for all $\omega' \in \Omega_{A\overline{B}}$ there is a $\omega \in \Omega_{AB}$ such that $\omega <_f \omega'$. Thus $f(\omega) \subsetneqq f(\omega')$, and because $\kappa(\omega)$ for a c-representation κ is defined by the sum of all impacts of negative literals in ω, it also holds that $\kappa(\omega) < \kappa(\omega')$ for $\kappa \in O$. Thus $A \mathrel|\joinrel\sim^{\kappa} B$ holds for all $\kappa \in O$, implying that $\mathrel|\joinrel\sim_{\mathcal{R}_n}^{<_f} \subseteq \mathrel|\joinrel\sim_{\mathcal{R}_n}^{O}$.

To show the other direction, we assume that $A \mathrel|\joinrel\not\sim_{\mathcal{R}_n}^{<_f} B$ and show that $A \mathrel|\joinrel\not\sim_{\mathcal{R}_n}^{O} B$. If $A \mathrel|\joinrel\not\sim_{\mathcal{R}_n}^{<_f} B$, then there is a world $\omega' \in \Omega_{A\overline{B}}$, such that for all worlds $\omega \in \Omega_{AB}$ $\omega \not<_f \omega'$ holds. If $\omega' <_f \omega$, then $\kappa(\omega') \leqslant \kappa(\omega)$ for every c-representation κ with $\kappa \models \mathcal{R}_n$ and therefore $A \mathrel|\joinrel\not\sim_{\mathcal{R}_n}^{O} B$. If $\omega' \not<_f \omega$ then for all worlds $\omega \in \Omega_{AB}$ $f(\omega') \not\subseteq f(\omega)$ and $f(\omega) \not\subseteq f(\omega')$, and we use Proposition 7 by setting $\Omega_V = \Omega_{AB}$ and construct a c-representation κ such that $\kappa(\omega') \leqslant \kappa(\omega)$ for all $\omega \in \Omega_{AB}$. Thus, $min\{\kappa(\omega) \mid \omega \models A\overline{B}\} \leqslant min\{\kappa(\omega) \mid \omega \models AB\}$, implying $A \mathrel|\joinrel\not\sim_{\kappa} B$ and therefore $A \mathrel|\joinrel\not\sim_{\mathcal{R}_n}^{O} B$.   □

Since both $<_f$ and $<_O$ are orderings of worlds and $\mathrel|\joinrel\sim_{\mathcal{R}_n}^{<_f}$ and $\mathrel|\joinrel\sim_{\mathcal{R}_n}^{<_O}$ are defined in the same way, it is now straightforward to show that $\mathcal{O}(CR(\mathcal{R}_n))$ is merged order compatible for any $\mathcal{R}_n$.

Proposition 9. *For $n > 1$, $\mathcal{O}(CR(\mathcal{R}_n))$ is merged order compatible for $\mathcal{R}_n$.*

Proof. To show that $O = \mathcal{O}(CR(\mathcal{R}_n))$ is merged order compatible for $\mathcal{R}_n$, we need to show that $\mathrel|\joinrel\sim_{\mathcal{R}_n}^{O} \subseteq \mathrel|\joinrel\sim_{\mathcal{R}_n}^{<_O}$. Since we already know $\mathrel|\joinrel\sim_{\mathcal{R}_n}^{O} = \mathrel|\joinrel\sim_{\mathcal{R}_n}^{<_f}$ (Proposition 8), due to Proposition 2 it suffices to show that $<_{f-} <_O$. If $\omega <_f \omega'$, then $f(\omega) \subsetneqq f(\omega')$. As was already pointed out in the proof of Proposition 8, this means that for all c-representations κ we have $\kappa(\omega) < \kappa(\omega')$ and thus $\omega <_O \omega'$. We now have $\mathrel|\joinrel\sim_{\mathcal{R}_n}^{O} = \mathrel|\joinrel\sim_{\mathcal{R}_n}^{<_f} = \mathrel|\joinrel\sim_{\mathcal{R}_n}^{<_O}$ and $\mathcal{O}(CR(\mathcal{R}_n))$ is merged order compatible.   □

Since we do not make use of impacts $\eta_i > n - 1$, the proofs of Propositions 8 and 9 also work for $O = \mathcal{O}(CR^{n-1}(\mathcal{R}_n))$, implying:

Proposition 10. *For $n > 1$, $\mathcal{O}(CR^{n-1}(\mathcal{R}_n))$ is merged order compatible for $\mathcal{R}_n$.*

These results now enable us to prove that $n - 1$ is sufficient for $\mathcal{R}_n$, implying that the inference relation induced by the solutions of $CR^{n-1}(\mathcal{R}_n)$ is equal to the skeptical inference over all c-representations for $\mathcal{R}_n$.

Proposition 11. *For $n > 1$, $CR^{n-1}(\mathcal{R}_n)$ is sufficient for $\mathcal{R}_n$.*

Proof. We need to show that $\mathrel|\joinrel\sim_{\mathcal{R}_n}^{\mathcal{O}(CR^{n-1}(\mathcal{R}_n))} = \mathrel|\joinrel\sim_{\mathcal{R}_n}^{\mathcal{O}(CR(\mathcal{R}_n))}$. Since both $\mathcal{O}(CR^{n-1}(\mathcal{R}_n))$ and $\mathcal{O}(CR(\mathcal{R}_n))$ are merged order compatible, it suffices to show that the inference cores are equal, i.e. $\lfloor \mathrel|\joinrel\sim_{\mathcal{R}_n}^{\mathcal{O}(CR^{n-1}(\mathcal{R}_n))} \rfloor = \lfloor \mathrel|\joinrel\sim_{\mathcal{R}_n}^{\mathcal{O}(CR(\mathcal{R}_n))} \rfloor$.

It is easy to see that if a pair of possible worlds (ω, ω') is in $\lfloor \mathrel|\joinrel\sim_{\mathcal{R}_n}^{\mathcal{O}(CR(\mathcal{R}_n))} \rfloor$, then it is also in $\lfloor \mathrel|\joinrel\sim_{\mathcal{R}_n}^{\mathcal{O}(CR^{n-1}(\mathcal{R}_n))} \rfloor$ since $\mathrel|\joinrel\sim_{\mathcal{R}_n}^{\mathcal{O}(CR^{n-1}(\mathcal{R}_n))}$ allows for possibly more inferences. To show the other direction, we assume that $(\omega, \omega') \notin \lfloor \vdash \frac{\mathcal{O}(CR(\mathcal{R}_n))}{\mathcal{R}_n} \rfloor$ and show that $(\omega, \omega') \notin \lfloor \vdash \frac{\mathcal{O}(CR^{n-1}(\mathcal{R}_n))}{\mathcal{R}_n} \rfloor$.

If (ω, ω') is not in the inference core of the unbounded skeptical c-inference, it means that there is a c-representation κ in which $\kappa(\omega) \geqslant \kappa(\omega')$. If $f(\omega') \subseteq f(\omega)$, then for $\vec{\eta} = (1, \ldots, 1)$ it holds that $\kappa_{\vec{\eta}}(\omega) \geqslant \kappa_{\vec{\eta}}(\omega')$. If $f(\omega) \subseteq f(\omega')$, then there is no c-representation κ such that $\kappa(\omega) \geqslant \kappa(\omega')$, contradicting the assumption. If neither $f(\omega') \subseteq f(\omega)$ nor $f(\omega) \subseteq f(\omega')$ holds, the precondition of Proposition 7 is met for ω' and $\Omega_V = \{\omega\}$, and we can construct a c-representation κ in $\mathcal{O}(CR^{n-1}(\mathcal{R}_n))$ such that $\kappa(\omega) \geqslant \kappa(\omega')$; hence $(\omega, \omega') \notin \lfloor \vdash \frac{\mathcal{O}(CR^{n-1}(\mathcal{R}_n))}{\mathcal{R}_n} \rfloor$. $\square$

6   Conclusions and Further Work

We introduced the notion of inference core of a nonmonotonic inference relation taking only so called base conditionals into account. By showing that a set of ranking models is merged order compatible, we can reduce the question of equality of inference relations to equivalence of inference cores. We illustrated arising differences between the set of all ranking models of a knowledge base $\mathcal{R}$ and the set of all c-representations of $\mathcal{R}$, and we applied our approach to skeptical c-inference for proving that for certain knowledge bases a maximal impact of $|\mathcal{R}| - 1$ is sufficient to fully capture the behavior of skeptical c-inference.

In our current work, we employ the concepts of inference cores and merged order compatibility for extending our investigations on sufficient upper bounds for $CR(\mathcal{R})$ to more general kinds of knowledge bases, and for addressing the open problems of characterizing knowledge bases whose set of c-representations is merged order compatible or whether this property holds for all knowledge bases. This goes along with finding a suitable characterization of merged order compatible sets of ranking models, and exploring relationships to approaches employing e.g. possibilistic or probabilistic semantics.

References

1. Beierle, C., Eichhorn, C., Kern-Isberner, G.: Skeptical inference based on C-representations and its characterization as a constraint satisfaction problem. In: Gyssens, M., Simari, G. (eds.) FoIKS 2016. LNCS, vol. 9616, pp. 65–82. Springer, Cham (2016). doi:10.1007/978-3-319-30024-5_4
2. Beierle, C., Kutsch, S.: Regular and sufficient bounds of finite domain constraints for skeptical c-inference. In: Benferhat, S., Tabia, K., Ali, M. (eds.) Proceedings of the 30th International Conference on Industrial, Engineering, Other Applications of Applied Intelligent Systems (IEA/AIE-2017). LNAI, vol. 10350. Springer, Heidelberg (2017)
3. Beierle, C., Eichhorn, C., Kern-Isberner, G., Kutsch, S.: Properties of skeptical c-inference for conditional knowledge bases and its realization as a constraint satisfaction problem, (2017, submitted)
4. Benferhat, S., Dubois, D., Prade, H.: Possibilistic and standard probabilistic semantics of conditional knowledge bases. J. Logic Comput. **9**(6), 873–895 (1999)

5. de Finetti, B.: La prévision, ses lois logiques et ses sources subjectives. Ann. Inst. H. Poincaré **7**(1), 1–68 (1973). ed. Kyburg, H., Smokler, H.E.: English Translation in Studies in Subjective Probability, pp. 93–158. Wiley, New York (1974)
6. Goldszmidt, M., Pearl, J.: Qualitative probabilities for default reasoning, belief revision, and causal modeling. Artif. Intell. **84**(1–2), 57–112 (1996)
7. Kern-Isberner, G.: Conditionals in Nonmonotonic Reasoning and Belief Revision. LNAI, vol. 2087. Springer, Heidelberg (2001)
8. Kern-Isberner, G.: A thorough axiomatization of a principle of conditional preservation in belief revision. Ann. Math. Artif. Intell. **40**(1–2), 127–164 (2004)
9. Pearl, J.: System Z: a natural ordering of defaults with tractable applications to nonmonotonic reasoning. In: Proceedings of the 3rd Conference on Theoretical Aspects of Reasoning About Knowledge (TARK 1990), pp. 121–135. Morgan Kaufmann Publisher Inc., San Francisco (1990)
10. Spohn, W.: Ordinal conditional functions: a dynamic theory of epistemic states. In: Harper, W., Skyrms, B. (eds.) Causation in Decision, Belief Change, and Statistics, II, pp. 105–134. Kluwer Academic Publishers (1988)

A Transformation System for Unique Minimal Normal Forms of Conditional Knowledge Bases

Christoph Beierle[1]($\boxtimes$), Christian Eichhorn[2], and Gabriele Kern-Isberner[2]

[1] Department of Computer Science, University of Hagen,
58084 Hagen, Germany
`christoph.beierle@fernuni-hagen.de`
[2] Department of Computer Science, TU Dortmund,
44221 Dortmund, Germany

Abstract. Conditional knowledge bases consisting of sets of conditionals are used in inductive nonmonotonic reasoning and can represent the defeasible background knowledge of a reasoning agent. For the comparison of the knowledge of different agents, as well as of different approaches to nonmonotonic reasoning, it is beneficial if these knowledge bases are as compact and straightforward as possible. To enable the replacement of a knowledge base $\mathcal{R}$ by a simpler, but equivalent knowledge base $\mathcal{R}'$, we propose to use the notions of elementwise equivalence or model equivalence for conditional knowledge bases. For elementwise equivalence, we present a terminating and confluent transformation system on conditional knowledge bases yielding a unique normal form for every $\mathcal{R}$. We show that an extended version of this transformation system takes model equivalence into account. For both transformation system, we prove that the obtained normal forms are minimal with respect to subset inclusion and the corresponding notion of equivalence.

1 Introduction

Defeasible Conditionals "If A then usually B" and conditional knowledge bases consisting of finite sets of such conditionals play a major role in nonmonotonic reasoning, as they are used to formalize the background knowledge of intelligent agents. A short, compact and straightforward normal form of these knowledge bases is desirable, not only to allow us to compare the knowledge of different agents, but also to store this knowledge in a form that is easily understandable and to compare different approaches of nonmonotonic reasoning. Additionally, the number of conditionals in a knowledge base is an important factor in the computational complexity of approaches that generate an epistemic state inductively on top of conditional knowledge bases. Thus, knowledge bases which contain no unnecessary conditionals may lead to significantly reduced computational efforts required when dealing with knowledge bases in these approaches.

This article extends the work presented in the short paper [2] in several directions and is organized as follows: After briefly recalling the required background in Sect. 2, we propose to use the notion of elementwise equivalence or

© Springer International Publishing AG 2017
A. Antonucci et al. (Eds.): ECSQARU 2017, LNAI 10369, pp. 236–245, 2017.
DOI: 10.1007/978-3-319-61581-3_22

model equivalence for conditional knowledge bases (Sect. 3) to enable the replacement of a knowledge base $\mathcal{R}$ by a simpler, but equivalent knowledge base $\mathcal{R}'$. In Sect. 4 we present a set of naturally arising transformation rules on conditional knowledge bases, develop a terminating and confluent transformation system for elementwise equivalence, yielding a unique normal form for every conditional knowledge base, and prove that two knowledge bases are elementwise equivalent iff they have the same conditional normal form. In Sect. 5, we show that an extended version of this transformation system takes model equivalence into account. We prove that both transformation systems yield normal forms that are minimal with respect to the corresponding notion of equivalence and subset inclusion. Section 6 concludes the paper and points out future work.

2 Background: Conditionals and OCFs

Let $\Sigma = \{V_1, ..., V_m\}$ be a propositional alphabet. A *literal* is the positive (v_i) or negated $(\overline{v}_i)$ form of a propositional variable V_i. From these we obtain the propositional language $\mathcal{L}$ as the set of formulas of Σ closed under negation $\neg$, conjunction $\wedge$, and disjunction $\vee$, as usual; for formulas $A, B \in \mathcal{L}$, $A \Rightarrow B$ denotes the material implication and stands for $\neg A \vee B$. For shorter formulas, we abbreviate conjunction by juxtaposition (i.e., AB stands for $A \wedge B$), and negation by overlining (i.e., $\overline{A}$ is equivalent to $\neg A$). Let Ω denote the set of possible worlds over $\mathcal{L}$; Ω will be taken here simply as the set of all propositional interpretations over $\mathcal{L}$ and can be identified with the set of all complete conjunctions over Σ. For $\omega \in \Omega$, $\omega \models A$ means that the propositional formula $A \in \mathcal{L}$ holds in the possible world ω.

A *conditional* $(B|A)$ with $A, B \in \mathcal{L}$ encodes the defeasible rule "if A then usually B" and is a trivalent logical entity with the evaluation [4,6]:

$$\llbracket (B|A) \rrbracket_\omega = \begin{cases} \textit{true} & \text{iff } \omega \models AB \quad \text{(verification)} \\ \textit{false} & \text{iff } \omega \models A\overline{B} \quad \text{(falsification)} \\ \textit{undefined} & \text{iff } \omega \models \overline{A} \quad \text{(not applicable)} \end{cases}$$

A *knowledge base* $\mathcal{R} = \{(B_1|A_1), ..., (B_n|A_n)\}$ is a finite set of such conditionals.

An *Ordinal Conditional Function* (OCF, ranking function) [9] is a function $\kappa : \Omega \to \mathbb{N}_0 \cup \{\infty\}$ that assigns to each world $\omega \in \Omega$ an implausibility rank $\kappa(\omega)$, that is, the higher $\kappa(\omega)$, the more surprising ω is. OCFs have to satisfy the normalization condition that there has to be a world that is maximally plausible, i.e., the preimage of 0 cannot be empty, formally $\kappa^{-1}(0) \neq \emptyset$. The rank of a formula A is defined by $\kappa(A) = \min\{\kappa(\omega) \mid \omega \models A\}$.

An OCF κ *accepts* a conditional $(B|A)$ (denoted by $\kappa \models (B|A)$) iff the verification of the conditional is less surprising than its falsification, i.e., iff $\kappa(AB) < \kappa(A\overline{B})$. This can also be understood as a nonmonotonic inference relation between the premise A and the conclusion B: We say that A κ-*entails* B (written $A \mathrel{|\!\sim}^\kappa B$) if and only if κ accepts the conditional $(B|A)$, formally

$$\kappa \models (B|A) \quad \text{iff} \quad \kappa(AB) < \kappa(A\overline{B}) \quad \text{iff} \quad A \mathrel{|\!\sim}^\kappa B. \tag{1}$$

The acceptance relation in (1) is extended as usual to a set $\mathcal{R}$ of conditionals by defining $\kappa \models \mathcal{R}$ iff $\kappa \models (B|A)$ for all $(B|A) \in \mathcal{R}$. This is synonymous to saying that $\mathcal{R}$ is *admissible* with respect to $\mathcal{R}$ [5]. A knowledge base $\mathcal{R}$ is *consistent* iff there exists an OCF κ such that $\kappa \models \mathcal{R}$.

Example 1 ($\mathcal{R}_{car}$). Let $\Sigma = \{C, E, F\}$ be an alphabet where C indicates whether something is a car (c), or not ($\overline{c}$), E indicates whether something is an e-car (e), or not ($\overline{e}$), and F indicates whether something needs fossil fuel (f), or not ($\overline{f}$). Let $\mathcal{R}_{car} = \{r_1, r_2, r_3, r_4, r_5, r_6, r_7\}$ be a knowledge base using Σ with:

$r_1 : (f|c)$ *"Usually cars need fossil fuel."*
$r_2 : (\overline{f}|e)$ *"Usually e-cars do not need fossil fuel."*
$r_3 : (c|e)$ *"E-cars usually are cars."*
$r_4 : (e|e\overline{f})$ *"E-cars that do not need fossil fuel usually are e-cars."*
$r_5 : (e\overline{f}|e)$ *"E-cars usually are e-cars that do not need fossil fuel."*
$r_6 : (\overline{e}|\top)$ *"Usually things are no e-cars."*
$r_8 : (cf \vee \overline{c}f|ce \vee c\overline{e})$ *"Things that are cars and e-cars or cars but not e-cars are cars that need fossil fuel or are no cars but need fossil fuel."*

This knowledge base is consistent: For instance, a ranking model κ for $\mathcal{R}_{car}$ is

ω	cef	$ce\overline{f}$	$c\overline{e}f$	$c\overline{e}\overline{f}$	$\overline{c}ef$	$\overline{c}e\overline{f}$	$\overline{c}\overline{e}f$	$\overline{c}\overline{e}\overline{f}$
$\kappa(\omega)$	2	1	0	1	4	2	0	0

with, e.g., $\kappa \models (\overline{f}|e)$ because $\kappa(e\overline{f}) = 1 < 2 = \kappa(ef)$ and $\kappa \models (\overline{e}|\top)$ because $\kappa(\overline{e}) = 0 < 1 = \kappa(e)$.

3   Model Based Equivalences

With the acceptance relation between ranking functions and knowledge bases, we now can define the set of ranking models of a knowledge base.

Definition 1 (ranking models). *Let $\mathcal{R} = \{(B_1|A_1), \ldots, (B_n|A_n)\}$ be a finite conditional knowledge base. The set of* ranking models *of $\mathcal{R}$ is the set of OCFs that are admissible with respect to $\mathcal{R}$, formally $Mod(\mathcal{R}) = \{\kappa | \kappa \models \mathcal{R}\}$.*

The notion of inconsistency gives us a possibility to determine whether every ranking model of a knowledge base accepts a given conditional:

Proposition 1 ([5]). *Let $\mathcal{R} = \{(B_1|A_1), \ldots, (B_n|A_n)\}$ be a finite conditional knowledge base. A conditional $(B|A)$ with $AB \not\equiv \bot$ is accepted by every ranking model $\kappa \in Mod(\mathcal{R})$ if and only if $\mathcal{R} \cup \{(\overline{B}|A)\}$ is inconsistent.*

Definition 2 (model equivalence). *Let $\mathcal{R}, \mathcal{R}'$ be knowledge bases. $\mathcal{R}$ and $\mathcal{R}'$ are* model equivalent, *denoted $\mathcal{R} \equiv_{mod} \mathcal{R}'$, iff $Mod(\mathcal{R}) = Mod(\mathcal{R}')$.*

By definition, the model set of an inconsistent knowledge base is empty, so all inconsistent knowledge bases are equivalent. We introduce the special knowledge base $\diamond$ that is inconsistent by definition; thus $\diamond \equiv_{mod} \mathcal{R}$ for every $\mathcal{R}$ with $Mod(\mathcal{R}) = \emptyset$; for instance, $\{(\bot|\top)\} \equiv_{mod} \diamond$. The idea of elementwise equivalence is that each piece of knowledge (i.e. conditional) in one knowledge base directly corresponds to a piece of knowledge in the other knowledge base.

Definition 3 (elementwise equivalence). *Let $\mathcal{R}$, $\mathcal{R}'$ be knowledge bases.*

- $\mathcal{R}$ *is an elementwise equivalent sub-knowledge base of $\mathcal{R}'$, denoted by $\mathcal{R} \ll_{ee}$ $\mathcal{R}'$, iff for every conditional $(B'|A') \in \mathcal{R}'$ that is not self-fulfilling (i.e. $A' \not\models B'$) there is a conditional $(B|A) \in \mathcal{R}$ such that $Mod(\{(B|A)\}) = Mod(\{(B'|A')\})$.*
- $\mathcal{R}$ *and $\mathcal{R}'$ are* strictly elementwise equivalent *iff $\mathcal{R} \ll_{ee} \mathcal{R}'$ and $\mathcal{R}' \ll_{ee} \mathcal{R}$.*
- $\mathcal{R}$ *and $\mathcal{R}'$ are* elementwise equivalent, *denoted by $\mathcal{R} \equiv_{ee} \mathcal{R}'$, iff either both $\mathcal{R}$ and $\mathcal{R}'$ are inconsistent, or both $\mathcal{R}$ and $\mathcal{R}'$ are consistent and strictly elementwise equivalent.*

Thus, two inconsistent knowledge bases are also elementwise equivalent according to Definition 3, e.g. $\{(B|A), (\overline{B}|A)\} \equiv_{ee} \{\{(B'|A'), (\bot|\top)\}$, enabling us to avoid cumbersome case distinctions when dealing with sets of consistent and inconsistent knowledge bases. We illustrate model equivalence and elementwise equivalence with the following excerpts of Example 1:

Example 2. Let $\mathcal{R}'_{car} = \{r_1, r_2, r_3, r_7\}$ and $\mathcal{R}''_{car} = \{r_1, r_2, r_3\}$ be knowledge bases with conditionals from Example 1. Since $f \equiv ef \vee \overline{c}f$ and $c \equiv ce \vee c\overline{e}$, for every OCF κ we have $\kappa(cf) = \kappa((ce \vee c\overline{e}) \wedge (ef \vee \overline{c}f))$ and likewise $\kappa(c\overline{f}) = \kappa((ce \vee c\overline{e}) \wedge \neg(ef \vee \overline{c}f))$ and hence $\kappa \models r_3$ if and only if $\kappa \models r_7$. Therefore, for every $\kappa \models \mathcal{R}''_{car}$ we also have $\kappa \models \mathcal{R}'_{car}$, and vice versa, which gives us $\mathcal{R}'_{car} \equiv_{mod} \mathcal{R}''_{car}$. For the same reason we have $Mod(\{r_3\}) = Mod(\{r_7\})$ (and, trivially, $Mod(\{r_i\}) = Mod(\{r_i\})$ for all $i \in \{1, 2, 3, 7\}$), which gives us $\mathcal{R}'_{car} \ll_{ee} \mathcal{R}''_{car}$ and $\mathcal{R}''_{car} \ll_{ee} \mathcal{R}'_{car}$ and hence $\mathcal{R}'_{car} \equiv_{ee} \mathcal{R}''_{car}$. So $\mathcal{R}'_{car}$ and $\mathcal{R}''_{car}$ are both model equivalent and also elementwise equivalent.

4 Normal Forms for Elementwise Equivalence

Similar to formulas in propositional logic, it is often advantageous to consider only conditional knowledge bases that are in a standardized normal form. In the following, we will develop rules for transforming a knowledge base toward a normal form. For this, we will use a function Π that assigns to a knowledge base $\mathcal{R}$ an ordered partition $\Pi(\mathcal{R}) = (\mathcal{R}_0, \ldots, \mathcal{R}_m)$ such that all conditionals in $\mathcal{R}_i$, $1 \leqslant i \leqslant m$, are tolerated by the set $\bigcup_{j=i}^{m} \mathcal{R}_j$ [8]; if no such partition exists, we extend Π by defining $\Pi(\mathcal{R}) = \diamond$. Thus, $\mathcal{R}$ is consistent iff $\Pi(\mathcal{R}) \neq \diamond$ [8].

For propositional formulas over a propositional alphabet Σ, there are various ways of defining a normal form such that precisely semantically equivalent formulas are mapped to the same normal form, using e.g. disjunctions of worlds or selected shortest formulas. In order to abstract from a particular choice, for the rest of this paper we assume a function ν that maps a propositional formula A to a unique normal form $\nu(A)$ such that $A \equiv A'$ iff $\nu(A) = \nu(A')$.

Using Π and ν, the transformation system $\mathcal{T}$ is given in Fig. 1:

(SF) removes a conditional $(B|A)$ if $A \models B$ since such a conditional is self-fulfilling because it can not be falsified by any world.

(SF) *self-fulfilling* : $\dfrac{\mathcal{R} \cup \{(B|A)\}}{\mathcal{R}}$ $A \models B,\ A \not\equiv \bot$

(DP) *duplicate* : $\dfrac{\mathcal{R} \cup \{(B|A),\ (B'|A')\}}{\mathcal{R} \cup \{(B|A)\}}$ $A \equiv A',\ B \equiv B'$

(CE) *conditional equivalence* : $\dfrac{\mathcal{R} \cup \{(B|A),\ (B'|A')\}}{\mathcal{R} \cup \{(B|A)\}}$ $AB \equiv A'B',\ A\overline{B} \equiv A'\overline{B'}$

(PN) *propositional normal form* : $\dfrac{\mathcal{R} \cup \{(B|A)\}}{\mathcal{R} \cup \{(\nu(B)|\nu(A))\}}$ $A \neq \nu(A)$ or $B \neq \nu(B)$

(CN) *conditional normal form* : $\dfrac{\mathcal{R} \cup \{(B|A)\}}{\mathcal{R} \cup \{(AB|A)\}}$ $B \not\equiv AB$

(CC) *counter conditional* : $\dfrac{\mathcal{R} \cup \{(B|A),\ (\overline{B}|A)\}}{\diamond}$

(SC) *self-contradictory* : $\dfrac{\mathcal{R} \cup \{(B|A)\}}{\diamond}$ $AB \equiv \bot$

(IC) *inconsistency* : $\dfrac{\mathcal{R}}{\diamond}$ $\mathcal{R} \neq \diamond, \Pi(\mathcal{R}) = \diamond$

Fig. 1. Transformation rules $\mathcal{T}$ for conditional knowledge bases

(DP) removes a conditional $(B'|A')$ which is a duplicate of a conditional $(B|A)$ under propositional equivalences of A and A' and of B and B'.
(CE) removes a conditional that is conditionally equivalent to another one.
(PN) propositionally normalizes antecedent and consequent of a conditional.
(CN) transforms a conditional $(B|A)$ to its conditional normal form by sharpening its consequent to the conjunction with its antecedent.
(CC) transforms a knowledge base containing both a conditional $(B|A)$ and its counter conditional $(\overline{B}|A)$ into the inconsistent knowledge base $\diamond$.
(SC) transforms a knowledge base containing a conditional that can not be verified by any world into the inconsistent knowledge base $\diamond$.
(IC) transforms an inconsistent knowledge base into $\diamond$.

We illustrate $\mathcal{T}$ transforming the knowledge base in the running example to a reduced, more compact form.

Example 3 ($\mathcal{T}(\mathcal{R}_{car})$). Consider the knowledge base $\mathcal{R}_{car}$ from Example 1.

(SF) In $\mathcal{R}_{car}$, r_4 is self-fulfilling since $ef \models e$, hence the application of (SF) yields $\mathcal{R}_{car}^{(SF)} = \mathcal{R}_{car} \setminus \{r_4\}$.
(DP) The conditionals r_1 and r_7 are duplicates since $c \equiv ce \vee c\overline{e}$ and $f \equiv (cf \vee \overline{c}f)$. So applying (DP) to $\mathcal{R}_{car}$ gives us $\mathcal{R}_{car}^{(DP)} = \mathcal{R}_{car} \setminus \{r_7\}$.
(CE) We have $e\overline{f} \equiv ee\overline{f}$ and $ef \equiv e \wedge (\overline{e} \vee f)$, therefore r_2 and r_5 are conditionally equivalent; applying (CE) to $\mathcal{R}_{car}$ yields $\mathcal{R}_{car}^{(CE)} = \mathcal{R}_{car} \setminus \{r_5\}$.

(*PN*) The conditional r_1 is equivalent to r_7 but shorter, so let us assume the shorter formula as propositional normal form. With ν being a function that converts a propositional formula to this normal form, applying (*PN*) to $\mathcal{R}_{car}$ gives us the same results as (*DP*), that is, $\mathcal{R}_{car}^{(PN)} = \mathcal{R}_{car} \setminus \{r_7\}$.

(*SC*) The knowledge base $\mathcal{R}_{car}$ contains no self-contradictory conditional; hence, (*SC*) can not be applied to $\mathcal{R}_{car}$.

Applying $\mathcal{T}$ exhaustively and in arbitrary sequence to $\mathcal{R}_{car}$ gives us the knowledge base $\mathcal{R}_{car}^{\mathcal{T}} = \mathcal{T}(\mathcal{R}_{car}) = \{r_1, r_2, r_3, r_6\}$.

Note that $\mathcal{T}$ is not a minimal set of transformation rules. For instance, (*DP*) is redundant since the effect of removing a conditional $(B'|A')$ as a duplicate of $(B|A)$ could also be achieved by applying (*PN*) to both conditionals, thereby mapping them both to the same normalized conditional in the resulting knowledge base. Similarly, (*CC*) and (*SC*) are redundant since these cases are also covered by the more general transformation rule (*IC*). However, our objective here is not to present a minimal set of rules, but a set of more or less naturally arising transformation rules.

Proposition 2. $\mathcal{T}$ *is terminating.*

Proof. The rules (*SF*), (*DP*), (*CE*), and (*IC*) all remove at least one conditional, (*PN*) and (*CN*) can be applied at most once to any conditional, and also (*CC*), (*SC*), and (*IC*) all remove at least one conditional. Hence, $\mathcal{T}$ is terminating. $\square$

Proposition 3 (*$\mathcal{T}$ correct*). *Let* $\mathcal{T}(\mathcal{R})$ *be the knowledge base obtained from* $\mathcal{R}$ *by exhaustively applying* $\mathcal{T}$ *to* $\mathcal{R}$*. Then* $\mathcal{R} =_{mod} \mathcal{T}(\mathcal{R})$*.*

Proof. We prove the proposition by showing that each single rule is correct.

- (*SF*) is correct since $(B|A)$ with $A \models B$ is verified by every OCF.
- (*DP*) is correct since $A \equiv A'$, $B \equiv B'$ implies that $\kappa \models (B|A)$ iff $\kappa \models (B'|A')$ for every OCF κ.
- (*CE*) is correct since $AB \equiv A'B'$, $A\overline{B} \equiv A'\overline{B'}$ implies that $\kappa \models (B|A)$ iff $\kappa \models (B'|A')$ for every OCF κ.
- (*PN*) is correct since for every OCF κ, we have $\kappa \models (B|A)$ iff $\kappa \models (\nu(B)|\nu(A))$.
- (*CN* is correct since for every OCF κ, we have $\kappa \models (B|A)$ iff $\kappa \models (AB|A)$.
- (*CC*) is correct since there is no OCF κ accepting both a conditional $(B|A)$ and its counter conditional $(\overline{B}|A)$.
- (*SC*) is correct since there is no OCF κ with $\kappa \models (B|A)$ if $AB \equiv \bot$.
- (*IC*) is correct since Π is a consistency test for any knowledge base $\mathcal{R}$. $\square$

While Proposition 3 states that $\mathcal{T}$ is correct with respect to model equivalence of knowledge bases, the following proposition shows that this is also the case with respect to the stricter notion of elementwise equivalence.

Proposition 4 (*$\mathcal{T}$ correct w.r.t. elementwise equivalence*). *Let* $\mathcal{T}(\mathcal{R})$ *be the knowledge base obtained from* $\mathcal{R}$ *by exhaustively applying* $\mathcal{T}$ *to* $\mathcal{R}$*. Then* $\mathcal{R} \equiv_{ee} \mathcal{T}(\mathcal{R})$*.*

Proof. According to the proof of Proposition 3, $\mathcal{R}$ is inconsistent iff $\mathcal{T}(\mathcal{R}) = \diamond$; thus, if $\mathcal{R}$ is inconsistent then $\mathcal{R} \equiv_{ee} \mathcal{T}(\mathcal{R})$. So let $\mathcal{R}$ be consistent. Then $\mathcal{T}(\mathcal{R})$ has been obtained from $\mathcal{R}$ by a finite number of applications of (SF), (DP), (CE), (PN), and (CN). For these applications we observe:

- (SF) preserves elementwise equivalence since for self-fulfilling conditionals no counterpart in the other knowledge base is required.
- (DP) and (CE) preserve elementwise equivalence since in both cases, we obviously have $Mod(\{(B|A)\}) = Mod(\{(B'|A')\})$ and thus $\{(B|A), (B'|A')\} \equiv_{ee} \{(B|A)\}$.
- (DP) and (CN) preserve elementwise equivalence since $Mod(\{(B|A)\}) = Mod(\{(\nu(B)|\nu(A))\})$ and since $Mod(\{(B|A)\}) = Mod(\{(AB|A)\})$. $\square$

$\mathcal{T}$ is an extended version of the transformation system NF presented in [2]. While NF is not confluent [1,2,7], the next proposition proves that $\mathcal{T}$ is confluent.

Proposition 5. $\mathcal{T}$ *is confluent.*

Proof. Since $\mathcal{T}$ is terminating, local confluence of $\mathcal{T}$ implies confluence of $\mathcal{T}$; local confluence of $\mathcal{T}$ in turn can be shown by ensuring that for every critical pair obtained form superpositioning two left hand sides of rules in $\mathcal{T}$ reduces to the same knowledge base [1,7]:

Any critical pair obtained from (CC), (SC), or (IC) and a rule in $\mathcal{T}$ reduces to $\diamond$ since all rules preserve the consistency status af a knowledge base.

Any critical pair obtained from (SF) with $\mathcal{T} \setminus \{(CC), (SC), (IC)\}$ reduces to the same knowledge base since a self-fulfilling conditional replaced by a transformation rule in $\{(DP), (CE), (PN), (CN)\}$ is still self-fulfilling. Furthermore, any critical pair involving (PN) can obviously be reduced to the same knowledge base; this observation also holds for (CN).

Thus, we are left with critical pairs obtained from (DP) and (CE). Consider

$$\mathcal{R}_0 = \mathcal{R} \cup \{(B|A), (B'|A'), (B''|A'')\}.$$

If (DP) can be applied to $\mathcal{R}_0$ at $\{(B|A), (B'|A')\}$ we get $\mathcal{R}_1 = \mathcal{R} \cup \{(B|A), (B''|A'')\}$, and if (CE) can be applied to $\mathcal{R}_0$ at $\{(B'|A'), (B''|A'')\}$ we get $\mathcal{R}_2 = \mathcal{R} \cup \{(B|A), (B'|A')\}$. The used applicability of (DP) to $\mathcal{R}_0$ ensures $A \equiv A', B \equiv B'$; hence (DP) can be applied to $\mathcal{R}_2$, yielding $\mathcal{R}_3 = \mathcal{R} \cup \{(B|A)\}$. The used applicability of (CE) to $\mathcal{R}_0$ ensures $A'B' \equiv A''B'', A'\overline{B'} \equiv A''\overline{B''}$; thus, we also have $AB \equiv A''B'', A\overline{B} \equiv A''\overline{B''}$ so that (CE) can be applied to $\mathcal{R}_1$, yielding $\mathcal{R} \cup \{(B|A)\} = \mathcal{R}_3$. Hence, $\mathcal{T}$ reduces both $\mathcal{R}_1$ and $\mathcal{R}_2$ to $\mathcal{R}_3$. Similarly, the other critical pairs obtained from (DP) and (CE) can be shown to be reducible to the same knowledge base. $\square$

$\mathcal{T}$ is not only confluent, but it also yields a knowledge base that is minimal when taking elementwise equivalence into account.

Proposition 6 ($\mathcal{T}$ minimizing w.r.t. elementwise equivalence). *For all knowledge bases $\mathcal{R}$ we have $\mathcal{T}(\mathcal{R}) = \diamond$ iff $\mathcal{R}$ is inconsistent, and if $\mathcal{R}$ is consistent, then for all knowledge bases $\mathcal{R}'$ it holds that:*

$$\mathcal{R}' \subsetneqq \mathcal{T}(\mathcal{R}) \quad \text{implies} \quad \mathcal{R}' \not\equiv_{ee} \mathcal{R} \tag{2}$$

Proof. For inconsistent $\mathcal{R}$, the proof follows from Proposition 4, so let $\mathcal{R}$ be consistent and $\mathcal{R}' \subsetneq \mathcal{T}(\mathcal{R})$. Proposition 4 implies $\mathcal{R} \equiv_{ee} \mathcal{T}(\mathcal{R})$; hence, it suffices to show $\mathcal{R}' \not\equiv_{ee} \mathcal{T}(\mathcal{R})$. If we assume the contrary, $\mathcal{R}' \equiv_{ee} \mathcal{T}(\mathcal{R})$, then $\mathcal{R}' \subsetneq \mathcal{T}(\mathcal{R})$ implies that there must be two different conditionals $(B_1|A_1), (B_2|A_2) \in \mathcal{T}(\mathcal{R})$ and a conditional $(B|A) \in \mathcal{R}'$ such that $Mod(\{(B_1|A_1)\}) = Mod(\{(B|A)\})$ and $Mod(\{(B_1|A_1)\}) = Mod(\{(B|A)\})$ and hence $Mod(\{(B_1|A_1)\}) = Mod(\{(B_2|A_2)\})$. This requires $A_1 \equiv A_2$, $A_1 B_1 \equiv A_2 B_2$, and $A_1 \overline{B_1} \equiv A_2 \overline{B_2}$, implying that (CE) could be applied to $(B_1|A_1), (B_2|A_2)$, a contradiction to our assumptions. Thus, $\mathcal{R}' \not\equiv_{ee} \mathcal{R}$.   □

Propositions 2–6 ensure that applying $\mathcal{T}$ to a knowledge base $\mathcal{R}$ always yields the unique normal form $\mathcal{T}(\mathcal{R})$ that is elementwise equivalent to $\mathcal{R}$ and minimal with respect to set inclusion.

Definition 4 (conditional normal form). *A knowledge base $\mathcal{R}$ is in* conditional normal form *iff $\mathcal{R} = \mathcal{T}(\mathcal{R})$.*

Thus, for every knowledge base $\mathcal{R}$, its conditional normal form is uniquely determined. Moreover, $\mathcal{T}$ provides a convenient test for the elementwise equivalence of knowledge bases.

Proposition 7 (elementwise equivalence). *Two knowledge bases $\mathcal{R}, \mathcal{R}'$ are elementwise equivalent iff the have the same conditional normal form, i.e.:*

$$\mathcal{R} \equiv_{ee} \mathcal{R}' \quad iff \quad \mathcal{T}(\mathcal{R}) = \mathcal{T}(\mathcal{R}') \tag{3}$$

5 Normal Forms for Model Equivalence

While $\mathcal{T}$ is correct with respect to both model equivalence and elementwise equivalence, it is minimizing for elementwise equivalence (Proposition 4), but it is not minimizing when taking model equivalence into account.

Example 4 (T not minimizing for model equivalence). We illustrate that $\mathcal{T}$ is not minimizing with respect to model equivalence using the running example with the knowledge bases $\mathcal{R}_{car}$ and $\mathcal{R}_{car}^{\mathcal{T}}$. We already illustrated that $\mathcal{R}_{car}^{\mathcal{T}}$ can be obtained from $\mathcal{R}_{car}$ by exhaustive application of the rules of $\mathcal{T}$ in Example 3, i.e. $\mathcal{T}(\mathcal{R}_{car}) = \mathcal{R}_{car}^{\mathcal{T}}$. But $\mathcal{R}_{car}^{\mathcal{T}}$ is not minimal with respect to set inclusion when taking model equivalence into account: Consider the knowledge base $\mathcal{R}_{car}' = \{r_1, r_2, r_3\} \subsetneq \mathcal{R}_{car}^{\mathcal{T}} = \{r_1, r_2, r_3, r_6\}$. We have $\mathcal{R}_{car}' \cup \{(e|\top)\} \equiv_{mod} \diamond$ and thus Proposition 1 gives us $\kappa \models (\overline{e}|\top)$ for all $\kappa \models \mathcal{R}_{car}'$. Therefore, since $r_6 = (\overline{e}|\top)$, every ranking model of $\mathcal{R}_{car}'$ is also a ranking model of $\mathcal{R}_{car}^{\mathcal{T}}$, thus $\mathcal{R}_{car}' \equiv_{mod} \mathcal{R}_{car}^{\mathcal{T}}$.

This example motivates the following extension of $\mathcal{T}$:

Definition 5 ($\mathcal{T}_2$). *$\mathcal{T}_2$ is the transformation system $\mathcal{T}$ extended by the rule:*

(RC) *redundant conditional* : $\dfrac{\mathcal{R} \cup \{(B|A)\}}{\mathcal{R}}$ $\Pi(\mathcal{R} \cup \{(\overline{B}|A)\}) = \diamond$

Since (RC) removes a conditional $(B|A)$ from a knowledge base $\mathcal{R} \cup \{(B|A)\}$ if every model of $\mathcal{R}$ accepts $(B|A)$, we immediately have $\mathcal{R} \cup \{(B|A)\} \equiv_{mod} \mathcal{R}$. Together with the properties for $\mathcal{T}$ shown above, we get:

Proposition 8 ($\mathcal{T}_2$ terminating and correct w.r.t. model equivalence). $\mathcal{T}_2$ *is terminating, and for all knowledge bases* $\mathcal{R}$, *we have* $\mathcal{R} \equiv_{mod} \mathcal{T}_2(\mathcal{R})$.

In contrast to $\mathcal{T}$, $\mathcal{T}_2$ is not confluent as the choice of where the transformation rule (RC) is applied may influence the result as illustrated by the following example.

Example 5 ($\mathcal{T}_2$ not confluent). Consider the knowledge base $\mathcal{R} = \{r_1, r_2, r_3, r_4\}$ with $r_1 = (f|c)$, $r_2 = (c|e)$, $r_3 = (\overline{f}|e)$, and $r_4 = (\overline{f}|ce)$. Let $\overline{r_3} = (f|e)$ and $\overline{r_4} = (f|ce)$. Then $\Pi(\{r_1, r_2, r_3\} \cup \{\overline{r_4}\}) = \diamond$, and hence applying (RC) to $\mathcal{R}$ at r_4 yields $\mathcal{R}_1 = \{r_1, r_2, r_3\}$. Furthermore, $\Pi(\{r_1, r_2, r_4\} \cup \{\overline{r_3}\}) = \diamond$, and hence applying (RC) to $\mathcal{R}$ at r_3 yields $\mathcal{R}_2 = \{r_1, r_2, r_4\}$. Applying $\mathcal{T}_2$ yields

$$R_1' = \mathcal{T}_2(R_1) = \{(\nu(cf)|\nu(c)), (\nu(ce)|\nu(e)), (\nu(e\overline{f})|\nu(e))\} \tag{4}$$

$$R_2' = \mathcal{T}_2(R_2) = \{(\nu(cf)|\nu(c)), (\nu(ce)|\nu(e)), (\nu(ce\overline{f})|\nu(ce))\} \tag{5}$$

and since $\mathcal{R}_1' \neq \mathcal{R}_2'$, these two knowledge bases are two different normal forms for $\mathcal{R}$ under $\mathcal{T}_2$.

On the other hand, $\mathcal{T}_2$ is minimizing when taking model equivalence into account.

Proposition 9 ($\mathcal{T}_2$ minimizing w.r.t. model equivalence). *For all knowledge bases* $\mathcal{R}$ *we have* $\mathcal{T}_2(\mathcal{R}) = \diamond$ *iff* $\mathcal{R}$ *is inconsistent, and if* $\mathcal{R}$ *is consistent, then for all knowledge bases* $\mathcal{R}'$ *it holds that:*

$$\mathcal{R}' \subsetneq \mathcal{T}_2(\mathcal{R}) \quad implies \quad \mathcal{R}' \not\equiv_{mod} \mathcal{R} \tag{6}$$

Proof. As in Proposition 6, we are left to prove the case for a consistent $\mathcal{R}$. So let $\mathcal{R}$ be consistent and $\mathcal{R}' \subsetneq \mathcal{T}_2(\mathcal{R})$. Proposition 8 implies $\mathcal{R} \equiv_{mod} \mathcal{T}_2(\mathcal{R})$; hence, it suffices to show $\mathcal{R}' \not\equiv_{mod} \mathcal{T}_2(\mathcal{R})$. If we assume the contrary, $\mathcal{R}' \equiv_{mod} \mathcal{T}_2(\mathcal{R})$, then $\mathcal{R}' \subsetneq \mathcal{T}_2(\mathcal{R})$ implies that there must be conditionals $(B_1|A_1), \ldots, (B_n|A_n) \in \mathcal{T}_2(\mathcal{R})$, $n \geqslant 1$, such that $\mathcal{R}' \cup \{(B_1|A_1), \ldots, (B_n|A_n)\} = \mathcal{T}_2(\mathcal{R})$ with $Mod(\{\mathcal{R}' \cup \{(B_1|A_1), \ldots, (B_n|A_n)\}\}) = Mod(\{\mathcal{T}_2(\mathcal{R})\})$. This implies that (RC) could be applied to $\mathcal{T}_2(\mathcal{R})$ at $(B_1|A_1)$, contradicting our assumptions. Thus, $\mathcal{R}' \not\equiv_{mod} \mathcal{R}$. $\square$

6   Conclusions and Future Work

In this paper we proposed notions of elementwise and model equivalence for conditional knowledge bases, enabling the replacement of a knowledge base $\mathcal{R}$ by an equivalent knowledge base $\mathcal{R}'$. Based on these notions, we presented the terminating and confluent transformation system $\mathcal{T}$ that for every knowledge base yields a minimal and unique conditional normal form with respect to element-wise equivalence, providing a straightforward test for elementwise equivalence of knowledge bases. We extended this system to the transformation system $\mathcal{T}_2$ that takes model equivalence into account. Both systems yield set-inclusion minimal knowledge bases with respect to the corresponding notion of equivalence. Note that we used OCFs as an exemplary model for knowledge bases. Both $\mathcal{T}$ and $\mathcal{T}_2$ should also apply to probabilistic or possibilistic [3] settings, it remains to show that the resulting knowledge bases respect the given semantics.

In our ongoing work, we are studying the practical consequences that result from using normalized knowledge bases instead of their non-normalized versions. Both transformation systems $\mathcal{T}$ and $\mathcal{T}_2$ are model preserving, so the normal forms obtained by these system can be used for all inference relations that take all or a single model into account. We are currently investigating to which extent this also applies to inference relations that are defined upon a set of preferred models.

Acknowledgment. This work was supported by DFG-Grant KI1413/5-1 to Gabriele Kern-Isberner as part of the priority program "New Frameworks of Rationality" (SPP 1516). Christian Eichhorn is supported by this Grant. We thank the anonymous reviewers for their valuable hints and comments.

References

1. Baader, F., Nipkow, T.: Term Rewriting and All That. Cambridge University Press, Cambridge (1998)
2. Beierle, C., Eichhorn, C., Kern-Isberner, G.: On transformations and normal forms of conditional knowledge bases. In: Benferhat, S., Tabia, K., Ali, M. (eds.) Proceedings of the 30th International Conference on Industrial, Engineering, Other Applications of Applied Intelligent Systems (IEA/AIE-2017). LNAI, vol. 10350, pp. 488–494. Springer, Heidelberg (2017)
3. Dubois, D., Prade, H.: Possibility theory and its applications: where do we stand? In: Kacprzyk, J., Pedrycz, W. (eds.) Springer Handbook of Computational Intelligence, pp. 31–60. Springer, Heidelberg (2015). doi:10.1007/978-3-662-43505-2_3
4. de Finetti, B.: La prévision, ses lois logiques et ses sources subjectives. Ann. Inst. H. Poincaré **7**(1), 1–68 (1937). English translation in Kyburg, H., Smokler, H.E. (eds.): Studies in Subjective Probability, pp. 93–158. Wiley, New York (1974)
5. Goldszmidt, M., Pearl, J.: Qualitative probabilities for default reasoning, belief revision, and causal modeling. Artif. Intell. **84**(1–2), 57–112 (1996)
6. Kern-Isberner, G.: Conditionals in Nonmonotonic Reasoning and Belief Revision - Considering Conditionals as Agents. LNCS, vol. 2087. Springer, Heidelberg (2001)
7. Knuth, D.E., Bendix, P.B.: Simple word problems in universal algebra. In: Leech, J. (ed.) Computational Problems in Abstract Algebra, pp. 263–297. Pergamon Press (1970)
8. Pearl, J.: System Z: a natural ordering of defaults with tractable applications to nonmonotonic reasoning. In: Parikh, R. (ed.) Proceedings of the 3rd Conference on Theoretical Aspects of Reasoning about Knowledge (TARK 1990), pp. 121–135. Morgan Kaufmann Publishers Inc., San Francisco (1990)
9. Spohn, W.: The Laws of Belief: Ranking Theory and Its Philosophical Applications. Oxford University Press, Oxford (2012)

On Boolean Algebras of Conditionals and Their Logical Counterpart

Tommaso Flaminio[1], Lluis Godo[2(✉)], and Hykel Hosni[3]

[1] Dipartimento di Scienze Teoriche e Applicate, Università dell'Insubria,
Via Mazzini 5, 21100 Varese, Italy
tommaso.flaminio@uninsubria.it
[2] Artificial Intelligence Research Institute (IIIA - CSIC),
Campus de la Univ. Autònoma de Barcelona s/n, 08193 Bellaterra, Spain
godo@iiia.csic.es
[3] Department of Philosophy, University of Milan,
Via Festa del Perdono 7, 20122 Milano, Italy
hykel.hosni@unimi.it

Abstract. This paper sheds a novel light on the longstanding problem of investigating the logic of conditional events. Building on the framework of Boolean algebras of conditionals previously introduced by the authors, we make two main new contributions. First, we fully characterise the atomic structure of these algebras of conditionals. Second, we introduce the *logic of Boolean conditionals* (LBC) and prove its completeness with respect to the natural semantics induced by the structural properties of the atoms in a conditional algebra as described in the first part. In addition we outline the close connection of LBC with *preferential consequence relations*, arguably one of the most appreciated systems of non-monotonic reasoning.

Keywords: Conditionals events · Uncertain reasoning · Boolean algebra of conditionals · Non-monotonic reasoning

1 Introduction

Conditionals play a fundamental role both in qualitative and in quantitative uncertain reasoning, see e.g. [1,6,7,9,12,14,15]. In the former, conditionals constitute the core focus of non-monotonic reasoning [8,10,11]. In quantitative uncertain reasoning, conditionals are central both for conditional probability, and more generally, for conditional uncertainty measures [5].

This paper builds on [4], where a Boolean algebra structure for conditionals was proposed with the goal of clarifying the relationship between conditional probabilities and simple probabilities on conditional events. The approach of considering (measure-free) conditionals as Boolean objects departs from previous ones in the literature where conditionals are mainly considered as three-valued objects, proposing different definitions for the operations between conditionals, see e.g. [2,3,7,9,13]. For a comparison, the reader may consult [4, Sect. 3.1].

© Springer International Publishing AG 2017
A. Antonucci et al. (Eds.): ECSQARU 2017, LNAI 10369, pp. 246–256, 2017.
DOI: 10.1007/978-3-319-61581-3_23

Intuitively, a Boolean algebra of conditional events, a BAC algebra for short, is an algebra built up over a Boolean algebra of plain events $A = (A, \wedge, \vee, \neg, \top, \bot)$ in which we allow *basic conditionals*, i.e. syntactic objects of the form $(a \mid b)$ with $a \in A$ and $b \in A' = A \backslash \{\bot\}$, to be freely combined with the usual Boolean operations, subject to the following plausible constraints (recall that in any Boolean algebra A, if $a, b \in A$, then $a \leq b$ iff $a \wedge b - a$):

- the conditional $(\top \mid \top)$ will be the top element of the algebra, while $(\bot \mid \top)$ will be the bottom;
- given $b \in A'$, we want the set of conditionals $\{(a \mid b) : a \in A\}$ to be the domain of a Boolean subalgebra, and in particular when $b = \top$, then we want this subalgebra to be isomorphic to A;
- in a conditional $(a \mid b)$ we can equivalently replace the consequent a by $a \wedge b$, that is, we require the conditionals $(a \mid b)$ and $(a \wedge b \mid b)$ to be equivalent;
- if $a \leq b \leq c$ then the result of conjunctively combining the conditionals $(a \mid b)$ and $(b \mid c)$ yields the conditional $(a \mid c)$.

This last condition captures a form of restricted transitivity and is clearly inspired by the chain rule of conditional probabilities: $P(a \mid b) \cdot P(b \mid c) = P(a \mid c)$ whenever $a \leq b \leq c$.

The purpose of this paper is to put BAC algebras on firm logical footing. To this end, we make two main new contributions. First, we fully characterise the atomic structure of BAC algebras, a problem that was left open in [4]. This is done in Sect. 3. Second, in Sect. 4, we introduce the *logic of Boolean conditionals* (LBC) and prove its completeness with respect to the natural semantics induced by the structural properties of the atoms in a conditional algebra. In Sect. 5 we conclude with a result to the effect that LBC is indeed a *preferential consequence relation*, in the sense of the well-known System P, see e.g. [10,11].

2 Boolean Algebras of Conditionals

Boolean algebras of conditionals, introduced and investigated in [4], are built as follows. Let $A = (A, \wedge, \vee, \neg, \top, \bot)$ be a Boolean algebra and let $A' = A \backslash \{\bot\}$.

The construction starts with considering the Boolean algebra of terms *freely generated* by the pairs $(a, b) \in A \times A'$, that will be denoted

$$\mathcal{F}(A \times A') = (\mathcal{F}(A \times A'), \wedge^*, \vee^*, \neg^*, \top^*, \bot^*).$$

Notice that all pairs $(a, b) \in A \times A'$ are such that $b > \bot$. This is motivated by the fact that we are avoiding, in this paper, to consider counterfactual conditionals. Since we want conditionals to satisfy a number of properties, what we do is to consider the greatest subalgebra of $\mathcal{F}(A \times A')$ where these properties hold. Technically speaking, we define it as a quotient algebra by a suitable congruence relation. Namely, consider the following elements in $\mathcal{F}(A \times A')$, where we use $\delta(c, c')$ to denote the element $c \leftrightarrow^* c'$ (of $\mathcal{F}(A \times A')$) for any $c, c' \in \mathcal{F}(A \times A')$:

(t1) $\delta((y, y), \top^*)$ for every $y \in A'$,

(t2) $\delta((x,y) \wedge^* (z,y), (x \wedge z, y))$ for every $x, z \in A$ and $y \in A'$,
(t3) $\delta(\neg^*(x,y), (\neg x, y))$ for every $x \in A$, $y \in A'$,
(t4) $\delta((x \wedge y, y), (x, y))$ for every $x \in A$, $y \in A'$,
(t5) $\delta((x,z), (x,y) \wedge^* (y,z))$ for every $x \in A$ and $y, z \in A'$ such that $x \leq y \leq z$.

Let $\mathfrak{C}$ be the filter of $\mathcal{F}(A \times A')$ generated by the set of all the instances of the above terms (t1-t5). We hence define the congruence relation $\equiv_{\mathfrak{C}}$ in the following way: $c \equiv_{\mathfrak{C}} c'$ if $\delta(c, c') \in \mathfrak{C}$.

Definition 1. *For every Boolean algebra A, we say that the quotient algebra $\mathcal{C}(A) = \mathcal{F}(A \times A')/_{\equiv_{\mathfrak{C}}}$ is the Boolean algebra of conditionals of A, the BAC algebra of A for short.*

Since $\mathcal{C}(A)$ is a quotient of a free Boolean algebra, it is a Boolean algebra as well. The elements of $\mathcal{C}(A)$ are in fact equivalence classes of elements of $\mathcal{F}(A \times A')$. We will denote the equivalence class of a pair $(a, b) \in A \times A'$ by $(a \mid b)$. Therefore, in $\mathcal{C}(A)$ we have *basic conditionals* of the form $(a \mid b)$ and *compound conditionals*, that is, those elements in $\mathcal{C}(A)$ that are algebraic terms definable in the language of Boolean algebras, modulo the identification induced by $\mathfrak{C}$. The operations on $\mathcal{C}(A)$ are denoted as follows, with the obvious interpretation

$$\mathcal{C}(A) = (\mathcal{C}(A), \sqcap_{\mathfrak{C}}, \sqcup_{\mathfrak{C}}, \neg_{\mathfrak{C}}, \perp_{\mathfrak{C}}, \top_{\mathfrak{C}}).$$

In order to simplify our notation we will omit the subscript $\mathfrak{C}$ whenever this leads to no ambiguity.

Notice that, the above conditions (t1)–(t5) used to define the quotient algebras $\mathcal{C}(A)$ automatically imply that, in any $\mathcal{C}(A)$, the following equations on conditionals will always hold: $(y \mid y) = \top, (x \mid y) \sqcap (z \mid y) = (x \wedge y \mid z), \neg(x \mid y) = (\neg x \mid y), (x \wedge y \mid y) = (x \mid y)$, and $(x \mid y) \sqcap (y \mid z) = (x \mid z)$ whenever $x \leq y \leq z$.

Observe as well, that the basic conditionals of $\mathcal{C}(A, A')$ need not be closed under meets and joins, though in some particular cases they can be, for instance as a consequence of (t2) and (t3), ensuring that the conjunction of two basic conditionals with the same antecedent is a basic conditional, and the negation of a basic conditional, is basic as well. In any algebra $\mathcal{C}(A)$, as in any Boolean algebra, the induced order relation, also denoted $\leq$, is defined as $(x \mid y) \leq (z \mid k)$ iff $(x \mid y) \sqcap (z \mid k) = (x \mid y)$.

Space constraints do not allow us to delve any further into BAC algebras. We refer the interested reader to [4] for further details, and limit ourselves to collect the properties required in this paper in the following proposition.

Proposition 1. *In every BAC algebra $\mathcal{C}(A)$, for every $x, z \in A$ and $y \in A'$, the following properties hold:*

(e1) $(y \mid y) = (\top \mid y) = \top$ *and* $(\neg y \mid y) = (\perp \mid y) = \perp$.
(e2) $(x \mid y) \sqcap (z \mid y) = (x \wedge z \mid y)$, *and hence, if* $x \wedge z = \perp$ *then* $(x \mid y) \sqcap (z \mid y) = \perp$.
(e3) *if* $z \in A'$ *and* $x \leq y \leq z$ *then* $(x \mid z) = (x \mid y) \sqcap (y \mid z)$.
(e4) $(x \mid y) \sqcup (z \mid y) = (x \vee z \mid y)$ *and* $\neg(x \mid y) = (\neg x \mid y)$.
(e5) $(x \wedge y \mid \top) \leq (x \mid y) \leq (\neg y \vee x \mid \top)$.

$(e6)$ $(x \mid \top) \sqcap (y \mid x) \leq (y \mid \top)$.
$(e7)$ $(x \wedge y \mid \top) \leq (x \mid y)$ and $(\neg x \wedge y \mid \top) \leq \neg(x \mid y)$.
$(e8)$ if $z \in A'$ then $(x \mid y) \sqcap (x \mid z) \leq (x \mid y \vee z)$.

Interesting readings of some of the above properties are the following. $(e1), (e2)$ and $(e4)$ force that for any $z \in A'$, the set $A \mid z = \{(a \mid z) \mid a \in A\}$ is (the domain of) a Boolean subalgebra of $\mathcal{C}(A)$. On the other hand, $(e5)$ shows that conditionals are stronger than material implications but weaker than conjunctions. Also, from a logical point of view, $(e6)$ states a form of modus ponens with conditionals. $(e7)$ states that, whenever y unconditionally holds true, a conditional $(x \mid y)$ holds true if x unconditionally holds true, while $(x \mid y)$ holds false otherwise. Finally, $(e8)$ corresponds to the so-called OR property, typical in nonmonotonic systems, see Sect. 5.

3   The Atoms of a Finite Algebra of Conditionals

We now move on to the investigation of the atoms of a BAC algebra $\mathcal{C}(A)$ for a *finite* A. To this end it is worth remembering that an element α of a Boolean algebra A is an *atom* of A iff when α covers $\bot$, that is, $\bot < \alpha$ and if $\bot \leq \beta \leq \alpha$ then either $\beta = \bot$ or $\beta = \alpha$. Note that every algebra $\mathcal{C}(A)$ is finite whenever A is. Indeed, if A is finite, $\mathcal{F}(A \times A')$ is finite as well, since the variety of Boolean algebras is locally finite. Thus, $\mathcal{C}(A)$ is finite and hence atomic. In the following, we write $Atom(B)$ to denote the set of atoms of any Boolean algebra B.

For the characterization theorem, we need some preliminary results. The following properties are immediate consequences of Proposition 1 $(e3)$. In what follows A will always denote a finite Boolean algebra such that $|Atom(A)| = n$.

Lemma 1. *Let $x, y, z \in A$. The following properties hold:*

1. *If $x \leq y \leq z$, then $(x \mid z) \leq (x \mid y)$; in particular $(x \mid \top) \leq (x \mid y)$.*
2. *If $x \wedge z = \bot$ and $\bot < x \leq y$, then $(x \mid \top) \sqcap (z \mid y) = \bot$.*

Now we can prove the following interesting results.

Proposition 2. *Let $i \leq n - 1$ and define $Seq_i(A)$ to be the set of sequences of length i of pairwise distinct atoms of A. Then the set*

$$Part_i(\mathcal{C}(A)) =$$
$$\{(\beta_1 \mid \top) \sqcap (\beta_2 \mid \neg\beta_1) \sqcap \ldots \sqcap (\beta_i \mid \neg\beta_1 \wedge \ldots \wedge \neg\beta_{i-1}) \mid \langle \beta_1, \beta_2, \ldots, \beta_i \rangle \in Seq_i(A)\}$$

is a partition of $\mathcal{C}(A)$, that is, $\bigsqcup Part_i(\mathcal{C}(A)) = \top$ and for any distinct $C, D \in Part_i(\mathcal{C}(A))$, $C \sqcap D = \bot$.

Proof. (1) The case $i = 1$ is easy as $Seq_1(A) = \{\langle \alpha \rangle \mid \alpha \in Atom(A)\}$, and it is clear that $\bigsqcup_\alpha (\alpha \mid \top) = (\bigvee_\alpha \alpha \mid \top) = \top$.
(2) Suppose the claim is true for $i - 1$, that is, $\bigsqcup Part_{i-1}(\mathcal{C}(A)) = \top$. Consider then a sequence $\overline{\beta} = \langle \beta_1, \ldots, \beta_{i-1} \rangle \in Seq_{i-1}$ and its corresponding compound conditional $H_{\overline{\beta}} = (\beta_1 \mid \top) \sqcap \ldots \sqcap (\beta_{i-1} \mid \neg\beta_1 \wedge \ldots \wedge \neg\beta_{i-2})$. By hypothesis, we know that $\bigsqcup_{\overline{\beta} \in Seq_{i-1}} H_{\overline{\beta}} = \bigsqcup Part_{i-1}(\mathcal{C}(A)) = \top$.

Let $D(\overline{\beta}) = Atom(A) \setminus \{\beta_1, \dots \beta_{i-1}\}$ be the set of $n - i + 1$ atoms disjoint from $\{\beta_1, \dots \beta_{i-1}\}$. Then it is clear that $\bigsqcup_{\beta \in D(\overline{\beta})} (\beta \mid \neg\beta_1 \wedge \dots \wedge \neg\beta_{i-1}) = \top$, and thus $H_{\overline{\beta}} = \bigsqcup_{\beta \in D(\overline{\beta})} H_{\overline{\beta}} \sqcap (\beta \mid \neg\beta_1 \wedge \dots \wedge \neg\beta_{i-1})$.

Therefore, since we can do this for every sequence $\overline{\beta} \in Seq_{i-1}$, we finally get that

$$\bigsqcup_{\overline{\beta} \in Seq_{i-1}} H_{\overline{\beta}} = \bigsqcup_{\overline{\beta} \in Seq_{i-1}} \left(\bigsqcup_{\beta \in D(\overline{\beta})} H_{\overline{\beta}} \sqcap (\beta \mid \neg\beta_1 \wedge \dots \wedge \neg\beta_{i-1}) \right) = \bigsqcup_{\overline{\delta} \in Seq_i} H_{\overline{\delta}} = \bigsqcup Part_i(\mathcal{C}(A))$$

Thus we have proved that $\bigsqcup Part_i(\mathcal{C}(A)) = \top$.   □

Let A be a Boolean algebra with n atoms. We denote by $Seq(A)$ the set of sequences $\overline{\alpha} = \langle \alpha_1, \alpha_2, \dots, \alpha_{n-1} \rangle$ of $n-1$ pairwise distinct atoms of A. Moreover, for every such a sequence $\overline{\alpha} \in Seq(A)$, let us consider the compound conditional

$$\omega_{\overline{\alpha}} = (\alpha_1 \mid \top) \sqcap (\alpha_2 \mid \neg\alpha_1) \sqcap \dots \sqcap (\alpha_{n-1} \mid \neg\alpha_1 \wedge \dots \wedge \neg\alpha_{n-2}),$$

or equivalently, $\omega_{\overline{\alpha}} = (\alpha_1 \mid \top) \sqcap (\alpha_2 \mid \alpha_2 \vee \dots \vee \alpha_n) \sqcap \dots \sqcap (\alpha_{n-1} \mid \alpha_{n-1} \vee \alpha_n)$.

Theorem 1. *The set of the atoms of $\mathcal{C}(A)$ is $Atom(\mathcal{C}(A)) = \{\omega_{\overline{\alpha}} : \overline{\alpha} \in Seq(A)\}$. As a consequence, $|Atom(\mathcal{C}(A))| = n!$ and $|\mathcal{C}(A)| = 2^{n!}$.*

Proof. We have to prove the following two conditions:

(i) For any $\omega_{\overline{\alpha}} \in Atom(\mathcal{C}(A))$, $\omega_{\overline{\alpha}} > \bot$. First of all, observe that, looking at the way the set $Atom(\mathcal{C}(A))$ is defined, if $\omega_{\overline{\alpha}} = \bot$ for some $\omega_{\overline{\alpha}} \in Atom(\mathcal{C}(A))$ then, by a symmetry argument it would be the case that every $\omega_{\overline{\beta}} \in Atom(\mathcal{C}(A))$ would also be $\bot$. Second, let us show that $\bigsqcup Atom(\mathcal{C}(A)) = \top$. Indeed, note that $Atom(\mathcal{C}(A)) = Part_{n-1}(\mathcal{C}(A))$, and thus this directly follows from Proposition 2 when taking $i = n - 1$. Therefore, we conclude that $\omega_{\overline{\alpha}} > \bot$ for every $\omega_{\overline{\alpha}} \in Atom(\mathcal{C}(A))$.

(ii) For any $\omega_{\overline{\alpha}} \in Atom(\mathcal{C}(A))$, there is no $D \in \mathcal{C}(A)$ such that $\bot < D < \omega_{\overline{\alpha}}$. It is enough to show that, for any element $(\gamma \mid b) \in \mathcal{C}(A) \setminus \{\bot\}$, with $\gamma \in Atom(A)$, we have that either $\omega_{\overline{\alpha}} \sqcap (\gamma \mid b) = \bot$ or $\omega_{\overline{\alpha}} \sqcap (\gamma \mid b) = \omega_{\overline{\alpha}}$ itself. Since $\gamma \in Atom(A)$, then $\gamma = \alpha_i$ for some $1 \le i \le n$. Then we have two cases: either $b = \alpha_i \vee \dots \vee \alpha_n$, and in that case $\omega_{\overline{\alpha}} \sqcap (\gamma \mid b) = \omega_{\overline{\alpha}}$, or otherwise b is of the form $b = \alpha_i \vee \alpha_k \vee a$, for some $k < i$. Then, in the latter case, we have $(\gamma \mid b) \sqcap (\alpha_k \mid \alpha_k \vee \dots \vee \alpha_n) = (\alpha_i \mid \alpha_i \vee \alpha_k \vee a) \sqcap (\alpha_k \mid \alpha_k \vee \dots \vee \alpha_n) \le (\alpha_i \mid \alpha_i \vee \alpha_k) \sqcap (\alpha_k \mid \alpha_k \vee \alpha_i) = \bot$, whence $(\gamma \mid b) \sqcap \omega_{\overline{\alpha}} = \bot$ as well.   □

Example 1. Let A be the Boolean algebra of 3 atoms $\{\alpha_1, \alpha_2, \alpha_3\}$ and 8 elements. Theorem 1 tells us that the atoms of the algebra $\mathcal{C}(A)$ are as follows:

$$Atom(\mathcal{C}(A)) = \{(\alpha_i \mid \top) \sqcap (\alpha_j \mid \neg\alpha_i) : i, j = 1, 2, 3 \text{ and } i \ne j\}.$$

Therefore, the algebra $\mathcal{C}(A)$, depicted in Fig. 1, has six atoms $\{x_1, \dots, x_6\}$ and $2^6 = 64$ elements. In particular, we have that $x_1 = (\alpha_1 \mid \top) \sqcap (\alpha_2 \mid \neg\alpha_1)$, $x_2 = (\alpha_1 \mid \top) \sqcap (\alpha_3 \mid \neg\alpha_1)$, $x_3 = (\alpha_2 \mid \top) \sqcap (\alpha_1 \mid \neg\alpha_2)$, $x_4 = (\alpha_2 \mid \top) \sqcap (\alpha_3 \mid \neg\alpha_2)$, $x_5 = (\alpha_3 \mid \top) \sqcap (\alpha_1 \mid \neg\alpha_3)$ and $x_6 = (\alpha_3 \mid \top) \sqcap (\alpha_2 \mid \neg\alpha_3)$.

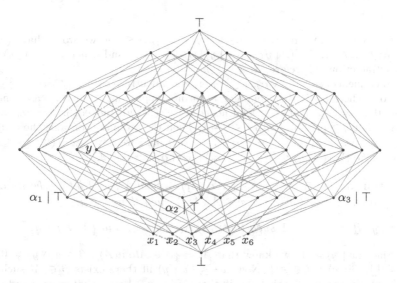

Fig. 1. The algebra of conditionals $\mathcal{C}(A)$, when $|Atom(A)| = 3$, the atoms of which are $x_1, x_2, x_3, x_4, x_5, x_6$. The element $y = (\alpha_1 \mid \neg\alpha_3)$ is $x_1 \sqcup x_2 \sqcup x_5$.

Let us consider the conditional $y = (\alpha_1 \mid \neg\alpha_3)$. Obviously, $y = \bigsqcup\{x_i : x_i \le y\}$ and, thanks to part (2) of Lemma 1, it is easy to see that, indeed, $y = x_1 \sqcup x_2 \sqcup x_5$.

As for a further explicative example, notice that $x_1 \sqcup x_2 = ((\alpha_1 \mid \top) \sqcap (\alpha_2 \mid \neg\alpha_1)) \sqcup ((\alpha_1 \mid \top) \sqcap (\alpha_3 \mid \neg\alpha_1)) = (\alpha_1 \mid \top) \sqcap (\alpha_2 \vee \alpha_3 \mid \neg\alpha_1) = (\alpha_1 \mid \top) \sqcap (\neg\alpha_1 \mid \neg\alpha_1) = (\alpha_1 \mid \top)$. Analogously $(\alpha_2 \mid \top) = x_3 \sqcup x_4$ and $(\alpha_3 \mid \top) = x_5 \sqcup x_6$.

We close this section by characterising the atoms which are below a given non-trivial basic conditional (i.e. different from $\top$).

Lemma 2. *Let* $\overline{\alpha} = \langle \alpha_1, \alpha_2, \ldots, \alpha_{n-1} \rangle \in Seq(A)$, *and let* $\omega_{\overline{\alpha}}$ *be its corresponding atom, i.e.* $\omega_{\overline{\alpha}} = (\alpha_1 \mid \top) \sqcap (\alpha_2 \mid \neg\alpha_1) \sqcap (\alpha_3 \mid \neg\alpha_1 \wedge \neg\alpha_2) \sqcap \ldots \sqcap (\alpha_{n-1} \mid \neg\alpha_1 \wedge \ldots \wedge \neg\alpha_{n-2})$. *Further let* $\beta \in Atom(A)$ *and* $y \in A$ *such that* $\beta < y$. *Then,*

$$\omega_{\overline{\alpha}} \le (\beta \mid y) \text{ iff } \exists\, i \le n-1 \text{ such that } \alpha_i = \beta \text{ and } \neg\alpha_1 \wedge \ldots \wedge \neg\alpha_{i-1} \ge y.$$

Proof. If $\alpha_i = \beta$ and $\neg\alpha_1 \wedge \ldots \wedge \neg\alpha_{i-1} \ge y$, then clearly $(\alpha_i \mid \neg\alpha_1 \wedge \ldots \wedge \neg\alpha_{i-1}) \le (\beta \mid y)$ and hence $\omega_{\overline{\alpha}} \le (\beta \mid y)$ as well. As for the other direction, we have two possibilities:

(i) There is $i \le n-1$ such that $\alpha_i = \beta$. If $i = 1$, then since $\top \ge y$, the condition is fulfilled. Then assume $i > 1$ and $\omega_{\overline{\alpha}} \sqcap (\beta \mid y) = \omega_{\overline{\alpha}}$, we want to prove that $\neg\alpha_1 \wedge \ldots \wedge \neg\alpha_{i-1} \ge y$. Indeed, we have:
 – If $\alpha_1 \le y$, we would have $(\alpha_1 \mid \top) \sqcap (\beta \mid y) \le (\alpha_1 \mid y) \sqcap (\alpha_i \mid y) = \bot$, and hence $\omega_{\overline{\alpha}} \sqcap (\beta \mid y) = \bot$, contradiction. Therefore $\alpha_1 \le \neg y$.
 – If $\alpha_2 \le y$, since $\alpha_1 \le \neg y$, we would have $(\alpha_2 \mid \neg\alpha_1) \sqcap (\beta \mid y) \le (\alpha_2 \mid y) \sqcap (\beta \mid y) = \bot$, and hence $\omega_{\overline{\alpha}} \sqcap (\beta \mid y) = \bot$, contradiction. Therefore, $\alpha_2 \le \neg y$.

– ...
- If $\alpha_{i-1} \leq y$, since $\alpha_1 \leq \neg y$, $\alpha_2 \leq \neg y$, ..., $\alpha_{i-2} \leq \neg y$, we would have $(\alpha_{i-1} \mid \neg\alpha_1 \wedge \ldots \wedge \alpha_{i-2}) \sqcap (\beta \mid y) \leq (\alpha_2 \mid y) \sqcap (\beta \mid y) = \bot$, and hence $\omega_{\overline{\alpha}} \sqcap (\beta \mid y) = \bot$, contradiction. Therefore, $\alpha_{i-1} \leq \neg y$.

As consequence, $\alpha_1 \vee \ldots \vee \alpha_{i-1} \leq \neg y$ or, equivalently, $\neg\alpha_1 \wedge \ldots \wedge \neg\alpha_{i-1} \geq y$.

(ii) $\beta = \alpha_n$, where α_n is the remaining atom not appearing $\overline{\alpha}$. In this case, one can show that $\omega_{\overline{\alpha}} \sqcap (\beta \mid y) = \bot$, and hence $\omega_{\overline{\alpha}} \not\leq (\beta \mid y)$. Indeed, if $\beta = \alpha_n < y$, it means that $y \geq \alpha_i \vee \alpha_n$, with $i \leq n-1$. Then the atom $\omega_{\overline{\alpha}}$ contains the conjunct $(\alpha_i \mid \neg\alpha_1 \wedge \ldots \neg\alpha_{i-1}) = (\alpha_i \mid \alpha_i \vee \ldots \vee \alpha_n)$, and we have $\omega_{\overline{\alpha}} \sqcap (\beta \mid y) \leq (\alpha_i \mid \alpha_i \vee \ldots \vee \alpha_n) \sqcap (\beta \mid y) \leq (\alpha_i \mid \alpha_i \vee \alpha_n) \sqcap (\beta \mid \alpha_i \vee \alpha_n) = \bot$.   □

Proposition 3. *Let $\overline{\alpha} = \langle \alpha_1, \alpha_2, \ldots, \alpha_{n-1} \rangle \in Seq(A)$, and let $\omega_{\overline{\alpha}}$ be its corresponding atom. Then, for any $x, y \in A \setminus \{\bot\}$ such that $x \not\geq y$.*

$$\omega_{\overline{\alpha}} \leq (x \mid y) \text{ iff } \exists\, i \leq n-1 \text{ such that: (1) } \alpha_i \leq x \wedge y, \text{ and (2) } \forall\, j < i, \neg\alpha_j \geq y.$$

Proof. Since $(x \mid y) \neq \bot$, we know that $At = \{\beta \in Atom(A) : \beta \leq x \wedge y\} \neq \emptyset$, and $(x \mid y) = \bigsqcup\{(\beta \mid y) : \beta \in At\}$. Now, $\omega_{\overline{\alpha}} \leq (x \mid y)$ iff there exists $\beta \in At$ such that $\omega_{\overline{\alpha}} \leq (\beta \mid y)$. By Lemma 2, this holds iff there is $i \leq n-1$ such that $\alpha_i = \beta$ and for all $j < i$, $\neg\alpha_j \geq y$.   □

4   Towards a Logic for Conditionals

In this section we define first steps towards a logic to reason with conditionals whose semantics is in accordance with the notion of BAC algebras as described above. Let $\mathcal{L}$ be the classical propositional logic language, built from a finite set of propositional variables $p_1, p_2, \ldots p_m$. Based on $\mathcal{L}$, we define the language $\mathcal{CL}$ of conditionals, in the style of e.g. [8], by the following stipulations:

- Atomic conditional formulas are expressions $(\varphi \mid \psi)$, where $\varphi, \psi \in \mathcal{L}$, and such that $\psi \nvdash \bot$. Atomic conditional formulas are in $\mathcal{CL}$.
- Further, if $\Phi, \Psi \in \mathcal{CL}$, then $\neg\Phi, \Phi \wedge \Psi, \Phi \vee \Psi \in \mathcal{CL}$.[1]

Definition 2. *The* Logic of Boolean conditionals *(LBC for short) has the following axioms and rules, where $\vdash_{PL}$ denotes classical derivability:*

(PL) Axioms and rule of classical propositional logic for $\mathcal{CL}$ formulas
(A1) $(\psi \mid \psi)$
(A2) $\neg(\varphi \mid \psi) \leftrightarrow (\neg\varphi \mid \psi)$
(A3) $(\varphi \mid \psi) \wedge (\delta \mid \psi) \leftrightarrow (\varphi \wedge \delta \mid \psi)$
(A4) $(\varphi \mid \psi) \leftrightarrow (\varphi \wedge \psi \mid \psi)$
(A5) $(\varphi \mid \psi) \leftrightarrow (\varphi \mid \chi) \wedge (\chi \mid \psi)$, if $\vdash_{PL} \varphi \rightarrow \chi$ and $\vdash_{PL} \chi \rightarrow \psi$
(R1) from $\vdash_{PL} \varphi \rightarrow \psi$ derive $(\varphi \mid \chi) \rightarrow (\psi \mid \chi)$
(R2) from $\vdash_{PL} \chi \leftrightarrow \psi$ derive $(\varphi \mid \chi) \leftrightarrow (\varphi \mid \psi)$

The notion of proof in LBC, $\vdash_{LBC}$, is defined as usual.

---

[1] We use the same symbols for connectives in $\mathcal{L}$ and in $\mathcal{CL}$ without danger of confusion.

The above axiomatic system is clearly inspired on the key properties of BAC algebras, and indeed we can prove a tight relation with them. We shall write $\mathbb{L}$ to denote the Lindenbaum-Tarski algebra for the propositional language $\mathcal{L}$. For $\Phi, \Psi \in \mathcal{CL}$, we write $\Phi \equiv \Psi$ if $\vdash_{LBC} \Phi \leftrightarrow \Psi$. (R1) and (R2) ensures that if $\vdash_{PL} \varphi \leftrightarrow \varphi'$ and $\vdash_{PL} \psi \leftrightarrow \psi'$, then $(\varphi \mid \psi) \equiv (\varphi' \mid \psi')$. The following holds.

Proposition 4. *The Lindenbaum-Tarski algebra for the language of conditionals $\mathcal{CL}$, i.e. the quotient algebra $\mathcal{CL}/_{\equiv}$, is a BAC algebra, namely $\mathcal{CL}/_{\equiv} = \mathcal{C}(\mathbb{L})$.*

Semantics and completeness

The guiding idea in defining a semantics for $\mathcal{CL}$ is that the evaluations of $\mathcal{CL}$-formulas should be in one-to-one correspondence with the atoms of the algebra $\mathcal{CL}/_{\equiv}$. Recall from Sect. 3 that the atoms of $\mathcal{C}(A)$ are of the form

$$(\alpha_1 \mid \top) \sqcap (\alpha_2 \mid \neg\alpha_1) \sqcap (\alpha_3 \mid \neg\alpha_1 \wedge \neg\alpha_2) \sqcap \ldots \sqcap (\alpha_{n-1} \mid \neg\alpha_1 \wedge \ldots \wedge \neg\alpha_{n-2}),$$

where $\alpha_1, \ldots \alpha_{n-1}$ are atoms of original algebra A, that is, the α_i's must correspond to maximal elementary conjunctions of literals, or equivalently to evaluations of $\mathcal{L}$. In the following let Ω be the set of (classical) interpretations for the propositional language $\mathcal{L}$, i.e. $\Omega = \{w : Var \to \{0, 1\}\}$. Note that if there are m propositional variables, then $|\Omega| = 2^m$. Therefore, the idea is to define $\mathcal{CL}$-evaluations as sequences $e = (w_1, \ldots, w_{2^m})$, of pair-wise distinct 2^m interpretations $w_1, \ldots, w_{2^m} \in \Omega$, and to stipulate that a $\mathcal{CL}$-evaluation e makes true a conditional $(\varphi \mid \psi)$ when the 'atomic' formula determined by e, $(w_1 \mid \top) \wedge (\overline{w_2} \mid \neg\overline{w_1}) \wedge \ldots \wedge (\overline{w_{n-1}} \mid \neg\overline{w_1} \wedge \ldots \wedge \neg\overline{w_{n-2}})$, is 'below' $(\varphi \mid \psi)$, where $n = 2^m$ and $\overline{w_i}$ denotes the maximal elementary conjunction of $\mathcal{L}$-literals that are true under $w_i \in \Omega$. Therefore, taking into account Proposition 3, we propose the following definition of $\mathcal{CL}$-evaluations.

Definition 3. *A $\mathcal{CL}$-evaluation is a sequence $e = (w_1, w_2, \ldots, w_n)$ of n pairwise distinct $w_1, \ldots, w_n \in \Omega$. The corresponding truth-evaluation of formulas of $\mathcal{CL}$ is as follows:*

- *for atomic $\mathcal{CL}$-formulas: $e(\varphi \mid \psi) = 1$ if $w_i \models \varphi$ for the lowest index i such that $w_i \models \psi$, and $e(\varphi \mid \psi) = 0$ otherwise.*
- *for compound $\mathcal{CL}$-formulas: e is extended using Boolean truth-functions.*

The corresponding notion of consequence is as expected: for any set of $\mathcal{CL}$-formulas $\Gamma \cup \{\Phi\}$, $\Gamma \models_{LBC} \Phi$ if, for every $\mathcal{CL}$-evaluation e such that $e(\Psi) = 1$ for all $\Psi \in \Gamma$, then $e(\Phi) = 1$.

In order to prove completeness for LBC, we need some preliminary results. For every valuation h of the Lindenbaum algebra $\mathcal{CL}/_{\equiv}$ into the 2-element Boolean algebra, i.e. for every Boolean homomorphism $h : \mathcal{CL}/_{\equiv} \to \{0, 1\}$, let ω^h be the unique atom in $\mathcal{CL}/_{\equiv}$ such that $h(\omega^h) = 1$. Also, let $\overline{\alpha_h} = (\alpha_1, \ldots, \alpha_{n-1})$ (for $\alpha_i \in Atom(\mathbb{L})$) be such that $\omega^h = \omega_{\overline{\alpha_h}}$. Further, for every α_i, let w_i^h be the unique evaluation in Ω such that $w_i^h(\alpha_i) = 1$. Then, we write $\Lambda(h) = (w_1^h, \ldots, w_{n-1}^h)$.

Lemma 3. $\Lambda(h)$ *is a $\mathcal{CL}$-evaluation, and for each $\mathcal{CL}$-formula Φ, $h(\Phi) = 1$ iff $\Lambda(h)(\Phi) = 1$.*

Proof. Let $h : \mathcal{CL}/_{\equiv} \rightarrow \{0, 1\}$ be a valuation as above. By construction, it is clear that $\Lambda(h)$ is a $\mathcal{CL}$-evaluation. Now we prove, by induction on the structure of the formula Φ, that $\Lambda(h)(\Phi) = 1$ iff $h(\Phi) = 1$. The interesting case is when Φ is an atomic conditional $(\varphi \mid \psi)$ such that $(\varphi \mid \psi) \not\equiv \top$. Then, $h(\Phi) = 1$ iff $\omega^h \leq (\varphi \mid \psi)$ iff (by Proposition 3) there is $i \leq n - 1$ such that $w_i^h(\varphi) = w_i^h(\psi) = 1$ and $w_l^h(\psi) = 0$ for all $l < i$, and thus, iff $\Lambda(h) = (w_1^h, \ldots, w_{n-1}^h)$ is such that a $\Lambda(h)(\Phi) = 1$. $\qquad\square$

Now the soundness and completeness of LBC easily follows from the above.

Theorem 2 (soundness and completeness). *LBC is sound and complete w.r.t. $\mathcal{CL}$-evaluations, i.e. $\vdash_{LBC} = \models_{LBC}$.*

Proof. Soundness is easy. As for completeness, assume that $\Gamma \not\vdash_{LBC} \Phi$. Thus there exists a homomorphism $h : \mathcal{CL}/_{\equiv} \rightarrow \{0, 1\}$ such that $h(\gamma) = 1$ for all $\gamma \in \Gamma$, and $h(\Phi) = 0$. Thus, by Lemma 3, $\Lambda(h)$ is a $\mathcal{CL}$-evaluation such that $\Lambda(h)(\gamma) = 1$ for every $\gamma \in \Gamma$ and $\Lambda(h)(\Phi) = 0$, i.e. $\Gamma \not\models_{LBC} \Phi$. $\qquad\square$

5   Relation to Non-monotonic Reasoning Models

Conditionals possess an implicit non-monotonic behaviour. Given a conditional $(\varphi \mid \psi)$, it does not follow in general that we can freely strengthen its antecedent, i.e. in general, $(\varphi \mid \psi) \not\vdash_{LBC} (\varphi \mid \psi \wedge \chi)$. For instance, φ, ψ, χ can be such that $\varphi \wedge \psi \not\models \bot$ while $\varphi \wedge \psi \wedge \chi \models \bot$. Actually, and not very surprisingly, the logic $\vdash_{LBC}$ satisfies the analogues of KLM-properties which characterize the well-known system P of preferential entailment [10,11].

Lemma 4. $\vdash_{LBC}$ *satisfies the following properties:*

Reflexivity: $\vdash_{LBC} (\varphi \mid \varphi)$
Left logical equivalence: if $\models_{PL} \varphi \leftrightarrow \psi$ then $(\chi \mid \varphi) \vdash_{LBC} (\chi \mid \psi)$
Right weakening: if $\models_{PL} \varphi \rightarrow \psi$ then $(\varphi \mid \chi) \vdash_{LBC} (\psi \mid \chi)$
Cut: $(\varphi \mid \psi) \wedge (\chi \mid \varphi \wedge \psi) \vdash_{LBC} (\chi \mid \psi)$
OR: $(\varphi \mid \psi) \wedge (\varphi \mid \chi) \vdash_{LBC} (\varphi \mid \psi \vee \chi)$
AND: $(\varphi \mid \psi) \wedge (\delta \mid \psi) \vdash_{LBC} (\varphi \wedge \delta \mid \psi)$
Cautious Monotony: $(\varphi \mid \psi) \wedge (\chi \mid \psi) \vdash_{LBC} (\chi \mid \varphi \wedge \psi)$.

Proof. Reflexivity, Left Logical Equivalence, Right Weakening and *AND* correspond to (A1), (R2), (R1), and (A3) of LBC, respectively. The other cases are proved as follows.

Cut: by (A4), $(\chi \mid \varphi \wedge \psi) \wedge (\varphi \mid \psi)$ is equivalent to $(\chi \wedge \varphi \wedge \psi \mid \varphi \wedge \psi) \wedge (\varphi \wedge \psi \mid \psi)$, and by (A5), it is equivalent to $(\chi \wedge \varphi \wedge \psi \mid \psi)$, and by (R1) this clearly implies $(\chi \mid \psi)$.
Cautious Monotony: by (A3), $(\varphi \mid \psi) \wedge (\chi \mid \psi)$ is equivalent to $(\varphi \wedge \chi \mid \psi)$, which by (A 4) is in turn equivalent $(\varphi \wedge \chi \wedge \psi \mid \psi)$, and by (A5) implies $(\varphi \wedge \chi \wedge \psi \mid \varphi \wedge \psi)$, which by (A3) is equivalent to $(\chi \mid \varphi \wedge \psi)$.

OR: $(\varphi \mid \psi) \wedge (\varphi \mid \chi)$ is equivalent to $[(\varphi \mid \psi) \wedge (\varphi \mid \chi) \wedge (\psi \mid \psi \vee \chi)] \vee [(\varphi \mid \psi) \wedge (\varphi \mid \chi) \wedge (\chi \mid \psi \vee \chi)]$, and this implies $[(\varphi \wedge \psi \mid \psi) \wedge (\psi \mid \psi \vee \chi)] \vee [(\varphi \wedge \chi \mid \chi) \wedge (\chi \mid \psi \vee \chi)]$, that is equivalent to $(\varphi \wedge \psi \mid \psi \vee \chi) \vee (\varphi \wedge \chi \mid \psi \vee \chi)$, which finally implies $(\varphi \mid \psi \vee \chi)$. $\square$

Now, let us fix a set of (atomic) conditional statements K, and let us define the consequence relation associated to K: $\varphi \hspace{1pt}\vdash_K \psi$ if $K \vdash_{LBC} (\psi \mid \varphi)$. Our last proposition is easily derived from the previous lemma.

Proposition 5. $\hspace{1pt}\vdash_K$ *is a preferential consequence relation.*

It can also be proved, but we omit this owing to space constraints, that Rational Monotonicity is also satisfied by LBC. Though this property is far more controversial than Cautious Monotonicity, it has been argued for in a number of circumstances.

6 Concluding Remarks

This paper deepens the investigation on Boolean algebras of conditional events that we began in [4]. Here, we have presented a full description of the atomic structure of these algebras and, based on this, we have defined a corresponding logic, LBC, to reason with conditionals. Moreover, we have shown tight connections of this logic with preferential nonmonotonic consequence relations.

Our previous work [4] was motivated by investigating the relationship of conditional probabilities on a Boolean algebra A and simple probabilities on the algebras of conditional events $\mathcal{C}(A)$. Our conjecture is that for any conditional probability on A there is a simple probability on $\mathcal{C}(A)$ agreeing on basic conditionals. Based on our new results, a close investigation on probabilities and other uncertainty measures on BAC algebras is the object of our future research.

Acknowledgments. We are thankful to the anonymous reviewers. Flaminio and Godo acknowledge partial support by the Spanish FEDER/MINECO project TIN2015-71799-C2-1-P.

References

1. Adams, E.W.: What is at stake in the controversy over conditionals. In: Kern-Isberner, G., Rödder, W., Kulmann, F. (eds.) WCII 2002. LNCS, vol. 3301, pp. 1–11. Springer, Heidelberg (2005). doi:10.1007/11408017_1
2. Calabrese, P.: An algebraic synthesis of the foundations of logic and probability. Inf. Sci. **42**, 187–237 (1987)
3. Dubois, D., Prade, H.: Measure-free conditioning, probability and non-monotonic reasoning. In: Proceedings of IJCAI 1989, vol. 2, pp. 1110–1114 (1989)
4. Flaminio, T., Godo, L., Hosni, H.: On the algebraic structure of conditional events. In: Destercke, S., Denoeux, T. (eds.) ECSQARU 2015. LNCS, vol. 9161, pp. 106–116. Springer, Cham (2015). doi:10.1007/978-3-319-20807-7_10
5. Friedman, N., Halpern, J.Y.: Plausibility measures and default reasoning. J. ACM **48**(4), 648–685 (2001)

6. Gilio, A., Sanfilippo, G.: Conditional random quantities and compounds of conditionals. Stud. Logica **102**(4), 709–729 (2014)
7. Goodman, I.R., Nguyen, H.T., Walker, E.A.: Conditional Inference and Logic for Intelligent Systems - A Theory of Measure-free Conditioning. North-Holland, Amsterdam (1991)
8. Halpern, J.Y.: Defining relative likelihood in partially ordered structures. J. Artif. Intell. Res. **7**, 1–24 (1997)
9. Kern-Isberner, G.: Conditionals in Nonmonotonic Reasoning and Belief Revision. Lecture Notes in Artificial Intelligence, vol. 2087. Springer, Heidelberg (2001)
10. Lehmann, D., Magidor, M.: What does a conditional knowledge base entail? Artif. Intell. **55**(1), 1–60 (1992)
11. Makinson, D.: Bridges From Classical to Non-monotonic Logic. College Publications, London (2005)
12. Makinson, D.: Conditional probability in the light of qualitative belief change. In: Hosni, H., Montagna, F. (eds.) Probability, Uncertainty and Rationality. Edizioni della Normale (2010)
13. Milne, P.: Bruno de Finetti and the logic of conditional events. Br. J. Philos. Sci. **48**(2), 195–232 (1997)
14. Nguyen, H.T., Walker, E.A.: A history and introduction to the algebra of conditional events and probability logic. IEEE Trans. Syst. Man Cybern. **24**(12), 1671–1675 (1994)
15. Schay, G.: An algebra of conditional events. J. Math. Anal. Appl. **24**, 334–344 (1968)

A Semantics for Conditionals with Default Negation

Marco Wilhelm[✉], Christian Eichhorn, Richard Niland,
and Gabriele Kern-Isberner

Department of Computer Science, TU Dortmund, Dortmund, Germany
marco.wilhelm@tu-dortmund.de

Abstract. Ranking functions constitute a powerful formalism for non-monotonic reasoning based on qualitative conditional knowledge. Conditionals are formalized defeasible rules and thus allow one to express that certain individuals or subclasses of some broader concept behave differently. More precisely, in order to model these exceptions by means of ranking functions, it is necessary to state that they behave contrarily with respect to the considered property. This paper proposes conditionals with default negation which instead enable a knowledge engineer to formalize exceptions without giving more specific information. This is useful when a subclass behaves indifferent towards a certain property, or the knowledge engineer wants to exclude a certain subclass because she is not aware of its behavior. Based on this novel type of conditionals, we further present and discuss a nonmonotonic inference formalism.

1 Introduction

Qualitative uncertain reasoning is often based on default rules of the form "if A holds, then typically B follows", representing semantically meaningful relationships between A and B that may serve as guidelines for rational decision making. Such rules are called *conditionals* and are formally written as $(B|A)$. Conditionals are different from material implications $A \Rightarrow B$ in that they are trivalent logical structures [3] that cannot be interpreted truth functionally but need richer epistemic structures to be evaluated. *Ranking functions*, also called *ordinal conditional functions* [10], provide a most convenient way for evaluating conditionals by assigning a degree of implausibility to formulas. Therewith, a conditional $(B|A)$ is accepted by a ranking function iff its verification $A \wedge B$ is more plausible than its falsification $A \wedge \neg B$. Based on this methodology, it is possible to express and reason about subclasses of individuals that behave differently to some broader concept, like the well-known penguins that form an exceptional subclass of non-flying birds. In order to model these exceptions by means of ranking functions, it is however necessary to state that they behave contrarily regarding the considered property. Therefore, this approach fails to address scenarios in which a subclass behaves *indifferently* towards a property of their superclass, as in the following example.

© Springer International Publishing AG 2017
A. Antonucci et al. (Eds.): ECSQARU 2017, LNAI 10369, pp. 257–267, 2017.
DOI: 10.1007/978-3-319-61581-3_24

Example 1. Hybrid electric cars (hybrids) form a subclass of cars with an electric engine. While one would generally attribute not having an additional gasoline engine to electric cars, one would not do so for hybrids. More particularly, it is unbeknownst whether said additional engine is a gasoline engine or, for instance, diesel-powered.

In this paper, we propose a novel knowledge representation and inference formalism which is based on *conditionals with default negation*. Using these extended conditionals leaves it open whether they apply to certain subclasses, and thus allows one to model exceptions such as the one in Example 1. Conditionals with default negation are ordinary conditionals enriched with default negated formulas within their premises that act as disqualifiers. Default negation, also called *negation as failure* [2], is an attenuated form of negation that is used to derive not D from the failure to derive D. In our framework, a conditional is *blocked* when a default negated formula D within its premise is assumed to hold. As a consequence, the conditional remains unconsidered when drawing inferences. We show that this approach leads to a novel inference relation that is convenient for our purpose. Although both conditionals and default negation are well-known in nonmonotonic reasoning, individually, (see, e.g., [2,4,5,8]), to our knowledge there is no work published yet which combines both concepts.

The rest of the paper is organized as follows: After preliminary remarks on conditionals and ranking functions, we give some examples that illustrate the benefit of expanding conditionals by default negation. Formal definitions as well as a discussion of our inference relation based on knowledge bases containing conditionals with default negation follow afterwards. Finally, we conclude.

2   Preliminaries

Let $\mathcal{L} = \mathcal{L}(\Sigma)$ be the set of *(propositional) formulas* over a finite set of *atoms* Σ built in the usual way using the connectives $\wedge$ (conjunction), $\vee$ (disjunction), and $\neg$ (negation). To shorten expressions, we write $A \wedge B$ as AB, $\neg A$ as $\overline{A}$, $\overline{A} \vee B$ as $A \Rightarrow B$, and $\top$ instead of $A \vee \overline{A}$ for $A, B \in \mathcal{L}$. A *literal* $\dot{a}$ is either the atom a or its negation $\overline{a}$. A *world* ω is a complete conjunction of literals, i.e., every atom occurs in ω exactly once, either positive or negated. Thus, the set of all worlds Ω corresponds to the complete set of interpretations of $\mathcal{L}$, and we say that a world ω is a *model* of a formula A, written $\omega \models A$, iff $\mathcal{I}(A) = \texttt{true}$ where $\mathcal{I}$ is the interpretation associated with ω. A set of formulas $\mathcal{F}$ entails a formula G, written $\mathcal{F} \models G$, iff every model of $\mathcal{F}$ is also a model of G. A *conditional* $(B|A)$ built upon formulas $A, B \in \mathcal{L}$ is a formalization of the defeasible rule "if A, then typically B" and leads to the trivalent evaluation $[\![(B|A)]\!]_\omega = 1$ iff $\omega \models AB$ (verification), $[\![(B|A)]\!]_\omega = 0$ iff $\omega \models A\overline{B}$ (falsification), and $[\![(B|A)]\!]_\omega = u$ iff $\omega \models \overline{A}$ (non-applicability) with respect to some world ω. A *knowledge base* $\mathcal{R} = (\mathcal{F}_\mathcal{R}, \mathcal{B}_\mathcal{R})$ consists of a finite set of formulas $\mathcal{F}_\mathcal{R}$ and a finite set of conditionals $\mathcal{B}_\mathcal{R}$. The set $\mathcal{F}_\mathcal{R}$ represents known *facts* that are certain, while conditionals in $\mathcal{B}_\mathcal{R}$ represent *beliefs* that are plausible. We call

$\Omega(\mathcal{R}) = \{\omega \in \Omega \mid \forall F \in \mathcal{F}_{\mathcal{R}} : \omega \models F\}$ the set of *possible worlds* with respect to (the facts in) $\mathcal{R}$. Thus, the worlds that contradict at least one fact from $\mathcal{F}_{\mathcal{R}}$ are declared as *impossible* and therefore excluded from $\Omega(\mathcal{R})$. The semantics of knowledge bases is defined via *ranking functions*. A ranking function, also called *ordinal conditional function* [10], is a mapping $\kappa : \Omega \to \mathbb{N}_0^\infty$ with $\kappa^{-1}(0) \neq \varnothing$, assigning to every world a degree of implausibility. The higher the rank $\kappa(\omega)$ is, the more implausible the world ω is believed to be, such that worlds with a rank of 0 are the most plausible worlds, and those with a rank of ∞ are completely disbelieved. A ranking function κ is called a *model of a conditional* $(B|A)$, written $\kappa \models (B|A)$, iff $\kappa(AB) < \kappa(A\overline{B})$, whereby the rank of a formula A is defined by $\kappa(A) = \min(\{\kappa(\omega) \mid \omega \models A\})$. κ is a *model of a knowledge base* $\mathcal{R}$ iff κ is a model of every conditional in $\mathcal{B}_{\mathcal{R}}$ and $\kappa^{-1}(\infty) = \Omega \setminus \Omega(\mathcal{R})$. Models of knowledge bases represent *epistemic states*. A knowledge base $\mathcal{R}$ is called *consistent* iff it has at least one model. The consistency of $\mathcal{R}$ can be characterized by the notion of *tolerance* [5]. A conditional $(B|A) \in \mathcal{B}_{\mathcal{R}}$ is tolerated by a set of conditionals $\mathcal{P} \subseteq \mathcal{B}_{\mathcal{R}}$ iff there is a possible world ω verifying $(B|A)$ without falsifying any of the conditionals in $\mathcal{P}$. Therewith, $\mathcal{R}$ is consistent iff there is an ordered partition $\mathfrak{P}(\mathcal{R}) = (\mathcal{P}_0, \mathcal{P}_1, \ldots, \mathcal{P}_k)$ of $\mathcal{B}_{\mathcal{R}}$ such that each conditional in $\mathcal{P}_m$ is tolerated by $\bigcup_{l=m}^k \mathcal{P}_l$ for $0 \leq m \leq k$. When choosing the sets $\mathcal{P}_i$ for $i = 0, \ldots, k$ to be maximal with respect to set inclusion (beginning from $\mathcal{P}_0$), one obtains a unique tolerance partition $\mathfrak{P}^Z(\mathcal{R})$. The superscript Z is in reference to the well-known System Z [9] from which the concept of tolerance partitions arises, and therefore, we call $\mathfrak{P}^Z(\mathcal{R})$ the *Z-partition* (of $\mathcal{B}_{\mathcal{R}}$ and thus) of $\mathcal{R}$. Eventually, ranking functions yield nonmonotonic inference relations. To this end, let κ be a model of the consistent knowledge base $\mathcal{R}$, and let A and B be formulas. Then, B can be inferred from A with respect to κ, written $A \mathrel{|\!\sim}_\kappa B$, iff $\kappa \models (B|A)$.

It is reasonable to assign the same rank to possible worlds with the same *conditional structure* [7], i.e., $\kappa(\omega)$ should equal $\kappa(\omega')$ if ω and ω' verify, falsify, and do not apply to the same conditionals in $\mathcal{B}_{\mathcal{R}}$, as these possible worlds are indistinguishable based on their behavior towards $\mathcal{R}$. This *indifference property* is not guaranteed by ranking functions in general, but it is a shared property of the *c-representations* [6] and of ranking functions which fulfill System Z.

Definition 1 (c-Representation). *A c-representation $\kappa_c(\mathcal{R})$ of a consistent knowledge base $\mathcal{R}$ with $\mathcal{B}_{\mathcal{R}} = \{(B_1|A_1), \ldots, (B_n|A_n)\}$ is a ranking function*

$$\kappa_c(\mathcal{R})(\omega) = \begin{cases} \sum_{i=1,\ldots,n,\ \omega \models A_i\overline{B_i}} \kappa_i^-, & \textit{iff } \omega \in \Omega(\mathcal{R}), \\ \infty, & \textit{iff } \omega \in \Omega \setminus \Omega(\mathcal{R}) \end{cases}$$

where the $\kappa_i^- \in \mathbb{N}_0$ for $i = 1, \ldots, n$ are impact values for falsifying the respective conditionals, and which have to be chosen such that

$$\kappa_i^- > \min_{\substack{\omega \in \Omega(\mathcal{R}) \\ \omega \models A_i B_i}} \left\{ \sum_{\substack{i \neq j \\ \omega \models A_j \overline{B_j}}} \kappa_j^- \right\} - \min_{\substack{\omega \in \Omega(\mathcal{R}) \\ \omega \models A_i \overline{B_i}}} \left\{ \sum_{\substack{i \neq j \\ \omega \models A_j \overline{B_j}}} \kappa_j^- \right\}, \tag{1}$$

which guarantees that all c-representations of $\mathcal{R}$ are indeed models of $\mathcal{R}$. We will focus on the unique *Z-c-representation* $\kappa_c^Z(\mathcal{R})$ induced by the Z-partition $\mathfrak{P}^Z(\mathcal{R})$ which exists for every consistent knowledge base $\mathcal{R}$ (cf. [7]). The impact values of $\kappa_c^Z(\mathcal{R})$ are defined as follows (ibid.): For every $\mathcal{P}_m$ in $\mathfrak{P}$, starting with $\mathcal{P}_0$, and for every $(B_i|A_i) \in \mathcal{P}_m$,

$$\kappa_i^- = \min_{\substack{\omega \in \Omega(\mathcal{R}) \text{ with } \omega \models A_i B_i \text{ and} \\ \forall (B|A) \in \bigcup_{l=m}^k \mathcal{P}_l : \omega \not\models A\overline{B}}} \left\{ \sum_{\substack{(B_j|A_j) \in \bigcup_{l=0}^{m-1} \mathcal{P}_l \\ \omega \models A_j \overline{B_j}}} \kappa_j^- \right\} + 1. \tag{2}$$

The recursive definition of the impact values (2) of $\kappa_c^Z(\mathcal{R})$ results in significantly lower computational costs in contrast to the general definition of the impact values (1). Furthermore, we will see that Z-c-representations are very useful when reasoning with conditionals with default negation as they coherently induce the required models of reducts. Before we introduce conditionals with default negation, we discuss the need for them by means of some illustrating examples.

3 Representing Exceptions with Conditionals

Conditionals in combination with c-representations are convenient to represent and reason with many kinds of defeasible rules. For example, it is possible to express that a class of individuals shows a typical property, whereas a certain subclass behaves contrarily towards this property.

Example 2. We consider the knowledge base $\mathcal{R} = (\{e \Rightarrow c\}, \{(r|c), (\neg r|e)\})$ stating that cars with an electric engine are, in particular, cars; cars typically have a long range; and cars with an electric engine typically do not have a long range. Commonsense deliberations tell us that from these conditionals we should be able to infer that cars without an electric engine do have a long range whereas electricity driven cars do not.

In Example 2, the exceptional behavior of electric driven cars with respect to a car's range is modelled by stating that cars with an electric engine behave contrarily to prototypical cars and thus do not have a long range. However, in some cases, one might want to treat a subclass as exceptional not because its members behave contrarily but because they behave indifferently towards a certain property of the superclass, as for instance in Example 1. Another reason for treating a subclass as exceptional is the lack of information on whether the subclass shares a property of its superclass.

Example 3. Typically, cars have a bad carbon footprint. Electric driven cars are a subclass of cars, and it is reasonable to say that they form an exception as one might not want to commit to an estimate of the car's carbon footprint when that of the original power source is unknown.

Altogether, one might want to exclude subclasses from sharing a property without stating anything about the subclasses except for their exceptionality. Therefore, it is necessary to extend the previously introduced framework of conditionals and ranking functions. The following example confirms this assessment by proving that there is no ranking function that can represent the epistemic state induced by the given assertions.

Example 4. We continue Example 1 and consider the set of atoms $\Sigma = \{e, g, h\}$ that denote whether a car has an electric engine or a gasoline engine, respectively whether it is a hybrid electric car (hybrid). A ranking function κ which shall model the assertion that cars with an electric engine are attributed not having a gasoline engine has to satisfy $\kappa(e\overline{g}) < \kappa(e.g.)$ $(\star)$. Further, since hybrids are a subclass of cars with an electric engine, we need $\kappa(\omega) = \infty$ for all worlds ω that contradict the formula $h \Rightarrow e$, i.e., $\kappa(\overline{e}gh) = \kappa(\overline{e}\,\overline{g}h) = \infty$ (†). In addition, we want to model that cars with an electric engine are hybrids with a gasoline engine more likely than non-hybrids without a gasoline engine, such that the restriction $\kappa(egh) < \kappa(e\overline{g}\overline{h})$ (‡) is satisfied as well. As $\kappa(e\overline{g}h) < \kappa(e.g.)$ $(\star)$ is a consequence from $(\star)$ and (‡), it inevitably follows that hybrids typically do not have a gasoline engine, i.e., $\kappa \models (\overline{g}|h)$:

$$\kappa(\overline{g}h) \leq \kappa(e\overline{g}h) \overset{(\star)}{<} \kappa(e.g.) \leq \kappa(cgh) \overset{(\dagger)}{=} \kappa(gh).$$

This obviously contradicts the desire for neither treating the presence nor the abscence of a gasoline engine as an attribute of hybrids.

In the next section, we introduce the concept of conditionals with default negation which proves to be an extension of the presented framework that is capable of dealing with all the exceptions discussed in this section.

4   Conditionals with Default Negation

Syntactically, conditionals with default negation are conditionals that additionally contain a default-negated set of *disqualifiers* $\mathcal{D}$ within their premise. Such a term not $\mathcal{D}$ is assumed to be true as long as no disqualifier $D \in \mathcal{D}$ is known to hold. In this paper, we use the notation not $\mathcal{D}$ as an abbreviation for the set of terms not $D_1, \ldots,$ not D_k, where $\mathcal{D} = \{D_1, \ldots, D_k\}$. In analogy to similar concepts from answer set programming [4], the semantics of a knowledge base containing conditionals with default negation is based on *reducts* of the knowledge base that are free of default negation. These reducts are built with respect to a formula S and describe the reasoner's focus on a concrete situation she is reasoning about. If one of a conditional's disqualifiers is a consequence of S, thereby falsifying the default negation not $\mathcal{D}$, the respective conditional is *blocked* and not contained in the reduct. As a consequence, the conditional is not considered when drawing inferences. Due to the concept of default-negation-free reducts, many concepts from reasoning with ordinary conditionals carry over to reasoning with conditionals with default negation.

For a formal definition of conditionals with default negation, we recall $\mathcal{L}$ to be a propositional language.

Definition 2 (Conditional with Default Negation). *Let* $A, B \in \mathcal{L}$, *and let* $\mathcal{D} \subseteq \mathcal{L}$ *be a finite set of formulas. Then,* $(B|A, \mathsf{not}\ \mathcal{D})$ *is called a* conditional with default negation. *If* $\mathcal{D}$ *is empty, then* $(B|A, \mathsf{not}\ \mathcal{D})$ *equals a conditional without default negation, and we write* $(B|A)$ *instead of* $(B|A, \mathsf{not}\ \varnothing)$.

Conditionals with default negation have the informal meaning "if A, then typically B, unless any $D \in \mathcal{D}$ holds". A *knowledge base (with default negation)* $\mathcal{R} = (\mathcal{F}_{\mathcal{R}}, \mathcal{B}_{\mathcal{R}})$ consists of a finite set of formulas $\mathcal{F}_{\mathcal{R}}$ and a finite set of conditionals with default negation $\mathcal{B}_{\mathcal{R}}$. We use the same notation and naming as for ordinary knowledge bases because knowledge bases without default negation can easily be expressed as knowledge bases with (empty) default negation.

Definition 3 (Reduct). *The* reduct $\mathcal{R}^S = (\mathcal{F}_{\mathcal{R}}, \mathcal{B}_{\mathcal{R}}^S)$ *of* $\mathcal{R}$ *by some formula* $S \in \mathcal{L}$ *is the knowledge base* $\mathcal{R}$ *with its set of conditionals* $\mathcal{B}_{\mathcal{R}}$ *being replaced by*

$$\mathcal{B}_{\mathcal{R}}^S = \Big\{ (B|A) \mid (B|A, \mathsf{not}\ \mathcal{D}) \in \mathcal{B}_{\mathcal{R}} \quad and \quad \forall D \in \mathcal{D} : \{S\} \cup \mathcal{F}_{\mathcal{R}} \not\models D \Big\}.$$

Thus, $\mathcal{B}_{\mathcal{R}}^S$ is obtained from $\mathcal{B}_{\mathcal{R}}$ by removing all conditionals that have a disqualifier $D \in \mathcal{D}$ with $\{S\} \cup \mathcal{F}_{\mathcal{R}} \models D$, and subsequently omitting all default negations in the remaining conditionals. Informally, $\mathcal{B}_{\mathcal{R}}^S$ represents those ordinary conditional beliefs that are plausible when S is *assumed* to be true. Note that this neither implies that S is *believed* to be true nor that S is *factually* true. The aim behind determining the reduct $\mathcal{R}^S$ is to be able to reason about the specific scenario in which S is true, regardless of its (relative) plausibility, by adapting the general belief $\mathcal{B}_{\mathcal{R}}$ to the assumption of S.

Example 5. By using a conditional with default negation, namely $(\overline{g}|e, \mathsf{not}\ \{h\})$, we are now able to formalize that cars with electric engines typically do not have gasoline engines unless they are hybrids, such that no estimation about hybrids having a gasoline engine is made (cf. Example 1). In total, we consider the knowledge base $\mathcal{R} = (\{h \Rightarrow e\}, \{(\overline{g}|e, \mathsf{not}\ \{h\}), (gh|e(gh \vee \overline{g}\overline{h}))\})$ over the set of atoms $\Sigma = \{e, g, h\}$ with the meanings

- $h \Rightarrow e$ $\quad\quad\quad\quad\ \widehat{=}$ $\underline{h}$ybrids are, in particular, cars with $\underline{e}$lectric engines,
- $(\overline{g}|e, \mathsf{not}\ \{h\})$ $\quad \widehat{=}$ cars with $\underline{e}$lectric engines typically do not have gasoline engines unless they are $\underline{h}$ybrids,
- $(gh|e(gh \vee \overline{g}\overline{h}))$ $\widehat{=}$ cars with $\underline{e}$lectric engines are more likely to be $\underline{h}$ybrids with a $\underline{g}$asoline engine than non-hybrids without.

The only two distinct reducts of $\mathcal{R}$ are $\mathcal{R}^S = (\{h \Rightarrow e\}, \{(gh|e(gh \vee \overline{g}\overline{h}))\})$ for S with $S \models h$ and $\mathcal{R}^{S'} = (\{h \Rightarrow e\}, \{(\overline{g}|e), (gh|e(gh \vee \overline{g}\overline{h}))\})$ for S' with $S' \not\models h$.

Although there are usually infinitely many formulas to reduce a given knowledge base $\mathcal{R}$ by, there are only finitely many distinct reducts of $\mathcal{R}$ as two formulas $S_1, S_2 \in \mathcal{L}$ may lead to the same reduct, i.e., $\mathcal{R}^{S_1} = \mathcal{R}^{S_2}$, which holds iff $\mathcal{B}_{\mathcal{R}}^{S_1} = \mathcal{B}_{\mathcal{R}}^{S_2}$. More precisely, the number of different reducts of $\mathcal{R}$ is obviously restricted by 2^n where $n = |\mathcal{B}_{\mathcal{R}}|$. Note that this upper bound is not tight.

Before we define the semantics of knowledge bases with default negation, we want to explicitly exclude from our investigations a specific class of conditionals. A conditional $(B|A, \text{not } \mathcal{D})$ is called *self-blocking with respect to a knowledge base* $\mathcal{R} = (\mathcal{F}_{\mathcal{R}}, \mathcal{B}_{\mathcal{R}})$ iff $\{A\} \cup \mathcal{F}_{\mathcal{R}} \models D$ for some $D \in \mathcal{D}$. The formula A in the premise of a self-blocking conditional can never be satisfied in the context of $\mathcal{F}_{\mathcal{R}}$ without satisfying any of the conditional's disqualifiers $D \in \mathcal{D}$, too. This means that the conditional is a vacuous assertion. Therefore, self-blocking conditionals are not of a reasoner's interest and we can disregard them without any loss. We call a knowledge base that is free of such self-blocking conditionals a *self-blocking-free knowledge base*, and consider only this kind of knowledge base in the following.

Now we have covered all necessary prerequisites to define the concept of models of knowledge bases with default negation.

Definition 4 (Model of a Self-blocking-free Knowledge Base). *Let $\mathcal{R}$ be a self-blocking-free knowledge base, let $\mathfrak{R}(\mathcal{R})$ be the set of all reducts of $\mathcal{R}$, and let $\mathcal{K}(\Omega)$ be the set of all ranking functions. A mapping $\eta : \mathfrak{R}(\mathcal{R}) \to \mathcal{K}(\Omega)$ is called a* model *of $\mathcal{R}$, written $\eta \models \mathcal{R}$, iff $\eta(\mathcal{R}^S) \models \mathcal{R}^S$ for every $\mathcal{R}^S \in \mathfrak{R}(\mathcal{R})$.*

In plain words, η is a model of $\mathcal{R}$ iff η maps every reduct of $\mathcal{R}$ to a ranking function that models this specific reduct. Further, a self-blocking-free knowledge base $\mathcal{R}$ is called *consistent* iff it has at least one model. Consistency can be checked easily following the next proposition.

Proposition 1. *A self-blocking-free knowledge base $\mathcal{R}$ is consistent iff its reduct $\mathcal{R}^\top$ is consistent.*

Proof. By definition $\mathcal{R}$ is consistent iff it has at least one model, which is the case iff every reduct of $\mathcal{R}$ is consistent. As for every formula S the set $\mathcal{B}_{\mathcal{R}}^S$ is a subset of $\mathcal{B}_{\mathcal{R}}^\top$ and subsets of consistent knowledge bases (without default negations) are always consistent, it suffices to verify the consistency of the reduct $\mathcal{R}^\top$. $\square$

Example 6. The knowledge base $\mathcal{R}$ from Example 5 is self-blocking-free (since $\{e, h \Rightarrow e\} \not\models h$) and consistent (as $\mathcal{R}^\top = (\{h \Rightarrow e\}, \{(\bar{g}|e), (gh|e(gh \vee \bar{g}\bar{h}))\})$ is consistent). The images of the model η of $\mathcal{R}$ with $\eta(\mathcal{R}^\top) = \kappa_c^Z(\mathcal{R}^\top)$ and $\eta(\mathcal{R}^h) = \kappa_c^Z(\mathcal{R}^h)$, i.e., the models of the reducts of $\mathcal{R}$, are shown in Table 1.

Table 1. Z-c-Representations $\kappa_c^Z(\mathcal{R}^\top)$ of $\mathcal{R}^\top = (\{h \Rightarrow e\}, \{(\bar{g}|e), (gh|e(gh \vee \bar{g}\bar{h}))\})$ as well as $\kappa_c^Z(\mathcal{R}^h)$ of $\mathcal{R}^h = (\{h \Rightarrow e\}, \{(gh|e(gh \vee \bar{g}\bar{h}))\})$.

ω	$\kappa_c^Z(\mathcal{R}^\top)(\omega)$	$\kappa_c^Z(\mathcal{R}^h)(\omega)$	ω	$\kappa_c^Z(\mathcal{R}^\top)(\omega)$	$\kappa_c^Z(\mathcal{R}^h)(\omega)$
egh	1	0	$\bar{e}gh$	∞	∞
$eg\bar{h}$	1	0	$\bar{e}g\bar{h}$	0	0
$e\bar{g}h$	0	0	$\bar{e}\bar{g}h$	∞	∞
$e\bar{g}\bar{h}$	2	1	$\bar{e}\bar{g}\bar{h}$	0	0

Based on the notion of models of knowledge bases with default negation, we are now able to define a nonmonotonic inference relation.

Definition 5 (Nonmonotonic Inference Relation). *Let $\mathcal{R}$ be a consistent knowledge base, let η be a model of $\mathcal{R}$, and let $A, B \in \mathcal{L}$ be formulas. Then, B can be inferred from A with respect to η, written $A\!\approx_\eta B$, iff $A\!\!\mid\!\sim_{\eta(\mathcal{R}^A)} B$, i.e., iff $\eta(\mathcal{R}^A) \models (B|A)$.*

The idea behind the inference $A\!\approx_\eta B$ is as follows: We assume A to be true and check if B is more plausible than $\overline{B}$ in the presence of A. If so, we infer B from A, and thus, $A\!\approx_\eta B$ holds. To this end, the assumption that A is true allows us to evaluate the default negations in $\mathcal{R}$ with respect to the trueness of $\mathcal{F}_\mathcal{R} \cup \{A\}$, i.e., we build the reduct $\mathcal{R}^A$ in which only those conditionals remain that are not blocked by A verifying one of their disqualifiers. Reasoning then reduces to reasoning with a knowledge base without default negation, namely the model $\eta(\mathcal{R}^A)$ of $\mathcal{R}^A$, and common techniques can be used to check whether B is more plausible than $\overline{B}$ in the presence of A, i.e., iff $\eta(\mathcal{R}^A) \models (B|A)$.

Example 7. We recall the model η with $\eta(\mathcal{R}^\top) = \kappa_c^Z(\mathcal{R}^\top)$ and $\eta(\mathcal{R}^h) = \kappa_c^Z(\mathcal{R}^h)$ of the knowledge base $\mathcal{R} = (\{h \Rightarrow e\}, \{(\overline{g}|e, \mathsf{not}\ \{h\}), (gh|e(gh \vee \overline{g}\overline{h}))\})$ from Example 6. As $\mathcal{R}^e = \mathcal{R}^\top$ and due to $\kappa_c^Z(\mathcal{R}^\top)(e\overline{g}) = 0 < 1 = \kappa_c^Z(\mathcal{R}^\top)(e.g.)$, it holds that $\kappa_c^Z(\mathcal{R}^e) \models (\overline{g}|e)$. Hence, we infer that cars with an electric engine typically do not have a gasoline engine. However, we can neither infer that hybrids do have a gasoline engine nor that they do not have since there is no preference either way: $\kappa_c^Z(\mathcal{R}^h)(gh) = 0 = \kappa_c^Z(\mathcal{R}^h)(\overline{g}h)$. Therefore, we can now obtain the desired inference behavior (cf. Example 4) unlike when reason based on ordinary knowledge bases.

The inference relation $\approx_\eta$ fulfills a wide range of inference properties, including *reflexivity* (REF), *and* (AND), *modus ponens in the consequence* (MPC), *right weakening* (RW), and *left logical equivalence* (LLE), as stated in the next proposition (cf. [8] for a discussion of the inference properties).

Proposition 2. *Let $\mathcal{R} = (\mathcal{F}_\mathcal{R}, \mathcal{B}_\mathcal{R})$ be a consistent knowledge base, and let η be a model of $\mathcal{R}$. Then, $A\!\approx_\eta A$ (REF) and*

$$
\begin{array}{llll}
A\!\approx_\eta B & \text{and} & A \approx_\eta C & \text{imply} & A \approx_\eta BC, & \text{(AND)} \\
A\!\approx_\eta B \Rightarrow C & \text{and} & A \approx_\eta B & \text{imply} & A \approx_\eta C, & \text{(MPC)} \\
B \models C & \text{and} & A \approx_\eta B & \text{imply} & A \approx_\eta C, & \text{(RW)} \\
A \equiv B & \text{and} & A \approx_\eta C & \text{imply} & B \approx_\eta C. & \text{(LLE)}
\end{array}
$$

Proof. The proofs are purely technical, so, due to spatial restrictions, we only present the proof of (LLE): Due to $A \equiv B$, it holds that $\mathcal{R}^A = \mathcal{R}^B$, and $A\!\approx_\eta C$ holds iff $\eta(\mathcal{R}^A) \models (C|A)$ iff $\eta(\mathcal{R}^B) \models (C|A)$ iff $\eta(\mathcal{R}^B) \models (C|B)$ iff $B\!\approx_\eta C$. Hence, (LLE) holds. Other proofs can be furnished in a similar fashion since $\approx_\eta$ essentially inherits its properties from $\mid\!\sim_\kappa$.

However, $\mathrel{\approx}_\eta$ does not satisfy the entirety System P [1], as the inference properties *cumulative transitivity* (CUT), i.e., $A\mathrel{\approx}_\eta B$ and $AB\mathrel{\approx}_\eta C$ imply $A\mathrel{\approx}_\eta C$, and *cautious monotonicity* (CM), i.e., $A\mathrel{\approx}_\eta B$ and $A\mathrel{\approx}_\eta C$ imply $AB\mathrel{\approx}_\eta C$, do *not* hold in general. The non-fulfillment of (CUT) and (CM) is an unpleasant property of nonmonotonic inference relations and thus of $\mathrel{\approx}_\eta$. However, the fulfillment of both (CUT) and (CM) can be guaranteed by weak additional assumptions (e.g., when claiming $\mathcal{R}^A = \mathcal{R}^{AB}$). Until now, any two images of a model η of a consistent knowledge base $\mathcal{R}$, i.e., the models of any two distinct reducts of $\mathcal{R}$, can be chosen completely independently. The following example shows that this freedom of choice can lead to undesirable behavior.

Table 2. All c-representations $\kappa_c(\mathcal{R}^\top)$ of $\mathcal{R}^\top = \{\varnothing, \{(s|dr), (\overline{s}|\overline{d}), (d|s)\}\}$ and $\kappa_c(\mathcal{R}^{\overline{r}})$ of $\mathcal{R}^{\overline{r}} = \{\varnothing, \{(s|dr), (\overline{s}|\overline{d})\}\}$. In the first case, the impact values have to be chosen such that $\kappa_1^- > 0$ and $\kappa_2^- + \kappa_3^- > 0$. In the second case, $\kappa_1^{-\prime} > 0$ and $\kappa_2^{-\prime} > 0$ have to hold.

ω	$\kappa_c(\mathcal{R}^\top)(\omega)$	$\kappa_c(\mathcal{R}^{\overline{r}})(\omega)$	ω	$\kappa_c(\mathcal{R}^\top)(\omega)$	$\kappa_c(\mathcal{R}^{\overline{r}})(\omega)$
drs	0	0	$\bar{d}rs$	$\kappa_2^- + \kappa_3^-$	$\kappa_2^{-\prime}$
$dr\bar{s}$	κ_1^-	$\kappa_1^{-\prime}$	$\bar{d}r\bar{s}$	0	0
$d\bar{r}s$	0	0	$\bar{d}\bar{r}s$	$\kappa_2^- + \kappa_3^-$	$\kappa_2^{-\prime}$
$d\bar{r}\bar{s}$	0	0	$\bar{d}\bar{r}\bar{s}$	0	0

Example 8. We consider the set of atoms $\Sigma = \{d, r, s\}$ and the knowledge base $\mathcal{R} = \{\varnothing, \{(s|dr), (\overline{s}|\overline{d}), (d|s, \text{not } \{\overline{r}\})\}\}$ with the meanings

- $(s|dr)$ $\;\widehat{=}\;$ if the housetop is damaged and it rains, then typically the floor inside is soaked,
- $(\overline{s}|\overline{d})$ $\;\widehat{=}\;$ if the housetop is not damaged, then typically the floor inside is not soaked,
- $(d|s, \text{not } \{\overline{r}\})$ $\;\widehat{=}\;$ if the floor inside is soaked, then typically the housetop is damaged, unless it does not rain.

$\mathcal{R}$ has two distinct reducts, depending on whether $\overline{r}$ is assumed to hold or not. Table 2 shows the schemata of all possible c-representations $\kappa_c(\mathcal{R}^\top)$ and $\kappa_c(\mathcal{R}^{\overline{r}})$ of these both reducts. A reasoner now is allowed to assign to the possible world $\omega = dr\bar{s}$ different ranks of implausibility, namely κ_1^- and $\kappa_1^{-\prime}$, depending on the reduct used to establish her epistemic state. However, in both cases the implausibility of $\omega = dr\bar{s}$ only depends on the impact value of the first conditional $(w|ds)$ which can be chosen regardless of the absence or presence of the third conditional, $(d|s, \text{not } \{\overline{r}\})$. Thus, the rank of implausibility should be the same for ω in both cases, i.e., $\kappa_1^- = \kappa_1^{-\prime}$ should hold.

Hence, Example 8 advises the reasoner to use a common meta strategy when determining the models of the different reducts of $\mathcal{R}$. An obvious strategy is to choose the respective Z-c-representations.

Definition 6 (Z-c-Mapping). *Let $\mathcal{R}$ be a consistent knowledge base. We call the model η of $\mathcal{R}$ with $\eta(\mathcal{R}^S) = \kappa_c^Z(\mathcal{R}^S)$ for every reduct $\mathcal{R}^S$ of $\mathcal{R}$ the Z-c-mapping of $\mathcal{R}$. We denote the Z-c-mapping of $\mathcal{R}$ with $\eta_c^Z(\mathcal{R})$.*

The Z-c-mapping $\eta_c^Z(\mathcal{R})$ of $\mathcal{R}$ is unique as Z-c-representations are unique, and hence, $\eta_c^Z(\mathcal{R})$ is predetermined by the knowledge base $\mathcal{R}$ itself.

Example 9. The Z-c-mapping of $\mathcal{R}$ from Example 8 can be obtained from the c-representation schemata given in Table 2 by instantiating all impact values $\kappa_1^-, \kappa_2^-, \kappa_3^-$ as well as $\kappa_1^{-\prime}, \kappa_2^{-\prime}$ with 1. Thus, $\kappa_1^- = \kappa_1^{-\prime}$ holds as desired.

Another example of a Z-c-mapping is given in Example 6. It remains an open question if Z-c-mappings show more inference properties than those presented in Proposition 2.

5   Conclusion

In this paper, we introduced a novel semantics for conditionals with default negation that allows for drawing nonmonotonic inferences. Basically, conditionals with default negation enable a reasoner to formulate conditional rules that typically hold unless a certain disqualifier applies. If so, the respective conditional remains unconsidered when drawing inferences from the reasoner's knowledge base. In our approach, the validity of the disqualifier is checked with respect to a conditional query, and thus, the reasoner is able to incorporate assumptions into her query and even to draw inferences based on implausible beliefs. Further, we argued that default negations truly extend the concept of conditionals as we were able to capture epistemic states by models of knowledge bases with default negation that cannot be reproduced using ordinary conditionals and ranking functions. In future work, we want to further investigate the properties of our proposed inference relation, particularly for specific subclasses of knowledge bases, formulate complexity results, and extend our approach to reasoning with probabilistic conditionals with default negation.

Acknowledgments. This research was supported by the DFG research unit FOR 1513 on "Hybrid Reasoning for Intelligent Systems" and the DFG grant KI1413/5-1 to Prof. Kern-Isberner as part of the priority program "New Frameworks of Rationality" (SPP 1516).

References

1. Adams, E.W.: The Logic of Conditionals: An Application of Probability to Deductive Logic. Springer, New York (1975)
2. Clark, K.L.: Negation as Failure. In: Gallaire, H., Minker, J. (eds.) Logic and Data Bases. Springer, New York (1978)
3. de Finetti, B.: Theory of Probability. Wiley, New York (1974)
4. Gelfond, M.: Answer sets. In: Lifschitz, V., van Hermelen, F., Porter, B. (eds.) Handbook of Knowledge Representation. Elsevier, San Diego (2008)

5. Goldszmidt, M., Pearl, J.: Qualitative probabilities for default reasoning, belief revision, and causal modeling. Artif. Intell. **84**, 57–112 (1996)
6. Kern-Isberner, G.: Conditionals in Nonmonotonic Reasoning and Belief Revision: Considering Conditionals As Agents. Springer, New York (2001)
7. Kern-Isberner, G.: A thorough axiomatization of a principle of conditional preservation in belief revision. Ann. Math. Artif. Intell. **40**, 127–164 (2004)
8. Lehmann, D.J., Magidor, M.: What does a conditional knowledge base entail? Artif. Intell. **55**, 1–60 (1992)
9. Pearl, J.: System Z: A natural ordering of defaults with tractable applications to default reasoning. In: Proceedings of the 3rd TARK Conference (1990)
10. Spohn, W.: The Laws of Belief: Ranking Theory and Its Philosophical Applications. Oxford University Press, Oxford (2012)

Credal Sets, Credal Networks

Incoherence Correction and Decision Making Based on Generalized Credal Sets

Andrey G. Bronevich[1,2]([⊠]) and Igor N. Rozenberg[1,2]

[1] National Research University Higher School of Economics,
Myasnitskaya 20, 101000 Moscow, Russia
brone@mail.ru, I.Rozenberg@gismps.ru
[2] JSC Research, Development and Planning Institute
for Railway Information Technology, Automation and Telecommunication,
Orlikov Per.5, Building 1, 107996 Moscow, Russia

Abstract. While making decisions we meet different types of uncertainty. Recently the concept of generalized credal set has been proposed for modeling conflict, imprecision and contradiction in information. This concept allows us to generalize the theory of imprecise probabilities giving us possibilities to process information presented by contradictory (incoherent) lower previsions. In this paper we propose a new way of introducing generalized credal sets: we show that any contradictory lower prevision can be represented as a convex sum of non-contradictory and fully contradictory lower previsions. In this way we can introduce generalized credal sets and apply them to decision problems. Decision making is based on decision rules in the theory of imprecise probabilities and the contradiction-imprecision transformation that looks like incoherence correction.

Keywords: Contradictory (incoherent) lower previsions · Decision making · Generalized credal sets · Incoherence correction

1 Introduction

Recently the extension of imprecise probabilities based on generalized credal sets has been proposed [3,4]. By the classical theory of imprecise probabilities [1,2, 6,8] we can model two types of uncertainty: conflict associated with probability measures and imprecision (non-specificity) linked with the choice of a probability measure among possible alternatives. Generalized credal sets allow us also to model contradiction when the avoiding sure loss condition is not fulfilled. Each upper generalized credal set consists of special plausibility functions, conceived as lower probabilities, whose bodies of evidence consist of singletons and certain event. The part consisting of singletons models conflict in information and the part described by a certain event models contradiction.

In our previous research [3,4] we have shown how we can work with contradictory lower and upper previsions based on generalized credal sets, we introduce

© Springer International Publishing AG 2017
A. Antonucci et al. (Eds.): ECSQARU 2017, LNAI 10369, pp. 271–281, 2017.
DOI: 10.1007/978-3-319-61581-3_25

the construction like natural extension in the classical theory of imprecise probabilities, we describe conditions when generalized credal sets generate models based on usual imprecise probabilities.

In the paper we show how generalized credal sets can be used for correcting incoherent information and how they can be applied to decision problems.

2   Monotone Measures: Basic Definitions and Notations

Let $X = \{x_1, ..., x_n\}$ be a finite set of elementary events, and let 2^X be the algebra of all subsets of X. A set function $\mu : 2^X \to [0, 1]$ is called a *monotone measure* if $\mu(\emptyset) = 0$, $\mu(X) = 1$ and $\mu(A) \leqslant \mu(B)$ for any $A, B \in 2^X$ such that $A \subseteq B$. A monotone measure μ is

- a *probability measure* if $\mu(A \cup B) = \mu(A) + \mu(B)$ for any $A, B \in 2^X$ such that $A \cap B = \emptyset$;
- a *belief function* if there is a set function $m : 2^X \to [0, 1]$ called the *basic belief assignment* (bba) with $m(\emptyset) = 0$ and $\sum_{B \in 2^X} m(B) = 1$ such that $\mu(A) = \sum_{B \subseteq A} m(B)$.

In the sequel M_{mon} denotes the set of all monotone measures on 2^X; M_{pr} denotes the set of all probability measures on 2^X; and M_{bel} denotes the set of all belief functions on 2^X.

We define on M_{mon} the following operations and relations:

- $\mu = a\mu_1 + (1 - a)\mu_2$ for $\mu_1, \mu_2 \in M_{mon}$ and $a \in [0, 1]$ if $\mu(A) = a\mu_1(A) + (1 - a)\mu_2(A)$ for all $A \in 2^X$;
- $\mu_1 \leqslant \mu_2$ for $\mu_1, \mu_2 \in M_{mon}$ if $\mu_1(A) \leqslant \mu_2(A)$ for all $A \in 2^X$;
- μ^d is the dual of μ if $\mu^d(A) = 1 - \mu(A^c)$ for all $A \in 2^X$, where A^c is the complement of A.

Let $Bel \in M_{bel}$ with bba m, then

- Bel^d is called a *plausibility function*;
- a set $B \in 2^X$ is called a *focal element* if $m(B) > 0$;
- the set of all focal elements is called the *body of evidence*;
- a belief function is called *categorical* if its body of evidence contains one focal element $B \in 2^X$. This set function is denoted by $\eta_{\langle B \rangle}$ and can be computed as $\eta_{\langle B \rangle}(A) = \begin{cases} 1, & B \subseteq A, \\ 0, & B \not\subseteq A. \end{cases}$;
- Any $Bel \in M_{bel}$ with bba m can represented as a convex sum of categorical belief functions $Bel = \sum_{B \in 2^X} m(B)\eta_{\langle B \rangle}$.

Assume that M is an arbitrary subset of M_{mon}, then $M^d = \{\mu^d | \mu \in M\}$. In such a way M_{bel}^d denotes the set of all plausibility functions on 2^X.

3   Credal Sets, Lower and Upper Previsions

In the following any $P \in M_{pr}$ can be represented as a point $(P(\{x_1\}), ..., P(\{x_n\}))$ in $\mathbb{R}^n$. By definition [1,6], a *credal set* $\mathbf{P}$ is a non-empty subset of M_{pr}, which is convex and closed. Convexity of $\mathbf{P}$ means that $P_1, P_2 \in \mathbf{P}$ and $a \in [0,1]$ implies that $aP_1 + (1 - a)P_2 \in \mathbf{P}$, and $\mathbf{P}$ is closed as a subset of $\mathbb{R}^n$. A model based on credal sets is one of the most general models of imprecise probabilities. We can describe credal sets using lower and upper previsions. Let K be a set of all real-valued functions $f : X \to \mathbb{R}$ on X. Then any $f \in K$ can be viewed as a random variable for a fixed $P \in M_{pr}$ and we can compute its expectation defined by $E_P(f) = \sum_{x \in X} f(x)P(\{x\})$. Let K' be an arbitrary subset of K, then any functional $\underline{E} : K' \to \mathbb{R}$ is called a *lower prevision* if each value $\underline{E}(f), f \in K'$, is viewed as a lower bound of expectation of the random variable f. This lower prevision is called non-contradictory (or it avoids sure loss) iff it defines the credal set

$$\mathbf{P}(\underline{E}) = \{P \in M_{pr} | \forall f \in K' : E_P(f) \geqslant \underline{E}(f)\} \tag{1}$$

Otherwise, when the set $\mathbf{P}(\underline{E})$ is empty, the lower prevision is called *contradictory* (or *incoherent*). Analogously, upper previsions are defined. Any functional $\bar{E} : K' \to \mathbb{R}$ is called an *upper prevision* if its values are viewed as upper bounds of expectations. It is non-contradictory (or it avoids sure loss) iff it defines the credal set

$$\mathbf{P}(\bar{E}) = \{P \in M_{pr} | \forall f \in K' : E_P(f) \leqslant \bar{E}(f)\},$$

and it incurs sure loss otherwise. Models of uncertainty based on upper and lower previsions are equivalent. It follows from the fact that every lower prevision $\underline{E} : K' \to \mathbb{R}$ and the corresponding upper prevision

$$\bar{E}(f) = -\underline{E}(-f), \quad -f \in K',$$

define the same credal set. The central role in reasoning based on imprecise probabilities plays the natural extension. Let $\underline{E} : K' \to \mathbb{R}$ be an non-contradictory lower prevision and $\mathbf{P}(\underline{E})$ be the credal set defined by formula (1), then the *natural extension* of $\underline{E}$ is a functional

$$\underline{E}'(f) = \inf_{P \in \mathbf{P}(\underline{E})} E_P(f), \quad f \in K'.$$

A lower prevision $\underline{E}$ is called *coherent* if $\underline{E}(f) = \underline{E}'(f)$ for all $f \in K'$. Analogously the natural extension of non-contradictory upper previsions is defined and coherent upper previsions are introduced. Monotone measures can be considered as special models of lower and upper previsions. In this case $K' = \{1_A\}_{A \in 2^X}$, where 1_A is the characteristic function of the set A, i.e. $\mu(A) = \underline{E}(1_A), A \in 2^X$, can be viewed as a set function. A monotone measure μ is called a lower probability if its values give us lower bounds of probabilities. It is non-contradictory if it defines the credal set $\mathbf{P}(\mu) = \{P \in M_{pr} | \mu \leqslant P\}$. We can define analogously the natural extension of non-contradictory lower previsions and the family of coherent lower probabilities. In the same way we define upper probabilities that give us upper bounds of probabilities, the natural extension of non-contradictory upper probabilities and coherent upper probabilities.

Remark 1. Obviously, $\min_{x \in X} f(x) \leqslant E_P(f) \leqslant \max_{x \in X} f(x)$ for any $P \in M_{pr}$ and $f \in K$. Thus, without decreasing generality we can assume that values $\underline{E}(f)$ of any lower prevision $\underline{E} : K' \to \mathbb{R}$ should be not larger than $\max_{x \in X} f(x)$, i.e. $\underline{E}(f) \leqslant \max_{x \in X} f(x)$ for any $f \in K'$. Analogously, we will assume that $\bar{E}(f) \geqslant \min_{x \in X} f(x)$ for any upper prevision $\bar{E} : K' \to \mathbb{R}$ and $f \in K'$, This assumption will be used later without mentioning about it.

4   Generalized Credal Sets for Describing Contradictory Lower Previsions

Assume that we have estimates $\hat{p}(x_i)$, $i = 1, ..., n$, of probabilities, but unfortunately $\sum_{i=1}^{n} \hat{p}(x_i) \neq 1$. What should we do? One can say that the available information is defective and it is not possible to use it. But if the value $\varepsilon = |\sum_{i=1}^{n} \hat{p}(x_i) - 1|$ is small, then this conclusion seems to be not useful. Otherwise we should correct $\hat{p}(x_i)$. Assume that $\sum_{i=1}^{n} \hat{p}(x_i) < 1$, then the correction can be done by adding to each $\hat{p}(x_i)$ a value $\alpha_i \geqslant 0$ such that $\sum_{i=1}^{n} (\hat{p}(x_i) + \alpha_i) = 1$. Thus, uncertainty can be modeled by the set of probability distributions

$$\left\{ (p(x_1), ..., p(x_n)) \,|\, p(x_i) \geqslant \hat{p}(x_i), i = 1, ..., n, \sum_{i=1}^{n} p(x_i) = 1 \right\}.$$

Observe that in this case values $\hat{p}(x_i)$ looks like lower bounds of probabilities, but this does not follow from the problem statement. To avoid ambiguity we should decide whether $\hat{p}(x_i)$ give us lower or upper bounds of probabilities. Lower probabilities have been intensively investigated in the theory of imprecise probabilities and they describe two types of uncertainty: conflict associated with probability measures and non-specificity linked with the choice of a probability measure among possible alternatives. If values $\hat{p}(x_i)$ are viewed as upper probabilities then we say that the available information incurs sure loss or it is contradictory.

Let us analyze the above model in detail. If $\hat{p}(x_i) = 0$, $i = 1, ..., n$, and $\hat{p}(x_i)$ are viewed as lower bounds of probabilities, then the set

$$\left\{ (p(x_1), ..., p(x_n)) \,|\, p(x_i) \geqslant 0, i = 1, ..., n, \sum_{i=1}^{n} p(x_i) = 1 \right\}$$

contains all possible probability distributions or probability measures on 2^X. Thus, in such a case, values $\hat{p}(x_i) = 0$, $i = 1, ..., n$, describe the situation of complete ignorance. We will describe this situation by a vacuous belief function $\eta_{\langle X \rangle}$ viewed as lower probability. Analogously, if $\hat{p}(x_i) = 0$, $i = 1, ..., n$, are viewed as upper bounds of probabilities, then we can describe contradiction by the set of all probability measures M_{pr}, or by $\eta_{\langle X \rangle}$ viewed as an upper probability. This situation can be understood as the case of full contradiction.

Although we can describe contradiction and non-specificity by the set of probability measures there is a principal difference between these two types of uncertainty. Non-specificity means that we don't know exactly what kind of probability model should be chosen among possible alternatives, but contradiction means that we have some deficiency in estimating probabilities. The last problem appears when we try to use simultaneously different probabilistic models for analyzing statistical data or to aggregate pieces of evidence from separate sources of information.

Let us remind the notion of contradiction from usual logic. Let we have a set of axioms $A_1,...,A_m$, and if we use the set-theoretical model, then any A_i can be represented as a subset of a finite set X. Then this system of axioms is contradictory iff $A_1 \cap ... \cap A_m = \emptyset$. In logic we can infer from the contradictory system of axiom that any conclusion is true. This situation can be described by the contradictory lower probability

$$\eta_X^d(A) = \begin{cases} 1, A \neq \emptyset, \\ 0, A = \emptyset. \end{cases}$$

Thus, the case of full contradiction can be described by any lower probability $\mu \in M_{mon}$ such that $\mu(A_1) = ... = \mu(A_m) = 1$ and $A_1 \cap ... \cap A_m = \emptyset$. In general the case of full contradiction can be described by the following definition.

Definition 1. The information described by a lower prevision $\underline{E} : K' \to \mathbb{R}$ is *fully contradictory* iff $\underline{E}$ can not be represented as a convex sum $\underline{E}(f) = a\underline{E}^{(1)}(f) + (1-a)\underline{E}^{(2)}(f)$ of a non-contradictory lower prevision $\underline{E}^{(1)} : K' \to \mathbb{R}$, and a (contradictory) lower prevision $\underline{E}^{(2)} : K' \to \mathbb{R}$ for some $a \in (0,1]$.

Lemma 1. *A lower prevision $\underline{E} : K' \to \mathbb{R}$ is fully contradictory iff for any $a \in (0,1]$ the lower prevision $\underline{E}'(f) = \frac{1}{a} \left(\underline{E}(f) - (1-a) \max_{x \in X} f(x) \right)$, $f \in K'$, is contradictory.*

Lemma 2. *If the set of contradictory previsions on K' is not empty, then the lower prevision $\hat{\underline{E}}(f) = \max_{x \in X} f(x)$, $f \in K'$, is fully contradictory.*

Remark 2. It is possible to choose K' such that every lower prevision is non-contradictory. In this case $\hat{\underline{E}}$ is also a non-contradictory lower prevision. Because the aim of the paper is to deal with contradictory information, in the next we will assume that K' is chosen providing the lower prevision $\hat{\underline{E}}$ to be fully contradictory.

Let $\underline{E} : K' \to \mathbb{R}$ be a lower prevision. Then by Lemma 1 and Lemma 2 (see also Remark 2) it can be always represented as a convex sum

$$\underline{E}(f) = a\underline{E}^{(1)}(f) + (1-a)\underline{E}^{(2)}(f), \tag{2}$$

where $\underline{E}^{(1)}$ is a non-contradictory lower prevision and a lower prevision $\underline{E}^{(2)}$ is fully contradictory. If $a \in (0,1]$, then by Lemma 1 $\underline{E}^{(2)}$ can be chosen to be equal to $\hat{\underline{E}}$. If the lower prevision $\underline{E}$ is fully contradictory, then $\underline{E}^{(2)} = \underline{E}$, $a = 0$, and we can take a non-contradictory lower prevision $\underline{E}^{(1)}$ arbitrarily. We see that the largest value of a characterizes the amount of contradiction in $\underline{E}$. Thus, we can introduce the following definition.

Definition 2. Let $\underline{E} : K' \to \mathbb{R}$ be a lower prevision and let A be the set of all possible values $a \in [0,1]$, for which the representation (2) exists for some non-contradictory lower prevision $\underline{E}^{(1)}$ and a fully contradictory lower prevision $\underline{E}^{(2)}$. Then the *amount of contradiction* is defined by $Con(\underline{E}) = 1 - \sup\{a | a \in A\}$.

Obviously, by Definition 2 $Con(\underline{E}) = 0$ iff $\underline{E}$ is a non-contradictory lower prevision, and $Con(\underline{E}) = 1$ iff $\underline{E}$ is fully contradictory. Let us introduce new concepts, which will help us to simplify the computation of $Con(\underline{E})$. Consider monotone measures on 2^X of the type

$$P = a_0 \eta^d_{\langle X \rangle} + \sum_{i=1}^{n} a_i \eta_{\langle \{x_i\} \rangle}, \tag{3}$$

where $\sum_{i=0}^{n} a_i = 1$, $a_i \geqslant 0$, $i = 0, ..., n$, and P is viewed as a lower probability. Such a P can be represented also as $P = a_0 \eta^d_{\langle X \rangle} + (1 - a_0)P'$, where $\eta^d_{\langle X \rangle}$ is a fully contradictory lower probability and P' is a probability measure defined by $P' = \frac{1}{1-a_0} \sum_{i=1}^{n} a_i \eta_{\langle \{x_i\} \rangle}$ for $a_0 \neq 1$. We can extend P to the lower prevision on the set of all functions in K by

$$\underline{E}_P(f) = a_0 \max_{x \in X} f(x) + \sum_{i=1}^{n} a_i f(x_i).$$

Again $\underline{E}_P$ can be represented as a convex sum of fully contradictory lower prevision $\hat{\underline{E}}$ and linear prevision $E_{P'}$, i.e. $\underline{E}_P(f) = a_0 \hat{\underline{E}}(f) + (1 - a_0)E_{P'}(f)$ for all $f \in K$. We will denote by M_{cpr} the set of all monotone measures defined by (3).

Lemma 3. *Let $P = a_0 \eta^d_{\langle X \rangle} + \sum_{i=1}^{n} a_i \eta_{\langle \{x_i\} \rangle}$ be in M_{cpr}. Then $Con(P) = a_0$.*

We will identify each $P \in M_{cpr}$ from (3) with a point $(a_1, ..., a_n)$ in $\mathbb{R}^n$. Let $P_1, P_2 \in M_{cpr}$ and $P_i = (a_1^{(i)}, ..., a_n^{(i)})$, $i = 1, 2$, then $P_1 \leqslant P_2$ iff $a_k^{(1)} \geqslant a_k^{(2)}$, $k = 1, ..., n$. Clearly, such P_1 and P_2 can describe the same information, but P_2 is a lower probability with higher contradiction.

Definition 3. A subset $\mathbf{P}$ of M_{cpr} is called an *upper generalized credal set* (UG-credal set) if

(1) $P_1 \in \mathbf{P}$, $P_2 \in M_{cpr}$, and $P_1 \leqslant P_2$ implies $P_2 \in \mathbf{P}$;

(2) $P_1, P_2 \in \mathbf{P}$ implies $aP_1 + (1-a)P_2 \in \mathbf{P}$ for every $a \in [0,1]$;

(3) $\mathbf{P}$ is a closed set if we consider it as a subset of $\mathbb{R}^n$.

We will describe any lower prevision $\underline{E} : K' \to \mathbb{R}$ by a UG-credal set $\mathbf{P}$ defined by

$$\mathbf{P} = \{P \in M_{cpr} | \forall f \in K' : E(f) \leqslant \underline{E}_P(f)\}. \tag{4}$$

Remark 3. Obviously, the set defined by (4) is not empty, because it always contains the measure $\eta^d_{\langle X \rangle}$.

Proposition 1. *Let $\underline{E} : K' \to \mathbb{R}$ be a lower prevision, and let $\mathbf{P}$ be its corresponding UG-credal set defined by (4). Then*

$$Con(\underline{E}) = \inf \{Con(P) | P \in \mathbf{P}\}. \tag{5}$$

5   Decision Making Based on Contradictory Lower Previsions

Assume that $\underline{E} : K' \to \mathbb{R}$ is a lower prevision and $Con(\underline{E}) = b$. If $b = 1$ then $\underline{E}$ is fully contradictory and $\underline{E}$ does not contain useful information. Therefore, this case is identical to the case of complete ignorance. Let $b < 1$, then for any $a \in (0, 1-b]$ the lower prevision $\underline{E}$ can be represented as $\underline{E}(f) = a\underline{E}^{(1)}(f) + (1-a)\underline{E}^{(2)}(f)$, $f \in K'$, where $\underline{E}^{(1)}$ is a non-contradictory and $\underline{E}^{(2)}$ is a fully contradictory lower prevision. Obviously, decision making should be based on information in $\underline{E}^{(1)}$. Notice also that decreasing parameter a we get information in $\underline{E}^{(1)}$ more imprecise. Therefore, it makes a sense taking $a = 1 - b$. It is also possible to choose $\underline{E}^{(2)} = \hat{E}$. After this choice the above representation can be rewritten as $\underline{E}(f) = (1-b)\underline{E}^{(1)}(f) + b\hat{E}(f)$, $f \in K'$.

Assume that a non-contradictory lower prevision $\underline{E}^{(1)}$ defines the credal set $\mathbf{P}' = \left\{P \in M_{pr} | \forall f \in K' : \underline{E}^{(1)}(f) \leqslant E_P(f)\right\}$. Then taking in account that $\hat{\underline{E}}$ describes the case of full contradiction, we can describe $\underline{E}$ by a credal set $\mathbf{P}''$ represented as a convex sum of two credal sets in which the first is $\mathbf{P}'$ and the second describes the case of complete ignorance, i.e.

$$\mathbf{P}'' = \{(1-b)P_1 + bP_2 | P_1 \in \mathbf{P}', P_2 \in M_{pr}\}. \tag{6}$$

The following proposition shows how the above set $\mathbf{P}''$ can be found based on UG-credal sets.

Proposition 2. *Let $\underline{E} : K' \to \mathbb{R}$ be a lower prevision, $Con(\underline{E}) = b$, and let $\mathbf{P}$ be its corresponding UG-credal set. Then*

$$\mathbf{P}'' = \{P' \in M_{pr} | \exists P \in \mathbf{P} : Con(P) = b, P' \leqslant P\}. \tag{7}$$

The above transformation $\underline{E} : K' \to \mathbb{R}$ of a contradictory lower prevision to the non-contradictory information presented by the credal set $\mathbf{P}''$ can be considered as incoherence correction in which full contradiction is transformed to complete ignorance. After this transformation we can use known models of decision making considered in imprecise probabilities. In our paper we will consider the decision rule justified in many works (e.g. [1,8].).

We will identify each decision with a function in K. Assume that available information is described by a credal set $\mathbf{P}'' \subseteq M_{pr}$. Then decision $f_2 \in K$ is at least preferable as decision $f_1 \in K$ ($f_1 \preccurlyeq f_2$) if $E_{P'}(f_1) \leqslant E_{P'}(f_2)$ for every $P' \in \mathbf{P}''$. This rule can be rewritten as $f_1 \preccurlyeq f_2$ if $\underline{E}_{\mathbf{P}''}(f_2 - f_1) \geqslant 0$, where $\underline{E}_{\mathbf{P}''}(f) = \inf\limits_{P \in \mathbf{P}''} E_P(f)$, $f \in K$.

Lemma 4. *Let we use notations as in formula (6). Then the expression for $\underline{E}_{\mathbf{P}''}(f)$ can be transformed to*

$$\underline{E}_{\mathbf{P}''}(f) = (1 - b)\underline{E}_{\mathbf{P}'}(f) + b \min\limits_{x \in X} f(x),$$

where $\underline{E}_{\mathbf{P}'}(f) = \inf\limits_{P \in \mathbf{P}'} E_P(f)$.

Let us consider the computational scheme by which this decision rule can be realized. A function $f \in K$ is called *normalized from above* if $\max\limits_{x \in X} f(x) = 0$. The following lemma shows how we can normalize functions for a given lower prevision.

Lemma 5. *Let $\underline{E} : K' \to \mathbb{R}$ be a lower prevision. Consider the set $K'' = \left\{ \bar{f} = f - \max\limits_{x \in X} f(x) | f \in K' \right\}$ of normalized from above functions. Then a lower prevision $\underline{E}' : K'' \to \mathbb{R}$ defines the same UG-credal set as $\underline{E}$ if $\underline{E}'(\bar{f}) = \underline{E}(f) - \max\limits_{x \in X} f(x)$ for all $f \in K'$.*

Clearly the above lemma allows us to assume that functions in K', on which a lower prevision $\underline{E}$ is defined, are normalized from above.

Proposition 3. *Let K' be a finite subset of normalized functions from above in K and let $\underline{E} : K' \to \mathbb{R}$ be a lower prevision. Then $Con(\underline{E}) = \max\{0, b\}$, where b is the solution of the following linear programming problem:*

$$b = 1 - \sum_{i=1}^{n} a_i \to \min,$$

$$\begin{cases} \sum\limits_{i=1}^{n} a_i f_k(x_i) \geqslant \underline{E}(f_k), & f_k \in K', \\ a_i \geqslant 0, & i = 1, ..., n, \end{cases}$$

Proposition 4. *Let K' be a finite subset of normalized functions from above in K and let $\underline{E} : K' \to \mathbb{R}$ be a lower prevision with $Con(\underline{E}) = b$. Then $c = (1-b)\underline{E}_{\mathbf{P}'}(f)$ for any $f \in K$ is the solution of the following linear programming problem:*

$$c = \sum_{i=1}^{n} a_i f(x_i) \to \min,$$

$$\begin{cases} \sum_{i=1}^{n} a_i f_k(x_i) \geqslant \underline{E}(f_k), \quad f_k \in K', \\ \sum_{i=1}^{n} a_i = 1 - b, \quad a_i \geqslant 0, \quad i = 1, ..., n. \end{cases}$$

Example 1. Let we have two pieces of evidence. The first says that the probability that it will be sunny tomorrow is higher or equal than 0.3. The second says that the probability of rain is higher or equal than 0.8. We can describe this information by the states of the world: $x_1 := sunny$, $x_2 := rain$, and denote $X = \{x_1, x_2\}$. Then we have $\underline{E}\left(1_{\{x_1\}}\right) = 0.3$, $\underline{E}\left(1_{\{x_2\}}\right) = 0.8$. For using our computational scheme functions $1_{\{x_1\}}$ and $1_{\{x_2\}}$ should be normalized from above. Doing it we get functions $f_1 = 1_{\{x_1\}} - 1_X$ and $f_2 = 1_{\{x_2\}} - 1_X$ with $\underline{E}(f_1) = -0.7$ and $\underline{E}(f_2) = -0.2$. Then the amount of contradiction can be computed by solving the following linear programming problem:

$$b = 1 - a_1 - a_2 \to \min$$

$$\begin{cases} -a_2 \geqslant -0.7, \\ -a_1 \geqslant -0.2, \\ a_1, a_2 \geqslant 0. \end{cases}$$

Thus, $b = 0.1$. Assume that we need to compute $c = (1-b)\underline{E}_{\mathbf{P}'}(f)$ for some $f \in K$. Then c can be computed by solving the following linear programming problem:

$$c = a_1 f(x_1) + a_2 f(x_2) \to \min,$$

$$\begin{cases} -a_2 \geqslant -0.7, \\ -a_1 \geqslant -0.2, \\ a_1 + a_2 = 0.9, \quad a_1, a_2 \geqslant 0. \end{cases}$$

Thus, $c = 0.2f(x_1) + 0.7f(x_2)$. In this case by Lemma 4 $\underline{E}_{\mathbf{P}''}(f) = 0.2f(x_1) + 0.7f(x_2) + 0.1 \min_{x \in X} f(x)$. Assume, for example, that we have two decisions: $g_1 :=$ go to the park; $g_2 :=$ go to the theater; defined by $g_1(x_1) = 3$, $g_1(x_2) = -1$, $g_2(x_1) = 1$, $g_2(x_2) = 1$. Then

$$\underline{E}_{\mathbf{P}''}(g_2 - g_1) = 0.2 \cdot (-2) + 0.7 \cdot 2 + 0.1 \cdot (-2) = 0.8 > 0,$$

i.e. decision g_2 is more preferable than decision g_1.

6   The Comparison with Previous Works

Incoherence correction has been considered in the papers by A. Capotorti and others (see [5] and references therein), and in the work [7] by E. Quaeghebeur. The main idea described in [5] is to use distances between incoherent lower prevision and the set of all possible coherent previsions, i.e. the best approximation is to use the closest coherent lower prevision to the available assessments. Among possible distances (divergences) are L_1- and L_2-distances, the logarithmic Bregman divergence, the discrepancy measure. In [7] the correction is produced by the lower envelope of maximal coherent lower previsions, which are lower than a given incoherent lower prevision.

Let us compare the incoherence correction based on generalized credal sets and the mentioned above approaches. Assume that $\mu \in M_{mon}$ is an upper envelope of the set of probability measures $\mathbf{P}$, i.e.

$$\mu(A) = \sup_{P \in \mathbf{P}} P(A), \quad A \in 2^X,$$

but it is viewed as a lower probability. Obviously, μ is a contradictory lower probability if $\mathbf{P}$ contains at least two different probability measures. If we apply methods from [5], then we choose some optimal approximation $P \in \mathbf{P}$ of μ. Thus, using this correction we cannot take in account that information is contradictory - every two decisions are comparable. If we use the approach considered in [7], then obviously after correction we get the coherent lower probability

$$\mu^d(A) = \inf_{P \in \mathbf{P}} P(A), \quad A \in 2^X.$$

Although sometimes corrections based on our approach and this one give us the same result (this is fulfilled for Example 1), but in some cases they can give us sufficiently different results, when, for example, we choose $\mathbf{P}$ such that μ is a fully contradictory lower probability and $\mathbf{P} \neq M_{pr}$. In this case, by our approach μ does not give us useful information, but the Quaeghebeur's approach supposes that μ contains some useful information that seems to be not correct.

References

1. Augustin, T., Coolen, F.P.A., de Cooman, G., Troffaes, M.C.M. (eds.): Introduction to Imprecise Probabilities. Wiley, New York (2014)
2. Bronevich, A.G., Klir, G.J.: Measures of uncertainty for imprecise probabilities: an axiomatic approach. Int. J. Approx. Reason. **51**, 365–390 (2010)
3. Bronevich, A.G., Rozenberg, I.N.: The generalization of the the conjunctive rule for aggregating contradictory sources of information based on generalized credal sets. In: Augustin, T., Doria, S., Miranda, E., Quaeghebeur, E. (eds.) Proceedings of the 9th International Symposium on Imprecise Probability: Theories and Applications, pp. 67–76. Aracne Editrice, Rome (2015)
4. Bronevich, A.G., Rozenberg, I.N.: The extension of imprecise probabilities based on generalized credal sets. In: Ferraro, M.B., Giordani, P., Vantaggi, B., Gagolewski, M., Gil, M.A., Grzegorzewski, P., Hryniewicz, O. (eds.) Advances in Intelligent Systems and Computing. 456, pp. 87–94. Springer Verlag, Berlin (2017)

5. Brozzi, A., Capotorti, A., Vantaggi, B.: Incoherence correction strategies in statistical matching. Int. J. Approx. Reason. **53**, 1124–1136 (2012)
6. Klir, G.J.: Uncertainty and Information: Foundations of Generalized Information Theory. Wiley-Interscience, Hoboken (2006)
7. Quaeghebeur, E.: Characterizing coherence, correcting incoherence. Int. J. Approx. Reason. **56**(Part B), 208–233 (2015)
8. Walley, P.: Statistical Reasoning with Imprecise Probabilities. Chapman and Hall, London (1991)

Reliable Knowledge-Based Adaptive Tests by Credal Networks

Francesca Mangili, Claudio Bonesana, and Alessandro Antonucci$^{(\boxtimes)}$

Istituto Dalle Molle di Studi sull'Intelligenza Artificiale, Lugano, Switzerland
{francesca,claudio,alessandro}@idsia.ch

Abstract. An *adaptive* test is a computer-based testing technique which adjusts the sequence of questions on the basis of the estimated ability level of the test taker. We suggest the use of *credal networks*, a generalization of Bayesian networks based on sets of probability mass functions, to implement adaptive tests exploiting the knowledge of the test developer instead of training on databases of answers. Compared to Bayesian networks, these models might offer higher expressiveness and hence a more reliable modeling of the qualitative expert knowledge. The counterpart is a less straightforward identification of the information-theoretic measure controlling the question-selection and the test-stopping criteria. We elaborate on these issues and propose a sound and computationally feasible procedure. Validation against a Bayesian-network approach on a benchmark about German language proficiency assessments suggests that credal networks can be reliable in assessing the student level and effective in reducing the number of questions required to do it.

1 Introduction

The use of communication and information technologies in education is actually growing. Both online (e.g., MOOCs) and classroom courses are urgently asking for more flexible and sophisticated e-learning and e-testing tools [1]. AI-based approaches such as intelligent tutoring systems, adapting the interaction with the student on the basis of his/her knowledge and/or psychological profile, represent an important direction to improve the quality of the (e-)learning experience [2].

Bayesian networks (BNs) [3] have been used to model the knowledge driving such intelligent systems [4]. However, collecting large sets of reliable data in educational domains may be difficult and time consuming (e.g., a course with few students, or taught for the first time), and the quantification should be based on expert knowledge only. To elicit a Bayesian network, an expert might face questions like: *"which is the probability of a student with a particular knowledge level giving the right answer to a question?"*. Giving sharp probabilities for questions of this kind can be problematic for an expert, whose knowledge is mostly qualitative (e.g., *"a right answer is very unlikely"*). Fuzzy linguistic approaches represent a viable, non-numerical, way to address these issues [5]. To stick within the probabilistic framework, verbal-numerical probability scales associated with sharp values [6] or intervals [7] have been also proposed.

© Springer International Publishing AG 2017
A. Antonucci et al. (Eds.): ECSQARU 2017, LNAI 10369, pp. 282–291, 2017.
DOI: 10.1007/978-3-319-61581-3_26

In this paper we show how to conjugate an interval-valued probabilistic elicitation of expert knowledge with the BN framework. This means to cope with a *credal network* (CN) [8], a generalization of BNs based on the imprecise probability theory [7], where local parameters are defined by set-valued probabilities. This simplifies the elicitation process and offers a more reliable handling of the related uncertainty. Moving from BNs to CNs implies two main issues: (i) numerical inferences will be interval-valued too, thus making debatable both the decision criterion [9] an the information measures [10] to adopt; and (ii) inference tasks in CNs typically belongs to higher complexity classes than their Bayesian counterparts [11]. Both these issues are addressed by defining a computationally feasible procedure based on CNs to be used for practical implementation of intelligent systems solely specified by expert knowledge. To the best of our knowledge this is the first attempt to perform e-testing with models of this kind.

We focus on the application of CNs to *computer adaptive testing* (CAT), i.e., an approach to e-testing that adjusts the sequence and the number of questions to the ability level of the test taker. CATs have the potential to make the test an individualised experience that challenges and does not discourage the test takers, as most of the questions are near their ability levels. Building upon *item response theory* [12], the common background underpinning CATs, graphical modeling (such as BNs and CNs) offers a powerful language for describing complex multivariate dependencies between skills and rich tasks. Several researchers have exploited the potential of BNs both in adaptive and non-adaptive educational assessment [13,14]. These authors focus on applications for which data are available to learn the model parameters. We regard this point as a serious limitation, possibly hindering CATs adoption by many teachers and instructors.

We start from a CAT procedure based on BNs that uses *entropy* as the information-theoretic measure driving the question selection and the stopping criteria (Sect. 2). Our goal is to improve this procedure by using CNs to better describe the pervasive uncertainty characterizing the model. A direct extension of the Bayesian framework to CNs would require the computation of bounds for the conditional entropy with respect to the CN specification. This corresponds to a non-linear non-convex optimization task. We therefore propose a number of simplifying assumptions to overcome this problem at the price of accepting sub-optimal question selection schemes (Sect. 3). The approach is tested on a real-world benchmark about German language proficiency assessment (Sect. 5). The results are promising: CAT based on CNs is effective in reducing the number of questions while maintaining a high accuracy in the evaluation and the approximations introduced do not compromise the procedure's effectiveness.

2 Adaptive Testing by Bayesian Networks

Skills modeling. We describe the knowledge level of a student as a collection of categorical variables, say $X := (X_1, \ldots, X_n)$, called *skills*. A joint *probability mass function* (PMF) $P(X)$ describes the uncertainty about the actual values of the skills. A compact specification of such multivariate model can be achieved by a BN [3]. This corresponds to: (i) a directed acyclic graph whose nodes are in one-to-one correspondence with the variables of $\boldsymbol{X}$; and (ii), for each $X_i \in \boldsymbol{X}$, a collection of conditional PMFs $P(X_i|\pi_{X_i})$, one for each value π_{X_i} of the joint variable Π_{X_i} denoting the *parents* (i.e., the immediate predecessors) of X_i. The *Markov condition* for BNs assumes every variable conditionally independent of its non-descendants non-parents given the parents. Accordingly, the joint PMF associated with a BN is such that $P(\boldsymbol{x}) := \prod_{i=1}^{n} P(x_i|\pi_{X_i})$, for each $\boldsymbol{x}$, where the values of x_i and π_{X_i} are those consistent with $\boldsymbol{x}$.

Questions modeling. The above joint probabilistic model describes the uncertainty about the skills of a student prior to his/her answers to the questions. To evaluate the student we formulate a number of *questions*, described as a collection of variables $\boldsymbol{Y} := (Y_1, \ldots, Y_m)$. We assume these variables to be Boolean, with the true value corresponding to the correct answer.[1] We call *background* of a question the set of skills "required" to answer it. This can be regarded as a conditional independence statement: given the background skills, the answer to the question is independent of the other skills and of the other questions. Following the Markov condition, this can be modeled by representing each question as a leaf node whose parents are the background skills. Such augmented graph requires the quantification, for each $Y_j \in \boldsymbol{Y}$, of a conditional PMF $P(Y_j|\pi_{Y_j})$ for each value π_{Y_j} of the background skills Π_{Y_j}. This procedure defines a BN over the skills and the questions, and hence a joint PMF $P(\boldsymbol{X}, \boldsymbol{Y})$.

Non-adaptive testing. Let $\boldsymbol{Y} = \boldsymbol{y}$ denote a student's answers to the test. In the above considered framework, the posterior knowledge about the skills is modeled by the joint PMF $P(\boldsymbol{X}|\boldsymbol{y})$. By running standard BN updating algorithms, the most probable level $\tilde{x}_i$ of skill X_i can be therefore evaluated as $\tilde{x}_i := \arg\max_{x_i} P(x_i|\boldsymbol{y})$, for each $X_i \in \boldsymbol{X}$. This reflects a non-adaptive, probabilistic approach to student evaluation.

Adaptive testing. To add adaptiveness to the above approach, every question should be chosen on the basis of the previous answers. As the goal is to gather information about the student skills, we evaluate the expected *information gain* (IG, i.e., the change in information entropy) associated with each possible new question, and pick the one maximizing this measure. The entropy of a BN over $\boldsymbol{X}$ can be computed as $H(\boldsymbol{X}) := \sum_{i=1}^{n} H(X_i|\Pi_{X_i})$ [3], where $H(X_i|\Pi_{X_i}) := \sum_{\pi_{X_i}} H(X_i|\pi_{X_i})P(\pi_{X_i})$ is the *conditional entropy* for X given its parents and $H(X_i|\pi_{X_i})$ is the entropy of the conditional PMF $P(X_i|\pi_{X_i})$.[2] Let $\boldsymbol{Y} = \boldsymbol{y}$ denote the answers to the questions already asked and $\boldsymbol{Y}'$ the set from which the next question should be picked. If the answer to every question $Y' \in \boldsymbol{Y}'$ would be

---

[1] Extension to non Boolean answers is trivial as all answers Y_i are *manifest* variables, and, thus, Y_i can be always regarded as a binary variable with the two values denoting the observed answer y_i and its negation [15].

[2] To have entropy levels between zero and one, we define the entropy of the PMF $P(X)$ as $H(X) := -\sum_x P(x) \log_b P(x)$, with b number of states of X.

known, and denoted by y', the question $\tilde{Y}' \in \mathbf{Y}'$ to choose would be the one leading to the largest IG. Yet, as the decision has to be made before the student's answer, conditional entropy should be considered instead, i.e.,

$$\tilde{Y}' := \arg \max_{Y' \in \mathbf{Y}'} [H(\mathbf{X}|\mathbf{y}) - H(\mathbf{X}|Y', \mathbf{y})] . \tag{1}$$

CATs should also decide when to stop asking questions. Again, entropy can be used as a measure to decide when the current evaluation is sufficiently informative, i.e., we stop the test if the skills entropy given the answers is below some threshold $\tilde{H}$. The overall approach is depicted in Fig. 1.

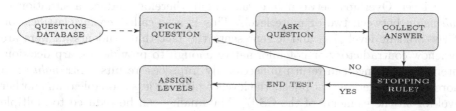

Fig. 1. CAT procedure

3   Adaptive Testing by Credal Networks

Credal sets and credal networks. A set of PMFs over X_i is called here *credal set* (CS) and denoted as $K(X_i)$. We always remove the inner points (i.e., those corresponding to convex combinations of the others) of a CS. CNs [16] are generalized BNs whose local PMFs are replaced by CSs. The BN defined in the previous section over the skills $\mathbf{X}$ and the questions $\mathbf{Y}$ becomes a CN if we replace with CSs the skill-to-skill and skill-to-question conditional PMFs. A joint CS $K(\mathbf{X}, \mathbf{Y})$ is consequently obtained as the collection of all the joint PMFs induced by BNs whose parameters take their values from the corresponding CSs, i.e.,

$$K(\mathbf{X}, \mathbf{Y}) := \left\{ P(\mathbf{X}, \mathbf{Y}) \Big|_{P(X_i|\pi_{X_i}) \in K(X_i|\pi_{X_i}), \ P(Y_j|\pi_{Y_j}) \in K(Y_j|\pi_{Y_j})}^{P(\mathbf{x},\mathbf{y}) := \prod_{i=1}^{n} P(x_i|\pi_{X_i}) \cdot \prod_{j=1}^{m} P(y_j|\pi_{Y_j}),} \right\} , \tag{2}$$

where the values of x_i, π_{X_i}, y_j, π_{Y_j} are those consistent with $\mathbf{x}$ and $\mathbf{y}$.

Expert knowledge modeling. For a reliable expert knowledge modeling, we use CSs induced by *probability intervals*. Qualitative judgments about the probability of a state are converted in interval constraints such as $l \leq P(x) \leq u$, with the interval $[l, u]$ capturing the expert knowledge behind the judgment in a more reliable way than a sharp assessment. The CS consistent with these constraints is eventually obtained by standard polyhedral algorithms. Verbal to interval-numeric scales such as that Table 2 are used. For instance, if for the probability of the true state of the Boolean variable Y the expert judgment is "very likely", the corresponding linear constraint is $.2 \leq P(Y = \text{true}) \leq .4$.

Non-adaptive testing. Given the answers $\boldsymbol{y}$ to the questions $\boldsymbol{Y}$, we evaluate the student as in the previous section by updating the marginal probabilities of each skill. With CNs, these posterior values are set-valued and their characterization can be provided by lower and upper bounds, say $\underline{P}(X_i|\boldsymbol{y})$ and $\overline{P}(X_i|\boldsymbol{y})$ for each $X_i \in \boldsymbol{X}$. CN updating algorithms can eventually compute these bounds. The task displays higher complexity than in the case of BNs (e.g., exact inference in non-binary singly-connected CNs is NP-hard [11]), but approximate techniques can be considered when exact inference is unfeasible [17].

To compare the posterior intervals and decide the actual level of the student we might adopt the (conservative) *interval dominance* criterion [9], which rejects a level if its upper probability is smaller than the lower probability of some other level. Overlaps between intervals might therefore induce a situation of *indecision* between two or more levels. This is a so-called *credal* classification of the student level [18], and it represents the fact that students answers are somehow contradictory or not informative enough to provide a sharp decision. Interval dominance can return unnecessarily imprecise results. *Maximality* is a more refined criterion that rejects the levels which are less probable than another level for all the elements of the CS [7]. Maximality can be reduced to multiple updating tasks on auxiliary binary leaf nodes defined for each pair of states [17].

Adaptive testing. To achieve CAT with CNs using entropy as measure of informativeness for PMFs, as in the BN approach of Sect. 2, computation of entropies should be extended to CSs. This topic has been the subject of much discussion [10]. A cautious approach [19] consists in taking the upper entropy $\overline{H}(\boldsymbol{X})$, i.e., the entropy of the most entropic PMF in the convex closure $\overline{K}(\boldsymbol{X})$ of $K(\boldsymbol{X})$. In our framework, we should, then, look for maximum values of conditional entropies, such as $\overline{H}(X_i|Y', \boldsymbol{y})$ or $\overline{H}(X_i|\Pi_{X_i})$, as conditional entropies are required to compute both: (i) the joint (unconditional) entropy $H(\boldsymbol{X})$ (and its posterior values); and (ii) the conditional entropies involved in the question selection in Eq. (1). By definition a conditional entropy is a convex combination (whose weights are the elements of a marginal PMF) of convex functions (the entropies). The objective function might, then, be non-convex, as the weights are also optimization variables.[3]

Then, to bypass this non-convex optimization task, we compute (i) by separately considering the entropies of each skill $X_i \in \boldsymbol{X}$. This is analogous to the marginal approach commonly considered in multi-label classification to minimize Hamming losses [20]. The issue (ii) is more challenging. We consider the following upper approximation of $\overline{H}(X_i|Y', \boldsymbol{y})$:

$$\overline{\overline{H}}(X_i|Y', \boldsymbol{y}) = \max_{P(y'|\boldsymbol{y}) \in \{\underline{P}(y'|\boldsymbol{y}), \overline{P}(y'|\boldsymbol{y})\}} \sum_{y' \in \{\text{true,false}\}} \overline{H}(X_i|\boldsymbol{y}, y')P(y'|\boldsymbol{y}), \quad (3)$$

---

[3] E.g., if $f(x)$ and $g(x)$ are convex functions of x, $h(x, y) := yf(x) + (1-y)g(x)$ is not convex even for $0 \le y \le 1$.

where the bounds of $P(y'|\boldsymbol{y})$ are obtained by standard CN updating algorithms. The problem thus reduces to the computation of upper entropies as

$$\overline{H}(X_i|\boldsymbol{y}) := \sup_{P(X_i|\boldsymbol{y})\in\overline{K}(X_i|\boldsymbol{y})} H(X_i|\boldsymbol{y}), \qquad (4)$$

where $\overline{K}(X_i|\boldsymbol{y})$ is the posterior CS after conditioning on the observed answers $\boldsymbol{y}$. If $\overline{K}(X_i|\boldsymbol{y})$ has a finite number of non-inner points, this is a linearly-constrained convex optimization whose solution typically corresponds to either the uniform PMF or a non-inner point on the frontier of $\overline{K}(X_i|\boldsymbol{y})$. A numerical solution can be easily found by a simple iterative approach in the special case of CS specified by probability intervals [19]. We have therefore computed the posterior lower and upper bounds of $P(X_i|\boldsymbol{y})$, and then maximized the entropy with respect to those bounds. The procedure induces an outer approximation of $\overline{K}(X_i|\boldsymbol{y})$, and hence the upper approximation of the maximum entropy $\overline{\overline{H}}(X|\boldsymbol{y}) \geq \overline{H}(X|\boldsymbol{y})$. Finally, to generalise Eq. (2) to CNs, we define the information gain provided by a question Y' for its background skill $X_{Y'}$ as $\overline{\overline{H}}(X_{Y'}|\boldsymbol{y}) - \overline{\overline{H}}(X_{Y'}|Y',\boldsymbol{y})$ and select the question $\tilde{Y}'$ leading to the maximum information gain, i.e.,

$$\tilde{Y}' := \arg\max_{Y'\in\boldsymbol{Y}'} \left[\overline{\overline{H}}(X_{Y'}|\boldsymbol{y}) - \overline{\overline{H}}(X_{Y'}|Y',\boldsymbol{y}) \right]. \qquad (5)$$

For the stopping criterion, as we do not consider the joint entropy over the skills, we separately require each $\overline{\overline{H}}(X_i|\boldsymbol{y})$ to be smaller than a threshold $\tilde{H}$. To be consistent with this choice, we remove from the set of questions to be selected, those whose background skills already satisfy this condition.

Note that the use of an outer approximation of the upper entropy affects only the question selection process (eventually making it sub-optimal), whereas it has no effect on the student evaluation given a set of answers.

4   Application to Language Assessment

Before the academic year begins, the students of the University of Applied Sciences and Arts of Southern Switzerland (SUPSI) are asked to take an online German language placement test with 95 questions. In years 2015 and 2016, the answers of 451 students to all the questions have been collected. This benchmark is used to simulate CATs based on BNs and CNs as described in Sects. 2 and 3.

Model elicitation. Four skills are assessed: *Wörtschatz* (X_1, vocabulary), *Kommunikation* (X_2, communication), *Hören* (X_3, listening), and *Lesen* (X_4, reading). For each skill the student is assigned to a knowledge level compliant with EU guidelines.[4] Levels A1, A2, B1, and B2 are considered, and skills are therefore modeled as quaternary variables. Teachers associate each question with a single skill, which is set as the unique background skill of the question. The

---

[4] http://www.coe.int/t/dg4/linguistic/Source/Framework_EN.pdf.

number of questions associated with $X_1/X_2/X_3/X_4$ is 26/24/30/15. The current evaluation method assigns levels by setting thresholds on the percentage γ of correct answers on each skill (A1 if $\gamma < 35\%$, A2 up to 55%, B1 up to 75%).[5]

We first elicit from the teachers the structure of the BN/CN graph over the skills. The result is a chain, which is augmented by leaf nodes modeling the questions, each having its background skill as single parent. Overall, a tree-shaped topology as in Fig. 2 is obtained. This makes exact inference in the BN fast, while in the CN a variable elimination might be slow (a minute for query in our setup). A faster approximate CN algorithm is therefore used [17].

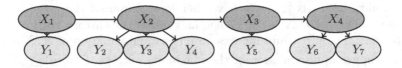

Fig. 2. A directed graph for CAT.

Teachers report their knowledge about the unconditional states of X_1 and the conditional states of X_i given X_{i-1}, for $i = 2, 3, 4$, as qualitative judgments (top of Table 1). To simplify the elicitation, the probabilities $P(X_i|X_{i-1})$ are given the same verbal judgment for all $i = 2, 3, 4$. A more detailed model could provide more accurate evaluations but it would be very hard for the domain expert to elicit it in a reliable way. Also, questions are divided by the teachers in three groups, corresponding to different difficulty levels. Questions in the same group are quantified in the same way, irrespective of their background skill, giving the judgments reported in the bottom part of Table 1.

For the CN, those judgements are translated in interval constraints for the corresponding events on the basis of the verbal-numerical scale in Table 2. Different probability intervals are considered for skills and questions as they refer to events of different type. For instance, when the expert considers "*impossible*" for an A1 level student to know the answer to a difficult question, the student is assigned a probability between .175 and .2 of answering correctly, as the questions offer only four choices plus the option of giving no answer. Notice that, by doing so, we are not anymore assuming that all questions in the same difficulty group share exactly the same conditional PMFs (as done by the BN model), as PMFs of different questions can vary independently in the given intervals. This seems a more sensible assumption than that of the precise model. For the BN, the PMFs corresponding to the centers of mass of the CSs defining the CN are used. Numerical inferences in the BN are consequently included in the intervals computed with the CN.

---

[5] These data as well as the software used for the simulations are freely available at http://ipg.idsia.ch/software.php?id=138.

Table 1. Expert judgements.

X_1	$P(X_1)$	$P(X_i\|X_{i-1})$	$X_{i-1} = $ A1	$X_{i-1} = $ A2	$X_{i-1} = $ B1	$X_{i-1} = $ B2
A1	*improbable*	$X_i=$A1	*fifty-fifty*	*uncertain*	*improbable*	*impossible*
A2	*uncertain*	$X_i=$A2	*uncertain*	*fifty-fifty*	*uncertain*	*improbable*
B1	*uncertain*	$X_i=$B1	*improbable*	*uncertain*	*fifty-fifty*	*uncertain*
B2	*improbable*	$X_i=$B2	*impossible*	*improbable*	*uncertain*	*fifty-fifty*

$P(Y = T\|X)$	$X = $ A1	$X = $ A2	$X = $ B1	$X = $ B2
Easy	*uncertain*	*fifty-fifty*	*expected*	*probable*
Medium	*improbable*	*uncertain*	*fifty-fifty*	*expected*
Difficult	*impossible*	*improbable*	*uncertain*	*fifty-fifty*

Table 2. A verbal-numerical scale for probability-intervals elicitation.

Judgement	*impossible*	*improbable*	*uncertain*	*fifty-fifty*	*expected*	*probable*
Skills	1–10%	10–20%	20–40%	30–50%	-	-
Questions	17.5–20%	22.5–25%	30–35%	60–65%	75–80%	95–97.5%

Experimental results. BN and CN methods in their non-adaptive (NA) and adaptive (AD) versions are considered. *Accuracy,* i.e., the proportion of stu-dents to whom the test assigns the same level of the current evaluation method, describes BN performances. This measure cannot be used for the set-valued out-puts of CN methods. In this case the u_{65} measure can provide a comparison with the accuracy [18]. If $\mathcal{L}$ is the set of levels assigned by the CN on a skill and L its cardinality, a *discounted* accuracy gives $1/L$ if $\mathcal{L}$ includes the true level and zero otherwise. The u_{65} is a concave reinforcement of this score based on risk-adverse arguments. Its underlying assumption is that acknowledging the indecision between more levels has larger utility than randomly choosing one of them (e.g., the teacher could set up further assessments in the undecided cases). Table 3 shows the NA comparison. In Fig. 3(left), the BN-NA accuracy is sepa-rately evaluated on the determinate (light bars) and indeterminate (dark bars) instances, i.e. those for which, respectively, a single level or multiple levels are returned by the CN model. On average, CN-NA returns single levels in 37.25% of the cases and, if this is not the case, an average of 2.36 levels (3.22 with interval dominance) are returned.

In the AD case we also track the average number of asked questions. Results are in Fig. 3(right). CN-AD (circles) is tested for different thresholds over the entropy (labels of the markers) against a version of BN-AD based on the joint entropy (triangles). Similar values are obtained by coping with mar-ginal entropies. We also allow the BN-AD method to return multiple levels by

Table 3. Non-adaptive tests results.

Algorithm	Average	X_1	X_2	X_3	X_4
BN-NA (acc)	63.09%	67.56%	60.85%	75.84%	48.10%
CN-NA (u_{65})	65.37%	67.71%	66.67%	70.33%	56.76%

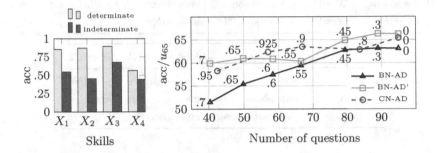

Fig. 3. Non-adaptive (left) and adaptive (right) tests performance.

maximizing the expected u_{65} utility over any possible set of levels. This variant is called BN-AD' and the corresponding u_{65} measure is reported (squares).

As a comment, CNs seem to identify hard-to-evaluate students as those for which multiple levels are provided. In fact, the agreement between the BN and the traditional tests is larger when the CN test is determinate. As a consequence, the CN u_{65} measure is, on average, larger than the BN accuracy. A limitation of the CN test is the large fraction of indeterminate evaluations. One can interpret this result as a lack of robustness of the BN model, as even small variations in the model specifications can result in different decisions. Results also show that, both BN-AD and CN-AD approaches reduce the number of questions asked without significantly affecting the accuracy. BN-AD performances are improved by the "credal" variant BN-AD'. The results becomes very similar to those of the CN-AD. Yet, the latter method appears to be a more principled and suitable approach for a direct modeling of qualitative expert knowledge.

5   Conclusions and Outlooks

A procedure for adaptive testing built solely on expert knowledge has been proposed based on credal networks. The procedure has been validated on a real dataset about a German language test. Results are promising, as the credal approach simplifies the model elicitation, recognizes when a sharp decision about the student level should not be made (that is, when the traditional and precise Bayesian evaluations disagree) and achieves an accuracy comparable to that of an indecisive Bayesian approach maximizing the expected u_{65} measure. However, the fraction of instances where CNs issue multiple levels remains rather large, therefore further research is needed to make CN-based CATs a viable solution for adaptive testing solely based on expert knowledge.

References

1. Pollard, E., Hillage, J.: Exploring e-learning. Inst. for Empl., Studies Brighton (2001)
2. Burns H., Luckhardt, C.A., Parlett, J.W., Redfield, C.L.: Intelligent Tutoring Systems: Evolutions in Design. Psychology Press (2014)
3. Koller, D., Friedman, N., Models, P.G.: Principles and Techniques. MIT Press, Cambridge (2009)
4. Almond, R.G., Mislevy, R.J., Steinberg, L., Yan, D., Williamson, D.: Bayesian Networks in Educational Assessment. Springer, New York (2015)
5. Badaracco, M., Martínez, L.: A fuzzy linguistic algorithm for adaptive test in intelligent tutoring system based on competences. Expert Syst. Appl. **40**(8), 3073–3086 (2013)
6. Renooij, S., Witteman, C.: Talking probabilities: communicating probabilistic information with words and numbers. Int. J. Approx. Reason. **22**(3), 169–194 (1999)
7. Walley, P.: Statistical Reasoning with Imprecise Probabilities. Chapman & Hall/CRC Monographs on Statistics & Applied Probability. Taylor & Francis (1991)
8. Piatti, A., Antonucci, A., Zaffalon, M.: Building knowledge-based systems by credal networks: a tutorial. In: Baswell, A.R. (ed.) Advances in Mathematics Research, vol. 11. Nova Science Publishers, New York (2010)
9. Troffaes, M.: Decision making under uncertainty using imprecise probabilities. Int. J. Approx. Reason. **45**(1), 17–29 (2007)
10. Klir, G., Wierman, M.: Uncertainty-Based Information: Elements of Generalized Information Theory. STUDFUZZ, vol. 15. Springer, Heidelberg (1999)
11. Mauá, D., de Campos, C., Benavoli, A., Antonucci, A.: Probabilistic inference in credal networks: new complexity results. J. Artif. Intell. Res. **50**, 603–637 (2014)
12. Hambleton, R.K., Swaminathan, H.: Item Response Theory: Principles and Applications, vol. 7. Springer Science & Business Media, New York (1985)
13. Vomlel, J.: Building adaptive tests using Bayesian networks. Kybernetika **40**(3), 333–348 (2004)
14. Plajner, M., Vomlel, J.: Bayesian network models for adaptive testing, arXiv preprint arXiv:1511.08488
15. Antonucci, A., Piatti, A.: Modeling unreliable observations in bayesian networks by credal networks. In: Godo, L., Pugliese, A. (eds.) SUM 2009. LNCS (LNAI), vol. 5785, pp. 28–39. Springer, Heidelberg (2009). doi:10.1007/978-3-642-04388-8_4
16. Cozman, F.G.: Credal networks. Artif. Intell. **120**, 199–233 (2000)
17. Antonucci, A., de Campos, C., Zaffalon, M., Huber, D.: Approximate credal network updating by linear programming with applications to decision making. Int. J. Approx. Reason. **58**, 25–38 (2014)
18. Zaffalon, M., Corani, G., Mauá, D.: Evaluating credal classifiers by utility-discounted predictive accuracy. Int. J. Approx. Reason. **53**(8), 1282–1301 (2012)
19. Abellan, J., Moral, S.: Maximum of entropy for credal sets. Int. J. Uncertain. Fuzz. **11**(05), 587–597 (2003)
20. Antonucci, A., Corani, G.: The multilabel naive credal classifier. Int. J. Approx. Reason. **83**, 320–336 (2016)

Decision Theory, Decision Making and Reasoning Under Uncertainty

Algorithms for Multi-criteria Optimization in Possibilistic Decision Trees

Nahla Ben Amor[1], Fatma Essghaier[1,2($\boxtimes$)], and Hélène Fargier[2]

[1] LARODEC, Le Bardo, Tunisia
nahla.benamor@gmx.fr, essghaier.fatma@gmail.com
[2] IRIT, Toulouse, France
fargier@irit.fr

Abstract. This paper raises the question of solving multi-criteria sequential decision problems under uncertainty. It proposes to extend to possibilistic decision trees the decision rules presented in [1] for non sequential problems. It present a series of algorithms for this new framework: *Dynamic Programming* can be used and provide an optimal strategy for rules that satisfy the property of monotonicity. There is no guarantee of optimality for those that do not—hence the definition of dedicated algorithms. This paper concludes by an empirical comparison of the algorithms.

Keywords: Possibility theory · Sequential decision problems · Multi-criteria decision making · Decision trees

1 Introduction

When information about uncertainty cannot be quantified in a probabilistic way, possibilistic decision theory is a natural field to consider [2–7]. Qualitative decision theory is relevant, among other fields, for applications to planning under uncertainty, where a suitable *strategy* (i.e. a set of conditional or unconditional decisions) is to be found, starting from a qualitative description of the initial world, of the available decisions, of their (perhaps uncertain) effects and of the goal to reach (see [8–10]). But up to this point, the evaluation of the strategies was considered in a simple, mono-criterion context, while it is often the case that several criteria are involved in the decision [11].

A theoretical framework has been proposed for multi-criteria/multi-agent (non sequential) decision making under possibilistic uncertainty [1,12]. In the present paper, we extend it to decision trees and we propose a detailed algorithmic study. After a refreshing on the background (Sect. 2), Sect. 3 presents our algorithms, and is completed, in Sect. 4, by an experimental evaluation.

2 Background

2.1 Multi-criteria Decision Making (MCDM) Under Uncertainty

Following Dubois and Prade's possibilistic approach of decision making under qualitative uncertainty, a non-sequential (i.e. one stage) decision can be seen

© Springer International Publishing AG 2017
A. Antonucci et al. (Eds.): ECSQARU 2017, LNAI 10369, pp. 295–305, 2017.
DOI: 10.1007/978-3-319-61581-3_27

as a possibility distribution[1] over a finite set of outcomes, called a (simple) *possibilistic lottery* [2]. Such a lottery is denoted $L = \langle \lambda_1/x_1, \ldots, \lambda_n/x_n \rangle$ where $\lambda_i = \pi_L(x_i)$ is the possibility that decision L leads to outcome x_i; this possibility degree can also be denoted by $L[x_i]$. In this framework, a decision problem is thus fully specified by a set of possibilistic lotteries on X and a utility function $u : X \mapsto [0,1]$. Under the assumption that the utility scale and the possibility scale are commensurate and purely ordinal, [2] proposes to evaluate each lottery by a qualitative, optimistic or pessimistic, global utility:

$$\text{Optimistic utility: } U^+(L) = \max_{x_i \in X} \min(\lambda_i, u(x_i)) \tag{1}$$

$$\text{Pessimistic utility: } U^-(L) = \min_{x_i \in X} \max(1 - \lambda_i, u(x_i)) \tag{2}$$

$U^+(L)$ is a mild version of the maximax criterion: L is good as soon as it is totally plausible that it gives a good consequence. On the contrary, the pessimistic index, $U^-(L)$ estimates the utility of an act by its worst possible consequence: its value is high whenever L gives good consequences in every "rather plausible" state.

This setting assumes a ranking of X by a *single* preference criterion, hence the use of a single utility function. When several criteria, say a set $Cr = \{1, \ldots, p\}$ of p criteria, have to be taken into account, u must be replaced by a vector $\boldsymbol{u} = \langle u_1, \ldots, u_p \rangle$ of utility functions u_j. If the criteria are not equally important, each j is equipped with a weight $w_j \in [0, 1]$ reflecting its importance.

In the absence of uncertainty, each decision leads to a unique consequence and the problem is a simple problem of qualitative MCDM aggregation; classically, such aggregation shall be either conjunctive (i.e. based on a weighted min) or disjunctive (i.e. based on a weighted max) - see [13] for more details about weighted min and weighted max aggregations.

In presence of uncertainty, the aggregation can be done *ex-ante* or *ex-post*:

- The *ex-ante* approach consists in computing the (optimistic or pessimistic) utility relative to each criterion j, and then performs the MCDM aggregation.
- The *ex-post* approach consists in first determining the aggregated utility (conjunctive or disjunctive) of each possible x_i; then the problem can be viewed as a mono-criterion problem of decision making under uncertainty.

Since the decision maker's attitude with respect to uncertainty can be either optimistic or pessimistic and the way of aggregating the criteria either conjunctive or disjunctive, [1,12] propose four *ex-ante* and four *ex-post* approaches:

$$U^{-\min}_{ante}(L) = \min_{j \in Cr} \; \max((1 - w_j), \min_{x_i \in X} \max(u_j(x_i), (1 - L[x_i]))) \tag{3}$$

$$U^{-\max}_{ante}(L) = \max_{j \in Cr} \; \min(w_j, \min_{x_i \in X} \max(u_j(x_i), (1 - L[x_i]))) \tag{4}$$

$$U^{+\min}_{ante}(L) = \min_{j \in Cr} \; \max((1 - w_j), \max_{x_i \in X} \min(u_j(x_i), L[x_i])) \tag{5}$$

---

[1] A possibility distribution π is a mapping from the universe of discourse to a bounded linearly ordered scale, typically by the unit interval [0, 1].

$$U_{ante}^{+\,max}(L) = \max_{j \in Cr} \; \min(w_j, \max_{x_i \in X} \min(u_j(x_i), L[x_i])) \tag{6}$$

$$U_{post}^{-\,min}(L) = \min_{x_i \in X} \; \max((1 - L[x_i]), \min_{j \in Cr} \max(u_j(x_i), (1 - w_j))) \tag{7}$$

$$U_{post}^{-\,max}(L) = \min_{x_i \in X} \; \max((1 - L[x_i]), \max_{j \in Cr} \min(u_j(x_i), w_j)) \tag{8}$$

$$U_{post}^{+\,min}(L) = \max_{x_i \in X} \; \min(L[x_i], \min_{j \in Cr} \max(u_j(x_i), (1 - w_j))) \tag{9}$$

$$U_{post}^{+\,max}(L) = \max_{x_i \in X} \; \min(L[x_i], \max_{j \in Cr} \min(u_j(x_i), w_j)) \tag{10}$$

In the notations above, the first (resp. second) sign denotes the attitude of the decision maker w.r.t. uncertainty (resp. the criteria). The $U_{ante}^{-\,min}$ utility for instance considers that the decision maker is pessimistic and computes the pessimistic utility of each criterion. Then the criteria are aggregated on a cautions basis: the higher is the satisfaction of the least satisfied of the important criteria, the better is the lottery. Using the same notations, $U_{post}^{-\,max}$ considers that a x_i is good as soon as one of the important criteria is satisfied: a max-based aggregation of the utilities is done, yielding a unique utility function $u()$ on the basis of which the pessimistic utility is computed. It should be noticed that the full pessimistic and full optimistic *ex-ante* utilities are equivalent to their *ex-post* counterparts [12], i.e. $U_{ante}^{-\,min} = U_{post}^{-\,min}$ and $U_{ante}^{+\,max} = U_{post}^{+\,max}$. But $U_{ante}^{-\,max}$ (resp. $U_{ante}^{+\,min}$) may differ from $U_{post}^{-\,max}$ (resp. from $U_{post}^{+\,min}$).

Example 1. *Consider two equally important criteria 1 and 2 ($w_1 = w_2 = 1$), and a lottery $L = \langle 1/x_a, 1/x_b \rangle$ leading to two equi possible consequences x_a and x_b such that x_a is good for 1 and bad for 2, and x_b is bad for 1 and good for 2: $u_1(x_a) = u_2(x_b) = 1$ and $u_2(x_a) = u_1(x_b) = 0$. It is easy to check that $U_{ante}^{+\,min}(L) = 0 \neq U_{post}^{+\,min}(L) = 1$.*

2.2   Possibilistic Decision Trees [14]

Decision trees provide an explicit modeling of sequential problems by representing, simply, all possible scenarios. A decision tree is a labeled tree $\mathcal{DT} = (\mathcal{N}, \mathcal{E})$ where $\mathcal{N} = \mathcal{D} \cup \mathcal{C} \cup \mathcal{LN}$ contains three kinds of nodes (see Fig. 1): $\mathcal{D}$ is the set of decision nodes (represented by squares); $\mathcal{C}$ is the set of chance nodes (represented by circles) and $\mathcal{LN}$ is the set of leaves. $Succ(N)$ denotes the set of children nodes of node N. For any $X_i \in \mathcal{D}, Succ(X_i) \subseteq \mathcal{C}$ i.e. a chance node (an action) must be chosen at each decision node. For any $C_i \in \mathcal{C}, Succ(C_i) \subseteq \mathcal{LN} \cup \mathcal{D}$: the set of outcomes of an action is either a leaf node or a decision node (and then a new action should be executed).

In the possibilistic context, leaves are labeled by utility degrees in the [0,1] scale and the uncertainty pertaining to the possible outcomes of each $C_i \in \mathcal{C}$, is represented by a *conditional possibility distribution* π_i on $Succ(C_i)$, such that $\forall N \in Succ(C_i), \pi_i(N) = \Pi(N|path(C_i))$ where $path(C_i)$ denotes all the value assignments of chance and decision nodes on the path from the root to C_i [14].

Solving a decision tree amounts at building a *complete strategy* that selects an action (a chance node) for each decision node: a strategy is a mapping $\delta : \mathcal{D} \mapsto \mathcal{C} \cup \{\perp\}$. $\delta(D_i) = \perp$ means that no action has been selected for D_i (δ is partial). Leaf nodes being labeled with utility degrees, the rightmost chance nodes can be seen as *simple possibilistic lotteries*. Then, each strategy δ can be viewed as a connected sub-tree of the decision tree and is identified with a *possibilistic compound lottery* L_δ, i.e. with a possibility distribution over a set of (simple or compound) lotteries. A compound lottery $\langle \lambda_1/L_1, ..., \lambda_k/L_k \rangle$ (and thus any strategy) can then be reduced into an equivalent simple lottery as follows[2] [2]:

$$Reduction(\langle \lambda_1/L_1, ..., \lambda_k/L_k \rangle) = \langle \max_{j=1,k}(\min(\lambda_1^j, \lambda_j))/u_1, ..., \max_{j=1,k}(\min(\lambda_n^j, \lambda_j))/u_n \rangle.$$

The pessimistic and optimistic utility of a strategy δ can then be computed on the basis of the reduction of L_δ: the utility of δ is the one of $Reduction(L_\delta)$.

3 Multi-criteria Optimization in Possibilistic Trees

Multi-criteria Possibilistic Decision Trees can now be defined: they are classical possibilistic decision trees, the leaves of which are evaluated according to several criteria - each leaf N is now labeled by a *vector* $u(N) = \langle u_1(N), ..., u_p(N) \rangle$ rather than by a single utility score (see Fig. 1). A strategy still leads to compound lottery, and can be reduced, thus leading in turn to a simple (but multi-criteria) lottery. We propose to base the comparison of strategies on the comparison, according to the rules O previously presented, of their reductions:

$$\delta_1 \succeq_O \delta_2 \quad \text{iff} \quad U_O(\delta_1) \geq U_O(\delta_2), \quad \text{where} \quad \forall \delta, U_O(\delta) = U_O(Reduction(L_\delta)) \quad (11)$$

Example 2. *Consider the tree of Fig. 1, involving two criteria that are supposed to be equally important and the strategy* $\delta(D_0) = C_1$, $\delta(D_1) = C_3$, $\delta(D_2) = C_5$. *It holds that* $L_\delta = \langle 1/L_{C_3}, 0.9/L_{C_5} \rangle$ *with* $L_{C_3} = \langle 0.5/x_a, 1/x_b \rangle$, $L_{C_5} = \langle 0.2/x_a, 1/x_b \rangle$. *Because* $Reduction(L_\delta) = \langle max(0.5, 0.2)/x_a, max(1, 0.9)/x_b \rangle = \langle 0.5/x_a, 1/x_b \rangle$, *we get* $U_{ante}^{+min}(\delta) = \min(\max\min(0.5, 0.3), \min(1, 0.6), \max(\min(0.5, 0.8) \min(1, 0.4))) = 0.5$.

The definition proposed by Eq. (11) is quite obvious but raises an algorithmic challenge: the set of strategies to compare is exponential w.r.t. the size of the tree which makes the explicit evaluation of the strategies not realistic. The sequel of the paper aims at providing algorithmic solutions to this difficulty.

3.1 Dynamic Programming as a tool for *ex-post* Utilities

Dynamic Programming [15] is an efficient procedure of strategy optimization. It proceeds by *backward induction*, handling the problem from the end (and in our case, from the leafs): the last decision nodes are considered first, and recursively

---

[2] Obviously, the reduction of a simple lottery is the simple lottery itself.

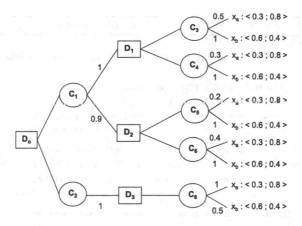

Fig. 1. A multi-criteria possibilistic decision tree

until the root is reached. This algorithm is sound and complete as soon as the decision rule leads to complete and transitive preferences and satisfies the principle of weak monotonicity,[3] that ensures that each sub strategy of an optimal strategy is optimal in its sub-tree. Hopefully, each of the *ex-post* criteria satisfy transitivity, completeness and weak monotonicity, because collapsing to either a classical U^- or a U^+ utility, which satisfy these properties [8,14]. The adaptation of Dynamic Programming to the *ex-post* rules is detailed in Algorithm 1.

In short, this algorithm aggregates the utility values of each leaf, and then builds an optimal strategy from the last decision nodes to the root of the tree, using the principle defined by [9,10] for classical (monocriterion) possibilistic decision trees.

3.2 Dynamic Programming for *ex-ante* Utilities?

The *ex-ante* variant of Dynamic Programming we propose is a little more tricky (see Algorithm 2). It keeps at each node a vector of p pessimistic (resp. optimistic) utilities, one for each criterion. The computation of the *ex-ante* utility can then be performed each time a decision is to be made. Recall that $U_{ante}^{-\min} = U_{post}^{-\min}$ and $U_{ante}^{+\max} = U_{post}^{+\max}$. Hence, for these two rules the optimization could also be performed by the *ex-post* algorithm. The two other rules, $U_{ante}^{-\max}$ and $U_{ante}^{+\min}$, unfortunately do not satisfy the monotonicity principle (see [1]). Hence, Algorithm 2 may provide a good strategy, but without any guarantee of optimality - it can be considered as an approximation algorithm in these two cases. Another approximation algorithm is the *ex-post* Algorithm described in the previous Section - even if it is not always the case, it often happens that $U_{post}^{-\max} = U_{ante}^{-\max}$ (resp. $U_{post}^{+\min} = U_{ante}^{+\min}$); if it is the case the solution provided by the *ex-post* Algorithm is optimal.

---

[3] Formally, $\succeq_O$ is said to be weakly monotonic iff whatever L, L' and L'', whatever (α, β) such that $\max(\alpha, \beta) = 1$: $L \succeq_O L' \Rightarrow \langle \alpha/L, \beta/L'' \rangle \succeq_O \langle \alpha/L', \beta/L'' \rangle$.

Algorithm 1. DynProgPost: *Ex-post* Dynamic Programming

Data: A Decision tree T, a node N in of T
Result: The value of the optimal strategy δ - δ is stored as a global variable
begin

    if $N \in \mathcal{LN}$ **then** // Leaf : MCDM aggregation

        for $i \in \{1, \ldots, p\}$ **do** $u_N \leftarrow (u_N \oplus (u_i \otimes \omega_i))$;

        // $\otimes = \min$, $\omega_i = w_i$, $\oplus = \max$ for disjunctive aggregation

        // $\otimes = \max$, $\omega_i = 1 - w_i$, $\oplus = min$ for conjunctive aggregation ;

    if $N \in \mathcal{C}$ **then** // Chance Node: compute the qualitative utility

        foreach $Y \in Succ(N)$ **do** $u_N \leftarrow (u_N \oplus (\lambda_Y) \otimes DynProgPost(Y))$;

        // $\otimes = \min$, $\lambda_Y = \pi(Y)$, $\oplus = \max$ for optimistic utility

        // $\otimes = \max$, $\lambda_Y = 1 - \pi(Y)$, $\oplus = \min$ for pessimistic utility

    if $N \in \mathcal{D}$ **then** // Decision node: determine the best decision

        $u^* \leftarrow 0$;

        foreach $Y \in Succ(N)$ **do**

            $u_Y \leftarrow DynProgPost(Y)$;

            if $u_Y \geq u^*$ **then** $\delta(N) \leftarrow Y$ and $u^* \leftarrow u_Y$;

    return u^*;

Algorithm 2. DynProgAnte: *Ex-ante* Dynamic Programming

Data: A Decision tree T, a node N in of T
Result: The value of the optimal strategy δ - δ is stored as a global variable
begin

    if $N \in \mathcal{LN}$ **then** // Leaf

        for $i \in \{1, \ldots, p\}$ **do** $\boldsymbol{u_N}[i] \leftarrow u_i$;

    if $N \in \mathcal{C}$ **then** // Chance Node: compute the utility vectors

        // Optimistic utility $\otimes = \min$, $\lambda_Y = \pi(Y)$, $\oplus = \max$, $\epsilon \leftarrow 0$

        // Pessimistic utility $\otimes = \max$, $\lambda_Y = 1 - \pi(Y)$, $\oplus = \min$, $\epsilon \leftarrow 1$

        for $i \in \{1, \ldots, p\}$ **do** $\boldsymbol{u_N}[i] \leftarrow \epsilon$;

        foreach $Y \in Succ(N)$ **do**

            $\boldsymbol{u_Y} \leftarrow DynProgAnte(Y)$;

            for $i \in \{1, \ldots, p\}$ **do** $\boldsymbol{u_N}[i] \leftarrow (\boldsymbol{u_N}[i] \oplus (\lambda_Y \otimes \boldsymbol{u_Y}[i]))$;

    if $N \in \mathcal{D}$ **then** // Decision node

        // Disjunctive MCDM: let $\otimes = \min$, $\omega_i = w_i$, $\oplus = \max$, $\epsilon \leftarrow 0$

        // Conjunctive MCDM: let $\otimes = \max$, $\omega_i = 1 - w_i$, $\oplus = min$, $\epsilon \leftarrow 1$

        $u^* \leftarrow 0$

        foreach $Y \in Succ(N)$ **do**

            $v_Y \leftarrow \epsilon$; $\boldsymbol{u_Y} \leftarrow DynProgAnte(Y)$;

            for $i \in \{1, \ldots, p\}$ **do** $v_Y \leftarrow v_Y \oplus (\boldsymbol{u_Y}[i] \otimes \omega_i)$;

            if $v_Y > u^*$ **then** $\delta(N) \leftarrow Y$ and $\boldsymbol{u_N} \leftarrow u_Y$;

    return $\boldsymbol{u_N}$;

3.3 Optimization of $U_{ante}^{-\max}$ by Multi Dynamic Programming

The lack of monotonicity of $U_{ante}^{-\max}$ is not dramatic, even when optimality must be guaranteed. With $U_{ante}^{-\max}$ indeed, we look for a strategy that has a good pessimistic utility U_j^- for *at least* one criterion j. This means that if it is possible to get for each j a strategy that optimizes U_j^- (and this can be done by Dynamic Programming, since the classical pessimistic utility is monotonic), the one with the highest value for $U_{ante}^{-\max}$ is globally optimal. Formally:

Proposition 1. $U_{ante}^{-\max}(L) = \max_{j=1,p} \min(w_j, U_j^-(L))$ where $U_j^-(L)$ is the pessimistic utility of L according to the sole criterion j.

Corollary 1. Let $\Delta^* = \{L_1^*, \ldots, L_p^*\}$ s.t. $\forall L$, $U_j^-(L_j^*) \geq U_j^-(L)$ and $L^* \in \Delta^*$. If $\max_{j=1,p} \min(w_j, U_j^-(L^*)) \geq \max_{j=1,p} \min(w_j, U_j^-(L_i^*)) \forall L_i^* \in \Delta^*$ then $U_{ante}^{-\max}(L^*) \geq U_{ante}^{-\max}(L), \forall L$.

Hence, the optimization problem can be solved by a series of p calls to a classical (monocriterion) pessimistic optimization. This is the principle of the Multi Dynamic Programming approach detailed by Algorithm 3.

Algorithm 3. MultiDynProg: right optimization of $U_{ante}^{-\max}$

Data: A tree T
Result: An Optimal strategy δ^* and its value u^*
begin
    $u^* = 0$; // Initialization
    for $i \in \{1, \ldots, p\}$ **do**
        $\delta_i = PesDynProg(T, i)$ // Call to classical possibilistic
        Dynamic Prog. [14] – returns an optimal strategy for U_i^-;
        $u_i = \max_{j=1\ldots p} \min(w_j, U_j^-(\delta_i))$;
        if $u_i > u^*$ **then** $\delta^* \leftarrow \delta_i$; $u^* \leftarrow u_i$;
    return u^*;

3.4 Right Optimization of $U_{ante}^{+\min}$: A Branch and Bound algorithm

Let us finally study the $U_{ante}^{+\min}$ utility. As previously said, it does not satisfy monotonicity and Dynamic Programming can provide a good strategy, but without any guarantee of optimality. To guarantee optimality, one can proceed by an implicit enumeration via a Branch and Bound algorithm, as done by [8] for Possibilistic Choquet integrals and by [16] for Rank Dependent Utility (both in the mono criterion case). The Branch and Bound procedure (see Algorithm 4) takes as argument a partial strategy δ and an upper bound of the $U_{ante}^{+\min}$ value of its best extension. It returns U^*, the $U_{ante}^{+\min}$ value of the best strategy found so far, δ^*. We can initialize δ^* with any strategy, e.g. the one provided by Algorithms 2 or 1. At each step of the Branch and Bound algorithm, the current partial strategy, δ, is developed by the choice of an action for some unassigned decision node. When several decision nodes are candidate, the one with the minimal rank (i.e. the former one according to the temporal order) is developed. The recursive procedure backtracks when either the current strategy is complete (then δ^* and U^* are updated) or proves to be worse than the current δ^* in any case. Function $UpperBound(D_0, \delta)$ provides an upper bound of the best completion of δ - in practice, it builds, for each criterion j, a strategy δ_j that maximizes U_j^+ (using [9,10]'s algorithm, which is linear). It then selects, among these strategies, the one with the highest $U_{ante}^{+\min}$. It is important to note that $UpperBound(D_0, \delta) = U_{ante}^{+\min}(\delta)$ when δ is complete. Whenever the value returned by $UpperBound(D_0, \delta)$ is lower or equal to U^*, the value of the best current strategy, the algorithm backtracks, yielding the choice of another action for the last considered decision node.

Algorithm 4. B&B algorithm for the optimization of $U_{ante}^{+,min}$

Data: A decision tree T, a (partial) strategy δ, an upper Bound U of $U_{ante}^{+,min}(\delta)$
Result: U^*: the $U_{ante}^{+,min}$ value of δ^* the best strategy found so far
begin
    if $\delta(D_0) = \perp$ **then** $\mathcal{D}_{pend} \leftarrow \{D_0\}$;
    else $D_{pend} \leftarrow \{D_i \in \mathcal{D} \text{ s.t. } \exists D_j, \delta(D_j) \neq \perp \text{ and } D_i \in Succ(\delta(D_j))\}$;
    if $\mathcal{D}_{pend} = \emptyset$ **then** // δ is a complete strategy
      | $\delta^* \leftarrow \delta$; $U^* \leftarrow U$;
    else
      | $D_{next} \leftarrow arg\ min_{D_i \in D_{pend}}\ i$;
      | **foreach** $C_i \in Succ(D_{next})$ **do**
        | $\delta(D_{next}) \leftarrow C_i$;
        | $U \leftarrow UpperBound(D_0, \delta)$;
        | **if** $U > U^*$ **then** $U^* \leftarrow B\&B(U, \delta)$;
    return U^*;

---

4   Experiments

Beyond the evaluation of the feasibility of the algorithms proposed, our experiments aim at evaluating to what extent the optimization of the problematic utilities, $U_{ante}^{-\max}$ and $U_{ante}^{+\min}$, can be approximated by Dynamic Programming.

The implementation has been done in Java, on a processor Intel Core $i7$ 2670 QMCPU, 2.2 GHz, 6 GB of RAM. The experiments were performed on complete binary decision trees. We have considered four sets of problems, the number of decisions to be made in sequence (denoted seq) varying from 2 to 6, with an alternation of decision and chance nodes: at each decision level l (i.e. odd level), the tree contains 2^{l-1} decision nodes followed by 2^l chance nodes.[4] In the present experiment, the number of criteria is set equal to 3. The utility values as well as the weights degrees are uniformly fired in the set $\{0, 0.1, 0.2, \ldots, 0.9, 1\}$. Conditional possibilities are chosen randomly in $[0, 1]$ and normalized. Each of the four samples of problems contains 1000 randomly generated trees.

Feasibility Analysis and Temporal Performances: Table 1 presents the execution time of each algorithm. Obviously, for each one, the CPU time increases with the size of the tree. But it remains affordable even for very big trees (1365 decisions). We can check that $U_{ante}^{-\max}$ (resp. $U_{ante}^{+\min}$) the approximation performed by *ex-post* Dynamic Programming is faster than the one performed by *ex-ante* Dynamic Programming, both being faster than the exact algorithm (Multi Dynamic Programming and Branch and Bound, respectively).

Table 1. Average CPU time, in milliseconds, of for each algorithms and for each rule, according the size of the tree (in number of decision nodes)

# decision nodes			5	21	85	341	1365
$U_{post}^{-\min}$ $U_{ante}^{-\min}$		Post Dyn. Prog	0.068	0.073	0.076	0.126	0.215
$U_{post}^{+\max}$ $U_{ante}^{+\max}$		Post Dyn. Prog	0.071	0.075	0.082	0.128	0.207
$U_{post}^{-\max}$		Post Dyn. Prog	0.068	0.083	0.090	0.140	0.235
$U_{post}^{+\min}$		Post Dyn. Prog	0.067	0.075	0.082	0.132	0.211
$U_{ante}^{-\max}$		Multi Dyn. Prog	0.172	0.203	0.247	0.295	1.068
$U_{ante}^{-\max}$		Ante Dyn. Prog	0.079	0.096	0.120	0.147	0.254
$U_{ante}^{+\min}$		Branch & Bound	0.576	1.012	1.252	1.900	5.054
$U_{ante}^{+\min}$		Ante Dyn. Prog	0.074	0.084	0.093	0.147	0.231

Quality of the Approximation: As previously mentioned the *ex-post* and the *ex-ante* Dynamic Programming algorithms are approximation algorithms for $U_{ante}^{-\max}$ and $U_{ante}^{+\min}$. The following experiments estimate the quality of these approximations. At this extent, we compute for each sample the success rate of the approximation algorithm considered, i.e. the number of trees for which the value provided by the approximation algorithm is actually optimal; then for the trees for which it fails to reach optimality, we report the average closeness value to $\frac{U_{Approx}}{U_{Exact}}$ where U_{Approx} is the utility of the strategy provided by the approximation algorithm and U_{Exact} is the optimal utility - the one of the solution by the exact algorithm (Branch and Bound for $U_{ante}^{+\min}$ and Multi Dynamic Programming for $U_{ante}^{-\max}$). The results are given in Table 2.

Clearly, *Ex-Post* Dynamic Programming provides a good approximation for $U_{ante}^{+\min}$ - its success rate decreases with the number of nodes but stay higher

---

[4] Hence, for a sequence length $seq = 2$ (resp. 3, 4, 5, 6), the number of decision nodes in each tree of the sample is equal to 5 (resp. 21, 85, 341, 1365).

Table 2. Quality of approximation of $U_{ante}^{-\max}$ and $U_{ante}^{+\min}$ by Dynamic Programming

# decision nodes		5	21	85	341	1365
% of success						
$U_{ante}^{-\max}$	Ante Dyn. Prog	17.3%	19%	22.1%	26.4%	31%
$U_{ante}^{-\max}$	Post. Dyn. Prog	15.4%	23.6%	30.7%	35.6%	40.4%
$U_{ante}^{+\min}$	Ante Dyn. Prog	87%	76.8%	68%	62.6%	59.6%
$U_{ante}^{+\min}$	Post Dyn. Prog	91.7%	90.8%	88.2%	86.7%	76%
Closeness value						
$U_{ante}^{-\max}$	Ante Dyn. Prog	0.522	0.56	0.614	0.962	0.981
$U_{ante}^{-\max}$	Post Dyn. Prog	0.473	0.529	0.556	0.58	0.62
$U_{ante}^{+\min}$	Ante Dyn. Prog	0.97	0.95	0.94	0.93	0.91
$U_{ante}^{+\min}$	Post Dyn. Prog	0.989	0.975	0.946	0.928	0.90

than 70%, and above all it has a very high closeness value (above 0.9); notice that it is always better than its *ex-ante* counterpart, in terms of success rate, of closeness and of CPU time. This is good news since it is polynomial while Branch and Bound, the exact algorithm, is exponential in the number of nodes. As to $U_{ante}^{-\max}$, none of the approximation algorithms is good. However, this is not so bad news since Multi Dynamic Programming, the exact algorithm is polynomial and has very affordable CPU time.

5   Conclusion

This paper proposes to extend to possibilistic decision trees the decision rules presented in [1] for non sequential problems. We show that, for the *ex-post* decision rules, as well as for U_{ante}^{+max} and U_{ante}^{-min}, the optimization can be achieved by Dynamic Programming. For $U_{ante}^{+\min}$ the optimization can be carried either by an exact but costly algorithm (Branch & Bound) or by an approximation one, (*ex-post* Dynamic Programming). For $U_{ante}^{-\max}$ we propose an exact algorithm (Multi Dynamic Programming) that performs better than Dynamic Programming. As future work, we would like to study the handling of several criteria in more sophisticated qualitative decision models such as possibilistic influence diagrams [14] or possibilistic Markov decision models [10].

References

1. Ben Amor, N., Essghaier, F., Fargier, H.: Solving multi-criteria decision problems under possibilistic uncertainty using optimistic and pessimistic utilities. In: Laurent, A., Strauss, O., Bouchon-Meunier, B., Yager, R.R. (eds.) IPMU 2014. CCIS, vol. 444, pp. 269–279. Springer, Cham (2014). doi:10.1007/978-3-319-08852-5_28
2. Dubois, D., Prade, H.: Possibility theory as a basis for qualitative decision theory. In: Proceedings of IJCAI 1995, pp. 1924–1930 (1995)

3. Dubois, D., Godo, L., Prade, H., Zapico, A.: Making decision in a qualitative setting: from decision under uncertainty to case-based decision. In: Proceedings of KR, pp. 594–607 (1998)
4. Giang, P.H., Shenoy, P.P.: A qualitative linear utility theory for Spohn's theory of epistemic beliefs. In: Proceedings of UAI, pp. 220–229 (2000)
5. Dubois, D., Prade, H., Sabbadin, R.: Decision theoretic foundations of qualitative possibility theory. EJOR **128**, 459–478 (2001)
6. Dubois, D., Fargier, H., Prade, H., Perny, P.: Qualitative decision theory: from savage's axioms to nonmonotonic reasoning. JACM **49**, 455–495 (2002)
7. Dubois, D., Fargier, H., Perny, P.: Qualitative decision theory with preference relations and comparative uncertainty: an axiomatic approach. Artif. Intell. **148**, 219–260 (2003)
8. Ben Amor, N., Fargier, H.: Possibilistic sequential decision making. Int. J. Approximate Reasoning **55**, 1269–1300 (2014)
9. Sabbadin, R., Fargier, H., Lang, J.: Towards qualitative approaches to multi-stage decision making. Int. J. Approximate Reasoning **19**, 441–471 (1998)
10. Sabbadin, R.: Empirical comparison of probabilistic and possibilistic Markov decision processes algorithms. In: Proceedings of ECAI, pp. 586–590 (2000)
11. Harsanyi, J.: Cardinal welfare, individualistic ethics, and interpersonal comparisons of utility. J. Polit. Econ. **63**, 309–321 (1955)
12. Ben Amor, N., Essghaier, F., Fargier, H.: Egalitarian collective decision making under qualitative possibilistic uncertainty: principles and characterization. In: Proceedings of AAAI, pp. 3482–3488 (2015)
13. Dubois, D., Prade, H.: Weighted minimum and maximum operations in fuzzy set theory. J. Inform. Sci. **39**, 205–210 (1986)
14. Garcias, L., Sabbadin, R.: Possibilistic influence diagrams. In: Proceedings of ECAI, pp. 372–376 (2006)
15. Bellman, R.: Dynamic Programming. Princeton University Press, New Jersey (1957)
16. Jeantet, G., Spanjaard, O.: Rank-dependent probability weighting in sequential decision problems under uncertainty. In: Proceedings of ICAPS, pp. 148–155 (2008)

Efficient Policies for Stationary Possibilistic Markov Decision Processes

Nahla Ben Amor[1]([⊠]), Zeineb EL khalfi[1,2]([⊠]), Hélène Fargier[2]([⊠]),
and Régis Sabaddin[3]([⊠])

[1] LARODEC, Le Bardo, Tunisie
nahla.benamor@gmx.fr, zeineb.khalfi@gmail.com
[2] IRIT, Toulouse, France
fargier@irit.fr
[3] INRA-MIAT, Toulouse, France
regis.sabbadin@inra.fr

Abstract. Possibilistic Markov Decision Processes offer a compact and tractable way to represent and solve problems of sequential decision under qualitative uncertainty. Even though appealing for its ability to handle qualitative problems, this model suffers from the *drowning effect* that is inherent to possibilistic decision theory. The present paper proposes to escape the drowning effect by extending to stationary possibilistic MDPs the lexicographic preference relations defined in [6] for non-sequential decision problems and provides a value iteration algorithm to compute policies that are optimal for these new criteria.

Keywords: Markov Decision Process · Possibility theory · Lexicographic comparisons · Possibilistic qualitative utilities

1 Introduction

The classical paradigm for sequential decision making under uncertainty is the one of expected utility-based *Markov Decision Processes* (MDP) [2,11], which assumes that the uncertain effects of actions can be represented by probability distributions and that utilities are additive. But the EU model is not tailored to problems where uncertainty and preferences are ordinal in essence. Alternatives to the EU-based model have been proposed to handle ordinal preferences/uncertainty. Remaining within the probabilistic, quantitative, framework while considering ordinal preferences has lead to *quantile-based* approaches [8,9,15,17,18]) Purely ordinal approaches to sequential decision under uncertainty have also been considered. In particular, possibilistic MDPs [1,4,12,13] form a purely qualitative decision model with an ordinal evaluation of plausibility and preference. In this model, uncertainty about the consequences of actions is represented by possibility distributions and utilities are also ordinal. The decision criteria are either the optimistic qualitative utility or its pessimistic counterpart [5]. However, it is now well known that possibilistic decision criteria suffer from the *drowning effect* [6]. Plausible enough bad or good consequences

© Springer International Publishing AG 2017
A. Antonucci et al. (Eds.): ECSQARU 2017, LNAI 10369, pp. 306–317, 2017.
DOI: 10.1007/978-3-319-61581-3_28

may completely blur the comparison between policies, that would otherwise be clearly differentiable. [6] have proposed *lexicographic refinements* of possibilistic criteria for the one-step decision case, in order to remediate the drowning effect. In this paper, we propose an extension of the lexicographic preference relations to stationary possibilistic MDPs.

The next Section recalls the background about possibilistic MDPs, including the drowning effect problem. Section 3 studies the lexicographic comparison of policies in finite horizon problems and presents a value iteration algorithm for the computation of lexi-optimal policies. Section 4 extends these results to the infinite-horizon case. Lastly, Section 5 reports experimental results. Proofs are omitted, but can be found in[1].

2 Possibilistic Markov decision process

2.1 Definition

A possibilistic Markov Decision Process (P-MDP) [12] is defined by:

- A finite set S of **states**.
- A finite set A of **actions**, A_s denotes the set of actions available in state s;
- A possibilistic transition function: each action $a \in A_s$ applied in state $s \in S$ is assigned a possibility distribution $\pi(.|s, a)$;
- A utility function μ: $\mu(s)$ is the intermediate satisfaction degree obtained in state s.

The uncertainty about the effect of an action a taken in state s is a possibility distribution $\pi(.|s, a) : S \to L$, where L is a qualitative ordered scale used to evaluate both possibilities and utilities (typically, and without loss of generality, $L = [0, 1]$): for any s', $\pi(s'|s, a)$ measures to what extent s' is a plausible consequence of a when executed in s and $\mu(s')$ is the utility of being in state s'. In the present paper, we consider stationary problems, i.e. problems in which states, the actions and the transition functions do not depend on the stage of the problem. Such a possibilistic MDP defines a graph, where states are represented by circles and are labelled by utility degrees and actions are represented by squares. An edge linking an action to a state denotes a possible transition and is labeled by the possibility of that state given the action is executed.

Example 1. Let us suppose that a "Rich and Unknown" person runs a startup company. Initially, s/he must choose between Saving money (*Sav*) or Advertising (*Adv*) and may then get Rich (*R*) or Poor (*P*) and Famous (*F*) or Unknown (*U*). In the other states, *Sav* is the only possible action. Figure 1 shows the stationary P-MDP that captures this problem, formally described as follows: $S = \{RU, RF, PU\}, A_{RU} = \{Adv, Sav\}, A_{RF} = \{Sav\},$ $A_{PU} = \{Sav\}, \pi(PU|RU, Sav) = 0.2, \pi(RU|RU, Sav) = 1; \pi(RF|RU, Adv) = 1; \pi(RF|RF, Sav) = 1, \pi(RU|RF, Sav) = 1, \mu(RU) = 0.5, \mu(RF) = 0.7,$ $\mu(PU) = 0.3.$

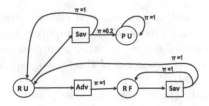

Fig. 1. A possibilistic stationary MDP

Solving a stationary MDP consists in finding a (stationary) policy, i.e. a function $\delta : S \rightarrow A_s$ which is optimal with respect to a decision criterion. In the possibilistic case, as in the probabilistic case, the value of a policy depends on the utility and on the likelihood of its trajectories. Formally, let Δ be the set of all policies encoded by a P-MDP. When the horizon is finite, each $\delta \in \Delta$ defines a list of scenarios called trajectories. Each trajectory is a sequence of states and actions $\tau = (s_0, a_0, s_1, \ldots, s_{E-1}, a_{E-1}, s_E)$.

To simplify notations, we will associate the vector $v_\tau = (\mu_0, \pi_1, \mu_1, \pi_2, \ldots, \pi_{E-1}, \mu_E)$ to each trajectory τ, where $\pi_{i+1} =_{def} \pi(s_{i+1}|s_i, a_i)$ and $\mu_i =_{def} \mu(s_i)$.

The possibility and the utility of τ given that δ is applied from s_0 are defined by:

$$\pi(\tau|s_0, \delta) = \min_{i=1..E} \pi(s_i|s_{i-1}, \delta(s_{i-1})) \quad \text{and} \quad \mu(\tau) = \min_{i=0..E} \mu(s_i) \qquad (1)$$

Two criteria, an optimistic and a pessimistic one, can then be used [5,13]:

$$u_{opt}(\delta, s_0) = \max_\tau \min\{\pi(\tau|s_0, \delta), \mu(\tau)\} \qquad (2)$$

$$u_{pes}(\delta, s_0) = \max_\tau \min\{1 - \pi(\tau|s_0, \delta), \mu(\tau)\} \qquad (3)$$

These criteria can be optimized by choosing, for each state, an action that maximizes the following counterparts of the Bellman equations [12]:

$$u_{opt}(s) = \max_{a \in A_s} \min\{\mu(s), \max_{s' \in S} \min(\pi(s'|s, a), u_{opt}(s'))\} \qquad (4)$$

$$u_{pes}(s) = \max_{a \in A_s} \min\{\mu(s), \min_{s' \in S} \max(1 - \pi(s'|s, a), u_{pes}(s'))\} \qquad (5)$$

This formulation is more general than the first one in the sense that it applies to both the finite and the infinite case. It has allowed the definition of a (possibilistic) *value iteration* algorithm which converges to an optimal policy in polytime $O(|S|^2 \cdot |A|^2 \cdot |L|)$ [12]. This algorithm proceeds by iterated modifications of a possibilistic value function $\tilde{Q}(s, a)$ which evaluates the "utility" (pessimistic or optimistic) of performing a in s.

---

[1] https://www.irit.fr/publis/ADRIA/PapersFargier/XKRU17MDP.pdf.

2.2   The Drowning Effect

Unfortunately, possibilistic utilities suffer from an important drawback called the *drowning effect*: plausible enough bad or good consequences may completely blur the comparison between acts that would otherwise be clearly differentiated; as a consequence, an optimal policy δ is not necessarily Pareto efficient - it may exist a policy δ' such that $u_{pes}(\delta'_s) = u_{pes}(\delta_s)$ while $\forall s, u_{pes}(\delta'_s) \succeq u_{pes}(\delta_s)$ and (ii) $\exists s, u_{pes}(\delta'_s) \succ u_{pes}(\delta_s)$ where δ_s (resp. δ'_s) is the restriction of δ (resp. δ') to the subtree rooted in s.

Example 2. The P-MDP of Example 1; it admits two policies δ and δ': $\delta(RU) = Sav; \delta(PU) = Stay; \delta(RF) = Sav; \delta'(RU) = Adv; \delta'(PU) = Stay; \delta'(RF) = Sav$. For horizon $E = 2$:

- δ has 3 trajectories: $\tau_1 = (RU, PU, PU)$ with $v_{\tau_1} = (0.5\ 0.2\ 0.3\ 1\ 0.3)$; $\tau_2 = (RU, RU, PU)$ with $v_{\tau_2} = (0.5\ 1\ 0.5\ 0.2\ 0.3)$; $\tau_3 = (RU, RU, RU)$ with $v_{\tau_3} = (0.5\ 1\ 0.5\ 1\ 0.5)$.
- δ' has 2 trajectories: $\tau_4 = (RU, RF, RF)$ with $v_{\tau_4} = (0.5\ 1\ 0.7\ 1\ 0.7)$; $\tau_5 = (RU, RF, RU)$ with $v_{\tau_5} = (0.5\ 1\ 0.7\ 1\ 0.5)$.

Thus $U_{opt}(\delta) = U_{opt}(\delta') = 0.5$. However δ' seems better than δ since it provides utility 0.5 for sure while δ provides a bad utility (0.3) in some non impossible trajectories (τ_1 and τ_2). τ_3 which is good and totally possible "drowns" τ_1 and τ_2: δ is considered as good as δ'.

2.3   Lexi-Refinements of Ordinal Aggregations

In ordinal (i.e. min-based and max-based) aggregation a solution to the drowning effect has been proposed, that is based on leximin and leximax comparisons [10]. It has then been extended to non-sequential decision making under uncertainty [6] and, in the sequential case, to decision trees [3]. Let us first recall the basic definition of these two preference relations. For any two vectors t and t' of length m built on L:

$$t \succeq_{lmin} t' \text{ iff } \forall i, t_{\sigma(i)} = t'_{\sigma(i)} \text{ or } \exists i^*, \forall i < i^*, t_{\sigma(i)} = t'_{\sigma(i)} \text{ and } t_{\sigma(i^*)} > t'_{\sigma(i^*)} \quad (6)$$

$$t \succeq_{lmax} t' \text{ iff } \forall i, t_{\mu(i)} = t'_{\mu(i)} \text{ or } \exists i^*, \forall i < i^*, t_{\mu(i)} = t'_{\mu(i)} \text{ and } t_{\mu(i^*)} > t'_{\mu(i^*)} \quad (7)$$

where, for any vector v (here, $v = t$ or $v = t'$), $v_{\mu(i)}$ (resp. $v_{\sigma(i)}$) is the ith best (resp. worst) element of v.

[6] have extended these procedures to the comparison of matrices built on L. Given a complete preorder $\trianglerighteq$ on vectors, it is possible to order the lines of the matrices (say, A and B) according to $\trianglerighteq$ and to apply an $lmax$ or an $lmin$ procedure:

$$A \succeq_{lmin(\trianglerighteq)} B \Leftrightarrow \forall j, a_{(\trianglerighteq,j)} \cong b_{(\trianglerighteq,j)} \text{ or } \exists i \text{ s.t. } \forall j > i, a_{(\trianglerighteq,j)} \cong b_{(\trianglerighteq,j)} \text{ and } a_{(\trianglerighteq,i)} \triangleright b_{(\trianglerighteq,i)} \quad (8)$$

$$A \succeq_{lmax(\trianglerighteq)} B \Leftrightarrow \forall j, a_{(\trianglerighteq,j)} \cong b_{(\trianglerighteq,j)} \text{ or } \exists i \text{ s.t.} \forall j < i, a_{(\trianglerighteq,j)} \cong b_{(\trianglerighteq,j)} \text{ and } a_{(\trianglerighteq,i)} \triangleright b_{(\trianglerighteq,i)} \quad (9)$$

where, for any $c \in (L^M)^N$, $c_{(\trianglerighteq,i)}$ is the ith largest sub-vector of c according to $\trianglerighteq$.

3   Lexicographic-Value Iteration for Finite Horizon P-MDPs

In (finite-horizon) possibilistic decision trees, the idea of [3] is to identify a strategy with the matrix of its trajectories, and to compare such matrices with a $\succeq_{lmax(lmin)}$ (resp. $\succeq_{lmin(lmax)}$) procedure for the optimistic (resp. pessimistic) case. We propose, in the following, a value iteration algorithm for the computation of such lexi-optimal policies in the finite (this Section) and infinite (Sect. 4) horizon cases.

3.1   Lexicographic Comparisons of Policies

Let E be the horizon of the P-MDP. A trajectory being a sequence of states and actions, a strategy can be viewed as a matrix where each line corresponds to a distinct trajectory. In the optimistic case each line corresponds to a vector $v_\tau = (\mu_0, \pi_1, \mu_1, \pi_2, \ldots, \pi_{E-1}, \mu_E)$ and in the pessimistic case to $w_\tau = (\mu_0, 1 - \pi_1, \mu_1, 1 - \pi_2, \ldots, 1 - \pi_{E-1}, \mu_E)$.

This allow us to define the comparison of trajectories and strategies by[2]:

$$\tau \succeq_{lmin} \tau' \text{ iff } (\mu_0, \pi_1, \ldots, \pi_E, \mu_E) \succeq_{lmin} (\mu'_0, \pi'_2, \ldots, \pi'_E, \mu'_E) \tag{10}$$

$$\tau \succeq_{lmax} \tau' \text{ iff } (\mu_0, 1 - \pi_1, \ldots, 1 - \pi_E, \mu_E) \succeq_{lmax} (\mu'_0, 1 - \pi'_1, \ldots 1 - \pi'_E, \mu'_E) \tag{11}$$

$$\delta \succeq_{lmax(lmin)} \delta' \text{ iff } \forall i, \ \tau_{\mu(i)} \sim_{lmin} \tau'_{\mu(i)}$$
$$or \ \exists i^*, \ \forall i < i^*, \tau_{\mu(i)} \sim_{lmin} \tau'_{\mu(i)} \ and \ \tau_{\mu(i^*)} \succ_{lmin} \tau'_{\mu(i^*)} \tag{12}$$

$$\delta \succeq_{lmin(lmax)} \delta' \text{ iff } \forall i, \ \tau_{\sigma(i)} \sim_{lmax} \tau'_{\sigma(i)}$$
$$or \ \exists i^*, \ \forall i < i^*, \tau_{\sigma(i)} \sim_{lmax} \tau'_{\sigma(i)} \ and \ \tau_{\sigma(i^*)} \succ_{lmax} \tau'_{\sigma(i^*)} \tag{13}$$

where $\tau_{\mu(i)}$ (resp. $\tau'_{\mu(i)}$) is the ith best trajectory of δ (resp δ') according to $\succeq_{lmin}$ and $\tau_{\sigma(i)}$ (resp. $\tau'_{\sigma(i)}$) is the ith worst trajectory of δ (resp δ') according to $\succeq_{lmax}$.

It is easy to show that we get efficient refinements of u_{opt} and u_{pes}.

Proposition 1. *If $u_{opt}(\delta) > u_{opt}(\delta')$ (resp. $u_{pes}(\delta) > u_{pes}(\delta')$) then $\delta \succ_{lmax(lmin)} \delta'$ (resp. $\delta \succ_{lmin(lmax)} \delta'$).*

Proposition 2. *Relations $\succeq_{lmin(lmax)}$ and $\succeq_{lmax(lmin)}$ are complete, transitive and satisfy the principle of strict monotonicity[3].*

---

[2] If a trajectory is shorter than E, neutral elements (0 for the optimistic case and 1 for the pessimistic one) are added at the end. If the policies have different numbers of trajectories, neutral trajectories (vectors) are added to the shortest one.

[3] A criterion O satisfies the principle of strict monotonicity iff: $\forall \delta, \delta', \delta'', \delta \succeq_O \delta' \iff \delta + \delta'' \succeq_O \delta' + \delta''$. $\delta + \delta''$ contains two disjoint sets of trajectories: the ones of δ and the ones of δ'' (and similarly for $\delta' + \delta''$). Then, adding or removing identical trajectories to two sets of trajectories does not change their comparison by $\succeq_{lmax(lmin)}$ (resp. $\succeq_{lmin(lmax)}$) - while it may transform a strict preference into an indifference if u_{opt} (resp. u_{pes}) were used.

Remark. We define the *complementary* MDP, $(S, A, \pi, \bar{\mu})$ of a given P-MDP (S, A, π, μ) where $\bar{\mu}(s) = 1 - \mu(s), \forall s \in S$. The complementary MDP simply gives complementary utilities. From the definitions of $\succeq_{lmax}$ and $\succeq_{lmin}$, we can check that:

Proposition 3. $\tau \succeq_{lmax} \tau' \Leftrightarrow \bar{\tau}' \succeq_{lmin} \bar{\tau}$ and $\delta \succ_{lmin(lmax)} \delta' \Leftrightarrow \bar{\delta}' \succeq_{lmax(lmin)} \bar{\delta}$.

where $\bar{\tau}$ and $\bar{\delta}$ are obtained by replacing μ with $\bar{\mu}$ in the trajectory/P-MDP.

Therefore, all results which we will prove in the following for $\succeq_{lmax(lmin)}$ also hold for $\succeq_{lmin(lmax)}$, if we take care to apply them to complementary strategies. Since considering $\succeq_{lmax(lmin)}$ involves less cumbersome expressions (no $1 - \cdot$), we will give the results for this criterion. Moreover, abusing notations slightly, we identify trajectories τ (resp. strategies) with their v_τ vectors (resp. matrices of v_τ vectors).

3.2   Basic Operations on Matrices of Trajectories

Before going further, we define some basic operations on matrices (typically, on $U(s)$ representing trajectories issued from s). For any matrix $U = (u_{ij})$ with n lines and m columns, $[U]_{l,c}$ denotes the restriction of U to its first l lines and first c columns.

Composition, $U \times (N_1, \ldots, N_a)$: Let U be a $a \times b$ matrix and $N_1, \ldots, N_a$ be a series of a matrices of dimension $n_i \times c$ (they all share the same number of columns). The composition of U with $(N_1, \ldots, N_a)$ denoted $U \times (N_1, \ldots, N_a)$ is a matrix of dimension $(\sum_{1 \leq i \leq a} n_i) \times (b + c)$. For any $i \leq a, j \leq n_j$, the $(\Sigma_{i' < i} n_{i'} + j)^{th}$ line of $U \times (N_1, \ldots, N_a)$ is the concatenation of the ith line of U and the jth line of N_i. The composition of $U \times (N_1, \ldots, N_a)$ is done in $O(n \cdot m)$ operations, where $n = \sum_{1 \leq i \leq a} n_i$ and $m = b + c$. The matrix $U(s)$ is typically the concatenation of the matrix $U = ((\pi(s'|s, a), \mu(s')), s' \in succ(s, a))$ with the matrices $N_{s'} = U(s')$.

Ordering Matrices $U^{lmaxlmin}$: Let U be a $n \times m$ matrix, $U^{lmaxlmin}$ is the matrix obtained by ordering the elements of the lines of U in increasing order and the lines of U according to $lmax$ (in decreasing order). The complexity of the operation depends on the sorting algorithm: if we use QuickSort then ordering the elements within a line is performed in $O(m \cdot log(m))$, and the inter-ranking of the lines is done in $O(n \cdot log(n) \cdot m)$ operations. Hence, the overall complexity in $O(n \cdot m \cdot log(n \cdot m))$.

Comparison of Ordered Matrices: Given two ordered matrices $U^{lmaxlmin}$ and $V^{lmaxlmin}$, we say that $U^{lmaxlmin} > V^{lmaxlmin}$ iff $\exists i, j$ such that $\forall i' < i, \forall j', U_{i',j'}^{lmaxlmin} = V_{i',j'}^{lmaxlmin}$ and $\forall j' < j, U_{i,j'}^{lmaxlmin} = V_{i,j'}^{lmaxlmin}$ and $U_{i,j}^{lmaxlmin} > V_{i,j}^{lmaxlmin}$. $U^{lmaxlmin} \sim V^{lmaxlmin}$ iff they are identical (comparison complexity: $O(n \cdot m)$).

3.3 Lexicographic-Value Iteration

In this section, we propose a value iteration algorithm (Algorithm 1 for the $lmax(lmin)$ variant; the $lmin(lmax)$ variant is similar) that computes a lexicographic optimal policy in a finite number of iterations. This algorithm is an iterative procedure that updates the utility of each state, represented by a finite matrix of trajectories, using the utilities of the neighboring states, until a halting condition is reached. At stage t, the procedure updates the utility of every states $s \in S$ as follows:

– For each $a \in A_s$, a matrix $Q(s, a)$ is built which evaluates the "utility" of performing a in s at stage t: this is done by combining $TU_{s,a}$ (comparison of the transition matrix $T_{s,a} = \pi(\cdot|s, a)$ and the utilities $\mu(s')$ of the states s' that may follows s when a is executed) with the matrices $U^{t-1}(s')$ of trajectories provided by these s'. The matrix $Q(s, a)$ is then ordered (the operation is made less complex by the fact that the matrices $U^{t-1}(s')$ have been ordered at $t-1$).
– The $lmax(lmin)$ comparison is performed on the fly to memorize the best $Q(s, a)$
– The value of s at t, $U^t(s)$, is the one given by the action $\delta^t(s) = a$ which provides the best $Q(s, a)$. U^t and δ^t are memorized (and U^{t-1} can be forgotten).

Algorithm 1. Lmax(lmin)-value iteration

Data: A possibilistic MDP and an horizon E
δ^*, the policy built by the algorithm, is a global variable
1   // δ a global variable starts as an empty set
Result: Computes and returns δ^* for MDP
2 **begin**
3    $t \leftarrow 0$;
4   **foreach** $s \in S$ **do** $U^t(s) \leftarrow ((\mu(s)))$;
5   **foreach** $s \in S$, $a \in A_s$ **do** $TU_{s,a} \leftarrow T_{s,a} \times ((\mu(s')), s' \in succ(s,a))$;
6   **repeat**
7      $t \leftarrow t + 1$;
8   **foreach** $s \in S$ **do**
9        $Q^* \leftarrow ((0))$;
10   **foreach** $a \in A$ **do**
11          $Future \leftarrow (U^{t-1}(s'), s' \in succ(s,a))$; // Gather the matrices provided by the successors of s;
12          $Q(s,a) \leftarrow (TU_{s,a} \times Future)^{lmaxlmin}$;
13   **if** $Q^* \leq_{lmaxlmin} Q(s,a)$ **then** $Q^* \leftarrow Q(s,a)$; $\delta^t(s) \leftarrow a$;
14        $U^t(s) \leftarrow Q^*(s, \delta^t(s))$
15   **until** $t == E$;
16    $\delta^*(s) \leftarrow argmax_a Q(s,a)$
17   **return** δ^*;

Proposition 4. *lmax(lmin)-Value iteration provides an optimal solution for* $\succeq_{lmaxlmin}$.

Time and space complexities of this algorithm are nevertheless expensive, since it eventually memorizes all the trajectories. At each step t its size may be about $b^t \cdot (2 \cdot t + 1)$, where b is the maximal number of possible successors of an action; the overall complexity of the algorithm is $O(|S| \cdot |A| \cdot |E| \cdot b^E)$, which is problematic. Notice now that, at any stage t and for any state s $[U^t(s)]_{1,1}$ (i.e. the top left value in $U^t(s)$) is precisely equal to $u_{opt}(s)$ at horizon t for the optimal strategy. We have seen that making the choices on this basis is not discriminant enough. On the other hand, taking the whole matrix is discriminant, but exponentially costly. Hence the idea of considering more than one line and one column, but less than the whole matrix - namely the first l lines and c columns of $U^t(s)^{lmaxlmin}$; hence the definition of the following preference:

$$\delta \succeq_{lmaxlmin,l,c} \delta' \text{ iff } [\delta^{lmaxlmin}]_{l,c} \geq [\delta'^{lmaxlmin}]_{l,c} \tag{14}$$

$\succeq_{lmaxlmin,1,1}$ corresponds to $\succeq_{opt}$ and $\succeq_{lmaxlmin,+\infty,+\infty}$ corresponds to $\succeq_{lmaxlmin}$.

The combinatorial explosion is due to the number of lines (because at finite horizon, the number of columns is bounded by $2 \cdot E + 1$), hence we shall bound the number of considered lines. The following proposition shows that this approach is sound:

Proposition 5. *For any* l, c $\delta \succ_{opt} \delta' \Rightarrow \delta \succ_{lmaxlmin,l,c} \delta'$.
For any l, c, l' *such that* $l' > l$, $\delta \succ_{lmaxlmin,l,c} \delta' \Rightarrow \delta \succ_{lmaxlmin,l',c} \delta'$.

Hence $\succ_{lmaxlmin,l,c}$ refines u_{opt} and the order over the strategies is refined for a fixed c when l increases. It tends to $\succ_{lmaxlmin}$ when $c = 2.E + 1$ and l tends to b^E.

Up to this point, the comparison by $\succeq_{lmaxlmin,l,c}$ is made on the basis of the first l lines and c columns of the *full* matrices of trajectories. This does obviously not reduce their size. The important following Proposition allows us to make the l, c reduction of the ordered matrices *at each step* (after each composition), and not only at the very end, thus keeping space and time complexities polynomial.

Proposition 6. *Let* U *be a* $a \times b$ *matrix and* $N_1, \ldots, N_a$ *be a series of* a *matrices of dimension* $a_i \times c$. *It holds that:*
$$[(U \times (N_1, \ldots, N_a))^{lmaxlmin}]_{l,c} = [(U \times ([N_1^{lmaxlmin}]_{l,c}, \ldots, [N_a^{lmaxlmin}]_{l,c}))^{lmaxlmin}]_{l,c}.$$

In summary, the idea of our Algorithm, that we call bounded lexicographic-value iteration (BL-VI) is to compute policies that are close to lexi-optimality, by keeping a sub matrix of each current value matrix - namely the first l lines and c columns. The algorithm is obtained by replacing line 12 of Algorithm 1, with:

$$\text{Line } 12' \; : \; Q(s,a) \leftarrow [(TU_{s,a} \times Future)^{lmaxlmin}]_{l,c};$$

Proposition 7. *Bounded lmax(lmin)-Value iteration provides a solution that is optimal for $\succeq_{lmaxlmin,l,c}$ and its time complexity is $O(|E| \cdot |S| \cdot |A| \cdot (l \cdot c) \cdot b \cdot log(l \cdot c \cdot b))$.*

In summary, this algorithm provides in polytime a strategy that is always as least as good as the one provided by u_{opt} (according to $lmax(lmin)$) and tends to lexi optimality when $c = 2 \cdot E + 1$ and l tends to b^E.

4 Lexicogaphic-Value Iteration for Infinite Horizon P-MDPs

In the infinite-horizon case, the comparison of matrices of trajectories by Eqs. (12) or (13) may not be enough to rank-order the policies. The length of the trajectories may be infinite, and their number infinite as well. This problem is well known in classical probabilistic MDP where a discount factor is used that attenuates the influence of later utility degrees - thus allowing the convergence of the algorithm [11]. On the contrary, classical P-MDPs do not need any discount factor and Value Iteration, based on the evaluation for $l = c = 1$, converges for infinite horizon P-MDPs [12].

In a sense, this limitation to $l = c = 1$ plays the role of a discount factor - which is too drastic; it is nevertheless possible to make the comparison using $\geq_{lmaxlmin,l,c}$. Let us denote $U^t(s)$ the matrix issued from s at horizon t when δ is executed. It holds that:

Proposition 8. $\forall l, c, \exists t$ *such that, forall* $t' > t$, $(U^t)_{l,c}^{lmaxlmin}(s) = (U^{t'})_{l,c}^{lmaxlmin}(s)$.

This means that from a given stage t, the value of a strategy is stable if computed with the bounded lmax(lmin) criterion. This criterion can thus be soundly used in the infinite-horizon case and bounded value iteration converges. To adapt the algorithm to the infinite case, we simply need to modify the halting condition at line 15 by:

$$Line15' : \mathbf{until}\ \left(U^t\right)_{l,c}^{lmaxlmin} == \left(U^{t-1}\right)_{l,c}^{lmaxlmin}.$$

Proposition 9. *Whatever l, c, Lmax(lmin)-Bounded Value iteration converges for infinite horizon P-MDPs.*

Proposition 10. *The overall complexity of Bounded lmax(lmin)-Value iteration algorithm is $O(|L| \cdot |S| \cdot |A| \cdot (l \cdot c) \cdot b \cdot log(l \cdot c \cdot b))$.*

5 Experiments

We now compare the performance of Bounded lexicographic value iteration (*BL-VI*) as an approximation of (unbounded) lexicographic value iteration (*UL-VI*), in the Lmax(lmin) variant. The two algorithms have been implemented in Java

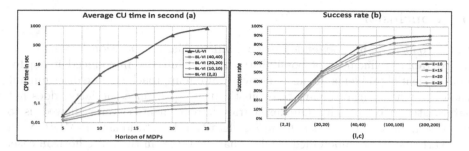

Fig. 2. Bounded lexicographic value iteration VS Unbounded lexicographic value iteration

and the experiments have been performed on an Intel Core i5 processor computer (1.70 GHz) with 8 GB DDR3L of RAM. We evaluate the performance of the algorithms by carrying out simulations on randomly generated P-MDPs with $|S| = 25$. The number of actions in each state is equal to 4. The output of each action is a distribution on two states randomly fired (i.e. the branching factor is equal to 2). The utility values are uniformly randomly fired in the set $L = \{0.1, 0.3, 0.5, 0.7, 1\}$. Conditional possibilities relative to decisions should be normalized. To this end, one choice is fixed to possibility degree 1 and the possibility degree of the other one is uniformly fired in L. For each experience, 100 P-MDPs are generated. The two algorithms are compared w.r.t. 2 measures: (i) CPU time and (ii) Pairwise success rate: *Success*, the percentage of optimal solutions provided by Bounded value iteration with fixed (l, c) w.r.t. the lmax(lmin) criterion in its full generality. The higher *Success*, the more important the effectiveness of cutting matrices with *BL-VI*; the lower this rate, the more important the drowning effect.

Figure 2 presents the average execution CPU time for the two algorithms. Obviously, for both *UL-VI* and *BL-VI*, the execution time increases with the horizon. Also, we observe that the CPU time of *BL-VI* increases according to the values of (l, c) but it remains affordable, as the maximal CPU time is lower than 1s for MDPs with 25 states and 4 actions when $(l, c) = (40, 40)$ and $E = 25$. Unsurprisingly, we can check that the *BL-VI* (regardless of the values of (l, c)) is faster than *UL-VI* especially when the horizon increases: the manipulation of l, c-matrices is obviously less expensive than the one of full matrices. The saving increases with the horizon.

As with the success rate, the results are described in Fig. 2. It appears that *BL-VI* provides a very good approximation especially when increasing (l, c). It provides the same optimal solution as the *UL-VI* in about 90% of cases, with an $(l, c) = (200, 200)$. Moreover, even when the success rate of *BL-VI* decreases (when E increases), the quality of approximation is still good: never less than 70% of optimal actions returned, with $E = 25$. These experiments conclude in favor of bounded value iteration: its approximated solutions are comparable in terms of quality for high (l, c) and increase when (l, c) increase, while it is much faster than the unbounded version.

6   Conclusion

In this paper, we have extended to possibilistic Markov Decision Processes the lexicographic refinement of possibilistic utilities initially introduced in [6] for non-sequential problems. It can be shown that our approach is more discriminant than the refinement of binary possibilistic utility [16] since the latter does not satisfy strict monotonicity. Our lexicographic refinements criteria allowed us to propose a *Lmax(lmin)-Value Iteration* algorithm for stationary P-MDPs with two variants: (i) an unbounded version that converges in the finite horizon case, but is unsuitable for infinite-horizon P-MDPs, since it generates matrices which size continuously increases with the horizon and (ii) a bounded version which has polynomial complexity. It bounds the size of the saved matrices and refines the possibilistic criteria, whatever the choice of the bounds. The convergence of this algorithm is shown for both the finite and the infinite horizon cases, and its efficiency has been observed experimentally even for low bounds.

There are two natural perspectives to this work. First, as far as the infinite horizon case is concerned, other types of lexicographic refinements could be proposed. One of these options could be to avoid the duplication of the set of transitions that occur several times in a single trajectory and consider only those which are observed. A second perspective of this work will be to define *reinforcement learning* [14] type algorithms for P-MDPs. Such algorithms would use samplings of the trajectories instead of full dynamic programming or quantile-based reinforcement learning approaches [7].

References

1. Bauters, K., Liu, W., Godo, L.: Anytime algorithms for solving possibilistic MDPs and hybrid MDPs. In: Gyssens, M., Simari, G. (eds.) FoIKS 2016. LNCS, vol. 9616, pp. 24–41. Springer, Cham (2016)
2. Bellman, R.: A Markovian decision process. J. Math. Mech. **6**, 679–684 (1957)
3. Ben Amor, N., El Khalfi, Z., Fargier, H., Sabbadin, R.: Lexicographic refinements in possibilistic decision trees. In: Proceedings ECAI 2016, pp. 202–208 (2016)
4. Drougard, N., Teichteil-Konigsbuch, F., Farges, J.L., Dubois, D.: Qualitative possibilistic mixed-observable MDPs. In: Proceedings UAI 2013, pp. 192–201 (2013)
5. Dubois, D., Prade, H.: Possibility theory as a basis for qualitative decision theory. In: Proceedings IJCAI 1995, pp. 1925–1930 (1995)
6. Fargier, H., Sabbadin, R.: Qualitative decision under uncertainty: back to expected utility. Artif. Intell. **164**, 245–280 (2005)
7. Gilbert, H., Weng, P.: Quantile reinforcement learning. In: Proceedings JMLR 2016, pp. 1–16 (2016)
8. Gilbert, H., Weng, P., Xu, Y.: Optimizing quantiles in preference-based Markov decision processes. In: Proceedings AAAI 2017, pp. 3569–3575 (2017)
9. Montes, I., Miranda, E., Montes, S.: Decision making with imprecise probabilities and utilities by means of statistical preference and stochastic dominance. Eur. J. Oper. Res. **234**(1), 209–220 (2014)
10. Moulin, H.: Axioms of Cooperative Decision Making. Cambridge University Press, Cambridge (1988)
11. Puterman, M.L.: Markov Decision Processes. Wiley, Hoboken (1994)

12. Sabbadin, R.: Possibilistic Markov decision processes. Eng. Appl. Artif. Intell. **14**, 287–300 (2001)
13. Sabbadin, R., Fargier, H.: Towards qualitative approaches to multi-stage decision making. Int. J. Approximate Reasoning **19**, 441–471 (1998)
14. Sutton, R.S., Barto, A.G.: Introduction to Reinforcement Learning. MIT Press, Cambridge (1998)
15. Szörényi, B., Busa-Fekete, R., Weng, P., Hüllermeier, E.: Qualitative multi-armed bandits: a quantile-based approach. In: Proceedings ICML 2015, pp. 1660–1668 (2015)
16. Weng, P.: Qualitative decision making under possibilistic uncertainty: toward more discriminating criteria. In: Proceedings UAI 2005, pp. 615–622 (2005)
17. Weng, P.: Markov decision processes with ordinal rewards: reference point-based preferences. In: Proceedings ICAPS 2011, pp. 282–289 (2011)
18. Yue, Y., Broder, J., Kleinberg, R., Joachims, T.: The k-armed dueling bandits problem. J. Comput. Syst. Sci. **78**(5), 1538–1556 (2012)

An Angel-Daemon Approach to Assess the Uncertainty in the Power of a Collectivity to Act

Giulia Fragnito, Joaquim Gabarro$^{(\boxtimes)}$, and Maria Serna

ALBCOM CS Department, Universitat Politècnica de Catalunya, Barcelona, Spain
giuliafragnito@yahoo.it, {gabarro,mjserna}@cs.upc.edu

Abstract. We propose the use of the angel-daemon ($a/ɒ$) framework to assess the Coleman's power of a collectivity to act under uncertainty in weighted voting games. In this framework uncertainty profiles describe the potential changes in the weights of a weighted game and fixes the spread of the weights' change. For each uncertainty profile a strategic $a/ɒ$ game can be considered. This game has two selfish players, the angel a and the daemon $ɒ$, a selects its action as to maximize the effect on the measure under consideration while $ɒ$ acts oppositely. Players a and $ɒ$ give a balance between the best and the worst. The $a/ɒ$ games associated to the Coleman's power are zero-sum games and therefore the expected utilities of all the Nash equilibria are the same. In this way we can asses the Coleman's power under uncertainty. Besides introducing the framework for this particular setting we analyse basic properties and make some computational complexity considerations. We provide several examples based in the evolution of the voting rules of the EU Council of Ministers.

Keywords: Weighted voting games · Coleman's power of a collectivity to act · Uncertainty profiles · Strategic games · Zero-sum games · EU Council of Ministers

1 Introduction

The distinction between *risk* and *uncertainty* has become increasingly important since [11] discussed it as we have imperfect knowledge of future events in our ever-changing world. Informally, *risk* can be measured by probabilities. In contrast, *uncertainty* refers to something where we cannot even gather the information required to figure out probabilities. However, in practice both are measured

G. Fragnito—Partially supported by a scholarship from Università degli Studi di Roma "La Sapienza". J. Gabarro and M. Serna are partially supported by funds from the Spanish Ministry for Economy and Competitiveness (MINECO) and the European Union (FEDER funds) under grant COMMAS (ref. TIN2013-46181-C2-1-R), and also by 2014 SGR 1034 (ALBCOM).

A. Antonucci et al. (Eds.): ECSQARU 2017, LNAI 10369, pp. 318–328, 2017.
DOI: 10.1007/978-3-319-61581-3_29

by historical standard deviation of the variable of interest [2,10]. This paper proposes an alternative, applying ideas from game theory and computer science.

The study of web applications is a field where uncertainty becomes unavoidable. The angel-daemon framework [9,18] provides a way to obtain numerical estimates of uncertainty in the execution of a Web service. In such a setting, the uncertainty is captured by an uncertainty profile describing a stressed environment for the execution of Web, or Cloud, applications. Uncertainty profiles provide a description of the perceived uncertain behaviour with respect to possible failing services or execution delays. That is, some sites can potentially misbehave but, a priori, we are uncertain about the specific sites that will do so. The model attempts to balance positive and negative aspects. Considering only positive aspects (minimizing damage) is usually too optimistic. In the opposite side, being pessimistic (maximizing damage) is also not realistic. When facing an intermediate situation with pessimistic and optimistic aspects. The framework considers two agents: *the angel* ($\mathfrak{a}$), dealing with the optimistic side; and the *daemon* ($\mathfrak{d}$), dealing with the pessimistic side. These agents act strategically in an associated angel-daemon zero-sum game. In this context, uncertain behaviours are identified with the Nash equilibria of the angel-daemon game and they are assessed by the value of the game. It is important to emphasize that the results in [9] are useful to analyse uncertain stable (or timeless) environments.

Another field where uncertainty becomes unavoidable is in the analysis of weighted voting systems. More specifically, the study of the power of the players inside a weighted voting game [17] is a well developed topic started by Lloyd Shapley in 1953 [16]. Also, the study of the uncertainty in such a weighted voting systems has been considered under theoretical [7] or practical [12] aspects. Less studied is the behaviour, as a whole, of a weighted voting game under uncertainty. In 1971 James Coleman [6] introduced the formal definition of the *power of a collectivity to act*, denoted here as Act. Practical applications appears yet in [6]. More recently, [1] (Table 4), it has been used to analyse the power of the EU Council of Ministers from 1958 to 2007. Its voting rules continuously change to adapt to EU enlargements. When considering a new voting criteria, there is uncertainty on the effects on the decision-making. Thus, a uncertainty assessment is convenient.

In this paper, we extend the $\mathfrak{a}/\mathfrak{d}$ framework to assess uncertainty in Act. We do that shaping uncertainty profiles $\mathcal{U}$ to deal with the a priori volatility in the weights of a voting game. Given an uncertainty profile $\mathcal{U}$, well adapted to Act, the $\mathfrak{a}/\mathfrak{d}$-game becomes a zero-sum game. Therefore, the value of the game divided by 2^n allows us to assess the Coleman's power to act under uncertainty, denoted as $\mathsf{Act}(\mathcal{U})$. Our objective is to analyse the viability of the model by first analysing some theoretical properties and secondly, using the simplicity of the model of weighted voting games, develop a series of scenarios in which we can test the adequacy of the approach.

2   Preliminaries

A *simple game* Γ is given by a tuple $(N, \mathcal{W})$ where N is a set of n players and $\mathcal{W}$ is a monotonic family of subsets of N [19]. In the context of simple games, the subsets of N are called *coalitions*. N is the *grand coalition* and $S \in \mathcal{W}$ is a *winning coalition*. Any subset of N which is not a winning coalition is called a *losing coalition*. The *Coleman's power of the collectivity to act* [6], is defined as $\mathsf{Act}(\Gamma) = \#\mathcal{W}/2^n$. This measure of collective power can be seen as the probability of the *yes* outcome assuming that all coalitions are equally like.

A *weighted voting game* [17] is a simple game defined by a tuple $\Gamma = \langle q; w_1, \ldots w_n \rangle$, where q is the *quota* and $w_i \in \mathbb{N}^+$ is the weight of player i, for all $1 \leq i \leq n$. The set of players is $N = [n] = \{1, \ldots, n\}$. Let $w(S) = \sum_{i \in S} w_i$ denote the weight of coalition S. The set of winning coalitions is $\mathcal{W}(\Gamma) = \{S \mid w(S) \geq q\}$. Therefore, the set of losing coalitions is $\mathcal{L}(\Gamma) = \{S \mid w(S) < q\}$. Let us consider the case where the weights of all the players are equal, $\Gamma = \langle q; w, \ldots, w \rangle$ with $n > 1$. We denote such a game as $\Gamma(n, q, w)$. Observe that, for $S \subseteq [n]$ it holds $w(S) = w|S|$ and $\mathcal{W}(\Gamma) = \{S \mid w|S| \geq q\}$. In the case of equal weights, to catch the straight majority of the total weight, it is needed to capture a weight strictly greater than $(nw)/2$. In such a case we cannot distinguish between the requirement of having the majority of the players from having the majority of the votes. In fact any quota between $w\lfloor \frac{n}{2} \rfloor + 1$ and $\lfloor \frac{nw}{2} \rfloor + 1$ defines the same game. We consider two basic families of weighted voting games. The *equal weight majority* on n players in which all the players have the same weight w, denote as $\Gamma(n, q, w)$ and the *equal weight majority* on n players game as $\Gamma(n, w) = \Gamma(n, w\lfloor n/2 \rfloor + 1, w)$. Observe that $\mathcal{W}(\Gamma(n, w)) = \mathcal{W}(\Gamma(n, 1))$ and $\mathsf{Act}(\Gamma(n, w)) = \mathsf{Act}(\Gamma(n, 1))$.

In order to study some uncertainty scenarios, we consider the weights of the different states at the Council of Ministers of the EU (now the Council of the EU) along the time [1]. In 1958 the founding members were Germany (DE), France (FR), Italy (IT), The Netherlands (NL), Belgium (BE), and Luxembourg (LU) listed in non-decreasing order of assigned weights. The Council of the EU at 1958 is summarized as $\Gamma_{\mathsf{EC6}} = \langle 12; 4, 4, 4, 2, 2, 1 \rangle$ where there are six players, the first one is DE, the second one FR, being the last one LU. We adopt the succinct notation $\Gamma_{\mathsf{EC6}} = \langle 12; 3{:}4, 2{:}2, 1{:}1 \rangle$ where notation 3:4 means "3 times weight 4". Note that, the total number of votes is $\#\mathcal{W}([6]) = 17$. The quota was defined by a qualified majority rule, QMR, of the 70.6% that is $q = 12 \approx 17 * (70.6/100) = 12.002$. Along the period 1958 to 2014 the number of states increases from the initial 6 to 9, 10, 12, 15, 25 and 27. The voting system changed to cover the new arrivals [1]. For instance, in 1973, three new states arrive; United Kingdom (UK), Denmark (DK) and Ireland (IE) having respectively 10, 3,3 votes. As in Sect. 6 we analyse the changes in Act when the weights of the six founding states are perturbed, we split the weights in two parts. The part corresponding to the founding six and the rest, $\Gamma_{\mathsf{EC9}} = \langle 41; 10, 10, 10, 5, 5, 2 \mid 10, 3, 3 \rangle = \langle 41; 3{:}10, 2{:}5, 1{:}2 \mid 1{:}10, 2{:}3 \rangle$. Following we give a list of the different voting systems taken from [1].

$$\Gamma_{\text{EC6}} = \langle 12; 3{:}4, 2{:}2, 1{:}1 \rangle$$
$$\Gamma_{\text{EC9}} = \langle 41; 3{:}10, 2{:}5, 1{:}2 \mid 1{:}10, 2{:}3 \rangle$$
$$\Gamma_{\text{EC10}} = \langle 45; 3{:}10, 2{:}5, 1{:}2 \mid 1{:}10, 1{:}5, 2{:}3 \rangle$$
$$\Gamma_{\text{EC12}} = \langle 54; 3{:}10, 2{:}5, 1{:}2 \mid 1{:}10, 1{:}8, 2{:}5, 2{:}3 \rangle$$
$$\Gamma_{\text{EC15}} = \langle 62; 3{:}10, 2{:}5, 1{:}2 \mid 1{:}10, 1{:}8, 2{:}5, 2{:}4, 3{:}3 \rangle$$
$$\Gamma_{\text{EC25}_1} = \langle 88; 3{:}10, 2{:}5, 1{:}2 \mid 1{:}10, 2{:}8, 4{:}5, 2{:}4, 8{:}3, 2{:}2 \rangle$$
$$\Gamma_{\text{EC25}_2} = \langle 232; 3{:}29, 1{:}13, 1{:}12, 1{:}4 \mid 1{:}29, 2{:}27, 4{:}12, 2{:}10, 5{:}7, 4{:}4, 1{:}3 \rangle$$
$$\Gamma_{\text{EC27}} = \langle 255; 3{:}29, 1{:}13, 1{:}12, 1{:}4 \mid 1{:}29, 2{:}27, 1{:}14, 4{:}12, 3{:}10, 5{:}7, 4{:}4, 1{:}3 \rangle$$

In all those games players 1 to 6 correspond to the six founding members in the same order as in the 1958 description. The parameters of those games and the Coleman's Power are the following.

Γ	Γ_{EC6}	Γ_{EC9}	Γ_{EC10}	Γ_{EC12}	Γ_{EC15}	Γ_{EC25_1}	Γ_{EU25_2}	Γ_{EC27}
$w([n])$	17	58	63	76	87	124	321	345
% of q	70.6	70.7	71.4	71.1	71.3	71	72.3	73.9
$\#\mathcal{W}(\Gamma)$	14	75	140	402	2549	1170000	1204448	2718774
$\text{Act}(\Gamma)$	0.2187	0.1464	0.1455	0.0981	0.0777	0.0348	0.0358	0.0202

Note that the power to act is quite small and roughly decreases along the time.

3   Uncertainty Profiles and $\mathfrak{a}/\mathfrak{d}$ Games

Let us move on to adapt the definition of *uncertainty profiles* [8,9]. They capture situations in which we have an approximate idea of the extension and nature of the perturbation but we are uncertain over the specific location where it will impact the system. A uncertainty profile is essentially based on three components: the set of players of Γ whose weights may be perturbed; the extent to which the perturbation can be applied; and the number of the components that can suffer the perturbation. As uncertainty may have both positive and negative effects, we have *angelic* and *daemonic* perturbations. The perturbation values are real numbers, so they can be either positive or negative.

Definition 1. A *uncertainty profile* is a tuple $\mathcal{U} = \langle \Gamma, \mathcal{A}, \mathcal{D}, \delta_a, \delta_{\mathfrak{d}}, b_a, b_{\mathfrak{d}} \rangle$ where $\Gamma = \langle q; w_1, \ldots, w_n \rangle$ is a weighted voting game; $\mathcal{A}, \mathcal{D} \subseteq [n]$ are the *sets of players* whose weights may be subject to angelic and daemonic perturbations, respectively; $\delta_a : \mathcal{A} \to \mathbb{Z}$ and $\delta_{\mathfrak{d}} : \mathcal{D} \to \mathbb{Z}$ represent the *strength* of the potential weight's perturbations; $b_a, b_{\mathfrak{d}} \in \mathbb{N}$ are such that $b_a \leq \#\mathcal{A}$ and $b_{\mathfrak{d}} \leq \#\mathcal{D}$ and they represent the *spread* of the angelic and daemonic perturbations.

The exerted perturbation follows joint actions (a, d), for $a \subseteq \mathcal{A}$, $d \subseteq \mathcal{D}$ with $\#a = b_a$ and $\#d = b_\mathfrak{d}$. The effects of a joint action is a perturbed game $\Gamma[a, d] = \langle q; w'_1, \ldots, w'_n \rangle$ defined as

$$w'_i = w_i + x_a(i)\delta_a(i) + x_d(i)\delta_\mathfrak{d}(i),$$

where $x_a(i) = 1$ if $i \in a$; 0 otherwise, and $x_d(i) = 1$ if $i \in d$; 0 otherwise. To ensure $w'_i \in \mathbb{N}^+$ we require that $|\delta_a(i)|, |\delta_\mathfrak{d}(i)|, |\delta_a(i) + \delta_\mathfrak{d}(i)| < w_i$. We consider also another way to define the perturbed game that we call *proportional quota* in contraposition to the *fixed quota* presented before. We define the perturbed game as $\Gamma_p[a, d] = \langle q', w'_1, \ldots, w'_n \rangle$ where $q' = \frac{q}{w([n])} w'([n])$. Observe that the proportion quota versus total weight is preserved in $\Gamma_p[a, d]$.

We are interested to know how the angelic and daemonic perturbations of a permissible joint action affect the number of winning coalitions through the perturbed games.

Example 1. Let $\mathcal{U} = \langle \Gamma_{\mathsf{EC6}}, \{2, 3\}, \{2, 3\}, \delta_a, \delta_\mathfrak{d}, 1, 1 \rangle$ where: The game is $\Gamma_{\mathsf{EC6}} = \langle 12; 4, 4, 4, 2, 2, 1 \rangle$, $\mathcal{A} = \mathcal{D} = \{2, 3\} = \{\mathsf{FR}, \mathsf{IT}\}$. We set the angelic perturbations so that they increase the number of votes, i.e., $\delta_a(2) = \delta_a(3) = 1$. By contrary, we set the daemonic perturbations so that such number is more strictly decreased, i.e., $\delta_\mathfrak{d}(2) = \delta_\mathfrak{d}(3) = -2$. Players' perturbed weights vary according to joint actions and we have four joint actions. The perturbed games in the fixed quota model are:

$$\Gamma_{\mathsf{EC6}}[\{\mathsf{IT}\}, \{\mathsf{FR}\}] = \langle 12; 4, 2, 5, 2, 2, 1 \rangle, \qquad \Gamma_{\mathsf{EC6}}[\{\mathsf{FR}\}, \{\mathsf{IT}\}] = \langle 12; 4, 5, 2, 2, 2, 1 \rangle,$$
$$\Gamma_{\mathsf{EC6}}[\{\mathsf{FR}\}, \{\mathsf{FR}\}] = \langle 12; 4, 3, 4, 2, 2, 1 \rangle, \qquad \Gamma_{\mathsf{EC6}}[\{\mathsf{IT}\}, \{\mathsf{IT}\}] = \langle 12; 4, 4, 3, 2, 2, 1 \rangle.$$

In the proportional quote model the total weight is 16 and $q' = \frac{12}{17} 16$.   □

Assessing uncertainty in a weighted voting game consists in evaluating the effects of a uncertainty profile. Such analysis is done by means of an associated strategic *zero-sum* $a/\mathfrak{d}$-*game*. Two agents are considered: the *angel* a, exerting the angelic perturbations of the uncertainty profile; and the *daemon* $\mathfrak{d}$, exerting the daemonic perturbations. Precisely, a attempts to maximize Act and $\mathfrak{d}$ acts in order to minimize it. As we are interested in the study of $\mathsf{Act} = \#\mathcal{W}/2^n$, we take as utility of the $a/\mathfrak{d}$-game $u_a = \#\mathcal{W}$. To avoid confusions, we use Γ to denote the underlying weighted voting game and G to denote zero-sum strategic $a/\mathfrak{d}$-game. For sake of simplicity we provide the definitions for the fixed quota model, similar definitions can be settled for the proportional quota model.

Definition 2. Given $\mathcal{U} = \langle \Gamma, \mathcal{A}, \mathcal{D}, \delta_a, \delta_\mathfrak{d}, b_a, b_\mathfrak{d} \rangle$, the associated *angel/daemon* (or $a/\mathfrak{d}$) game is $G(\mathcal{U}) = \langle \{a, \mathfrak{d}\}, A_a, A_\mathfrak{d}, u_a, u_\mathfrak{d} \rangle$. Game $G(\mathcal{U})$ has two players: the angel a and the daemon $\mathfrak{d}$. The player's actions are $A_a = \{a \subseteq \mathcal{A} \mid \#a = b_a\}$ and $A_\mathfrak{d} = \{d \subseteq \mathcal{D} \mid \#d = b_\mathfrak{d}\}$. For $(a, d) \in A_a \times A_\mathfrak{d}$ utilities are $u_a(a, d) = \#\mathcal{W}(\Gamma[a, d])$ and $u_\mathfrak{d}(a, d) = -u_a(a, d)$.

Notice that, in an $a/\mathfrak{d}$ game the set of strategy profiles is $A_a \times A_\mathfrak{d}$. a and $\mathfrak{d}$ choices of actions can be done probabilistically. Mixed strategies for

$\mathfrak{a}$ and $\mathfrak{d}$ are probability distributions $\alpha : A_{\mathfrak{a}} \to [0,1]$ and $\beta : A_{\mathfrak{d}} \to [0,1]$ respectively. A *mixed strategy* is a tuple (α, β) such that $u_{\mathfrak{p}}(\alpha, \beta) = \sum_{(a,d) \in A_{\mathfrak{a}} \times A_{\mathfrak{d}}} \alpha(a) u_{\mathfrak{p}}(a,d) \beta(d)$ for $\mathfrak{p} \in \{\mathfrak{a}, \mathfrak{d}\}$. Given $u_{\mathfrak{a}}(a,d) = \#\mathcal{W}(\Gamma[a,d])$ it makes sense to extend $\#\mathcal{W}$ to mixed strategies defining $\#\mathcal{W}(\alpha, \beta) = u_{\mathfrak{a}}(\alpha, \beta) = \sum_{(a,d) \in A_{\mathfrak{a}} \times A_{\mathfrak{d}}} \alpha(a) (\#\mathcal{W}(\Gamma[a,d])) \beta(d)$.

Let $\Delta_{\mathfrak{a}}$ and $\Delta_{\mathfrak{d}}$ denote the set of mixed strategies for $\mathfrak{a}$ and $\mathfrak{d}$, respectively. A pure strategy profile (a,d) is a special case of mixed strategy profile (α, β) in which $\alpha(a) = 1$ and $\beta(d) = 1$. A mixed strategy profile (α, β) is a *Nash equilibrium* if for any $\alpha' \in \Delta_{\mathfrak{a}}$ it holds $u_{\mathfrak{a}}(\alpha, \beta) \geq u_{\mathfrak{a}}(\alpha', \beta)$ and for any $\beta' \in \Delta_{\mathfrak{d}}$ it holds $u_{\mathfrak{d}}(\alpha, \beta) \geq u_{\mathfrak{d}}(\alpha, \beta')$. A *pure Nash equilibrium*, PNE, is a Nash equilibrium (a,d) where a and d are pure strategies.

It is well known that all Nash equilibria of a zero-sum game G have the same value $\nu(G)$ corresponding to the utility of the row player [13]. For an $\mathfrak{a}/\mathfrak{d}$ game $G(\mathcal{U})$ we have: $\nu(G(\mathcal{U})) = \max_{\alpha \in \Delta_{\mathfrak{a}}} \min_{\beta \in \Delta_{\mathfrak{d}}} \#\mathcal{W}(\alpha, \beta) = \min_{\beta \in \Delta_{\mathfrak{d}}} \max_{\alpha \in \Delta_{\mathfrak{a}}} \#\mathcal{W}(\alpha, \beta)$. Considering $\mathfrak{a}/\mathfrak{d}$ games we extend the definition of the Coleman's power to act to uncertainty profiles as follows.

Definition 3. Given Γ with n players, $\mathcal{U} = \langle \Gamma, \mathcal{A}, \mathcal{D}, \delta_{\mathfrak{a}}, \delta_{\mathfrak{d}}, b_{\mathfrak{a}}, b_{\mathfrak{d}} \rangle$ and $u_{\mathfrak{a}}(a,d) = \#\mathcal{W}(\Gamma[a,d])$. We define $\#\mathcal{W}(\mathcal{U}) = \nu(G(\mathcal{U}))$ being $G(\mathcal{U})$ the corresponding $\mathfrak{a}/\mathfrak{d}$-game. The *Coleman's power to act* of $\mathcal{U}$ is $\mathsf{Act}(\mathcal{U}) = \#\mathcal{W}(\mathcal{U})/2^n$.

When (α, β) is a Nash equilibrium of $G(\mathcal{U})$ the utility of the angel player verifies $u_{\mathfrak{a}}(\alpha, \beta) = \#\mathcal{W}(\mathcal{U}) = \sum_{(a,d) \in A_{\mathfrak{a}} \times A_{\mathfrak{d}}} \alpha(a)\beta(d)\#\mathcal{W}(\Gamma[a,d])$ and therefore, dividing by 2^n we get $\mathsf{Act}(\mathcal{U}) = \sum_{(a,d) \in A_{\mathfrak{a}} \times A_{\mathfrak{d}}} \alpha(a)\beta(d)\mathsf{Act}(\Gamma[a,d])$.

Example 2. We continue with $\mathcal{U}$ given in Example 1. As $\mathcal{A} - \mathcal{D} = \{\text{FR}, \text{IT}\} - \{2,3\}$ and $b_{\mathfrak{a}} = b_{\mathfrak{d}} = 1$ we have $A_{\mathfrak{a}} = A_{\mathfrak{d}} = \{\{\text{FR}\}, \{\text{IT}\}\}$. Then, for example, $u_{\mathfrak{a}}(\{\text{FR}\}, \{\text{FR}\}) = \#\mathcal{W}(\Gamma_{\text{EC6}}[\{\text{FR}\}, \{\text{FR}\}] = \#\mathcal{W}(\langle 12; 4, 3, 4, 2, 2, 1\rangle)$. Observe that a reordering of the weights does not change the number of winning coalitions, i.e., $\#\mathcal{W}(\langle 12; 4, 3, 4, 2, 2, 1\rangle) = \#\mathcal{W}(\langle 12; 2{:}4, 1{:}3, 2{:}2, 1{:}1\rangle)$. The $\mathfrak{a}/\mathfrak{d}$-game is described by the following utility matrix for $\mathfrak{a}$.

	{FR}	{IT}
{FR}	$\#\mathcal{W}(\langle 12; 2{:}4, 1{:}3, 2{:}2, 1{:}1\rangle) = 11$	$\#\mathcal{W}(\langle 12; 1{:}5, 1{:}4, 3{:}2, 1{:}1\rangle) = 12$
{IT}	$\#\mathcal{W}(\langle 12; 1{:}5, 1{:}4, 3{:}2, 1{:}1\rangle) = 12$	$\#\mathcal{W}(\langle 12; 2{:}4, 1{:}3, 2{:}2, 1{:}1\rangle) = 11$

There is only one Nash equilibrium with $\alpha(FR) = \beta(FR) = 1/2$ and $\alpha(IT) = \beta(IT) = 1/2$. Then $\#\mathcal{W}(\mathcal{U}) = 23/2$ and $\mathsf{Act}(\mathcal{U}) = 23/2^7 \approx 0.1796$ □

4   Majority Games with Equal Weights

We consider the case in which all the players have equal weight, $\Gamma(n, q, w)$ or $\Gamma(n, w)$. We start by perturbing two players with equal and opposite strengths. The *minimal egalitarian* profile is defined as $\mathcal{ME}(n, w, \delta, \mathcal{A}, \mathcal{D}) = \langle \Gamma(n,w), \mathcal{A}, \mathcal{D}, \delta_{\mathfrak{a}}, \delta_{\mathfrak{d}}, 1, 1 \rangle$, where $\delta_{\mathfrak{a}}(i) = \delta$, for $i \in \mathcal{A}$, and $\delta_{\mathfrak{d}}(i) = -\delta$, for $i \in \mathcal{D}$.

Lemma 1. *Let $n > 2$, $w > 1$, $0 < \delta < w$, and $\Gamma = \Gamma(n, w)$. Let $\mathcal{A}, \mathcal{D} \subseteq [n]$ and $\mathcal{U} = \mathcal{ME}(n, w, \delta, \mathcal{A}, \mathcal{D})$. Then, we have $\#\mathcal{W}(\Gamma[\{i\}, \{j\}]) = \#\mathcal{W}(\Gamma(w, n))$ if $j = i$, otherwise $\#\mathcal{W}(\Gamma[\{i\}, \{j\}]) = \#\mathcal{W}(\Gamma(n, w)) + \binom{n-2}{\lfloor n/2 \rfloor - 1}$.*

Proof. Observe that, when both $\mathfrak{a}$ and $\mathfrak{d}$ select a common player $\{i\}$, the two perturbations cancel and the so obtained game is the initial one. When $(a, d) = (\{i\}, \{j\})$ with $i \neq j$ the perturbed game is such that $w_i' = w + \delta$, $w_j' = w - \delta$ and the remaining players have weight w. Thus, the set of winning coalitions under $(\{i\}, \{j\})$, denoted by $\mathcal{W}(\Gamma[\{i\}, \{j\}])$, is given by the disjoint union $\{S \mid \#S \geq \lfloor n/2 \rfloor + 1\} \cup \{\{i\} \cup S \mid S \subseteq N\setminus\{i, j\}, \#S = \lfloor n/2 \rfloor - 1\}$. The number of winning coalitions follows by straightforward combinatorial arguments. □

Example 3. Take $n = 3$, $w = 2$ and $\delta = 1$. In such a case $w\lfloor n/2 \rfloor + 1 = 3$, $\Gamma(3, 2) = \langle 3; 2, 2, 2 \rangle$ and $\mathcal{W}(\Gamma(3, 2)) = \{\{1, 2\}, \{1, 3\}, \{2, 3\}, \{1, 2, 3\}\}$. Under the uncertainty profile $\mathcal{U} = \mathcal{ME}(n, w, \delta, \mathcal{A}, \mathcal{D})$, where $\mathcal{A} = \{1, 2\}$, $\mathcal{D} = \{1\}$, the winning coalitions are $\mathcal{W}(\Gamma[\{1\}, \{1\}]) = \mathcal{W}(\Gamma(3, 2))$ and $\mathcal{W}(\Gamma[\{2\}, \{1\}]) = \mathcal{W}(\Gamma(3, 2)) \cup \{\{2\}\}$. □

Theorem 1. *Let $n > 2$, $w > 1$, $0 < \delta < w$, and $\Gamma = \Gamma(n, w)$. Let $\mathcal{A}, \mathcal{D} \subseteq [n]$ and $\mathcal{U} = \mathcal{ME}(n, w, \delta, \mathcal{A}, \mathcal{D})$. Assume $\#\mathcal{A} > 0$, $\#\mathcal{D} > 0$. Then, if $\mathcal{A} = \mathcal{D}$, $\mathrm{PNE}(\Gamma(\mathcal{U})) = \emptyset$, if $\mathcal{A} \neq \mathcal{D}$ and $\mathcal{A} \subseteq \mathcal{D}$, $\mathrm{PNE}(G(\mathcal{U})) = \{(\{i\}, \{i\}) \mid i \in \mathcal{A}\}$ and $\mathsf{Act}(\mathcal{U}) = \mathsf{Act}(\Gamma(n, w))$, otherwise $\mathrm{PNE}(G(\mathcal{U})) = \{(\{i\}, \{j\}) \mid i \in \mathcal{A}\setminus\mathcal{D}, j \in \mathcal{D}\}$ and $\mathsf{Act}(\mathcal{U}) = \mathsf{Act}(\Gamma(n, w)) + \frac{1}{2^n}\binom{n-2}{\lfloor n/2 \rfloor - 1}$.*

It is easy to see that all the previous results also hold in the proportional quota model. We conclude with an example in which the $\mathfrak{a}/\mathfrak{d}$ game has no PNE but exactly one NE.

Example 4. Let $n > 2$ and $w > 1$. Let $\Gamma = \Gamma(n, w)$ and $\mathcal{U} = \mathcal{ME}(n, w, 1, \{1, 2\}, \{1, 2\})$, by Theorem 1 we know that $G(\mathcal{U})$ has no PNE. As $b_{\mathfrak{a}} = b_{\mathfrak{d}} = 1$ we have $A_{\mathfrak{a}} = A_{\mathfrak{d}} = \{\{1\}, \{2\}\}$ and $\mathfrak{a}$'s payoff matrix in the $\mathfrak{a}/\mathfrak{d}$ game is the following.

	$\{1\}$	$\{2\}$
$\{1\}$	$\#\mathcal{W}(\Gamma(n, w))$	$\#\mathcal{W}(\Gamma(n, w)) + \frac{1}{2^n}\binom{n-2}{\lfloor n/2 \rfloor - 1}$
$\{2\}$	$\#\mathcal{W}(\Gamma(n, w)) + \frac{1}{2^n}\binom{n-2}{\lfloor n/2 \rfloor - 1}$	$\#\mathcal{W}(\Gamma(n, w))$

A straightforward computation shows that the unique (mixed) Nash equilibrium is $((1/2, 1/2), (1/2, 1/2))$. Therefore, $\mathsf{Act}(\mathcal{U}) = \mathsf{Act}(\Gamma(n, w)) + \frac{1}{2^{n-1}}\binom{n-2}{\lfloor n/2 \rfloor - 1}$. □

5   Computational Complexity Considerations

We refer the reader to [15] for the definition of complexity classes and to [3–5] for further computational results in the context of weighted voting games.

Observe that the number of losing coalitions in a weighted voting game corresponds to the number of solutions of a 0–1 Knapsack problem [15]. The counting version of 0–1 Knapsack is known to be a #P-complete problem, therefore computing the number of winning coalitions of a given weighted voting game is also #P-complete. The hardness result for Knapsack like problems can be be extended to decisional problems in our context.

Theorem 2. *Computing $\#\mathcal{W}(\Gamma)$, given Γ, is #P-complete. The following problems are NP-hard: deciding whether $\mathsf{Act}(\Gamma) \neq \mathsf{Act}(\Gamma')$, given Γ and Γ'; given $\mathcal{U}$ and a joint action $(a, d) \in A_a \times A_\eth$, deciding if d is a best response to a in $G(\mathcal{U})$; and given $\mathcal{U}$ associated to Γ, deciding whether $\mathsf{Act}(\mathcal{U}) \neq \mathsf{Act}(\Gamma)$.*

A slight modification in those reductions allows us to get similar complexity limits for the proportional quota model. The preceding results rely on the hardness of 0–1 Knapsack like problems. Although those problems are NP-hard, they do admit pseudo-polynomial time algorithms. The problems become solvable in time polynomial in n, provided the weights are polynomial in n. So, we can compute $\#\mathcal{W}(\Gamma)$ in pseudo-polynomial time using Dynamic Programming on an array of size $w([n])$ adapting the traditional 0–1 Knapsack algorithm. However, even when the weights are polynomial in n, the number of strategies in $G(\mathcal{U})$ can be exponential, in such a case the complexity of computing $\nu(G(\mathcal{U}))$ is still open.

6   A Study Based on the Council of the EU

When we face a uncertainty profile $\mathcal{U}$ in which, all the weights are polynomial in n and where the spread is constant, it is possible to compute efficiently a complete description of $G(\mathcal{U})$. Afterwards we compute $\nu(G(\mathcal{U}))$ solving a Linear Programming problem [14]. We use this approach to assess experimentally properties of Act in small games. We have computed $\mathsf{Act}(\mathcal{U})$ for several uncertainty profiles when the game is taken from the voting systems of the Council of the EU and uncertainty is limited to the weights of the six founder members. Our setting and results are summarized in Fig. 1.

For our first case study we selected the weights of the funding states in Γ_{EC6}, Γ_{EC27} and Γ_{EC12}. The differences in weights provide the perturbation functions δ_{12-27} and δ_{12-6} given in Fig. 1(a). From the point of view of Γ_{EC12}, a attempts to move forward the weights to those in Γ_{EC27} while $\eth$ wishes to move them back to Γ_{EC6}. Our aim is to evaluate the monotonicity of the assessment under severe weight perturbations when a and $\eth$ can act in the same set of nations in opposite directions. We assessed uncertainty profiles of the form $\mathcal{U}_{12}(b_a, b_\eth) = \langle \Gamma_{\mathsf{EC12}}, [6], [6], \delta_{12-27}, \delta_{12-6}, b_a, b_\eth \rangle$, for all possible values of $(b_a, b_\eth)$. In Fig. 1(b) we provide, for each combination of $(b_a, b_\eth)$, $\mathsf{Act}(\mathcal{U}_{12}(b_a, b_\eth))$ in the fixed quota model (left) and in the proportional quota model (right). As one can expect, the variability of Act is higher when the quota is fixed. In the fixed quota model, an increase of power by a ($\eth$) results in an increase (decrease) of Act. The quota is unchanged while some angelic players' weights are significantly increased.

This phenomena does not appear in the variable quota model as it can be seen, for example, when $b_a = 2$.

For our second case of study we selected some variations of the previous uncertainty profiles keeping a and $\eth$ with the same spread. We assessed uncertainty profiles of the form $\mathcal{U}_{12SD}(b) = \langle \Gamma_{EC12}, \{0, 1, 2\}, \{3, 4, 5\}, \delta_{12-27}, \delta_{12-6}, b, b \rangle$ and $\mathcal{U}_{12SI}(b) = \langle \Gamma_{EC12}, \{0, 1, 3\}, \{3, 4, 5\}, \delta_{12-27}, \delta_{12-6}, b, b \rangle$ and their reversed versions $\mathcal{U}_{12SDr}(b)$ and $\mathcal{U}_{12SIr}(b)$. With *reversed*, we mean that $\delta_a = \delta_{12-6}$ and $\delta_{\eth} = \delta_{12-27}$. In such a scenario, a tries to reduce the damage by decreasing Act as less as possible, $\eth$ acts in opposite way. The resulting assessment values are summarized in Fig. 1(c) for the two models. As one can expect, in the reversed case Act increases. Observe that by reversing the roles of a and $\eth$, the perturbed

Γ	DE	FR	IT	NL	BE	LU
EC6	4	4	4	2	2	1
EC12	10	10	10	5	5	2
EC27	29	29	29	13	12	4

δ	DE	FR	IT	NL	BE	LU
δ_{12-27}	19	19	19	8	7	2
δ_{12-6}	-6	-6	-6	-3	-3	-1
δ_1	1	1	1	1	1	1
δ_{-1}	-1	-1	-1	-1	-1	-1

(a) Some weighted voting games and some perturbation functions

FQ	0	1	2	3	4	5	6
0	0,098	0,049	0,016	0,002	0,000	0,000	0,000
1	0,405	0,339	0,262	0,177	0,145	0,115	0,105
2	0,573	0,513	0,449	0,385	0,348	0,319	0,311
3	0,666	0,636	0,603	0,566	0,537	0,512	0,509
4	0,722	0,697	0,668	0,637	0,608	0,578	0,575
5	0,762	0,737	0,711	0,682	0,658	0,637	0,631
6	0,773	0,749	0,723	0,695	0,679	0,654	0,645

PQ	0	1	2	3	4	5	6
0	0,098	0,098	0,095	0,088	0,092	0,101	0,104
1	0,151	0,156	0,158	0,158	0,163	0,172	0,183
2	0,182	0,175	0,164	0,155	0,160	0,167	0,172
3	0,170	0,163	0,165	0,156	0,161	0,166	0,172
4	0,155	0,152	0,151	0,141	0,143	0,150	0,156
5	0,141	0,142	0,136	0,134	0,131	0,133	0,141
6	0,134	0,131	0,125	0,124	0,118	0,127	0,131

(b) Assessment of profiles with $\delta_a = \delta_{12-27}$ and $\delta_{\eth} = \delta_{12-6}$, Γ_{EC12} for all b_a, $b_{\eth}$

FQ	1	2	3
ec12SD	0,381	0,516	0,602
ec12SDr	0,390	0,528	0,602
ec12SI	0,381	0,516	0,602
ec12SIr	0,198	0,432	0,602

PQ	1	2	3
ec12SD	0,163	0,180	0,148
ec12SDr	0,164	0,188	0,148
ec12SI	0,163	0,180	0,173
ec12SIr	0,111	0,167	0,173

(c) Assessing profiles with reversed roles

b	0	1	2	3
ec6	0,22	0,23	0,24	0,22
ec9	0,27	0,26	0,25	0,25
ec10	0,39	0,41	0,41	0,41
ec12	0,10	0,10	0,10	0,10

b	0	1	2	3
ec15	0,04	0,04	0,04	0,04
ec25-1	0,03	0,03	0,03	0,03
ec25-2	0,04	0,04	0,04	0,04
ec27	0,02	0,02	0,02	0,02

(d) Assessing minimal egalitarian profiles on the Council of the EU.

Fig. 1. Experimental results

games are the same but the Nash equilibria are different. As we can see in the tables, the $a/\eth$ approach provides as expected different assessments when the objectives change.

In our third case study we analyse the extent to which a fixed uncertainty model with unit perturbations affects the assessment of Actin the Council of the EU. We assessed uncertainty profiles $\mathcal{U}_x(b) = \langle \Gamma_{\text{ECx}}, \{0,1,2\}, \{3,4,5\}, \delta_1, \delta_{-1}, b, b \rangle$. The results are given in Fig. 1(d). Under such profiles the total weights of the players is preserved, thus the fixed and proportional models are equivalent. As one can expect, under so small perturbations Act slightly increases as the spread increases and does not present big variations.

7 Conclusions and Open Problems

We have extended the $a/\eth$ framework to tackle with the uncertainty in Coleman's power to act issued from the imprecisions on weights in voting games. We have provided an extension of the power measure to uncertainty profiles. We developed several properties and examples. Finally, we conducted an experimental study showing the monotonicity of the Power to act of the Council of the EU under uncertainty. Act changes proportionally to the strength and the spread of the perturbation. As expected, such variations are more obvious in the fixed model than in the proportional one.

We are working towards extending the framework to voting systems in which fluctuations of weights are due to a-priori uncertainty in the level of abstention. Other related topics merits to be studied. In [6] two other measures were considered. For a player i of a weighted voting game Γ. The *power to initiate action*, $\text{Initiate}_i(\Gamma)$, defined as $\#\{S \in \mathcal{L}(\Gamma) \mid S \cup \{i\} \in \mathcal{W}(\Gamma)\}/\#\mathcal{L}(\Gamma)$ which gives the likelihood that i turns a loosing coalition into a winning one. The *power to prevent action*, $\text{Prevent}_i(\Gamma)$, given by $\#\{S \in \mathcal{W}(\Gamma) \mid S\backslash\{i\} \in \mathcal{L}(\Gamma)\}/\#\mathcal{W}(\Gamma)$ which is the fraction of winning coalitions for which i is critical. It remains open how to address individual measures like the power to initiate or the power to prevent in an uncertain environment. This extension could allow a comparison with the probabilistic approach undertaken in [1]. In the same lines, uncertainty profiles well adapted to the Shapley values, will allow a comparison with probabilistic approaches like the one given in [7] and to much more practical approaches like those given in [1,12].

References

1. Antonakakis, N., Badinger, H., Reuter, W.H.: From Rome to Lisbon and beyond: Member states'power, efficiency, and proportionality in the EU Council of Ministers. Working Paper 175, Vienna University of Economics and Business, Department of Economics (2014)
2. Arratia, A.: Computational finance. An introductory course with R. Atlantis Press, Paris (2014)

3. Aziz, H.: Algorithmic and complexity aspects of simple coalitional games. Ph.D. thesis, University of Warwick (2009)
4. Chalkiadakis, G., Elkind, E., Wooldridge, M.: Computational aspects of cooperative game theory. Morgan & Claypool (2011)
5. Chalkiadakis, G., Wooldridge, M.: Weighted voting games. In: Brandt, F., Conitzer, V., Endriss, U., Lang, J., Procaccia, A. (eds.) Handbook of Computational Social Choice, pp. 377–394. Cambridge University Press, New York (2016)
6. Coleman, J.: Control of collectivities and the power of a collectivity to act. In: Lieberman, B. (ed.) Social choice, pp. 269–300. Gordon and Breach, reedited in Routledge Revivals, 2011(1971)
7. Fatima, S.S., Wooldridge, M., Jennings, N.R.: An analysis of the Shapley value and its uncertainty for the voting game. In: Gleizes, M.P., Kaminka, G.A., Nowé, A., Ossowski, S., Tuyls, K., Verbeeck, K. (eds.) EUMAS 2005, Belgium, December 7–8, 2005. pp. 480–481. Koninklijke Vlaamse Academie van Belie voor Wetenschappen en Kunsten (2005)
8. Gabarro, J., Serna, M.: Uncertainty in basic short-term macroeconomic models with angel-daemon games. Int. J. Data Anal. Tech. Strat. (2017, in press)
9. Gabarro, J., Serna, M., Stewart, A.: Analysing web-orchestrations under stress using uncertainty profiles. Comput. J. **57**(11), 1591–1615 (2014)
10. Hull, J.: Risk Management and Financial Institutions, 3rd edn. Pearson, Hoboken (2012)
11. Knight, F.: Risk, Uncertainty and Profit. Houghton Mifflin, Boston (1921)
12. Mielcova, E.: The uncertainty in voting power: the case of the Czech parliament 1996–2004. AUCO Czech Econo. Rev. **4**(2), 201–221 (2010)
13. von Neumann, J., Morgenstern, O.: Theory of Games and Economic Behavior, 60th Anniversary, Commemorative edn. Princeton University Press, Princeton and Oxford (1953)
14. Osborne, M.: An Introductions to Game Theory. Oxford University Press, New York and Oxford (2004)
15. Papadimitriou, C.: Computational Complexity. Addison-Wesley, Reading (1994)
16. Shapley, L.: A value for n-person games. In: Kuhn, H., Tucker, A. (eds.) Contributions to the Theory of Games, vol. II, pp. 307–317. Princeton University Press, Princeton (1953). Included in Classics in Game Theory
17. Shapley, L.: Simple games: an outline of the descriptive theory. Syst. Res. Behav. Sci. **7**(1), 59–66 (1962)
18. Stewart, A., Gabarro, J., Keenan, A.: Uncertainty in the cloud: an angel-daemon approach to modelling performance. In: Destercke, S., Denoeux, T. (eds.) ECSQARU 2015. LNCS (LNAI), vol. 9161, pp. 141–150. Springer, Cham (2015). doi:10.1007/978-3-319-20807-7_13
19. Taylor, A., Zwicker, W.: Simple Games: Desirability Relations, Trading, Pseudoweightings. Princeton University Press, Princeton (1999)

Decision Theory Meets Linear Optimization Beyond Computation

Christoph Jansen$^{(\boxtimes)}$, Thomas Augustin, and Georg Schollmeyer

Department of Statistics, LMU, Munich, Germany
{christoph.jansen,augustin,georg.schollmeyer}@stat.uni-muenchen.de

Abstract. The paper is concerned with decision making under complex uncertainty. We consider the Hodges and Lehmann-criterion relying on uncertain classical probabilities and Walley's maximality relying on imprecise probabilities. We present linear programming based approaches for computing optimal acts as well as for determining least favorable prior distributions in finite decision settings. Further, we apply results from duality theory of linear programming in order to provide theoretical insights into certain characteristics of these optimal solutions. Particularly, we characterize conditions under which randomization pays out when defining optimality in terms of the Gamma-Maximin criterion and investigate how these conditions relate to least favorable priors.

Keywords: Linear programming · Decision making · Least favorable prior · Duality · Maximality · Imprecise probabilities · Gamma-maximin · Hodges & Lehmann

1 Introduction

Many problems arising in modern sciences, e.g. estimation and hypothesis testing in statistics or modeling an agent's preferences in economics, can be embedded in the formal framework of *decision theory under uncertainty*. However, as the specification of a *precise* (i.e. classical) probability measure on the space of uncertain states often turns out to be too restrictive from an applicational point of view, decision theory using *imprecise probabilities* (for a survey see, e.g., [12]) has become a more and more attractive modeling tool recently. For determining optimal decisions with respect to the complex decision criteria particularly (but not exclusively) arising in the context of the theory of imprecise probabilities, *linear programming theory* (see, e.g., [15]) often turns out to be well-suited: By embedding decision problems into this general optimization framework, one can draw on the whole theoretical toolbox of this well-investigated mathematical discipline. Particularly, this allows for a computational treatment of complex decision making problems in standard software (e.g. MATLAB or for statisticians R) and, therefore, helps in order to make the abstract theory applicable for practitioners. Accordingly, there exists plenty of literature on linear optimization driven algorithms for facing complex decision problems. Examples include [6,13]. A survey is given in [5].

© Springer International Publishing AG 2017
A. Antonucci et al. (Eds.): ECSQARU 2017, LNAI 10369, pp. 329–339, 2017.
DOI: 10.1007/978-3-319-61581-3_30

However, quite similar to characterizations of imprecise probabilities and natural extensions in [17, Chap. 4] and [14], the opportunities of using linear programming in decision theory are by far not exhausted by producing powerful algorithms (see [18, p. 402]). Instead, applying basic results on duality from linear programming theory (such as, e.g., the *complementary slackness* property, see, e.g., [15, Sect. 5.5]) can often provide theoretical insights on both the connection between different decision criteria and the specific properties shared by all optimal solutions with respect to a certain criterion.

The paper is structured as follows: In Sect. 2, we recall the classical model of finite decision theory as well as the extended version of the model allowing for randomized acts. In Sect. 3, we give a linear program for determining optimal randomized acts with respect to a decision criterion of Hodges and Lehmann which tries to cope with uncertain prior probabilistic information and investigate the corresponding dual programming problem. In Sect. 4, we consider the case of decision making under imprecise probabilistic information. Particularly, we present an algorithm for checking maximality of pure acts in one single linear program in Sect. 4.1 and use duality theory for deriving connections between least favorable prior distributions and the Gamma-Maximin criterion in Sect. 4.2. Finally, Sect. 5 is preserved for concluding remarks.

2   The Basic Model

Throughout the paper, we consider the standard model of *finite* decision theory: An *agent* (or *decision maker*) has to decide which *act* a_i to pick from a finite set $\mathbb{A} = \{a_1, \ldots, a_n\}$. However, the *utility* of the chosen act depends on which *state of nature* from a finite set $\Theta = \{\theta_1, \ldots, \theta_m\}$ corresponds to the true description of reality. Specifically, we assume that the utility of every pair $(a, \theta) \in \mathbb{A} \times \Theta$ can be evaluated by a *known* real-valued *cardinal utility function* $u : \mathbb{A} \times \Theta \to \mathbb{R}$. For simplicity, we will often use the notation $u_{ij} := u(a_i, \theta_j)$, where $i = 1, \ldots, n$ and $j = 1, \ldots, m$. The structure of the basic model and a running example repeatedly considered throughout the paper are visualized in Table 1. For every act $a \in \mathbb{A}$, the utility function u is naturally associated with a random variable $u_a : (\Theta, 2^\Theta) \to \mathbb{R}$ defined by $u_a(\theta) := u(a, \theta)$ for all $\theta \in \Theta$. Similarly, for every $\theta \in \Theta$, we can define a random variable $u^\theta : (\mathbb{A}, 2^\mathbb{A}) \to \mathbb{R}$ by setting $u^\theta(a) := u(a, \theta)$ for all $a \in \mathbb{A}$.

Depending on the context, we also allow for *randomized acts*, i.e. classical probability measures λ on $(\mathbb{A}, 2^\mathbb{A})$. Choosing λ is then interpreted as leaving your final decision to a random experiment which yields act a_i with probability $\lambda(\{a_i\})$. We denote the set of randomized acts on $(\mathbb{A}, 2^\mathbb{A})$ by $G(\mathbb{A})$.

The utility function u on $\mathbb{A} \times \Theta$ is then extended to a utility function $G(u)$ on $G(\mathbb{A}) \times \Theta$ by assigning each pair (λ, θ) the expectation of the random variable u^θ under the measure λ, i.e. $G(u)(\lambda, \theta) := \mathbb{E}_\lambda[u^\theta]$, which corresponds to the expectation of utility that choosing the randomized act λ will lead to, given θ is the true description of reality. Every *pure act* $a \in \mathbb{A}$ then can uniquely be identified with the *Dirac-measure* $\delta_a \in G(\mathbb{A})$, and we have $u(a, \theta) = G(u)(\delta_a, \theta)$

Table 1. Basic model (left) and running example with acts $\mathbb{A} = \{a_1, a_2, a_3\}$, states $\Theta = \{\theta_1, \ldots, \theta_4\}$ (right) and the credal set $\mathcal{M} := \{\pi : 0.3 \leqslant \pi(\{\theta_2\}) + \pi(\{\theta_3\}) \leqslant 0.7\}$ additionally considered in the Sects. 4.1 and 4.2.

$u(a_i, \theta_j)$	θ_1	$\cdots$	θ_m
a_1	$u(a_1, \theta_1)$	$\cdots$	$u(a_1, \theta_m)$
$\vdots$	$\vdots$	$\cdots$	$\vdots$
a_n	$u(a_n, \theta_1)$	$\cdots$	$u(a_n, \theta_m)$

$u(a_i, \theta_j)$	θ_1	θ_2	θ_3	θ_4
a_1	20	15	10	5
a_2	30	10	10	20
a_3	20	40	0	20

for all $(a, \theta) \in \mathbb{A} \times \Theta$. Again, for every $\lambda \in G(\mathbb{A})$ fixed, the extended utility function $G(u)$ is associated with a random variable $G(u)_\lambda$ on $(\Theta, 2^\Theta)$ by setting $G(u)_\lambda(\theta) := G(u)(\lambda, \theta)$ for all $\theta \in \Theta$. Finally, we refer to the triplet $(\mathbb{A}, \Theta, u)$ as the *(finite) decision problem* and to the triplet $(G(\mathbb{A}), \Theta, G(u))$ as the corresponding *randomized extension*.

Within this framework, our goal is to determine an *optimal* act (depending on the context, either randomized or pure). However, any appropriate definition of optimality depends on (what we assume about) the *mechanism generating the states of nature*. Here, traditional decision theory mainly covers two *extremes*: The mechanism follows a *known* probability measure π on $(\Theta, 2^\Theta)$ or it can be compared to a *game against an omniscient enemy*. In this cases optimality is almost unanimously defined by either *maximizing expected utility* with respect to π (also known as *Bayes-criterion*) or applying the *Maximin-criterion* (i.e. choosing an act that has maximal utility under the worst possible state of nature).

In contrast, defining optimality of acts becomes less obvious if the prior π is only *partially* known (case of *imprecise probabilities*) or there is uncertainty about the complete appropriateness of it (case of *uncertainty about precise probabilities*). The following sections are concerned with these two situations.

3 Handling Uncertain Precise Probabilistic Information: The Hodges and Lehmann-Criterion

Apart from the border cases of maximizing expected utility with respect to a precise prior π in the presence of perfect probabilistic information and the Maximin-criterion in complete absence of probabilistic information, classical decision theory tries to cope with decision making under uncertain probabilistic information, too: Anticipating ideas of *robust statistics*, Hodges and Lehmann proposed applying the Bayes-criterion only to such acts, whose worst possible utility does not fall below a certain amount of the Minimax utility (see [4]). Their idea is to utilize probabilistic information from previous experience while simultaneously distrusting the complete appropriateness of this information and restricting analysis to acts that are not too bad under the worst state. They also give the following alternative representation of their approach that has a different, intuitively more accessible, interpretation[1]: The decision maker is allowed to model his *degree of trust* in the prior by a parameter $\alpha \in [0,1]$. Specifically, if π is a probability measure on $(\Theta, 2^\Theta)$, a randomized act $\lambda^* \in G(\mathbb{A})$ is said to be *Hodges and Lehmann*-optimal w.r.t. π and α (short: $\Phi_{\pi,\alpha}$-optimal), if $\Phi_{\pi,\alpha}(\lambda^*) \geqslant \Phi_{\pi,\alpha}(\lambda)$ for all $\lambda \in G(\mathbb{A})$, where

$$\Phi_{\pi,\alpha}(\lambda) := (1 - \alpha) \cdot \min_\theta G(u)(\lambda, \theta) + \alpha \cdot \mathbb{E}_\pi \Big[G(u)_\lambda \Big] \tag{1}$$

Thus, the parameter α in (1) controls how the linear trade-off between expectation maximization w.r.t. π and applying the Maximin-criterion is actually made. The following Proposition 1 describes an algorithm for determining a randomized Hodges and Lehmann-optimal act for arbitrary pairs (π, α).[2]

Proposition 1. *Consider the linear programming problem*

$$(1 - \alpha) \cdot (w_1 - w_2) + \alpha \cdot \sum_{i=1}^n \mathbb{E}_\pi(u_{a_i}) \cdot \lambda_i \longrightarrow \max_{(w_1, w_2, \lambda_1, \ldots, \lambda_n)} \tag{2}$$

with constraints $(w_1, w_2, \lambda_1, \ldots, \lambda_n) \geqslant 0$ *and*

- $\sum_{i=1}^n \lambda_i = 1$
- $w_1 - w_2 \leqslant \sum_{i=1}^n u_{ij} \cdot \lambda_i$ *for all* $j = 1, \ldots, m$.

Then the following holds:

(i) *Every optimal solution* $(w_1^*, w_2^*, \lambda_1^*, \ldots, \lambda_n^*)$ *to (2) induces a* $\Phi_{\pi,\alpha}$-*optimal randomized act* $\lambda^* \in G(\mathbb{A})$ *by setting* $\lambda^*(\{a_i\}) := \lambda_i^*$.

(ii) *There always exists an* $\Phi_{\pi,\alpha}$-*optimal randomized act.* $\qquad\square$

By computing the dual linear program of the linear program given in Proposition 1, we receive the following Corollary. It can be interpreted as a method to construct priors that take the agent's *scepticism* about the prior probability π (expressed by the parameter α) into account.

Corollary 1. *Let* $\lambda^* \in G(\mathbb{A})$ *denote a* $\Phi_{\pi,\alpha}$-*optimal randomized act. Then, there exists a probability measure* $\mu_{\pi,\alpha}$ *on* $(\Theta, 2^\Theta)$ *and a pure act* $a^* \in \mathbb{A}$ *such that*

$$\Phi_{\pi,\alpha}(\lambda^*) = \mathbb{E}_{\mu_{\pi,\alpha}}[u_{a^*}] \tag{3}$$

Proof. The dual of the optimization problem (2) is given by:

$$z_1 - z_2 \longrightarrow \min_{(z_1, z_2, \sigma_1, \ldots, \sigma_m)} \tag{4}$$

with constraints $(z_1, z_2, \sigma_1, \ldots, \sigma_m) \geqslant 0$ *and*

---

[1] A further mathematical characterization from the viewpoint of Gamma-Maximity for certain imprecise probabilities is given in Footnote 3.

[2] The proofs of Propositions 1, 2 and 3 are straightforward and therefore left out.

- $\sum_{j=1}^{m} \sigma_j = 1 - \alpha$
- $z_1 - z_2 \geq \sum_{j=1}^{m} u_{ij} \cdot \sigma_j + \alpha \cdot \mathbb{E}_\pi(u_{a_i})$ for all $i = 1, \ldots, n$.

Let $(z_1^*, z_2^*, \sigma_1^*, \ldots, \sigma_m^*)$ denote an optimal solution to (4). Then the constraints guarantee that assigning $\mu_{\pi,\alpha}(\{\theta_j\}) := \alpha \cdot \pi(\{\theta_j\}) + \sigma_j^*$ for all $j = 1, \ldots, m$ induces a probability measure on $(\Theta, 2^\Theta)$ and that for all expectation maximal acts $a^* \in \mathbb{A}$ with respect to $\mu_{\pi,\alpha}$ it holds that $z_1^* - z_2^* = \mathbb{E}_{\mu_{\pi,\alpha}}[u_{a^*}]$. Further, by duality, we know that $z_1^* - z_2^*$ coincides with the optimal value of program (2) and, therefore, with $\Phi_{\pi,\alpha}(\lambda^*)$ where $\lambda^* \in G(\mathbb{A})$ denotes an Hodges and Lehmann-optimal randomized act. Thus, $\Phi_{\pi,\alpha}(\lambda^*) = \mathbb{E}_{\mu_{\pi,\alpha}}[u_{a^*}]$, as desired. $\square$

Running Example (Table 1): Let π denote the prior on $(\Theta, 2^\Theta)$ induced by $(0.2, 0.7, 0.05, 0.05)$ and let our trust in π be expressed by $\alpha = 0.35$. Resolving the linear programming problem from Proposition 1 gives the optimal solution $(8, 0, 0.8, 0, 0.2)$. Thus, a $\Phi_{\pi,0.35}$-optimal randomized act $\lambda^* \in G(\mathbb{A})$ is induced by $(0.8, 0, 0.2)$. Next, we can use Corollary 1 to compute $\mu_{\pi,0.35}$. An optimal solution of problem (4) is given by the vector $(11.78, 0, 0, 0, 0.6385, 0.0115)$, and thus the measure $\mu_{\pi,0.35}$ is induced by the vector $(0.070, 0.245, 0.656, 0.029)$.

4   Handling Imprecise Probabilistic Information: The Gamma-Maximin View

We now turn to decision criteria taking into account the uncertainty in the prior information in a more direct way: For modeling prior knowledge, instead of one classical probability, we consider polyhedral sets of probability measures that are a common tool in different theories of imprecise probabilities, like e.g. *linear partial information* ([7]), *credal sets* ([8]), *lower previsions* ([16]) or *interval probability* ([17]) as well as in robust statistics, like e.g. *ε-contamination models* (see [3, p. 12]). Particularly, we assume probabilistic information is expressed by a polyhedrical set $\mathcal{M}$ of probability measures on $(\Theta, 2^\Theta)$ of the form

$$\mathcal{M} := \left\{ \pi \mid \underline{b}_s \leqslant \mathbb{E}_\pi(f_s) \leqslant \overline{b}_s \; \forall s = 1, \ldots, r \right\} \tag{5}$$

where, for all $s = 1, \ldots, r$, we have $(\underline{b}_s, \overline{b}_s) \in \mathbb{R}^2$ such that $\underline{b}_s \leqslant \overline{b}_s$ and $f_s : \Theta \to \mathbb{R}$. Specifically, the available information is assumed to be describable by lower and upper bounds for the expected values of a finite number of random variables on the space of states. Clearly, if uncertainty is described by a set of probability measures, defining meaningful criteria for decision making strongly depend on the agent's *attitude towards ambiguity*, i.e. towards the non-stochastic uncertainty between the measures contained in $\mathcal{M}$. Accordingly, many competing criteria exist (see [12] for a survey or [2,8,16] for original sources). In the following sections, we present linear programming based results for a selection of such criteria, namely *Walley's maximality* and the *Gamma-Maximin* criterion. For the latter, we also investigate some connections to least favorable priors.

4.1   Checking Maximality of Pure Acts

The idea behind maximality of an act $a^* \in \mathbb{A}$ is quite simple: One repeatedly compares an act a^* pairwise to all other acts and checks whether there exists an element of the set $\mathcal{M}$ with respect to which u_{a^*} dominates the corresponding other act in expectation. Formally, an act $a^* \in \mathbb{A}$ is said to be $\mathcal{M}$-maximal, if

$$\forall\, a \in \mathbb{A}\ \exists\, \pi_a \in \mathcal{M}:\quad \mathbb{E}_{\pi_a}(u_{a^*}) \geqslant \mathbb{E}_{\pi_a}(u_a) \tag{6}$$

Naturally, the above definition extends to randomized acts. However, when also considering randomized acts, the criterion of $\mathcal{M}$-Maximality coincides (see [16, p. 163]) with another well-investigated criterion known from IP decision theory contributed to *Levi*: E-admissibility. For a detailed discussion of connections between the two criteria see [11]. An algorithm for determining the set of all randomized E-admissible acts has been introduced in [13]. However, for finite $\mathbb{A}$, being $\mathcal{M}$-Maximal is a strictly weaker condition and, therefore, needs to be checked separately from E-admissibility. Other approaches for doing so have already been proposed in [6]. Proposition 2 describes an algorithm for checking $\mathcal{M}$-Maximality of a pure act $a_z \in \mathbb{A}$ by solving one single linear program.

Proposition 2. *Let $(\mathbb{A}, \Theta, u)$ denote a finite decision problem and let $\mathcal{M}$ be of the form (5). Further, let $a_z \in \mathbb{A}$ be any act. Consider the linear program*

$$\sum_{i=1}^{n}\Big(\sum_{j=1}^{m}\gamma_{ij}\Big) \longrightarrow \max_{(\gamma_{11},\ldots,\gamma_{nm})} \tag{7}$$

with constraints $(\gamma_{11}, \ldots, \gamma_{nm}) \geqslant 0$ and

- $\sum_{j=1}^{m} \gamma_{ij} \leqslant 1$ *for all $i = 1, \ldots, n$*
- $\underline{b}_s \leqslant \sum_{j=1}^{m} f_s(\theta_j) \cdot \gamma_{ij} \leqslant \overline{b}_s$ *for all $s = 1, ..., r$, $i = 1, \ldots, n$*
- $\sum_{j=1}^{m}(u_{ij} - u_{zj}) \cdot \gamma_{ij} \leqslant 0$ *for all $i = 1, \ldots, n$.*

Then $a_z \in \mathbb{A}$ is $\mathcal{M}$-Maximal iff the optimal outcome of (7) equals n.   □

If $(\gamma_{11}^*, \ldots, \gamma_{nm}^*)$ is an optimal solution to problem (7) yielding an value of n, we can construct $\pi_{a_i} \in \mathcal{M}$ for which act a_z dominates act a_i in expectation by setting $\pi_{a_i}(\{\theta_j\}) := \gamma_{ij}$. The problem possesses $n(3 + r)$ constraints and nm decision variables. Determining the set of all maximal acts requires to solve n such linear programs. Compared to this, the algorithm based on pairwise comparisons of acts proposed in [6] here translates to solving $n^2 - n$ linear programs with m decision variables, however, with only $r + 2$ constraints.

Running Example (Table 1): Resolving the linear programming problem from Proposition 2 for every act a_1, a_2 and a_3 separately gives optimal value 3 for each of them. Thus, all available acts are $\mathcal{M}$-Maximal.

4.2   Gamma-Maximin and Least Favorable Priors

In this section, we first present a linear program for identifying a *least favorable prior distribution* from the credal set $\mathcal{M}$ under consideration. Afterwards, we investigate the dual of this linear program and, in this way, provide a connection between pure acts $a \in \mathbb{A}$ that maximize expected utility with respect to a least favorable prior and randomized acts $\lambda \in G(\mathbb{A})$ that are optimal with respect to the Gamma-Maximin criterion.

Before we proceed, some additional notation is needed: For a credal element $\pi \in \mathcal{M}$, let $B(\pi)$ denote the maximal expectation with respect to π that an act from $\mathbb{A}$ can yield (that is $B(\pi) = \mathbb{E}_{\pi}(u_{a^*})$, where $a^* \in \mathbb{A}$ maximizes expected utility with respect to π). The set of all acts $a \in \mathbb{A}$ that maximize expected utility with respect to π is denoted by $\mathbb{A}_{\pi}$. Further, we call a credal element $\pi^- \in \mathcal{M}$ a *least favorable prior (lfp)* from $\mathcal{M}$ iff $B(\pi^-) \leqslant B(\pi)$ holds for all $\pi \in \mathcal{M}$. Specifically, π^- is a lfp, if it yields the minimal best possible expected utility under all concurring elements on the credal set. Proposition 3 describes a linear program for determining a lfp from $\mathcal{M}$.

Proposition 3. *Let $(\mathbb{A}, \Theta, u)$ denote a decision problem and let $\mathcal{M}$ be of the form (5). Consider the linear program*

$$w_1 - w_2 \longrightarrow \min_{(w_1, w_2, \pi_1, \ldots, \pi_m)} \tag{8}$$

with constraints $(w_1, w_2, \pi_1, \ldots, \pi_m) \geqslant 0$ and

- $\sum_{j=1}^{m} \pi_j = 1$
- $\underline{b}_s \leqslant \sum_{j=1}^{m} f_s(\theta_j) \cdot \pi_j \leqslant \overline{b}_s$ *for all $s = 1, \ldots, r$*
- $w_1 - w_2 \geqslant \sum_{j=1}^{m} u_{ij} \cdot \pi_j$ *for all $i = 1, \ldots n$.*

Then the following holds:

(i) Every optimal solution $(w_1^*, \ldots, \pi_m^*)$ to (8) induces a least favorable prior $\pi^- \in \mathcal{M}$ by setting $\pi^-(\{\theta_j\}) := \pi_j^*$.

(ii) There always exists a least favorable prior. □

A lfp can be understood as a kind of "pignistic" probability, representing the decision problem under complex uncertainty in a way that is specific to the problem and the criterion under consideration, but in return gives the exact criterion value. This contrasts lfps from pignistic probabilities in Smets' spirit, who argued that a decision problem under complex uncertainty could be approached by distinguishing between a *credal level*, where the uncertain beliefs are to be expressed with all their ambiguity and scarceness by an imprecise probability (belief function in Smets' context), and a *decision level*, where eventually the imprecise probability is condensed into a traditional probability on which expected utility theory could be applied (see, e.g., [9,10], as well as, e.g., [1] for geometric techniques to represent belief functions by a single precise probability).

We now show some connections between least favorable priors and randomized Gamma-Maximin acts w.r.t. $\mathcal{M}$ ($\mathcal{M}$-Maximin). Recalling its definition, a randomized act $\lambda^* \in G(\mathbb{A})$ is said to be $\mathcal{M}$-*Maximin optimal* iff for all $\lambda \in G(\mathbb{A})$:

$$\underline{\mathbb{E}}_{\mathcal{M}}[G(u)_{\lambda^*}] \geqslant \underline{\mathbb{E}}_{\mathcal{M}}[G(u)_\lambda] \tag{9}$$

where $\underline{\mathbb{E}}_{\mathcal{M}}(X) := \min_{\pi \in \mathcal{M}} \mathbb{E}_\pi(X)$ for random variables $X : (\Theta, 2^\Theta) \to \mathbb{R}$.[3] It turns out that the linear program from Proposition 3 is *dual* to the one for determining a randomized $\mathcal{M}$-Maximin act described in [13, Sect. 3.2]. Together with *complementary slackness* (see, e.g., [15, Sect. 5.5]) from linear optimization theory, this allows to derive connections between lfps and the Gamma-Maximin.

Proposition 4. *Let* $(\mathbb{A}, \Theta, u)$ *denote a finite decision problem and let* $\mathcal{M}$ *be of the form (5). Then the following holds:*

(i) *If* π^- *is a lfp from* $\mathcal{M}$, *then for all optimal randomized* $\mathcal{M}$-*Maximin acts* $\lambda^* \in G(\mathbb{A})$ *we have* $\lambda^*(\{a\}) = 0$ *for all* $a \in \mathbb{A} \backslash \mathbb{A}_{\pi^-}$.

(ii) *Let* π^- *denote a lfp from* $\mathcal{M}$ *and let* $\lambda^* \in G(\mathbb{A})$ *denote a randomized* $\mathcal{M}$-*Maximin act. Then for all* $a \in \mathbb{A}_{\pi^-}$ *we have*

$$\mathbb{E}_{\pi^-}[u_a] = \underline{\mathbb{E}}_{\mathcal{M}}[G(u)_{\lambda^*}]$$

Proof. The dual programming problem of problem (8) is given by:

$$z_1 - z_2 + \sum_{s=1}^{r} (\underline{b}_s x_s - \overline{b}_s y_s) \longrightarrow \max_{(z_1, z_2, x_1, \dots, x_r, y_1, \dots, y_r, \lambda_1, \dots, \lambda_n)} \tag{10}$$

with constraints $(z_1, z_2, x_1, \dots, x_r, y_1, \dots, y_r, \lambda_1, \dots, \lambda_n) \geqslant 0$ and

- $\sum_{i=1}^{n} \lambda_i = 1$
- $z_1 - z_2 + \sum_{s=1}^{r} f_s(\theta_j)(x_s - y_s) \leq \sum_{i=1}^{n} u_{ij} \cdot \lambda_i$ for all $j = 1, \dots, m$.

The resulting linear program (10) is exactly the one for determining a randomized act $\lambda^* \in G(\mathbb{A})$ which is optimal with respect to the $\mathcal{M}$-Maximin criterion as proposed and proven in [13, Sect. 3.2]. We now can use standard results on duality and complementary slackness (see, e.g., [15, Chap. 5]) to proof the proposition:

---

[3] For the special case of an ε-contamination model (a.k.a. *linear-vacuous model*) of the form $\mathcal{M}_{(\pi_0, \varepsilon)} := \{(1 - \varepsilon)\pi_0 + \varepsilon\pi : \pi \in \mathcal{P}(\Theta)\}$, where $\mathcal{P}(\Theta)$ denotes the set of all probability measures on $(\Theta, 2^\Theta)$, $\varepsilon > 0$ is a fixed contamination parameter and $\pi_0 \in \mathcal{P}(\Theta)$ is the central distribution, Gamma-Maximin is mathematically closely related to the Hodges and Lehmann-criterion: For fixed $X : (\Theta, 2^\Theta) \to \mathbb{R}$ we have $\underline{\mathbb{E}}_{\mathcal{M}_{(\pi_0, \varepsilon)}}(X) = \min_{\pi \in \mathcal{P}(\Theta)}((1 - \varepsilon)\mathbb{E}_{\pi_0}(X) + \varepsilon\mathbb{E}_\pi(X)) = (1 - \varepsilon)\mathbb{E}_{\pi_0}(X) + \varepsilon \min_{\pi \in \mathcal{P}(\Theta)} \mathbb{E}_\pi(X) = (1 - \varepsilon)\mathbb{E}_{\pi_0}(X) + \varepsilon \min_{\theta \in \Theta} X(\theta)$. Thus, maximizing the lower expectation w.r.t. the ε-contamination model is equivalent to maximizing the Hodges and Lehmann-criterion with trust parameter $(1 - \varepsilon)$ and prior π_0.

Part (i): Let $\pi^- \in \mathcal{M}$ denote a lfp and let $a_z \in \mathbb{A}\backslash\mathbb{A}_{\pi^-}$. Then

$$(\max\{B(\pi^-),0\}, -\min\{B(\pi^-),0\}, \pi^-(\{\theta_1\}), \ldots, \pi^-(\{\theta_m\})) \qquad (11)$$

defines an optimal solution to (8) for which it holds that $B(\pi^-) > \mathbb{E}_{\pi^-}(u_{a_z})$. Thus, there exists an optimal solution to (8), for which the constraint $w_1 - w_2 \geq \sum_{j=1}^m u_{zj} \cdot \pi_j$ holds strictly and, therefore, the corresponding slack variable is strictly greater 0. Hence, by complementary slackness, the corresponding variable in the dual problem (10), that is λ_z, equals 0 for every optimal solution of problem (10). Finally, note that $\{\lambda_z^* : \lambda_z^* \text{ appears in optimal solution}\} = \{\lambda^*(\{a_z\}) : \lambda^* \in G(\mathbb{A}) \ \mathcal{M}\text{-Maximin optimal}\}$, since, as (implicitly) shown in [13, Sect. 3.2], every $\mathcal{M}$-Maximin optimal $\lambda^* \in G(\mathbb{A})$ induces an optimal solution to (10), namely

$$(z_1^*, z_2^*, x_1, \ldots, x_r^*, y_1^*, \ldots, y_r^*, \lambda^*(\{a_1\}), \ldots, \lambda^*(\{a_n\})) \qquad (12)$$

where $(z_1^*, z_2^*, x_1, \ldots, x_r^*, y_1^*, \ldots, y_r^*)$ denotes an optimal solution to a reduced version of problem (10) with $(\lambda_1, \ldots, \lambda_n) := (\lambda^*(\{a_1\}), \ldots, \lambda^*(\{a_n\}))$ fixed.

Part (ii): Let $\pi^- \in \mathcal{M}$ denote an lfp and $\lambda^* \in G(\mathbb{A})$ denote an $\mathcal{M}$-Maximin act. Use (11) and (12) to construct optimal solutions to (8) and (10). As the optimal value of (8) equals $B(\pi^-)$ and the optimal value of (10) equals $\mathbb{E}_{\mathcal{M}}[G(u)_{\lambda^*}]$, the result follows by the duality theorem. $\qquad\Box$

As an immediate consequence of Proposition 4 (i), we can specify a condition under which randomization cannot improve utility, if optimality is defined in terms of the Gamma-Maximin criterion. Specifically, we have the following corollary.

Corollary 2. *If there exists a lfp π^- from $\mathcal{M}$ such that $\mathbb{A}_{\pi^-} = \{a_z\}$ for some $z \in \{1, \ldots, n\}$, then $\delta_{a_z} \in G(\mathbb{A})$ is the unique randomized $\mathcal{M}$-Maximin act. Specifically, considering randomized acts is unnecessary in such situations.* $\qquad\Box$

Running Example (Table 1): Algorithm 8 leads to the optimal solution vector $(13, 0, 0, 0, 0.7, 0.3)$. Thus, a lfp π^- from $\mathcal{M}$ is induced by $(0, 0.7, 0.3, 0)$. Simple computation gives $\mathbb{A}_{\pi^-} = \{a_2\}$. Hence, according to Corollary 2, a_2 is the unique $\mathcal{M}$-Maximin act (even compared to randomized acts) with utility 13.

5   Summary and Concluding Remarks

We presented linear programming based approaches for determining optimal randomized acts and investigated what can be learned by dualizing these. Future research includes the following issues: If $\mathcal{M}$ is non-degenerated, i.e. $\pi(\{\theta\}) > 0$ for all $(\pi, \theta) \in \mathcal{M} \times \Theta$, the same holds for every lfp π^-. Since every π^- induces an optimal solution to (8), complementary slackness implies that all constraints of problem (10) are binding for every optimal solution. This gives a system of linear *equations* that have to be satisfied by every randomized $\mathcal{M}$-Maximin act. A natural question is: Under which conditions is this system sufficient to identify an optimal act without solving an optimization problem at all? A further interesting point is that algorithm (7) for checking maximality of an act a_z takes into account all other acts a_i in one linear program simultaneously. This could be used to modify the algorithm for finding maximal acts that are not too far from being E-admissible in the sense that the involved probabilities π_{a_i} that establish maximality of a_z differ not too much w.r.t. the L_1-norm which can be guaranteed by imposing further *linear* constraints.

Acknowledgement. The authors would like to thank the three anonymous referees for their helpful comments and their support.

References

1. Cuzzolin, F.: Two new Bayesian approximations of belief functions based on convex geometry. IEEE T. Syst. Man. Cy. B **37**, 993–1008 (2007)
2. Gilboa, I., Schmeidler, D.: Maxmin expected utility with non-unique prior. J. Math. Econ. **18**, 141–153 (1989)
3. Huber, P.: Robust Statistics. Wiley, New York (1981)
4. Hodges, J., Lehmann, E.: The use of previous experience in reaching statistical decisions. Ann. Math. Stat. **23**, 396–407 (1952)
5. Hable, R., Troffaes, M.: Computation. In: Augustin, T., Coolen, F., de Cooman, G., Troffaes, M. (eds.) Introduction to Imprecise Probabilities, pp. 329–337. Wiley, Chichester (2014)
6. Kikuti, D., Cozman, F., Filho, R.: Sequential decision making with partially ordered preferences. Artif. Intel. **175**, 1346–1365 (2011)
7. Kofler, E., Menges, G.: Entscheiden bei unvollständiger Information. Springer, Berlin (1976)
8. Levi, I.: The Enterprise of Knowledge: An Essay on Knowledge, Credal Probability, and Chance. MIT Press, Cambridge (1983)
9. Smets, P.: Decision making in the TBM: the necessity of the pignistic transformation. Int. J. Approx. Reason. **38**, 133–147 (2005)
10. Smets, P.: Decision making in a context where uncertainty is represented by belief functions. In: Srivastava, R., Mock, T. (eds.) Belief Functions in Business Decisions, pp. 17–61. Physica, Heidelberg (2002)
11. Schervish, M., Seidenfeld, T., Kadane, J., Levi, I.: Extensions of expected utility theory and some limitations of pairwise comparisons. In: Bernard, J.-M., Seidenfeld, T., Zaffalon, M. (eds.) Proceedings of ISIPTA 2003, pp. 496–510. Carleton Scientific, Waterloo (2003)
12. Troffaes, M.: Decision making under uncertainty using imprecise probabilities. Int. J. Approx. Reason. **45**, 17–29 (2007)
13. Utkin, L., Augustin, T.: Powerful algorithms for decision making under partial prior information and general ambiguity attitudes. In: Cozman, F., Nau, R., Seidenfeld, T. (eds.) Proceedings of ISIPTA 2005, pp. 349–358 (2005)
14. Utkin, L., Kozine, I.: Different faces of the natural extension. In: de Cooman, G., Fine, T., Seidenfeld, T. (eds.) Proceedings of ISIPTA 2001, pp. 316–323 (2001)
15. Vanderbei, R.: Linear Programming: Foundations and Extensions. Springer, New York (2014)
16. Walley, P.: Statistical Reasoning with Imprecise Probabilities. Chapman and Hall, London (1991)

17. Weichselberger, K.: Elementare Grundbegriffe einer allgemeineren Wahrschein-
lichkeitsrechnung I: Intervallwahrscheinlichkeit als umfassendes Konzept. Physica,
Heidelberg (2001)
18. Weichselberger, K.: Interval probability on finite sample spaces. In: Rieder, H. (ed.)
Robust Statistics, Data Analysis, and Computer Intensive Methods, pp. 391–409.
Springer, New York (1996)

Axiomatization of an Importance Index for Generalized Additive Independence Models

Mustapha Ridaoui[1(✉)], Michel Grabisch[1], and Christophe Labreuche[2]

[1] Paris School of Economics, Université Paris I - Panthéon-Sorbonne, Paris, France
{mustapha.ridaoui,michel.grabisch}@univ-paris1.fr
[2] Thales Research and Technology, Palaiseau, France
christophe.labreuche@thalesgroup.com

Abstract. We consider MultiCriteria Decision Analysis models which are defined over discrete attributes, taking a finite number of values. We do not assume that the model is monotonically increasing with respect to the attributes values. Our aim is to define an importance index for such general models, encompassing Generalized-Additive Independence models as particular cases. They can be seen as being equivalent to k-ary games (multichoice games). We show that classical solutions like the Shapley value are not suitable for such models, essentially because of the efficiency axiom which does not make sense in this context. We propose an importance index which is a kind of average variation of the model along the attributes. We give an axiomatic characterization of it.

Keywords: MultiCriteria decision analysis · k-ary game · Shapley value

1 Introduction

In MultiCriteria Decision Analysis (MCDA), a central question is to determine the importance of attributes or criteria. Suppose the preference of a decision maker has been represented by a numerical model. For interpretation and explanation purpose of the model, a basic requirement is to be able to assess the importance of each attribute. If this is easy for a number of elementary models (essentially additive ones), it becomes more challenging with complex models.

For models based on the Choquet integral w.r.t. a capacity or fuzzy measure (see a survey in [8]), it has been recognized since a long time ago that the Shapley value [15], a concept borrowed from game theory, is the adequate tool to quantify the importance of attributes.

Choquet integral-based models belong to the category of decomposable models, that is, where utility functions are defined on each attribute, and then are aggregated by some increasing function. In this paper, we depart from this kind of models and focus on models where there is no such separation of utilities among the attributes. Typically, the Generalized Additive Independence (GAI) model proposed by Fishburn [4,5] is of this type, since of the form $U(x) = \sum_{S \in \mathcal{S}} u_S(x_S)$, where $\mathcal{S}$ is a collection of subsets of N, the index set of

© Springer International Publishing AG 2017
A. Antonucci et al. (Eds.): ECSQARU 2017, LNAI 10369, pp. 340–350, 2017.
DOI: 10.1007/978-3-319-61581-3_31

all attributes. In this paper, however, we do not take advantage of this peculiar form, and consider a numerical model without particular properties, except that the underlying attributes are discrete, and thus take a finite number of values. Note that in many applications, especially in the AI field, this is the case, in particular for GAI models [1,2,6].

As far as we know, the question of the definition of an importance index for such a general case remains open. As we will explain, discrete models can be seen as k-ary capacities or more generally k-ary games [9] (also called multichoice games [11]), and thus it seems natural to take as importance index the various definitions of Shapley-like values for multichoice games existing in the literature. There is however a major drawback inherent to these values: they all satisfy the efficiency axiom, that is, the sum of the importance indices over all attributes is equal to $v(k_N)$, the value of the game when all attributes take the highest value. If this axiom is natural in a context of cooperative game, where the Shapley value defines a rational way to share among the players the total benefit $v(k_N)$ of the game, it has no justification in MCDA, especially if the model v is not monotone increasing.

The approach we propose here is inspired by the calculus of variations: we define the importance index of an attribute as the average variation of v (depicting the satisfaction of the decision maker) when the value of attribute i is increased by one unit. We propose an axiomatic definition, where the chosen axioms are close to those of the original Shapley value.

Section 2 recalls the basic concepts. Section 3 informally defines what is the aim of our importance index. The axiomatic characterization is presented in Sect. 4. The new index is then interpreted (Sect. 5).

2   Preliminaries

Throughout the paper, $N = \{1, \ldots, n\}$ is a finite set which can be thought as the set of attributes (in MCDA), players (in cooperative game theory), etc., depending on the application. In this paper, we will mainly focus on MCDA applications. Cardinality of sets will be often denoted by corresponding lower case letters, e.g., n for $|N|$, s for $|S|$, etc.

The set of all possible values taken by attribute $i \in N$ is denoted by L_i. As it is often the case in MCDA, we assume that these sets are finite, and we represent them by integer values, i.e., $L_i = \{0, 1, \ldots, k^i\}$. Alternatives are thus elements of the Cartesian product $L = \times_{i \in N} L_i$ and take the form $x = (x_1, x_2, \ldots, x_n)$ with $x_i \in L_i$, $i = 1, \ldots, n$. For $x, y \in L$, we write $x \leq y$ if $x_i \leq y_i$ for every $i \in N$. For $S \subseteq N$ and $x \in L$, x_S is the restriction of x to S. L_{-i} is a shorthand for $\times_{j \neq i} L_j$. For each $y_{-i} \in L_{-i}$, and any $\ell \in L_i$, (y_{-i}, ℓ_i) denotes the combined alternative x such that $x_i = \ell_i$ and $x_j = y_j, \forall j \neq i$. The vector $0_N = (0, \ldots, 0)$ is the null alternative of L, and $k_N = (k_1^1, \ldots, k_n^n)$ is the top element of L. 0_{-i} denotes the element of L_{-i} in which all coordinates are zero. We call vertex of L any element $x \in L$ such that x_i is either 0 or k^i, for each $i \in N$. We denote by $\Gamma(L) = \times_{i \in N} \{0, k^i\}$ the set of vertices of L. For each $x \in L$, we denote by

$S(x) = \{i \in N \mid x_i > 0\}$ the support of x, and by $K(x) = \{i \in N \mid x_i = k^i\}$ the kernel of x. Their cardinalities are respectively denoted by $s(x)$ and $k(x)$. We call *elementary cell* of L a unit cell $\times_{i \in N}\{x_i, x_i + 1\}$, for $x_i \in L_i \backslash k^i$.

We suppose to have a numerical representation $v : L \to \mathbb{R}$ of the preference of the decision maker (DM) over the set of alternatives in L. For the sake of generality, we do not make any assumption on v, except that $v(0_N) = 0$ (this is not a restriction, as most of numerical representations are unique up to a positive affine transformation). In particular, there is no assumption of monotonicity, that is, we do not assume that $x_1 \geq x_1', \ldots, x_n \geq x_n'$ implies $v(x_1, \ldots, x_n) \geq v(x_1', \ldots, x_n')$. Example 1 below illustrates that it is quite common to observe this lack of monotonicity.

Example 1. The level of comfort of humans depends on three main attributes: temperature of the air (X_1), humidity of the air (X_2) and velocity of the air (X_3). Then $v(x_1, x_2, x_3)$ measures the comfort level. One can readily see that v is not monotone in its three arguments. For x_2 and x_3 fixed, v is maximal for intermediate values of the temperature (typically around 23 °C). Similarly, the value of humidity maximizing v is neither too low nor too high. Finally, for x_1 relatively large, some wind is well appreciated, but not too much. Hence for any i, and supposing the other two attributes being fixed, there exists an optimal value $\widehat{\ell_i} \in L_i$ such that v is increasing in x_i below $\widehat{\ell_i}$, and then decreasing in x_i above $\widehat{\ell_i}$.

Although we will not use this specific form for v in the sequel, we mention as typical example of a model not necessarily satisfying monotonicity the Generalized Additive Independence (GAI) model, i.e., v is written as $v(x) = \sum_{S \in \mathcal{S}} v_S(x_S)$, where $\mathcal{S}$ is a collection of subsets of N [4,5]. This model has been widely used in AI [1,2,6].

For convenience, we assume from now on that all attributes have the same number of elements, i.e., $k^i = k$ for every $i \in N$ ($k \in \mathbb{N}$). Note that if this is not the case, k is set to $\max_{i \in N} k^i$, and we duplicate some elements of L_i when $k^i < k$. A fundamental observation is that when $k = 1$, v is nothing other than a *pseudo-Boolean function* $v : \{0,1\}^N \to \mathbb{R}$ vanishing on 0_N, or put otherwise via the identity between sets and their characteristic functions, a *(cooperative) game* (in characteristic form) $\mu : 2^N \to \mathbb{R}$, with $\mu(\varnothing) = 0$. A game v is *monotone* if $v(A) \leq v(B)$ whenever $A \subseteq B$. A monotonic game is called a *capacity* [3] or *fuzzy measure* [16]. For the general case $k \geq 1$, $v : L \to \mathbb{R}$ is called a *multichoice game* or *k-ary game* [11], and the numbers $0, 1, \ldots, k$ in L_i are seen as the level of activity of the players. By analogy with the classical case $k = 1$, a *k-ary capacity* is a monotone k-ary game, i.e., satisfying $v(x) \leq v(y)$ whenever $x \leq y$, for each $x, y \in L$ [9]. Hence, a k-ary capacity represents a preference on L which is increasing with the value of the attributes. We denote by $\mathcal{G}(L) = \{v : L \to \mathbb{R}, v(0_N) = 0\}$ the set of functions defined on L vanishing on 0_N. A function is called *single-peaked* if it has only one local maximum and no troughs.

By analogy with classical games, a unanimity game for k-ary game denoted u_x, for each $x \in L$ with $x \neq 0_N$ is defined by

$$u_x(y) = \begin{cases} 1, \text{ if } & y \geq x \\ 0, \text{ otherwise} \end{cases}$$

Note that the set of unanimity games forms a basis of the vector space of k-ary games. One advantage of unanimity games is that they are monotone. Hence this basis is relevant for k-ary capacities. In order to obtain a basis of k-ary games not necessarily made of monotone functions, we define for each $x \in L$ such that $x \neq 0_N$, the game δ_x by

$$\delta_x(y) = \begin{cases} 1, \text{ if } & y = x \\ 0, \text{ otherwise} \end{cases}$$

It is obvious that any k-ary game v can be written as

$$v = \sum_{\substack{x \in L \\ x \neq 0_N}} v(x)\delta_x. \tag{1}$$

3 Definition of an Importance Index: What Do We Aim at Doing?

We restrict ourselves to the MCDA setting and interpretation in this paper. When dealing with numerical representations of preference in MCDA, one of the primary concerns is to give an interpretation of the model in terms of importance of the attributes. When v is a capacity or a game ($k = 1$), or with continuous models extending capacities and games like the Choquet integral, the standard solution is to take the Shapley value, introduced by Shapley in the context of cooperative games [15]. A *value* is a function $\phi : \mathcal{G}(2^N) \rightarrow \mathbb{R}^N$ that assigns to every game μ a payoff vector $\phi(\mu)$. It is interpreted in the MCDA context as the vector of importance of the attributes. The value introduced by Shapley is one of the most popular, and is defined by:

$$\phi_i^{Sh}(\mu) = \sum_{S \subseteq N \setminus i} \frac{(n - s - 1)!s!}{n!}\left(\mu(S \cup i) - \mu(S)\right), \forall i \in N. \tag{2}$$

A standard property shared by many values in the literature is *efficiency*: $\sum_{i \in N} \phi_i(\mu) = \mu(N)$. This property is very natural in game theory, as $\mu(N)$ is the total benefit obtained from the cooperation of all players in N, and by efficiency the payoff vector $\phi(v)$ represents a sharing of this total benefit.

If the Shapley value has been widely used in MCDA with great success (see, e.g., [8]), it must be stressed that it was only in the case of monotonically increasing models, i.e., based on a capacity μ. In such cases, $\mu(N)$ is set to 1, the value of the best possible alternative, and the importance index of an attribute could be seen as a kind of contribution of that attribute to the best possible alternative. However if the model is not monotone increasing, such an interpretation fails. Hence, we are facing here a double difficulty: to propose a "value" both valid for $k \geq 1$ and for nonincreasing models. There have been many proposed values for multichoice games, e.g., Hsiao and Raghavan [11], van den Nouweland et al. [17], Klijn et al. [12], Peters and Zank [14] and Grabisch and Lange [10], etc. All of them satisfy the classical efficiency axiom.

We wish to capture in our importance index the impact of each attribute on the overall utility. Let us consider for illustration function δ_y with $k = 2$, $n = 3$ and $y = (2, 1, 1)$. Attribute 1 is non-decreasing and has a positive impact on the overall utility. Attribute 2 has neither a positive nor a negative impact on the overall utility, since $\delta_y(x)$ is going from value 0 (at $x_2 = 0$) to 1 (at $x_2 = 1$), and then decreasing to value 0 again (at $x_2 = 2$). Hence attribute 2 has globally neither a positive nor a negative impact. The same holds with attribute 3. Hence, denoting by $\phi(\delta_y)$ our importance index for that function, one shall have $\phi_1(\delta_y) > 0$, $\phi_2(\delta_y) = \phi_3(\delta_y) = 0$, so that the sum $\sum_{i \in N} \phi_i(v) > 0$ cannot be equal to $v(2, 2, 2) - v(0, 0, 0) = 0$, and hence ϕ does not satisfy efficiency. Rather, the index $\phi_i(v)$ shall measure the impact of attribute i on v, as the total variation on v if we increase the value of attribute i of one unit (going from value x_i to $x_i + 1$), when x is varying over the domain.

4   Axiomatization

We define in this section an importance index according to the ideas explained above, by using an axiomatic description. Our first three axioms are the same as those used by Shapley when characterizing his value in [15]: linearity, null player and symmetry. Our approach will follow Weber [18], who introduces the axioms one by one and at each step gives a characterization. Throughout this section, we consider a value as a mapping $\phi : \mathcal{G}(L) \to \mathbb{R}$.

We expect our importance index to have an exponential complexity of computation, as for the Shapley value. In order to reduce the complexity burden for some functions v having some particular properties, we assume that for GAI models, the value can be decomposed over each utility function v_S. Hence ϕ shall be additive: $\phi_i(v + v') = \phi_i(v) + \phi_i(v')$. Moreover, utilities are in MCDA invariant to positive linear transformations. Hence ϕ_i shall be homogeneous: $\phi_i(\alpha v) = \alpha \phi_i(v)$ for every $\alpha > 0$. Lastly the value shall be symmetric: $\phi_i(-v) = -\phi_i(v)$. The previous three properties yield linearity axiom **L**.

Linearity Axiom (L): ϕ is linear on $\mathcal{G}(L)$, i.e., $\forall v, w \in \mathcal{G}(L), \forall \alpha \in \mathbb{R}$,

$$\phi_i(v + \alpha w) = \phi(v) + \alpha \phi(w).$$

Proposition 1. *Under axiom (L), for all $i \in N$, there exists constants $a_x^i \in \mathbb{R}$, for all $x \in L$, such that $\forall v \in \mathcal{G}(L)$,*

$$\phi_i(v) = \sum_{x \in L} a_x^i v(x). \tag{3}$$

The proof of this result and the other ones are omitted due to space limitation.

The second axiom that characterizes the Shapley value in [18] is called the null player axiom. It says that a player $i \in N$ who brings no contribution (i.e., $\mu(S \cup i) = \mu(S), \forall S \subseteq N \backslash \{i\}$) should receive a zero payoff. This definition can be easily extended to $v \in \mathcal{G}(L)$ and to a MCDA setting.

Definition 1. A criterion $i \in N$ is said to be null for $v \in \mathcal{G}(L)$ if

$$v(x + 1_i) = v(x), \forall x \in L, x_i < k.$$

Remark 1. Let $i \in N$ be a null criterion for $v \in \mathcal{G}(L)$. we have,

$$\forall x \in L, v(x_{-i}, x_i) = v(x_{-i}, 0_i).$$

If an attribute is null w.r.t. a function $v \in \mathcal{G}(L)$, then this attribute has no influence on v, and hence the importance of this attribute shall be zero. We propose the following axiom.

Null Axiom (N): If a criterion i is null for $v \in \mathcal{G}(L)$, then $\phi_i(v) = 0$.

Proposition 2. *Under axioms (N) and (L), for all $i \in N$, there exists $b_x^i \in \mathbb{R}$, for all $x \in L$ with $x_i < k$, such that $\forall v \in \mathcal{G}(L)$,*

$$\phi_i(v) = \sum_{\substack{x \in L \\ x_i < k}} b_x^i \big(v(x + 1_i) - v(x)\big). \tag{4}$$

This proposition shows that ϕ_i is a linear combination of the added-values on v, going from value x_i to $x_i + 1$, over all x.

The classical symmetry axiom says that the numbering of the attributes has no influence on the value. It means that the computation of value should not depend on the numbering of the attributes.

Let σ be a permutation on N. For all $x \in L$, we denote $\sigma(x)_{\sigma(i)} = x_i$. For all $v \in \mathcal{G}(L)$, The function $\sigma \circ v$ is defined by $\sigma \circ v(\sigma(x)) = v(x)$.

Symmetry Axiom (S): For any permutation σ of N,

$$\phi_{\sigma(i)}(\sigma \circ v) = \phi_i(v), \forall i \in N.$$

Proposition 3. *Under axioms (N), (L) and (S), $\forall v \in \mathcal{G}(L), \forall i \in N$,*

$$\phi_i(v) = \sum_{\substack{x \in L \\ x_i < k}} b_{x_i; n_0, \dots, n_k} \big(v(x + 1_i) - v(x)\big),$$

where $b_{x_i; n_0, \dots, n_k} \in \mathbb{R}$, and n_j is the number of components of x_{-i} being equal to j.

This result means that the coefficients in front of the added-values on v, going from value x_i to $x_i + 1$, do not depend on the precise value of x, but only on the number of terms of x_{-i} taking values $0, 1, \ldots, k$.

Let us take for example $N = \{1, 2, 3\}$, $k = 2$ and $x = (0, 2, 2)$. We have,

$$b_x^1 = b_{0;0,0,2}, b_x^2 = b_{2;1,0,1}, \text{ and } b_x^3 = b_{2;1,0,1}.$$

The next axiom enables an easier computation of coefficients b_x^i while reducing their number.

Invariance Axiom (I): Let us consider two functions $v, w \in \mathcal{G}(L)$ such that, for all $i \in N$,

$$v(x + 1_i) - v(x) = w(x) - w(x - 1_i), \forall x \in L, x_i \notin \{0, k\}$$
$$v(x_{-i}, 1_i) - v(x_{-i}, 0_i) = w(x_{-i}, k_i) - w(x_{-i}, k_i - 1), \forall x_{-i} \in L_{-i}.$$

Then $\phi_i(v) = \phi_i(w)$.

Taking two functions v and w for which the differences $v(x + 1_i) - v(x)$ (measuring the added value of improving x of one unit on attribute i can be deduced from that of w just by shifting of one unit, then the mean importance of attribute i shall be the same for v and w. In other words, what is essential is the absolute value of the differences $v(x + 1_i) - v(x)$ and not the value x at which it occurs.

Proposition 4. *Under axioms (L), (N) and (I)*, $\forall v \in \mathcal{G}(L), \forall i \in N$,

$$\phi_i(v) = \sum_{x_{-i} \in L_{-i}} b_{x_{-i}}^i \big(v(x_{-i}, k_i) - v(x_{-i}, 0_i) \big).$$

Axiom **(I)** implies that we only need to look at the difference of v between the extreme value 0 and k. The evaluation on the intermediate elements of L_i do not count.

Proposition 5. *Under axioms (L), (N), (I) and (S)*, $\forall v \in \mathcal{G}(L), \forall i \in N$,

$$\phi_i(v) = \sum_{x_{-i} \in L_{-i}} b_{n(x_{-i})} \big(v(x_{-i}, k_i) - v(x_{-i}, 0_i) \big),$$

where $n(x_{-i}) = (n_0, n_1, \ldots, n_k)$ with n_j the number of components of x_{-i} being equal to j.

As explained in Sect. 3, we do not require that ϕ satisfies efficiency. In the context of game theory, $\phi_i^{Sh}(\mu)$ is the amount of money alloted to player i, so that relation $\sum_{i \in N} \phi_i^{Sh}(\mu) = \mu(N)$ means that all players share among themselves the total worth $\mu(N)$. We have no such interpretation in MCDA. By contrast, we interpret $\phi_i(v)$ as an overall added value when increasing the value of attribute i of one unit – thereby going from any point x to $(x_i + 1, x_{-i})$. Hence $\sum_{i \in N} \phi_i(v)$ can be interpreted as the overall added value when increasing simultaneously the value of all attributes of one unit – thereby going from any point x to $x + 1 = (x_1 + 1, \ldots, x_n + 1)$. For an arbitrary function v, there is a priori no particular property for the previous sum. We thus consider a very special case of functions following Example 1. These functions are single peaked. The simplest version of these functions is the family of functions δ_y. For those functions, we immediately see from Proposition 5 that $\phi_i(\delta_y) = 0$ for every i such that $y_i \neq 0, k$, as already mentioned in Sect. 3. Based on this remark, we should only bother on attributes which are equal to either 0 or k in y. We have therefore three cases (recall that $s(y), k(y)$ are the cardinalities of the support and kernel of y):

- $k(y) \neq 0$ and $s(y) = n$. Then $y - 1 \in L$ because no component of y is equal to 0, and we have $\delta_y(y) - \delta_y(y - 1) = 1$. Note that $\delta_y(x + 1) - \delta_y(x) = 0$ for any $x \neq y - 1$ and $x, x + 1 \in L$. Therefore, by the above argument, we have $\sum_{i \in N} \phi_i(\delta_y) = 1$.
- $k(y) = 0$ and $s(y) < n$. This is the dual situation: $y + 1 \in L$ because no component is equal to k, and we have $\delta_y(y + 1) - \delta_y(y) = -1$. Since $\delta_y(x + 1) - \delta_y(x) = 0$ for any other possible x, we get $\sum_{i \in N} \phi_i(\delta_y) = -1$.
- $k(y) \neq 0$ and $s(y) < n$. This time there are both components equal to 0 and to k in y. Therefore, neither $y + 1$ nor $y - 1$ belong to L, and for any possible $x \in L$ s.t. $x + 1 \in L$, we have $\delta_y(x + 1) - \delta_y(x) = 0$. Therefore, $\sum_{i \in N} \phi_i(\delta_y) = 0$.

To summarize, we shall write

$$\sum_{i \in N} \phi_i(\delta_y) = \begin{cases} +1 & \text{if } k(y) \neq \text{ and } s(y) = n \\ -1 & \text{if } k(y) = 0 \text{ and } s(y) < n \\ 0 & \text{else} \end{cases}$$

This can be written in the following compact form.

Restricted Efficiency Axiom (RE): For all $x \in L \backslash \{0_N\}$,

$$\sum_{i \in N} \phi_i(\delta_x) = \delta_x(x_{-i}, k_i) - \delta_x(x_{-j}, 0_j)$$

where, $i = argmax\ x$ and $j = argmin\ x$.

Note that the previous formula takes the form of standard efficiency $\sum_{i \in N} \phi_i^{Sh}(\mu) = \mu(N) - \mu(\emptyset)$. The final result is the following.

Theorem 1. *Under axioms (L), (N), (I), (S) and (RE), for all $v \in \mathcal{G}(L)$*

$$\phi_i(v) = \sum_{x_{-i} \in L_{-i}} \frac{(n - s(x_{-i}) - 1)! k(x_{-i})!}{(n + k(x_{-i}) - s(x_{-i}))!} \left(v(x_{-i}, k_i) - v(x_{-i}, 0_i) \right), \forall i \in N.$$

Our axiomatic characterization has in common with the Shapley value the satisfaction of axioms **(L)**, **(N)** and **(S)**. The other axioms **(I)** and **(RE)** are different.

We note that we have the following relation, for every $v \in \mathcal{G}(L)$

$$\sum_{i \in N} \phi_i(v) = \sum_{\substack{x \in L \\ x_j < k}} (v(x+1) - v(x)).$$

The right-hand side of this expression corresponds exactly to the interpretation provided above saying that $\sum_{i \in N} \phi_i(v)$ is the overall impact of going from any point x to $x+1$.

We apply Theorem 1 with the following example.

Example 2. Let $x \in L \backslash \{0_N\}$ and $i \in N$.
- The computation of ϕ_i w.r.t. δ_x gives,

$$\phi_i(\delta_{x_{-i},0}) = -\phi_i(\delta_{x_{-i},k}) = \frac{(n - s(x_{-i}) - 1)! k(x_{-i})!}{(n + k(x_{-i}) - s(x_{-i}))!}, \text{and } \phi_i(\delta_{x_{-i},x_i}) = 0, \forall x_i \notin \{0, k\}.$$

- The computation of ϕ_i w.r.t. u_x gives,

$$\phi_i(u_x) = \sum_{\substack{y_{-i} \in L_{-i} \\ y_{-i} \geq x_{-i}}} \frac{(n - s(x_{-i}) - 1)! k(x_{-i})!}{(n + k(x_{-i}) - s(x_{-i}))!}.$$

Let us take for exemple $N = \{1, 2, 3\}$, $k = 3$, and $x = (2, 0, 3)$, we have,

$$\phi_1(\delta_x) = 0, \phi_2(\delta_x) = -\frac{1}{2}, \phi_3(\delta_x) = \frac{1}{2},$$

$$\phi_1(u_x) = \frac{3}{2}, \phi_2(u_x) = 0, \phi_3(u_x) = \frac{9}{2}.$$

We note that $\phi_i(\delta_x)$ can be nul when the peak is attained inside the domain ($i = 1$), is strictly positive if the peak is attained at the maximal value k ($i = 3$), and is strictly negative if the peak is attained at 0 ($i = 2$). On the other hand, $\phi_i(u_x)$ is always non-negative as the unanimity function is non-decreasing.

5   Interpretation

We propose here an interpretation of ϕ in continuous spaces, that is, after extending v to the continuous domain $[0, k]^N$. We consider thus a function $U : [0, k]^N \to \mathbb{R}$ which extends v: $U(x) = v(x)$ for every $x \in L$. The importance of attribute i can be defined as (see [13, Proposition 5.3.3, p. 141])

$$Imp_i(U) = \int_{[0,k]^{n-1}} \Big(U(k_i, z_{-i}) - U(0_i, z_{-i}) \Big) dz_{-i} = \int_{[0,k]^n} \frac{\partial U}{\partial z_i}(z) \, dz.$$

In this formula, the local importance of attribute i for function U at point z is equal to $\frac{\partial U}{\partial z_i}(z)$. The index $Imp_i(U)$ appears as the mean of relative amplitude of the range of U w.r.t. attribute i, when the remaining variables take uniformly random values.

The most usual extension of v on $[0, k]^N$ is the Choquet integral with respect to k-ary capacities [7]. Let us compute Imp_i in this case. We write $Imp_i(U) = \sum_{x \in \{0,...,k-1\}^N} \int_{[x,x+1]^n} \frac{\partial U}{\partial z_i}(z)\, dz$. In $[x, x+1]^n$, U is equal to $v(x)$ plus the Choquet integral C_{μ_x} w.r.t. capacity μ_x defined by $\mu_x(S) = v((x+1)_S, x_{-S}) - v(x)$ for every $S \subseteq N$. By [13], $\int_{[0,1]^n} \frac{\partial C_{\mu_x}}{\partial z_i}(z)\, dz = \phi_i^{Sh}(\mu_x)$. Hence

$$Imp_i(U) = \sum_{x \in \{0,...,k-1\}^N} \phi_i^{Sh}(\mu_x). \tag{5}$$

We then obtain the following result.

Lemma 1. *If U is the Choquet integral w.r.t. k-ary capacity v, then $Imp_i(U) = \phi_i(v)$.*

Hence the counterpart of ϕ_i on continuous domains is the integrated local importance.

6   Conclusion and Related Works

We have proposed a new importance index for GAI models. It quantifies the impact of each attribute on the overall utility. According to the linearity, null criterion, symmetry and invariance properties, $\phi_i(v)$ takes the form of the sum over $x \in \{0,\dots, k-1\}^N$ of a value over the restriction of functions v on $\times_{i \in N}\{x_i, x_i + 1\}$ (see (5)). In our construction, the value at an elementary cell $\times_{i \in N}\{x_i, x_i + 1\}$ corresponds to the usual Shapley value.

We will explore in future work the possibility of the use of other values such as the Banzhaf value. We will also investigate other indices ϕ_i which measure the impact in absolute value of attribute i. In this case, $\phi_i(\delta_y)$ is not equal to zero when $0 < y_i < k$.

References

1. Bacchus, F., Grove, A.: Graphical models for preference and utility. In: Conference on Uncertainty in Artificial Intelligence (UAI), Montreal, Canada, pp. 3–10, July 1995
2. Braziunas, D., Boutilier, C.: Minimax regret based elicitation of generalized additive utilities. In: Proceedings of the Twenty-Third Conference on Uncertainty in Artificial Intelligence (UAI-07), Vancouver, pp. 25–32 (2007)
3. Choquet, G.: Theory of capacities. Annales de l'institut Fourier **5**, 131–295 (1953)
4. Fishburn, P.: Interdependence and additivity in multivariate, unidimensional expected utility theory. Int. Econ. Rev. **8**, 335–342 (1967)
5. Fishburn, P.: Utility Theory for Decision Making. Wiley, New York (1970)
6. Gonzales, C., Perny, P., Dubus, J.: Decision making with multiple objectives using GAI networks. Artif. Intell. J. **175**(7), 1153–1179 (2000)

7. Grabisch, M., Labreuche, C.: Capacities on lattices and k-ary capacities. In: International Conference of the Euro Society for Fuzzy Logic and Technology (EUSFLAT), Zittau, Germany, 10–12 September 2003

8. Grabisch, M., Labreuche, C.: A decade of application of the Choquet and Sugeno integrals in multi-criteria decision aid. Ann. Oper. Res. **175**, 247–286 (2010)

9. Grabisch, M., Labreuchem, C.: Capacities on lattices and k-ary capacities. In: 3rd International Conference of the European Society for Fuzzy Logic and Technology (EUSFLAT 2003), Zittau, Germany, pp. 304–307, September 2003

10. Grabisch, M., Lange, F.: Games on lattices, multichoice games and the Shapley value: a new approach. Math. Methods Oper. Res. **65**(1), 153–167 (2007)

11. Hsiao, C.R., Raghavan, T.E.S.: Shapley value for multi-choice cooperative games, I. Discussion paper of the University of Illinois at Chicago, Chicago (1990)

12. Klijn, F., Slikker, M., Zarzuelo, J.: Characterizations of a multi-choice value. Int. J. Game Theor. **28**(4), 521–532 (1999)

13. Marichal, J.-L.: Aggregation operators for multicriteria decision aid. Ph.D. thesis, University of Liège (1998)

14. Peters, H., Zank, H.: The egalitarian solution for multichoice games. Ann. Oper. Res. **137**(1), 399–409 (2005)

15. Shapley, L.: A value for n-person games. In: Kuhn, I.H., Tucker, A. (eds.) Contributions to the Theory of Games, II(28), pp. 307–317. Princeton University Press, Princeton (1953)

16. Sugeno, M.: Theory of fuzzy integrals and its applications. Ph.D. thesis, Tokyo Institute of Technology (1974)

17. van den Nouweland, A., Tijs, S., Potters, J., Zarzuelo, J.: Cores and related solution concepts for multi-choice games. Research Memorandum FEW 478, Tilburg University, School of Economics and Management (1991)

18. Weber, R.J.: Probabilistic values for games. In: Roth, A.E. (ed.) The Shapley Value: Essays in Honor of Lloyd S. Shapley, pp. 101–120. Cambridge University Press, Cambridge (1988)

Fuzzy Sets, Fuzzy Logic

Probability Measures in Gödel$_\Delta$ Logic

Stefano Aguzzoli[1]([✉]), Matteo Bianchi[2], Brunella Gerla[2], and Diego Valota[1]

[1] Department of Computer Science, Università degli Studi di Milano,
via Comelico 39/41, 20135 Milan, Italy
{aguzzoli,valota}@di.unimi.it

[2] Dipartimento di Scienze Teoriche e Applicate, Università degli Studi dell'Insubria,
via Mazzini 5, 21100 Varese, Italy
{brunella.gerla,matteo.bianchi}@uninsubria.it

Abstract. In this paper we define and axiomatise finitely additive probability measures for events described by formulas in Gödel$_\Delta$ (G$_\Delta$) propositional logic. In particular we show that our axioms fully characterise finitely additive probability measures over the free finitely generated algebras in the variety constituting the algebraic semantics of G$_\Delta$ as integrals of elements of those algebras (represented canonically as algebras of $[0, 1]$-valued functions), with respect to Borel probability measures.

Keywords: Probability measures in non-classical logics · Gödel propositional logic, Gödel$_\Delta$ propositional logic, Free algebras

1 Introduction

Probability theory over non-classical logics has been an increasingly interesting research topic for the last pair of decades. The literature on the subject amounts to several research articles and monograph chapters. One of the first papers connecting Łukasiewicz logic with probability theory was [11], in which the notion of state (finitely additive probability measure) over MV-algebras was introduced, and from then several directions have been pursued (see [9]). More recently, in [5,6] the notion of state was studied for Gödel-Dummett logic G, by showing also a connection with the defuzzification process of fuzzy systems, and in [4] for the logic of Nilpotent Minimum. In all such logics, formulas can be seen as functions taking values in the real interval $[0, 1]$, hence it is reasonable to expect that finitely additive probability measures over formulas behave as integrals with respect to some measure over $[0, 1]$. The logic G$_\Delta$ was introduced in [7]. It is obtained by expanding the language of G with a unary operator Δ that increases expressiveness: informally, we can say that Δ allows to make "fuzzy" statements "crisp", allowing to express the characteristic functions of 1-sets of fuzzy sets (see Eq. (2)). In this paper we introduce the notion of state for classes of logically equivalent formulas of G$_\Delta$. Indeed we show that classes of logically equivalent formulas of G$_\Delta$ can be represented as real valued functions, and that a definition of state over logically equivalent formulas can be given in such a way that it corresponds to integrals of such functions with respect to a

© Springer International Publishing AG 2017
A. Antonucci et al. (Eds.): ECSQARU 2017, LNAI 10369, pp. 353–363, 2017.
DOI: 10.1007/978-3-319-61581-3_32

measure over $[0, 1]$. One of the advantages of having Δ is that the functional representation and the axiomatisation of states can be slightly simplified, w.r.t. the one provided in [6] for G.

We further characterise states by combinatorial means, considering the dual space of G_Δ-algebras: such characterisation can be fruitfully adapted to other logics, and in the last section we shall briefly mention the case of Drastic Product logic.

2   Algebraic Notions

We assume that the reader is acquainted with many-valued logics in Hájek's sense, and with their algebraic semantics. We refer to [9,10] for any unexplained notion. We recall that MTL is the logic, on the language $\{\&, \wedge, \vee, \to, \neg, \bot, \top\}$, of all left-continuous t-norms and their residua, and that its associated algebraic semantics in the sense of Blok and Pigozzi [8] is the variety MTL of *MTL-algebras*, that is, prelinear, commutative, bounded, integral, residuated lattices [9]. In an MTL-algebra $\mathcal{A} = (A, *, \Rightarrow, \sqcap, \sqcup, \sim, 0, 1)$ the connectives $\&, \to$, $\wedge, \vee, \neg, \bot, \top$ are interpreted, respectively, by $*, \Rightarrow, \sqcap, \sqcup, \sim, 0, 1$. Totally ordered MTL-algebras are called MTL-chains. In every chain $\sqcap = \min$ and $\sqcup = \max$.

A logic L is the extension of MTL via a set of axioms $\{\varphi_i\}_{i \in I}$ if and only if $\mathbb{L}$ is the subvariety of MTL-algebras satisfying $\{\bar{\varphi}_i = 1\}_{i \in I}$, where $\bar{\varphi}_i$ is obtained from φ_i by replacing the connectives with the corresponding operations, and every propositional variable in φ with an individual variable.

Gödel logic G is axiomatised as MTL plus $\varphi \to (\varphi \& \varphi)$. The variety $\mathbb{G}$ of G-algebras is axiomatised as MTL plus $x \Rightarrow (x * x) = 1$ (see [9]). The operations of a G-chain are defined as follows.

$$x * y = \min\{x, y\} \qquad x \Rightarrow y = \begin{cases} 1 & \text{if } x \leq y, \\ y & \text{otherwise} \end{cases} \qquad \sim x = \begin{cases} 1 & \text{if } x = 0, \\ 0 & \text{otherwise.} \end{cases} \tag{1}$$

The logic G_Δ [7] is obtained by expanding the language with a new unary connective Δ. The corresponding variety $\mathbb{G}_\Delta$ is axiomatised as follows.

$$\Delta(x) \sqcup \sim\Delta(x) = 1. \tag{$\Delta 1$}$$

$$\Delta(x \sqcup y) \Rightarrow (\Delta(x) \sqcup \Delta(y)) = 1. \tag{$\Delta 2$}$$

$$\Delta(x) \Rightarrow x = 1. \tag{$\Delta 3$}$$

$$\Delta(x) \Rightarrow \Delta(\Delta(x)) = 1. \tag{$\Delta 4$}$$

$$\Delta(x \Rightarrow y) \Rightarrow (\Delta(x) \Rightarrow \Delta(y)) = 1. \tag{$\Delta 5$}$$

For every G_Δ-chain, the operation Δ has the following semantics:

$$\Delta(x) = \begin{cases} 1 & \text{if } x = 1, \\ 0 & \text{otherwise.} \end{cases} \tag{2}$$

$\mathbb{G}_\Delta$ is the variety generated by all the G_Δ-chains. Moreover, it is singly generated by the *standard* G_Δ-chain $[0, 1]_\Delta = ([0, 1], *, \min, \max, \Rightarrow, \sim, 0, 1, \Delta)$ where $* =$

min, while $\Rightarrow$ and $\sim$ are defined as in Eq. (1). Since in $\mathbb{G}$ and in $\mathbb{G}_\Delta$ the operations interpreting $\&$ and $\wedge$ always coincide, from now on we drop the first one from the signature.

Let $\mathcal{A}$ be an MTL algebra, then $\mathfrak{p} \subseteq A$ is a *filter* of $\mathcal{A}$ if for all $y \in A$, if there is x in $\mathfrak{p}$ such that $x \leq y$ then $y \in \mathfrak{p}$, and $x * y \in \mathfrak{p}$ for all $x, y \in \mathfrak{p}$. We call *proper* the filters $\mathfrak{p}$ such that $\mathfrak{p} \neq A$. If $\mathcal{A}$ is a $\mathbb{G}_\Delta$-algebra then a filter $\mathfrak{p}$ of $\mathcal{A}$ is a filter of its Gödel reduct $\bar{\mathcal{A}}$ further satisfying $x \in \mathfrak{p}$ implies $\Delta x \in \mathfrak{p}$.

Filters and congruences of $\mathcal{A}$ are in bijection. Indeed, $x\theta_\mathfrak{p} y$ if and only if $(x \Rightarrow y) \sqcap (y \Rightarrow x) \in \mathfrak{p}$, and $\mathfrak{p}_\theta = \{x \in A \mid x\theta 1\}$. By abuse of notation we write $A/\mathfrak{p}$ to denote $A/\theta_\mathfrak{p}$.

A filter $\mathfrak{p}$ of $\mathcal{A}$ is *prime* if it is proper and for all $x, y \in A$, either $x \Rightarrow y \in \mathfrak{p}$ or $y \Rightarrow x \in \mathfrak{p}$. The set $Spec(\mathcal{A})$ of all prime filters of $\mathcal{A}$ ordered by reverse inclusion is called the *prime spectrum* of $\mathcal{A}$. The inclusion-maximal elements of $Spec(\mathcal{A})$ are the *maximal filters* of $\mathcal{A}$, and they form the *maximal spectrum* $Max(\mathcal{A}) \subseteq Spec(\mathcal{A})$. For each $\mathfrak{p} \in Spec(\mathcal{A})$, $A/\mathfrak{p}$ is a chain.

Proposition 1 ([2], **Lemma 7.2**). *Every finite $\mathbb{G}_\Delta$-algebra $\mathcal{A}$ is a direct product of chains. That is, $\mathcal{A} \simeq \prod_{\mathfrak{p} \in Max(\mathcal{A})} A/\mathfrak{p}$, and $Max(\mathcal{A}) = Spec(\mathcal{A})$.*

We recall that in any lattice $\mathcal{A}$ an element a is *join-irreducible* if whenever $a = b \vee c$, then $a = b$ or $a = c$. In any finite lattice every element a is the join of all join-irreducible elements below a. An element a *covers* an element b if $b < a$, and for every $c \in \mathcal{A}$, if $b \leq c \leq a$, then either $c = b$ or $c = a$.

3 Functional Representation

Two formulas φ and ψ over n many distinct variables $x_1, x_2, \ldots, x_n$ are *logically equivalent* in $\mathbb{G}_\Delta$ iff both $\varphi \rightarrow \psi$ and $\psi \rightarrow \varphi$ are $\mathbb{G}_\Delta$ tautologies. As is well known, the *Lindenbaum algebra* formed by the set of all classes of *logically equivalent* formulas over the set of variables $\{x_1, \ldots, x_n\}$ equipped with the operations inherited from the connectives, constitutes a $\mathbb{G}_\Delta$-algebra, which in turn is (isomorphic with) the free n-generated $\mathbb{G}_\Delta$-algebra $\mathbf{F}_n(\mathbb{G}_\Delta)$.

Since the algebra $[0,1]_\Delta$ singly generates the whole variety $\mathbb{G}_\Delta$, from universal algebra we have that $\mathbf{F}_n(\mathbb{G}_\Delta)$ is isomorphic with the subalgebra of the algebra of all functions $f : [0,1]_\Delta^n \rightarrow [0,1]_\Delta$ generated by the projection functions $\overline{x_i} : (t_1, \ldots, t_n) \mapsto t_i$, for all $i \in \{1, 2, \ldots, n\}$. To fix notation, for each formula φ over $\{x_1, \ldots, x_n\}$ we shall write $\overline{\varphi}$ for the corresponding element of $\mathbf{F}_n(\mathbb{G}_\Delta)$, considered as a function $\overline{\varphi} : [0,1]^n \rightarrow [0,1]$. Moreover, we shall denote each operation of $\mathbf{F}_n(\mathbb{G}_\Delta)$ with the same symbol of the connective it interprets.

In this section we shall characterise the functions $\overline{\varphi}$ belonging to $\mathbf{F}_n(\mathbb{G}_\Delta)$.

Let $\approx$ be the binary relation on $[0,1]^n$ defined in the following way: given two n-tuples $\mathbf{u} = (u_1, \cdots, u_n), \mathbf{v} = (v_1, \cdots, v_n) \in [0,1]^n$ we set $\mathbf{u} \approx \mathbf{v}$ if and only if there is a permutation σ of $\{1, \ldots, n\}$ and a map $\prec : \{0, \ldots, n\} \rightarrow \{<, =\}$ such that (we write $\prec_i$ for $\prec(i)$)

$$0 \prec_0 u_{\sigma(1)} \prec_1 \cdots \prec_{n-1} u_{\sigma(n)} \prec_n 1 \text{ iff } 0 \prec_0 v_{\sigma(1)} \prec_1 \cdots \prec_{n-1} v_{\sigma(n)} \prec_n 1. \quad (3)$$

The relation $\approx$ is an equivalence relation. We denote by $[\mathbf{u}]$ the equivalence class of $\mathbf{u}$. The quotient set $[0,1]^n / \approx$ is hence a partition of $[0,1]^n$.

With each class $[\mathbf{u}]$, where $0 \prec_0 u_{\sigma(1)} \prec_1 \cdots \prec_{n-1} u_{\sigma(n)} \prec_n 1$, we associate a unique *ordered partition* $\rho_{\mathbf{u}} = Q_1 < \cdots < Q_h$ (i.e., a partition equipped with a total order among its blocks) of the set $\{\bot, x_1, \ldots, x_n, \top\}$ in the following way:

- $\bot \in Q_1$; $\top \in Q_h$; $h > 1$;
- if $\prec_i$ is = then $x_{\sigma(i)}$ and $x_{\sigma(i+1)}$ belong to the same Q_j;
- if $\prec_i$ is < and $x_{\sigma(i)} \in Q_j$ then $x_{\sigma(i+1)} \in Q_{j+1}$.

We call such an ordered partition a *Gödel n-partition*. There is a bijection between Gödel n-partitions and equivalence classes $[\mathbf{u}] \in [0,1]^n / \approx$. If $\rho = \rho_{\mathbf{u}}$ is a Gödel n-partition, we denote by D_ρ the associated equivalence class $[\mathbf{u}]$. We write Ω_n for the set of all Gödel n-partitions.

A n-variate G_Δ-function is a function $f : [0,1]^n \to [0,1]$ such that for every $\mathbf{u} \in [0,1]^n$ (equivalently, for any $\rho \in \Omega_n$) the restriction of f to $[\mathbf{u}]$ (equivalently, to D_ρ) is either equal to 0, or to 1, or to a projection function $\overline{x_i}$.

Theorem 1. *The elements of $\mathbf{F}_n(\mathbb{G}_\Delta)$ are exactly the n-variate G_Δ-functions.*

Proof. An easy induction on the complexity of a formula φ over $\{x_1, \ldots, x_n\}$ proves that $\overline{\varphi}$ is a G_Δ-function.

To prove the other direction we start by defining formulas whose associated function is the characteristic function of the set $[\mathbf{u}]$ for all $\mathbf{u} \in [0,1]^n$. Let us define the following derived connective:

$$x \lhd y = \Delta(x \to y) \land \neg\Delta(y \to x).$$

Note that when interpreted in $[0,1]$ we have $\overline{x \lhd y} = 1$ if $x < y$ and $\overline{x \lhd y} = 0$ otherwise. Let $\mathbf{u} = (u_1, \cdots, u_n)$ in $[0,1]^n$ with $0 \prec_0 u_{\sigma(1)} \prec_1 \cdots \prec_{n-1} u_{\sigma(n)} \prec_n 1$. To simplify notation, let us put $x_{\sigma(0)} = \bot$ and $x_{\sigma(n+1)} = \top$. Moreover, let $x \leftrightarrow y$ denote $(x \to y) \land (y \to x)$. For any $\rho = \rho_{\mathbf{u}} \in \Omega_n$, consider the formula

$$\chi_\rho = \bigwedge_{i=0}^{n} \delta_i,$$

where

$$\delta_i = \begin{cases} \Delta(x_{\sigma(i)} \leftrightarrow x_{\sigma(i+1)}) & \text{iff } \prec_i \text{ is } =, \\ x_{\sigma(i)} \lhd x_{\sigma(i+1)} & \text{iff } \prec_i \text{ is } < . \end{cases}$$

Then it is straightforward to check that $\overline{\chi_\rho}(\mathbf{v}) = 1$ iff $\mathbf{v} \approx \mathbf{u}$, while $\overline{\chi_\rho}(\mathbf{v}) = 0$ otherwise.

Let now $f : [0,1]^n \to [0,1]$ be a G_Δ-function. For each $\rho \in \Omega_n$, let y_ρ be the necessarily unique element of $\{\bot, x_1, \ldots, x_n, \top\}$ such that $\overline{y_\rho}$ coincides with f over the whole of D_ρ. Consider the formula

$$\varphi = \bigvee_{\rho \in \Omega_n} (\chi_\rho \land y_\rho).$$

Fix any point $\mathbf{u} \in [0,1]^n$. Then $\overline{\varphi}(\mathbf{u})$ coincides with $\overline{\chi_\rho \wedge y_\rho}(\mathbf{u})$ for the unique $\rho \in \Omega_n$ such that $\mathbf{u} \in D_\rho$, since $\overline{\chi_\tau}(\mathbf{u}) = 0$ for all $\rho \neq \tau \in \Omega_n$, while $\overline{\chi_\rho}(\mathbf{u}) = 1$. Whence, $\overline{\varphi}(\mathbf{u}) = \overline{y_\rho}(\mathbf{u})$, that is $\overline{\varphi}(\mathbf{u}) = f(\mathbf{u})$, since $\mathbf{u} \in D_\rho$. We have proved $\overline{\varphi} = f$. $\qquad\qquad\qquad\qquad\qquad\qquad\qquad\qquad\qquad\qquad\qquad\qquad\qquad\qquad\qquad$ $\square$

4 States on $\mathbf{F}_n(\mathbb{G}_\Delta)$

In this section we introduce the notion of *state* over a finitely generated free G_Δ-algebra, and prove our main result, stating that integrals of elements of such algebras with respect to Borel probability measures exactly correspond to our notion of states.

Definition 1. *A state on $\mathbf{F}_n(\mathbb{G}_\Delta)$ is a function $s: \mathbf{F}_n(\mathbb{G}_\Delta) \to [0,1]$ such that, for every $f, g \in \mathbf{F}_n(\mathbb{G}_\Delta)$:*

1. *$s(\bot) = 0$, $s(\top) = 1$;*
2. *$s(f \vee g) = s(f) + s(g) - s(f \wedge g)$;*
3. *If $f \leq g$ then $s(f) \leq s(g)$;*
4. *If $f \leq g$ and $s(g) = s(f)$ then $s(\Delta(g \to f)) = 1$.*

Note that, as in the case of Gödel logic [6], the following theorem shows that Definition 1 provides an axiomatisation à la Kolmogorov, where no explicit conditions on the connectives $\neg$, $\to$ and Δ are required, apart from item 4, which deals with the interaction of order, Δ and $\to$ (recall that $\neg\varphi$ is $\varphi \to \bot$).

Theorem 2. *The following hold.*

1. *If $s: \mathbf{F}_n(\mathbb{G}_\Delta) \to [0,1]^n$ is a state, there exists a Borel probability measure μ on $[0,1]^n$ such that*

$$\int_{[0,1]^n} f \, \mathrm{d}\mu = s(f), \text{ for every } f \in \mathbf{F}_n(\mathbb{G}_\Delta). \qquad (4)$$

2. *Viceversa, for any Borel probability measure μ on $[0,1]^n$, the function $s: \mathbf{F}_n(\mathbb{G}_\Delta) \to [0,1]$ defined by (4) is a state.*

Proof. We prove (1). Let s be a state, and consider a Gödel n-partition ρ and the region $D_\rho \subseteq [0,1]^n$. Throughout the proof we fix $x_0 = 0$ and $x_{n+1} = 1$. For any ρ such that $s(\overline{\chi_\rho}) \neq 0$, we define the element $\mathbf{z}_\rho \in [0,1]^n$ whose i-th component z_ρ^i is equal to

$$z_\rho^i = \frac{s(\overline{x_i} \wedge \overline{\chi_\rho})}{s(\overline{\chi_\rho})}.$$

We claim that $\mathbf{z}_\rho \in D_\rho$. Indeed consider $i, j \in \{0, \cdots, n+1\}$ and suppose first that for every element $(t_1, \ldots, t_n) \in D_\rho$, $t_i = t_j$. Hence $\overline{x_i} \wedge \overline{\chi_\rho} = \overline{x_j} \wedge \overline{\chi_\rho}$ and $z_\rho^i = z_\rho^j$. On the other hand, if $t_i < t_j$ for some i, j, then $\overline{x_i} \wedge \overline{\chi_\rho} < \overline{x_j} \wedge \overline{\chi_\rho}$ and, by 3. in Definition 1, $s(\overline{x_i} \wedge \overline{\chi_\rho}) \leq s(\overline{x_j} \wedge \overline{\chi_\rho})$. If it were $s(\overline{x_i} \wedge \overline{\chi_\rho}) = s(\overline{x_j} \wedge \overline{\chi_\rho})$, by

4. in Definition 1 it would be the case that $s(\Delta((\overline{x_j} \wedge \overline{\chi_\rho}) \to (\overline{x_i} \wedge \overline{\chi_\rho}))) = 1$. This is impossible by 1. in Definition 1, since $(\overline{x_j} \wedge \overline{\chi_\rho}) = \overline{\chi_\rho}$ and then $\Delta((\overline{x_j} \wedge \overline{\chi_\rho}) \to (\overline{x_i} \wedge \overline{\chi_\rho})) = \overline{1}$,

hence it must be that $s(\overline{x_i} \wedge \overline{\chi_\rho}) < s(\overline{x_j} \wedge \overline{\chi_\rho})$ and $z_\rho^i < z_\rho^j$. We can hence conclude that $z_\rho \in D_\rho$.

We are ready to define the discrete Borel probability measure μ determined by $\mu([0,1]^n \setminus \{z_\rho \mid \rho \in \Omega_n\}) = 0$ and $\mu(\{z_\rho\}) = s(\overline{\chi_\rho})$ for each $\rho \in \Omega_n$.

If $f \in \mathbf{F}_n(\mathbb{G}_\Delta)$, we have:

$$\int_{[0,1]^n} f \, d\mu = \sum_{\rho \in \Omega_n} \int_{D_\rho} f \, d\mu = \sum_{\rho \in \Omega_n} f(z_\rho) \mu(\{z_\rho\}) = \sum_{\rho \in \Omega_n} f(z_\rho) s(\overline{\chi_\rho}).$$

By Theorem 1, f is a G_Δ-function. Then on every region D_ρ the function f is equal to either 0 or 1 or a projection function $\overline{x_i}$. Let $\rho(i) \in \{0, 1, \cdots, n+1\}$ be such that $\overline{x_{\rho(i)}} = f \restriction D_\rho$. We hence have

$$\sum_{\rho \in \Omega_n} f(z_\rho) s(\overline{\chi_\rho}) = \sum_{\rho \in \Omega_n} z_\rho^{\rho(i)} s(\overline{\chi_\rho}) = \sum_{\rho \in \Omega_n} \frac{s(\overline{x_{\rho(i)}} \wedge \overline{\chi_\rho})}{s(\overline{\chi_\rho})} s(\overline{\chi_\rho}) = \sum_{\rho \in \Omega_n} s(\overline{x_{\rho(i)}} \wedge \overline{\chi_\rho}).$$

Since, for $\rho \neq \sigma$ we have $(\overline{x_{\rho(i)}} \wedge \overline{\chi_\rho}) \wedge (\overline{x_{\sigma(i)}} \wedge \overline{\chi_\sigma}) = \overline{1}$, by 2. in Definition 1 we have

$$\sum_{\rho \in \Omega_n} s(\overline{x_{\rho(i)}} \wedge \overline{\chi_\rho}) = s \left(\bigvee_{\rho \in \Omega_n} (\overline{x_{\rho(i)}} \wedge \overline{\chi_\rho}) \right) = s(f).$$

We now prove (2). It is easy to check that the function $s(f) = \int_{[0,1]^n} f \, d\mu$ satisfies properties 1., 2. and 3. of Definition 1, so we focus on the last property. Let $f, g \in \mathbf{F}_n(\mathbb{G}_\Delta)$ with $f \leq g$ and

$$\int_{[0,1]^n} g \, d\mu = \int_{[0,1]^n} f \, d\mu. \tag{5}$$

Let $A = \{x \in [0,1]^n \mid f(x) = g(x)\}$ and $B = [0,1]^n \setminus A = \{x \in [0,1]^n \mid f(x) < g(x)\}$. We have

$$\int_{[0,1]^n} f \, d\mu = \int_A f \, d\mu + \int_B f \, d\mu = \int_A g \, d\mu + \int_B f \, d\mu.$$

and

$$\int_{[0,1]^n} g \, d\mu = \int_A g \, d\mu + \int_B g \, d\mu.$$

Whence, by (5), $\int_B f \, d\mu = \int_B g \, d\mu$. Since in B we have $f < g$ it must be $\mu(B) = 0$, hence $\mu(A) = 1$ and, since $\Delta(g \to f)(x) = 1$ for every $x \in A$,

$$\int_{[0,1]^n} \Delta(g \to f) \, d\mu = \int_A \Delta(g \to f) \, d\mu + \int_B \Delta(g \to f) \, d\mu = \mu(A) = 1.$$

$\square$

In words, fixing a state on $\mathbf{F}_n(\mathbb{G}_\Delta)$ precisely amounts to integrating $\mathbb{G}_\Delta$-functions of n variables with respect to an appropriate Borel probability measure μ on $[0,1]^n$. The proof of the following corollary is omitted for lack of space.

Corollary 1. *The states of $\mathbf{F}_n(\mathbb{G}_\Delta)$ are precisely the convex combinations of finitely many truth value assignments.*

Example 3. For $n = 2$, the set of all Gödel partitions Ω_2 counts 11 elements. Consider $\rho_1 = \{0, x, y\} < \{1\}$, $\rho_2 = \{0\} < \{x, y\} < \{1\}$ and $\rho_3 = \{0\} < \{x\} < \{y\} < \{1\}$. Let s be the state on $\mathbf{F}_2(\mathbb{G}_\Delta)$ given by setting $s(\overline{\chi_{\rho_1}}) = 1/3$, $s(\overline{\chi_{\rho_2}}) = 1/6$, $s(\overline{\chi_{\rho_3}}) = 1/2$, $s(\overline{x \wedge \chi_{\rho_2}}) = 1/12$, $s(\overline{x \wedge \chi_{\rho_3}}) = 1/12$, $s(\overline{y \wedge \chi_{\rho_3}}) = 1/6$, $s(\overline{\chi_\sigma}) = 0$ for $\sigma \notin \{\rho_1, \rho_2, \rho_3\}$; all the other values of s are determined by the previous ones. Then we have three points $\mathbf{z}_{\rho_1} = (0,0)$, $\mathbf{z}_{\rho_2} = (1/2, 1/2)$ and $\mathbf{z}_{\rho_3} = (1/6, 1/3)$ on which we can define the discrete measure μ by setting $\mu(\{\mathbf{z}_{\rho_1}\}) = 1/3$, $\mu(\{\mathbf{z}_{\rho_2}\}) = 1/6$ and $\mu(\{\mathbf{z}_{\rho_3}\}) = 1/2$. Consider now the $\mathbb{G}_\Delta$-function f that is equal to 1 over D_{ρ_1}, it is equal to 0 over D_{ρ_2} and it is equal to $\overline{y}$ on D_{ρ_3} (the other values of f are not relevant to our example). Then

$$s(f) = s(\overline{\chi_{\rho_1}} \vee (\overline{y} \wedge \overline{\chi_{\rho_3}})) = s(\overline{\chi_{\rho_1}}) + s(\overline{y} \wedge \overline{\chi_{\rho_3}}) = \frac{1}{3} + \frac{1}{6} = \frac{1}{2}$$

and

$$\int_{[0,1]^2} f \, \mathrm{d}\mu = \sum_{i=1}^3 f(z_{\rho_i}) \mu(\{z_{\rho_i}\}) = 1 \cdot \frac{1}{3} + 0 \cdot \frac{1}{6} + \frac{1}{3} \cdot \frac{1}{2} = \frac{1}{2} = s(f).$$

5 A Dual Equivalence

In this section we recall from [2] a dual equivalence concerning finite $\mathbb{G}_\Delta$-algebras that will be useful for dealing with states with combinatorial tools.

The variety $\mathbb{G}_\Delta$ of $\mathbb{G}_\Delta$-algebras and their homomorphisms form a category. We write $(\mathbb{G}_\Delta)_{fin}$ for the full subcategory of $\mathbb{G}_\Delta$ whose objects have finite cardinality. Authors in [2] introduce a dual equivalence between $(\mathbb{G}_\Delta)_{fin}$ and a suitable combinatorial category by adapting the well-known dual categorical equivalence between finite G-algebras and finite forests (see [3,5] for details). In this section we briefly recall their results.

Lemma 1 ([2]). *Let $\mathcal{C}$ and $\mathcal{D}$ be G_Δ-chains, and let $h : \mathcal{C} \to \mathcal{D}$ be a homomorphism. Then h is injective.*

Let $h \colon \mathcal{A} \to \prod_{i \in I} \mathcal{C}_i$ be a homomorphism of G-algebras, where each $\mathcal{C}_i$ is a G-chain. Then h is called *chain-injective* if, given each projection $\pi_j \colon \prod_{i \in I} \mathcal{C}_i \to \mathcal{C}_j$, the homomorphism $\pi_j \circ h \colon \mathcal{A} \to \mathcal{C}_j$ is injective.

By Lemma 1 each homomorphism h of G_Δ-algebras is a chain-injective homomorphism between their Gödel reducts. These observation allows to prove a dual equivalence between $(\mathbb{G}_\Delta)_{fin}$ and the combinatorial category we are going to describe.

Given a poset $(P, \leq)$, a subset C of P is a *subchain* of P if it is totally ordered by the restriction of $\leq$. A subset $Q \subseteq P$ is *downward-closed* if $Q = \downarrow Q$, for $\downarrow Q = \{x \in P \mid \exists y \in Q, x \leq y\}$. Analogously, $Q \subseteq P$ is *upward-closed* if $Q = \uparrow Q$, for $\uparrow Q = \{x \in P \mid \exists y \in Q, y \leq x\}$. Given a poset P which is a disjoint union $C_1 \cup C_2 \cup \cdots \cup C_u$ of chains, we write $\mathcal{C}(P)$ for the multiset $\{C_1, C_2, \ldots, C_u\}$.

Let MC be the category whose objects are finite multisets of (nonempty) finite chains, and whose morphisms $h \colon C \to D$, are defined as follows. Display C as $\{C_1, \ldots, C_m\}$ and D as $\{D_1, \ldots, D_n\}$. Then $h = \{h_i\}_{i=1}^{m}$, where each h_i is an order preserving surjection $h_i \colon C_i \twoheadrightarrow D_j$ for some $j \in \{1, 2, \ldots, n\}$.

Theorem 4 ([2]). *The categories $(\mathbb{G}_\Delta)_{fin}$ and MC are dually equivalent.*

We remark that, for each $\mathcal{A} \in (\mathbb{G}_\Delta)_{fin}$, the poset $Spec(\bar{\mathcal{A}})$, that is, the prime spectrum of the G-algebra reduct of $\mathcal{A}$, ordered by reverse inclusion, is isomorphic with the poset of the join-irreducible elements of $\mathcal{A}$ ordered by restriction. As a matter of fact, each prime filter $\mathfrak{p}$ of $\bar{\mathcal{A}}$ is generated by a join-irreducible element a as $\mathfrak{p} = \{b \in \mathcal{A} \mid a \leq b\}$. On the other hand, each join-irreducible element of $\mathcal{A}$ singly generates a prime filter of $\bar{\mathcal{A}}$. Recall also that in general $Max(\mathcal{A}) = Spec(\mathcal{A}) \neq Spec(\bar{\mathcal{A}})$, and equality between the two prime spectra occurs if and only if $\mathcal{A}$ is a Boolean algebra. Further, it is clear that the poset $Spec(\bar{\mathcal{A}})$ is the disjoint union of chains, as $\mathcal{A}$ is a direct product of chains by Proposition 1.

The functor implementing one side of the dual equivalence is:

$$\mathsf{Spec}^\Delta : (\mathbb{G}_\Delta)_{fin} \to \mathsf{MC},$$

defined on objects as $\mathsf{Spec}^\Delta(\mathcal{A}) = \mathcal{C}(Spec(\bar{\mathcal{A}}))$, and on morphisms as follows. If $h : \mathcal{A} \to \mathcal{B}$ then $(\mathsf{Spec}^\Delta h) : \mathsf{Spec}^\Delta(\mathcal{B}) \to \mathsf{Spec}^\Delta(\mathcal{A})$ is given by $(\mathsf{Spec}^\Delta h)_i(\mathfrak{p}) = h^{-1}[\mathfrak{p}] \cap C_i$ for each $\mathfrak{p} \in Spec(\bar{\mathcal{B}})$ and each $i \in \{1, 2, \ldots, m\}$. The other side of the dual equivalence is given by the functor

$$\mathsf{Sub}^\Delta : \mathsf{MC} \to (\mathbb{G}_\Delta)_{fin},$$

defined on objects by the following prescriptions. Given $\{C\} \in \mathsf{MC}$, define $\Delta C = C$ and $\Delta D = \emptyset$ for each proper downward-closed subchain $D \subsetneq C$. For all downward-closed subchains $D_1, D_2 \subseteq C$, define $D_1 \to D_2 = C \backslash \uparrow (D_1 \setminus D_2)$ (that is, $D_1 \to D_2 = C$ if $D_1 \subseteq D_2$, and $D_1 \to D_2 = D_2$ otherwise). Further, define $\sim D_1 = C$ if $D_1 = \emptyset$ and $\sim D_1 = \emptyset$ otherwise. Then,

$$\mathsf{Sub}^\Delta(C) = (\{D \subseteq C \mid D = \downarrow D\}, \cup, \cap, \to, \sim, \emptyset, C, \Delta).$$

Whence, for any $C = \{C_1, C_2, \ldots, C_m\} \in \mathsf{MC}$, we define

$$\mathsf{Sub}^\Delta(C) = \prod_{i=1}^{m} \mathsf{Sub}^\Delta(C_i).$$

Given a morphism $f : C \to D$ in MC, to define the dual homomorphism $\mathsf{Sub}^\Delta(f) : \mathsf{Sub}^\Delta(D) \to \mathsf{Sub}^\Delta(C)$ in $(\mathbb{G}_\Delta)_{fin}$, we recall that if C is $\{C_1, \ldots, C_m\}$

and D is $\{D_1, \ldots, D_n\}$, then $f = \{f_i\}_{i=1}^m$ is composed by order preserving surjections $f_i \colon C_i \twoheadrightarrow D_{j(i)}$ for some $j \colon \{1, 2, \ldots, m\} \to \{1, 2, \ldots, n\}$. Then, given $E = (E_1, \ldots, E_n) \in \mathsf{Sub}^\Delta(D)$ we define $(\mathsf{Sub}^\Delta(f))(E) = (C_i \cap f_i^{-1}[E_{j(i)}])_{i=1}^m$.

6 A Combinatorial Way to States

In this section we shall use the dual equivalence proved in Sect. 5, to formulate states over $\mathbf{F}_n(\mathbb{G}_\Delta)$ in a purely combinatorial way. The combinatorial notion corresponding to *state* is the following definiton of *labeling*. For every $\mathfrak{p}, \mathfrak{q} \in \mathrm{Spec}^\Delta \mathbf{F}_n(\mathbb{G}_\Delta)$ we write $\mathfrak{p} \lesseqgtr \mathfrak{q}$ to mean that $\mathfrak{p}$ and $\mathfrak{q}$ are comparable, that is $\mathfrak{p} \leq \mathfrak{q}$ or $\mathfrak{q} \leq \mathfrak{p}$, in $\mathrm{Spec}(\overline{\mathbf{F}_n(\mathbb{G}_\Delta)})$. We shall use the same notation $f \lesseqgtr g$ to mean that two join-irreducible elements of $\mathbf{F}_n(\mathbb{G}_\Delta)$ are comparable. For any join-irreducible element $g \in \mathbf{F}_n(\mathbb{G}_\Delta)$, we shall write $\langle g \rangle$ for the prime filter of $\mathbf{F}_n(\mathbb{G}_\Delta)$ generated by g.

Definition 2. *A labeling l is a function $l \colon \mathrm{Spec}^\Delta \mathbf{F}_n(\mathbb{G}_\Delta) \to [0, 1]$, such that*

1. $\sum_{\mathfrak{p} \in \mathrm{Spec}^\Delta \mathbf{F}_n(\mathbb{G}_\Delta)} l(\mathfrak{p}) = 1$;
2. *If $l(\mathfrak{p}) = 0$ then $l(\mathfrak{q}) = 0$ for all $\mathfrak{q} \lesseqgtr \mathfrak{p}$.*

We will show that labelings and states are in bijection. Moreover, labelings can be used to compute states (compare with the analogous notion given for G-algebras in [5]). For each $f \in \mathbf{F}_n(\mathbb{G}_\Delta)$ let $J(f)$ be the set of all join-irreducible elements $g \subset \mathbf{F}_n(\mathbb{G}_\Delta)$ such that $g \leq f$. Note that $f = \bigvee_{g \in J(f)} g$. Moreover, observe that $J(\Delta f) = \{g \in J(\mathbf{F}_n(\mathbb{G}_\Delta)) \mid \forall h \lesseqgtr g, h \in J(f)\}$.

Theorem 5. *Let S_n be the collection of all states $s \colon \mathbf{F}_n(\mathbb{G}_\Delta) \to [0, 1]$, and let L_n be the collection of all labelings $l \colon \mathrm{Spec}^\Delta \mathbf{F}_n(\mathbb{G}_\Delta) \to [0, 1]$. Then, the map defined for every formula φ over the set of variables $\{x_1, \ldots, x_n\}$ by*

$$(S(l))(\overline{\varphi}) = \sum_{g \in J(\overline{\varphi})} l(\langle g \rangle)$$

is a bijective correspondence $S \colon L_n \to S_n$.

Proof. Notice that for each labeling $l \in L_n$ the map $S(l)$ clearly satisfies Definition 1 (1), (2), (3). To prove Definition 1 (4), we observe that if $f, g \in \mathbf{F}_n(\mathbb{G}_\Delta)$ are such that $f \leq g$ and $(S(l))(g) = (S(l))(f)$, then $J(f) \subseteq J(g)$, and $l(\langle h \rangle) = 0$ for any $h \in J(g) \backslash J(f)$. Whence, by the definition of labeling, and the isomorphism between the posets $\mathrm{Spec}(\overline{\mathbf{F}_n(\mathbb{G}_\Delta)})$ and $J(\mathbf{F}_n(\mathbb{G}_\Delta))$, we have $l(\langle k \rangle) = 0$ for all $k \in J(\mathbf{F}_n(\mathbb{G}_\Delta))$ comparable with h. Now, $(S(l))(\Delta(g \to f)) = \sum_{h \in H} l(\langle h \rangle)$, for $H = \{k \in J(\mathbf{F}_n(\mathbb{G}_\Delta)) \mid \forall h \lesseqgtr k, h \notin \uparrow(J(g) \backslash J(f))\}$. Then $(S(l))(\Delta(g \to f)) = 1 - \sum_{k \in K} l(\langle k \rangle)$ for $K = \{k \in J(\mathbf{F}_n(\mathbb{G}_\Delta)) \mid \exists h \lesseqgtr k, h \in \uparrow(J(g) \backslash J(f))\}$. Whence, $\sum_{k \in K} l(\langle k \rangle) = 0$, and we conclude that $(S(l))(\Delta(g \to f)) = 1$, and $S(l)$ maps L_n to S_n.

We now prove injectivity of S. If $l_1 \neq l_2$ are two different labelings, then there is a join-irreducible element g such that $l_1(\langle g \rangle) \neq l_2(\langle g \rangle)$ hence $(S(l_1))(g) \neq (S(l_2))(g)$ and $S(l_1) \neq S(l_2)$.

To prove surjectivity, we construct the map inverse to S. We define for every $s \in S_n$ the following labeling. For any join-irreducible element $g \in \mathbf{F}_n(\mathbb{G}_\Delta)$ let $h \in J(g) \cup \{\bot\}$ be the unique element such that g covers h, whence $J(g) \setminus J(h) = \{g\}$. Then we set $(L(s))(\langle g \rangle) = s(g) - s(h)$. It is straightforward to check that $S(L(s)) = s$ and $L(S(l)) = l$.   □

7   Drastic Product Algebras

Drastic Product algebras constitute the subvariety $\mathbb{DP}$ of $\mathbb{MTL}$ axiomatised by $x \sqcup \sim(x * x) = 1$. Let $\mathsf{MC}^\top$ be the non-full subcategory of MC whose morphisms $h \colon C \to D$ satisfy the following additional constraint: for each $i = 1, 2, \ldots, m$, if the target D_j of h_i is not isomorphic with $\mathbf{1} = \{*\}$, then $h_i^{-1}(\max D_j) = \{\max C_i\}$.

Theorem 6 ([1]). *$\mathsf{MC}^\top$ is dually equivalent to the category $\mathbb{DP}_{fin}$ of finite DP algebras and their homomorphisms.*

The above theorem implies that the category of finite DP algebras and their homomorphisms is equivalent to a non-full subcategory of finite $\mathbb{G}_\Delta$-algebras and their homomorphisms. Using this fact, we can adapt the results of Sect. 4 to axiomatise states over DP-algebras. Details will be given elsewhere.

References

1. Aguzzoli, S., Bianchi, M., Valota, D.: A note on drastic product logic. In: Laurent, A., Strauss, O., Bouchon-Meunier, B., Yager, R.R. (eds.) IPMU 2014. CCIS, vol. 443, pp. 365–374. Springer, Cham (2014). doi:10.1007/978-3-319-08855-6_37
2. Aguzzoli, S., Codara, P.: Recursive formulas to compute coproducts of finite Gödel algebras and related structures. In: 2016 IEEE International Conference on Fuzzy Systems (FUZZ-IEEE). pp. 201–208 (2016)
3. Aguzzoli, S., D'Antona, O.M., Marra, V.: Computing minimal axiomatizations in Gödel propositional logic. J. Log. Comput. **21**, 791–812 (2011)
4. Aguzzoli, S., Gerla, B.: Probability measures in the logic of nilpotent minimum. Stud. Log. **94**(2), 151–176 (2010)
5. Aguzzoli, S., Gerla, B., Marra, V.: De Finetti's no-Dutch-book criterion for Gödel logic. Stud. Logica **90**, 25–41 (2008)
6. Aguzzoli, S., Gerla, B., Marra, V.: Defuzzifying formulas in Gödel logic through finitely additive measures. In: 2008 IEEE International Conference on Fuzzy Systems (IEEE World Congress on Computational Intelligence), pp. 1886–1893 (2008)
7. Baaz, M.: Infinite-valued Gödel logics with 0-1-projections and relativizations. In: Hájek, P. (ed.) Gödel'96: Logical foundations of mathematics, computer science and physics–Kurt Gödel's legacy, Brno, Czech Republic, August 1996, proceedings. Lecture Notes in Logic, vol. 6, pp. 23–33. Springer-Verlag, Berlin (1996)

8. Blok, W., Pigozzi, D.: Algebraizable logics, Memoirs of The American Mathematical Society, vol. 77. American Mathematical Society (1989)
9. Cintula, P., Hájek, P., Noguera, C. (eds.): Handbook of Mathematical Fuzzy Logic, vol. 1, 2, 3. College Publications, London (2011)
10. Hájek, P.: Metamathematics of Fuzzy Logic, Trends in Logic, vol. 4. Kluwer Academic Publishers, Dordrecht (1998)
11. Mundici, D.: Averaging the truth-value in Łukasiewicz logic. Stud. Logica 55(1), 113–127 (1995)

Fuzzy Weighted Attribute Combinations Based Similarity Measures

Giulianella Coletti[1], Davide Petturiti[2], and Barbara Vantaggi[3(✉)]

[1] Dip. Matematica e Informatica, Università di Perugia, Perugia, Italy
giulianella.coletti@unipg.it
[2] Dip. Economia, Università di Perugia, Perugia, Italy
davide.petturiti@unipg.it
[3] Dip. S.B.A.I., Università di Roma "La Sapienza", Rome, Italy
barbara.vantaggi@sbai.uniroma1.it

Abstract. Some similarity measures for fuzzy subsets are introduced: they are based on fuzzy set-theoretic operations and on a weight capacity expressing the degree of contribution of each group of attributes. For such measures, the properties of dominance and T-transitivity are investigated.

Keywords: Fuzzy subset · Capacity · Similarity measure · T-transitivity

1 Introduction

In many contexts, a similarity measure can be (loosely) seen as a mathematical tool to express quantitatively what is in common between two "objects". Similarity measures expressing the degree of similarity of objects are adopted in many applications such as multi-criteria decision making, economics, finance, information retrieval, psychology, automatic classification, probability, statistics and data mining. Almost all the proposed similarity measures present in the literature take into account (at most) the significance value of different features individually, disregarding their mutual influence. This is true both in the classic (crisp) ambit, where each attribute can only be present or absent, as well as in the fuzzy one, where any attribute can be present with a degree $\alpha \in [0, 1]$.

On the other hand, in many fields (such as multi-criteria decision making [1,9,10,14], multi-attribute utility theory [11], cooperative game theory [18], text mining [16], to cite some) capacities and the Choquet integral are used to model and evaluate the measure of a specific concept (such as utility, power, coalition effort), taking into account the significance value of different features and their mutual (positive or negative) interactions. This is due to the fact that a weight of importance is attached to every subset of features.

Our aim is to propose similarity measures able to consider weights which can be interpreted as the "importance" of groups of attributes, in measuring similarity. For this, starting from the approach proposed in [2], where only positive

© Springer International Publishing AG 2017
A. Antonucci et al. (Eds.): ECSQARU 2017, LNAI 10369, pp. 364–374, 2017.
DOI: 10.1007/978-3-319-61581-3_33

interactions are taken into account, we define similarity measures based on a weight capacity and the Choquet integral, generalising the index proposed by Jaccard [12]. For these measures, we limit to a finite universe of fuzzy attributes, investigating dominance and T-transitivity properties, where T is a t-norm.

2   Preliminaries

We introduce capacities and briefly recall some well-known notions in fuzzy measure theory and their related integrals useful in this paper. Let $N = \{1, \ldots, n\}$ be a finite index set and denote with $\wp(N)$ its power set. We call *significance assessment* a function $\sigma : \wp(N) \to \mathbb{R}$ satisfying the following conditions:

(S1) $\sigma(\emptyset) = 0$;
(S2) $\sum_{\{i\} \subseteq B \subseteq A} \sigma(B) \geq 0$, for every $A \in \wp(N)$ and every $i \in A$.

Notice that requirement **(S2)** imposes, in particular, to assign a non-negative weight to singletons. The function σ is then used to compute the *weight capacity* $\mu : \wp(N) \to [0, +\infty)$ defined for every $A \in \wp(N)$ as:

$$\mu(A) = \sum_{B \subseteq A} \sigma(B).$$

By Proposition 2 in [4] it immediately follows that μ is a capacity [5], i.c., it satisfies:

(C1) $\mu(\emptyset) = 0$;
(C2) $A \subseteq B \Longrightarrow \mu(A) \leq \mu(B)$, for every $A, B \in \wp(N)$.

In the case the significance assessment σ fulfils the further normalization condition $\sum_{A \in \wp(N)} \sigma(A) = 1$, then μ is a *normalized capacity*, that is $\mu(N) = 1$.

Given a weight capacity μ on $\wp(N)$ and $X \in [0, 1]^N$, the *Choquet integral* of X with respect to μ is defined (see [8]) by

$$\mathbf{C}_\mu(X) = \sum_{i=1}^{n} [X(\pi(i)) - X(\pi(i-1))] \, \mu(\{\pi(i), \ldots, \pi(n)\}),$$

where π is a permutation of N such that $X(\pi(1)) \leq \ldots \leq X(\pi(n))$ and $X(\pi(0)) := 0$. In particular, if $X \in \{0, 1\}^N$ then X can be identified with a subset of N (still denoted with X) and so $\mathbf{C}_\mu(X) = \mu(X)$.

In the particular case σ ranges in the set of non-negative real numbers, the corresponding weight capacity μ is *totally monotone*, i.e., it satisfies, for every $n \geq 2$, the *n-monotonicity condition*, for every $A_1, \ldots, A_n \in \wp(N)$,

$$\mu\left(\bigcup_{i=1}^{n} A_i\right) \geq \sum_{\emptyset \neq I \subseteq \{1,\ldots,n\}} (-1)^{|I|+1} \mu\left(\bigcap_{i \in I} A_i\right).$$

If further $\sigma(A) = 0$ for every $A \in \wp(N)$ such that $|A| > 1$, then the corresponding weight capacity μ is *additive*, i.e., for every $A, B \in \wp(N)$ such that $A \cap B = \emptyset$, $\mu(A \cup B) = \mu(A) + \mu(B)$. Weighted means are particular cases of Choquet integrals for the case μ is additive.

3   Weighted Attribute Combinations Based Similarities

Here we assume that every object is described by a set of attributes indexed by the finite set $N = \{1, \ldots, n\}$, which can be only present or absent: any object description is thus regarded as a subset of N, which is identified with its *indicator function*, so, we simply denote it as a function $X : N \to \{0, 1\}$. Denote with $\mathcal{C} = \{0, 1\}^N$ the set of all (crisp) object descriptions.

We start from the similarity measure proposed by Jaccard in [12]

$$S_J(X, Y) = \frac{|X \cap Y|}{|X \setminus Y| + |Y \setminus X| + |X \cap Y|} = \frac{|X \cap Y|}{|X \Delta Y| + |X \cap Y|} = \frac{|X \cap Y|}{|X \cup Y|}, \quad (1)$$

where $|Z|$ stands for the cardinality of the set Z and all the attributes are considered equally important (or effective) in the evaluation of the similarity between two object descriptions.

The equalities in Eq. (1) continue to hold also if the cardinality measure is replaced by the sum of positive weights (of indicator functions) attached to each attribute (i.e., by the weighted mean) which quantify their "significance" in evaluating the similarity between two object descriptions.

As already discussed in multi-criteria decision making, taking into account interactions among attributes can make the model more effective. To reach this aim, the generalisation of (1) using the weighted mean in place of set cardinality is not sufficient. In fact, in order to consider weights satisfying (S1) and (S2) for groups of attributes, we need to use a proper non-additive capacity.

Example 1. Let us consider apartments in New York described by the following crisp attributes indexed by $N = \{1, 2, 3, 4\}$:

1: the apartment is located in a skyscraper;
2: the apartment has a terrace;
3: the apartment has a panoramic view;
4: the apartment is equipped with a lift;

and assign the following significance assessment σ and the corresponding μ (we remove braces and commas on subsets of N to save space):

$\wp(N)$	1	2	3	4	12	13	14	23	24	34	123	124	134	234	1234
σ	0.3	0.2	0.3	0.2	0.4	−0.1	−0.1	−0.1	0	0	−0.1	0	0	0	0
μ	0.3	0.2	0.3	0.2	0.9	0.5	0.4	0.4		0.4	0.5	0.9	1	0.6	0.6	1

Let us consider the pairs (X, Y) and (X', Y') reported below:

Subset	1 2 3 4
X	1 0 1 0
Y	0 0 1 1
X'	1 0 1 1
Y'	0 1 1 1

Subset	1 2 3 4
$X \cap Y$	0 0 1 0
$X \setminus Y$	1 0 0 0
$Y \setminus X$	0 0 0 1
$X \cup Y$	1 0 1 1

Subset	1 2 3 4
$X' \cap Y'$	0 0 1 1
$X' \setminus Y'$	1 0 0 0
$Y' \setminus X'$	0 1 0 0
$X \cup Y$	1 1 1 1

By using the above μ it follows

$$\frac{\mu(X \cap Y)}{\mu(X \setminus Y) + \mu(Y \setminus X) + \mu(X \cap Y)} = \frac{3}{8} < \frac{1}{2} = \frac{\mu(X' \cap Y')}{\mu(X' \setminus Y') + \mu(Y' \setminus X') + \mu(X' \cap Y')},$$

$$\frac{\mu(X \cap Y)}{\mu(X \Delta Y) + \mu(X \cap Y)} = \frac{3}{7} > \frac{5}{14} = \frac{\mu(X' \cap Y')}{\mu(X' \Delta Y') + \mu(X' \cap Y')},$$

$$\frac{\mu(X \cap Y)}{\mu(X \cup Y)} = \frac{1}{2} = \frac{1}{2} = \frac{\mu(X' \cap Y')}{\mu(X' \cup Y')}.$$

So, depending on the particular functional form chosen for (1) we reach completely different similarity orderings between the pairs (X, Y) and (X', Y'). ∎

Then, the generalization of (1) to fuzzy subsets is not obvious. For that we refer to the Choquet integral with respect to a non-necessarily additive measure. A possible choice is to compute a L^p-norm attribute-wise and then apply the Choquet integral to the vector of differences. Another possibility, as done in what follows, is to apply the fuzzy set-theoretic operations attribute-wise and then apply the Choquet integral.

Choosing the Choquet integral the equalities in Eq. (1) may not hold anymore: different choices of the expression of (1) can even invert the inequality of a comparative degree of similarity between two pairs of object descriptions, as shown in Example 1. For crisp sets the generalization through the Choquet integral leads to different similarities if μ is not additive, but each of them satisfies the *maximality condition*: for every $X, Y \in \mathcal{C}$ it holds $S(X, X) = S(Y, Y) \geq S(X, Y)$.

4   Fuzzy Weighted Attribute Combinations Based Similarities

Now, we assume that every object is described by a set of attributes indexed by the finite set $N = \{1, \ldots, n\}$, and that each one can be present with a different degree of membership: any object description is thus regarded as a fuzzy subset of N [17]. In order to avoid cumbersome notation, every fuzzy subset X of N is identified with its *membership function*, so, we simply denote it as a function $X : N \to [0,1]$. Denote with $\mathcal{F} = [0,1]^N$ the set of all possible fuzzy object descriptions and still with $\mathcal{C} = \{0,1\}^N$ the subset of crisp object descriptions.

We consider a t-norm T together with its dual t-conorm S and the complement $(\cdot)^c = 1 - (\cdot)$ to perform fuzzy set-theoretic operations. As usual (see [13]), we denote the main t-norms and t-conorms, for every $x, y \in [0,1]$, as

$$\begin{aligned}
T_M(x, y) &= \min\{x, y\}, & S_M(x, y) &= \max\{x, y\}, \\
T_P(x, y) &= x \cdot y, & S_P(x, y) &= x + y - x \cdot y, \\
T_L(x, y) &= \max\{x + y - 1, 0\}, & S_L(x, y) &= \min\{x + y, 1\}.
\end{aligned}$$

For every $X, Y \in \mathcal{F}$, we define $X \cap Y = T(X, Y)$, $X \setminus Y = T(X, Y^c)$, $Y \setminus X = T(Y, X^c)$, $X \Delta Y = S(X \setminus Y, Y \setminus X)$ and $X \cup Y = S(X, Y)$, where all operations are intended pointwise on the elements of N.

Different definitions of similarities have been given for fuzzy subsets [3,6,7] essentially based on the "common" and the "different" parts of the compared fuzzy subsets.

We introduce three classes of similarity measures $S_i^\mu : \mathcal{F}^2 \rightarrow [0,1]$, for $i = 1,2,3$, each parametrized by a weight capacity μ or, equivalently, by a significance assessment σ, defined, for every $X, Y \in [0,1]^N$, as:

$$S_1^\mu(X,Y) = \frac{\mathbf{C}_\mu(X \cap Y)}{\mathbf{C}_\mu(X \setminus Y) + \mathbf{C}_\mu(Y \setminus X) + \mathbf{C}_\mu(X \cap Y)}, \tag{2}$$

$$S_2^\mu(X,Y) = \frac{\mathbf{C}_\mu(X \cap Y)}{\mathbf{C}_\mu(X \Delta Y) + \mathbf{C}_\mu(X \cap Y)}, \tag{3}$$

$$S_3^\mu(X,Y) = \frac{\mathbf{C}_\mu(X \cap Y)}{\mathbf{C}_\mu(X \cup Y)}. \tag{4}$$

If the denominator (and so the numerator) of S_i^μ vanishes, we set $S_i^\mu(X,Y) := 0$ even if different generalisations could be introduced by considering refinements of the capacity μ or better by considering conditional capacities (we omit this discussion due to limit of space).

If μ is additive then the similarity S_3^μ is a special case of that introduced in [15] and coincides with the Jaccard similarity when μ is the Laplace measure (discrete uniform measure). Notice that, under an additive μ, the similarity measures S_1^μ, S_2^μ and S_3^μ coincide on $\mathcal{C}^2$, but are generally different on $\mathcal{F}^2 \setminus \mathcal{C}^2$. The choice among the different similarities could be based on the role played by the operations between fuzzy sets (so, by the chosen t-norm) and that played by the weight capacity: in S_1^μ and S_2^μ the role of the t-norm is more relevant than in S_3^μ.

Even if μ is additive, the maximality condition $S_i^\mu(X,X) \geq S_i^\mu(X,Y)$, for $i = 1,2$, may fail. For instance, when (T_M, S_M) or (T_P, S_P) are taken to perform fuzzy set-theoretic operations, it is sufficient to take a crisp set X and a proper fuzzy set Y. Instead, when (T_L, S_L) are used, maximality fails by taking a crisp set X and a proper fuzzy set Y pointwise less or equal than $\frac{1}{2}$.

Due to the freedom on the choice of μ (or, equivalently, σ), in general there is no dominance relation holding between S_1^μ, S_2^μ and S_3^μ, as shown in the following example.

Example 2. Let $N = \{1,2,3\}$ and take the weight capacities μ_1 and μ_2 defined on $\wp(N)$ as:

$\wp(N)$	$\emptyset$	$\{1\}$	$\{2\}$	$\{3\}$	$\{1,2\}$	$\{1,3\}$	$\{2,3\}$	N
μ_1	0	0.1	0.1	0.1	0.2	0.2	0.2	1
μ_2	0	0.25	0.25	0.5	0.5	0.5	0.5	1

Let T and S be any pair of dual t-norm and t-conorm and consider the crisp subsets of N below

Fuzzy subset	1	2	3
X		0	1	1
Y		1	1	0

If we take the weight capacity μ_1 then we get $\mathbf{C}_{\mu_1}(X \cap Y) = 0.1$, $\mathbf{C}_{\mu_1}(X \backslash Y) = 0.1$, $\mathbf{C}_{\mu_1}(Y \backslash X) = 0.1$, $\mathbf{C}_{\mu_1}(X \Delta Y) = 0.2$ and $\mathbf{C}_{\mu_1}(X \cup Y) = 1$, so, it holds

$$S_3^{\mu_1}(X,Y) = 0.1 < 0.3333 = S_1^{\mu_1}(X,Y) = S_2^{\mu_1}(X,Y).$$

On the other hand, if we take the weight capacity μ_2 then we get $\mathbf{C}_{\mu_2}(X \cap Y) = 0.25$, $\mathbf{C}_{\mu_2}(X \backslash Y) = 0.5$, $\mathbf{C}_{\mu_2}(Y \backslash X) = 0.25$, $\mathbf{C}_{\mu_2}(X \Delta Y) = 0.5$ and $\mathbf{C}_{\mu_2}(X \cup Y) = 1$, so, it holds

$$S_1^{\mu_2}(X,Y) = S_3^{\mu_2}(X,Y) = 0.25 < 0.3333 = S_2^{\mu_2}(X,Y). \qquad \blacksquare$$

Restricting S_1^μ, S_2^μ and S_3^μ on $\mathcal{C}^2$, if μ is *superadditive*, i.e., for every $A, B \in \wp(N)$ with $A \cap B = \emptyset$ it holds

$$\mu(A \cup B) \geq \mu(A) + \mu(B),$$

we have that $S_1^\mu(X,Y) \geq S_2^\mu(X,Y) \geq S_3^\mu(X,Y)$ for every $X, Y \in \mathcal{C}$. In analogy, if μ is *subadditive*, i.e., for every $A, B \in \wp(N)$ with $A \cap B = \emptyset$ it holds

$$\mu(A \cup B) \leq \mu(A) + \mu(B),$$

we have that $S_1^\mu(X,Y) \leq S_2^\mu(X,Y) \leq S_3^\mu(X,Y)$ for every $X, Y \in \mathcal{C}$.

Nevertheless, regarding S_1^μ, S_2^μ and S_3^μ as functions on the whole $\mathcal{F}^2$, superadditivity or subadditivity of μ do not determine any form of dominance as shown in the following example.

Example 3. Let $N = \{1, 2, 3\}$ and take the weight capacities μ_1 and μ_2 defined on $\wp(N)$ as:

$\wp(N)$	$\emptyset$	$\{1\}$	$\{2\}$	$\{3\}$	$\{1,2\}$	$\{1,3\}$	$\{2,3\}$	N
μ_1	0	0	0	0	0	0	0	1
μ_2	0	1	1	1	1	1	1	1

which are easily seen to be, respectively, superadditive and subadditive (actually, they are totally monotone and totally alternating). This implies that $S_1^{\mu_1}(X,Y) \geq S_2^{\mu_1}(X,Y) \geq S_3^{\mu_1}(X,Y)$ and $S_1^{\mu_2}(X,Y) \leq S_2^{\mu_2}(X,Y) \leq S_3^{\mu_2}(X,Y)$, for every $X, Y \in \mathcal{C}$.

Note that $\mathbf{C}_{\mu_1}(X) = \min_{i=1,\ldots,3} X(i)$ and $\mathbf{C}_{\mu_2}(X) = \max_{i=1,\ldots,3} X(i)$, for $X \in \mathcal{F}$.

Consider the fuzzy subsets of N below

Fuzzy subset	1	2	3
X	0.2	0.4	0.3
Y	0.9	0.7	0.8

If we take $T = T_M$ and $S = S_M$, simple computations show that

$$S_1^{\mu_1}(X,Y) = 0.2222 < S_3^{\mu_1}(X,Y) = 0.2857 < S_2^{\mu_1}(X,Y) = 0.6666,$$

$$S_1^{\mu_2}(X,Y) = 0.2666 < S_3^{\mu_2}(X,Y) = 0.4444 < S_2^{\mu_2}(X,Y) = 0.5714.$$

An analogous situation happens if fuzzy set-theoretic operations are executed taking $T = T_L$ and $S = S_L$, since it holds

$$S_3^{\mu_1}(X,Y) = 0.1 < S_1^{\mu_1}(X,Y) = 0.25 < S_2^{\mu_1}(X,Y) = 1,$$

$$S_3^{\mu_2}(X,Y) = 0.1 < S_1^{\mu_2}(X,Y) = 0.125 < S_2^{\mu_2}(X,Y) = 1.$$

∎

Now, we consider the T'-transitivity of the similarity measures S_1^μ, S_2^μ and S_3^μ, i.e., for every $X, Y, Z \in \mathcal{F}$,

$$S_i^\mu(X,Z) \geq T'(S_i^\mu(X,Y), S_i^\mu(Y,Z)),$$

where T' is a t-norm possibly different from the t-norm T used in the fuzzy set-theoretic operations. In the relevant literature, different authors maintain that T_M-transitivity is a too strong requirement [6,7], then weaker forms of transitivity must be required (essentially choosing a different t-norm).

The following example shows that the above measures are generally not T'-transitive, for any t-norm such that $T_L \leq T' \leq T_M$.

We recall that if a similarity measure is T'-transitive for some t-norm $T_L \leq T' \leq T_M$, then it is T''-transitive for any t-norm T'' such that $T_L \leq T'' \leq T'$ [7].

Example 4. Let N and μ_1 as in Example 3 and take the fuzzy subsets of N

Fuzzy subset	1	2	3
X	0.47	0.87	0.95
Y	0.46	0.99	0.56
Z	0.98	0.23	0.21

Taking $T = T_M$ and $S = S_M$ to perform fuzzy set-theoretic operations, we have

$$S_1^{\mu_1}(X,Z) = 0.75, \qquad S_2^{\mu_1}(X,Z) = 0.913043, \qquad S_3^{\mu_1}(X,Z) = 0.241379,$$
$$S_1^{\mu_1}(X,Y) = 0.884615, \qquad S_2^{\mu_1}(X,Y) = 0.978723, \qquad S_3^{\mu_1}(X,Y) = 0.978723,$$
$$S_1^{\mu_1}(Y,Z) = 0.875, \qquad S_2^{\mu_1}(Y,Z) = 0.954545, \qquad S_3^{\mu_1}(Y,Z) = 0.375,$$

so, for $i = 1, 2, 3$, we have that

$$S_i^{\mu_1}(X, Z) < T_L(S_i^{\mu_1}(X, Y), S_i^{\mu_1}(Y, Z)) \le T_M(S_i^{\mu_1}(X, Y), S_i^{\mu_1}(Y, Z)).$$

∎

Next we prove that if the weight capacity μ is additive, then the similarity measure S_3^μ is T_L-transitive.

Proposition 1. *If the weight capacity $\mu : \wp(N) \to [0, +\infty)$ is additive, then the similarity measure S_3^μ is T_L-transitive.*

Proof. For any $X, Y, Z \in [0, 1]^N$ it is sufficient to show that

$$S_3^\mu(X, Z) + 1 \ge S_3^\mu(X, Y) + S_3^\mu(Y, Z).$$

Note that

$$S_3^\mu(X, Y) = \frac{\mathbf{C}_\mu(X \cap Y)}{\mathbf{C}_\mu(X \cup Y)} \le \frac{\mathbf{C}_\mu(X \cap Y) + c}{\mathbf{C}_\mu(X \cup Y) + c}$$

for any non-negative constant c, in particular, for $c = \mathbf{C}_\mu(X \cup Y \cup Z) - \mathbf{C}_\mu(X \cup Y)$.
Analogously,

$$S_3^\mu(Y, Z) = \frac{\mathbf{C}_\mu(Y \cap Z)}{\mathbf{C}_\mu(Y \cup Z)} \le \frac{\mathbf{C}_\mu(Y \cap Z) + c'}{\mathbf{C}_\mu(Y \cup Z) + c'}$$

for $c' = \mathbf{C}_\mu(X \cup Y \cup Z) - \mathbf{C}_\mu(Y \cup Z)$.
Then,

$$\begin{aligned}
S_3^\mu(X, Y) + S_3^\mu(Y, Z) &\le \frac{\mathbf{C}_\mu(X \cap Y) + c}{\mathbf{C}_\mu(X \cup Y) + c} + \frac{\mathbf{C}_\mu(Y \cap Z) + c'}{\mathbf{C}_\mu(Y \cup Z) + c'} \\
&= \frac{\mathbf{C}_\mu(X \cap Y) + c}{\mathbf{C}_\mu(X \cup Y \cup Z)} + \frac{\mathbf{C}_\mu(Y \cap Z) + c'}{\mathbf{C}_\mu(X \cup Y \cup Z)} \\
&\le \frac{\mathbf{C}_\mu(X \cap Y \setminus Z) + \mathbf{C}_\mu(X \cap Z) + c}{\mathbf{C}_\mu(X \cup Z)} + \frac{\mathbf{C}_\mu(Y \cap Z) + c'}{\mathbf{C}_\mu(X \cup Z)} \\
&= 1 + S_3^\mu(X, Z).
\end{aligned}$$

□

5   A Paradigmatic Example

The following example is inspired to the main example in [9].

Example 5. We consider 3 students x, y, z evaluated with respect to 3 subjects: mathematics (1), physics (2) and literature (3), whose final marks are given on a scale from 0 to 20:
    The above evaluation vectors determine three fuzzy subsets of $N = \{1, 2, 3\}$ by rescaling the marks to range in $[0, 1]$:

Student	1	2	3
x	18	16	10
y	10	12	18
z	14	15	15

Fuzzy subset	1	2	3
X	0.9	0.8	0.5
Y	0.5	0.6	0.9
Z	0.7	0.75	0.75

$\wp(N)$	$\emptyset$	$\{1\}$	$\{2\}$	$\{3\}$	$\{1,2\}$	$\{1,3\}$	$\{2,3\}$	N
μ	0	0.45	0.45	0.3	0.5	0.9	0.9	1

It is common knowledge that "usually" students good at mathematics are also good at physics, and vice versa. Thus, to weigh the subjects on which the students are evaluated we use the capacity $\mu : \wp(N) \rightarrow [0,1]$ given below:

Notice that the capacity μ is neither superadditive nor subadditive since:

$$\mu(\{1,2\}) = 0.5 < 0.45 + 0.45 = \mu(\{1\}) + \mu(\{2\}),$$
$$\mu(\{1,3\}) = 0.9 > 0.45 + 0.3 = \mu(\{1\}) + \mu(\{3\}),$$
$$\mu(\{2,3\}) = 0.5 > 0.45 + 0.3 = \mu(\{2\}) + \mu(\{3\}).$$

Taking $T = T_M$ and $S = S_M$, the values of S_1^μ, S_2^μ and S_3^μ are reported in Table 1.

Thus, denoting with $\precsim_i$ the weak order (with asymmetric part $\prec_i$) induced by the similarity measure S_i^μ on $\{X,Y,Z\}^2$, for $i = 1, 2, 3$, we obtain:

$$\begin{matrix}(X,Y)\\(Y,X)\end{matrix} \prec_1 \begin{matrix}(Y,Z)\\(Z,Y)\end{matrix} \prec_1 \begin{matrix}(X,Z)\\(Z,X)\end{matrix} \prec_1 (Y,Y) \prec_1 (X,X) \prec_1 (Z,Z),$$

$$\begin{matrix}(X,Y)\\(Y,X)\end{matrix} \prec_2 \begin{matrix}(Y,Z)\\(Z,Y)\end{matrix} \prec_2 \begin{matrix}(X,Z)\\(Z,X)\end{matrix} \prec_2 (Y,Y) \prec_2 (Z,Z) \prec_2 (X,X),$$

$$\begin{matrix}(X,Y)\\(Y,X)\end{matrix} \prec_3 \begin{matrix}(X,Z)\\(Z,X)\end{matrix} \prec_3 \begin{matrix}(Y,Z)\\(Z,Y)\end{matrix} \prec_3 \begin{matrix}(X,X)\\(Y,Y)\\(Z,Z)\end{matrix}.$$

Table 1. Values of S_1, S_2 and S_3 on $\{X,Y,Z\}^2$

	X	Y	Z
X	0.5538	0.4866	0.5298
Y	0.4866	0.5354	0.5281
Z	0.5298	0.5281	0.5775

	X	Y	Z
X	0.7680	0.5266	0.6368
Y	0.5266	0.6974	0.6318
Z	0.6368	0.6318	0.7322

	X	Y	Z
X	1	0.6124	0.7591
Y	0.6124	1	0.8034
Z	0.7591	0.8034	1

This shows that, generally, the three similarity measures induce different weak orders, moreover, disregarding reflexive pairs, S_1^μ and S_2^μ select X and Z as the most similar, while S_3^μ selects Y and Z. ∎

6 Conclusions

The application of similarity measures whose evaluation is based on the Choquet integral requires the prior identification of a weight capacity μ or, equivalently, a significance assessment σ. The choice of μ or σ deeply impacts on the similarity orderings induced by the proposed measures. The elicitation of μ or σ by a field expert seems to be the most "natural" procedure, in agreement to what happens in multi-criteria decision analysis [9]. Nevertheless, in this context a learning procedure in the line of that in [2] can be envisaged.

Acknowledgment. This work was partially supported by INdAM-GNAMPA through the Project 2015 U2015/000418 and the Project 2016 U2016/000391 and by the Italian Ministry of Education, University and Research, under grant 2010FP79LR_003.

References

1. Angilella, S., Greco, S., Lamantia, F., Matarazzo, B.: The application of fuzzy integrals in multicriteria decision making. Eur. J. Oper. Res. **158**(3), 734–744 (2004)
2. Baioletti, M., Coletti, G., Petturiti, D.: Weighted attribute combinations based similarity measures. In: Greco, S., Bouchon-Meunier, B., Coletti, G., Fedrizzi, M., Matarazzo, B., Yager, R.R. (eds.) IPMU 2012. CCIS, vol. 299, pp. 211–220. Springer, Heidelberg (2012). doi:10.1007/978-3-642-31718-7_22
3. Bouchon-Meunier, B., Coletti, G., Lesot, M.-J., Rifqi, M.: Towards a conscious choice of a fuzzy similarity measure: a qualitative point of view. In: Hüllermeier, E., Kruse, R., Hoffmann, F. (eds.) IPMU 2010. LNCS (LNAI), vol. 6178, pp. 1–10. Springer, Heidelberg (2010). doi:10.1007/978-3-642-14049-5_1
4. Chateauneuf, A., Jaray, J.Y.: Some characterizations of lower probabilities and other monotone capacities through the use of möbius inversion. Math. Soc. Sci. **17**(3), 263–283 (1989)
5. Choquet, G.: Theory of capacities. Ann. Inst. Fourier **5**, 131–295 (1953)
6. De Baets, B., Janssens, S., Meyer, H.D.: On the transitivity of a parametric family of cardinality-based similarity measures. Int. J. Approximate Reasoning **50**(1), 104–116 (2009)
7. De Baets, B., Meyer, H.D.: Transitivity-preserving fuzzification schemes for cardinality-based similarity measures. Eur. J. Oper. Res. **160**(3), 726–740 (2005)
8. Denneberg, D.: Non-Additive Measure and Integral, Series B: Mathematical and Statistical Methods, vol. 27. Kluwer Academic Publishers, Dordrecht (1994)
9. Grabisch, M.: The application of fuzzy integrals in multicriteria decision making. Eur. J. Oper. Res. **69**(3), 279–298 (1995)
10. Grabisch, M.: Fuzzy integral in multicriteria decision making. Fuzzy Sets Syst. **89**(3), 445–456 (1996)
11. Grabisch, M., Kojadinovic, I., Meyer, P.: A review of methods for capacity identification in choquet integral based multi-attribute utility theory: applications of the kappalab R package. Eur. J. Oper. Res. **186**(2), 766–785 (2008)

12. Jaccard, P.: Nouvelles recherches sur la distribution florale. Bull. Soc. Vaud. Sci. Nat. **44**, 223–270 (1908)
13. Klement, E., Mesiar, R., Pap, E.: Triangualr Norms, vol. 8. Kluwer Academic Publishers, Dordrecht (2000)
14. Marichal, J.: An axiomatic approach of the discrete choquet integral as a tool to aggregate interacting criteria. IEEE Trans. Fuzzy Syst. **8**(6), 800–807 (2000)
15. Scozzafava, R., Vantaggi, B.: Fuzzy inclusion and similarity through coherent conditional probability. Fuzzy Sets Syst. **160**(3), 292–305 (2009)
16. Wilbik, A., Keller, J.M., Alexander, G.: Similarity evaluation of sets of linguistic summaries. Int. J. Intell. Syst. **27**, 226–238 (2012)
17. Zadeh, L.A.: Fuzzy sets. Inf. Control **8**, 338–353 (1965)
18. Yu, X., Zhang, Q.: An extension of cooperative fuzzy games. Fuzzy Sets Syst. **161**, 1614–1634 (2010)

Online Fuzzy Temporal Operators for Complex System Monitoring

Jean-Philippe Poli[✉], Laurence Boudet, Bruno Espinosa,
and Laurence Cornez

CEA, LIST, Data Analysis and System Intelligence Laboratory,
91191 Gif-sur-Yvette cedex, France
{jean-philippe.poli,laurence.boudet,bruno.espinosa,
laurence.cornez}@cea.fr

Abstract. Online fuzzy expert systems can be used to process data and
event streams, providing a powerful way to handle their uncertainty and
their inaccuracy. Moreover, human experts can decide how to process the
streams with rules close to natural language. However, to extract high
level information from these streams, they need at least to describe the
temporal relations between the data or the events.

In this paper, we propose temporal operators which relies on the
mathematical definition of some base operators in order to characterize trends and drifts in complex systems. Formalizing temporal relations
allows experts to simply describe the behaviors of a system which lead
to a break down or an ineffective exploitation. We finally show an experiment of those operators on wind turbines monitoring.

1 Introduction

Complex systems are now equipped with hundreds of sensors which deliver continuous signals. Sensors provide either measurements at a dynamic or constant
sampling rate (i.e. data streams, e.g. connected thermometers), either events
whenever they are detected (i.e. event streams, e.g. presence detectors). Such
streams are generally processed, filtered and combined to get higher level information. These operations can be applied to predictive maintenance of complex
systems.

Predictive maintenance consists in monitoring an engineering system in order
to detect changes in its exploitation and prevent damages. Having a continuous
report of in-service systems allows an optimal use of it, the avoidance of important damages and early-stage failure detection. Moreover, it changes the organization of maintenance services by replacing scheduled and periodic maintenance
and by minimizing the involvement of operators.

Artificial intelligence plays an important role in predictive maintenance [4]
and provides system-specific solutions : signal processing and statistical learning techniques have been successfully applied to obtain a type of damage or
a type of risk. Predictive maintenance mainly relies on data from process sensors (temperature, pressure, etc.) and test sensors (vibration, acoustic, humidity,

A. Antonucci et al. (Eds.): ECSQARU 2017, LNAI 10369, pp. 375–384, 2017.
DOI: 10.1007/978-3-319-61581-3_34

etc.) [3]. In order to better handle the sensors inaccuracy and the uncertainty in the assessment of the system's state, fuzzy logic has been applied to predictive maintenance [8,12].

Our work consists in developing an online fuzzy expert system which can take data or event streams as input. The goal is to reinforce the expressivity of such systems to let experts author their own rules with complex fuzzy relations. Gathering the knowledge of different experts can be a suitable approach to predictive maintenance, avoiding some difficulties of the techniques described formerly:

– no past data are needed to build the models;
– the decision can be explained through the trace of activated rules.

In the case of predictive maintenance, the rules consist in detecting patterns in time-series which lead to a damage. Numerous authors [1,2,11] state temporal relations are a prerequisite to describe such patterns. One can distinguish different approaches. On the one hand, fuzzy temporal relations [2,5,10] can be used to describe the temporality of events but are not always relevant for online causal reasoning. On the other hand, some papers suggest to linguistically describe time-series [6,7] using fuzzy natural logic, specifically on complete time-series, i.e. in an offline way.

In this article, we remind 3 base fuzzy temporal relations which are then combined into more complex relations. The compositional paradigm we use allows to create new intuitive relations because they combined simple operators. The new operators are the first of a series of temporal operators which can be used to describe time-series. In our work, we make the following assumptions:

– sensors give correct timestamps: there is no uncertainty in the acquisition timestamps, but we take into account the vagueness in the relations between the timestamps;
– sensors values are fuzzified to both manipulate linguistic terms and manage their inaccuracy.

The article is organized as follows: the next section presents the previous work and the notations. The new temporal operators are described in Sect. 3. In Sect. 4, we describe their use by an application to wind turbine predictive maintenance. Finally, Sect. 5 draws the conclusions and perspectives of this work.

2   Previous Work

In our previous work, we introduced a compositional paradigm which consists in deriving specialized operators from base operators in the temporal domain [9]. In this paper, we take advantage from these operators to build new temporal ones for online characterization of time-series.

The temporal operators use two concepts to deal with event streams [9]. On the one hand, expiration is the faculty for a temporal expression to yell that its value has expired and must be re-evaluated. On the other hand, they are applied on a scope. A scope is a fuzzy set defined on a temporal domain, anchored at the present moment, and whose membership function gives the importance of a moment in this temporal domain. For instance, Fig. 1.c shows such a scope representing "the last 10 s". Both concepts ensure a satisfying computational cost and allow an online execution.

Let E be a fuzzy expression, $eval(E, t)$ be the value of E at time t. Let S be a fuzzy scope and μ_S its membership function. In the remainder of this paper, we will use the following temporal operators:

- The occurrence operator which indicates if an expression has a degree of fulfillment strictly greater than 0 throughout the scope :

$$Occ(E, S, t_{now}) = \bigvee_{t \in supp(S)} eval(E, t) \wedge \mu_S(t) \tag{1}$$

When its value is strictly greater than 0, it means that at least at one moment of the scope, the operand expression has been observed. It is a disjunction over all the moments t_i in the scope of conjunctions of the operand value at time t_i and the value of the scope membership function for t_i.

- The ratio operator which aggregates the different degrees of fulfillment of the operand expression E throughout a scope S:

$$Ratio(E, S, t_{now}) = \frac{\int_{t \in supp(S)} eval(E, t) \wedge \mu_S(t)}{\int_{t \in supp(S)} \mu_S(t)} \tag{2}$$

It aggregates the different values of the operand E on the scope S, divided by the area under the scope membership function. It is related to Zadeh's relative count applied on a fuzzy scope.

- The persistence operator which indicates if at each moment of S, the degree of fulfillment of E is strictly greater than 0:

$$StrictPers(E, S, t_{now}) = \neg Occ(\neg E, S, t_{now}). \tag{3}$$

It equals 0 if there exists a moment t_i in the scope S such as $eval(E, t_i) = 0$. This is why we called it "strict". To moderate its definition, we can either replace the Occ operator by the $Ratio$ inside its definition, or simply use $Ratio$ instead of $StrictPers$.

In the next section, we use these operators to define new temporal operators to both characterize trends of time-series and to compare two of them. Adopting an iterative approach, we first define the following operators and we will add new ones when are not sufficient anymore.

3   Signal Characterization Operators

To illustrate the behavior of the operators, we introduce some examples of input signals and parameters we will use throughout this section.

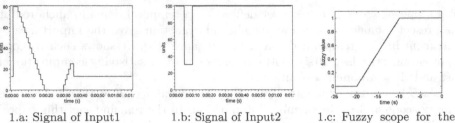

1.a: Signal of Input1   1.b: Signal of Input2   1.c: Fuzzy scope for the
                                                          last 10 seconds

Fig. 1. Examples of two signals and a fuzzy scope

For the sake of comprehension, Fig. 1(a) and (b) show two simple simulated signals. We will use these signals to illustrate the behavior of the operators and in the Sect. 4, we will use more realistic signals.

In the remainder of this section, without loss of generality, the operators are defined upon the *Ratio* operator. As a consequence of the use of the *Ratio* operator, those operators are considered tolerant. Thus, if at some moment the input signal is changing for a short while, the direct effect of its change is smoothed. If a more strict behavior is needed, it is possible to replace the *Ratio* by the *StrictPers* operator.

3.1 Growth, Decline and Variation

In predictive maintenance, it is important to be able to characterize drifts of some sensors, because it can lead to the detection of a damage. The goal here is to monitor the growth or the decline of an input value with operators such as:

$$input \; \langle adverb \rangle \; \text{decreases/increases throughout } S.$$

where adverb is a fuzzy set which represents, for example, "slowly" or "significantly" and S is a fuzzy scope.

To compute a degree of fulfillment for such relations, saving all the values in the scope is not necessary. We chose instead to compute the gradient between the two last samples and then to characterize its direction with a fuzzy set corresponding to the adverb. The fuzzy set is thus defined on a quarter of the trigonometric circle (top-right quadrant for the growth and bottom-right quadrant for the decline). Figures 2(a) and 3(a) show an example of membership functions for adverbs "slowly" and "significantly" applied respectively to the decline and the growth operator.

To aggregate the characterizations of the gradient over the scope, we can use the *Ratio*. Thus, the *Decreases* operator can be defined as:

$$Decreases(I, S, \mu_g, t_{now}) = Ratio(\mu_g(grad(I, t_{now})), S, t_{now}) \qquad (4)$$

where I is the real input of the system whose values change, *grad* is the direction of the gradient, and μ_g is the membership function of the adverb fuzzy set.

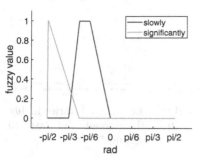

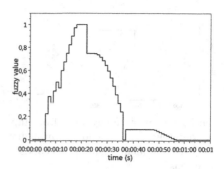

2.a: Examples of "slowly" and "significantly" membership functions

2.b: "Input1 significantly decreases
throughout the last 10 seconds"

Fig. 2. Examples of membership functions for the adverbs of the *Decreases* operator
and result on Input 1

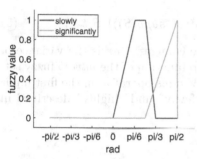

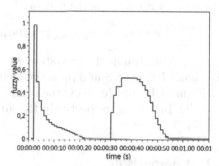

3.a: Examples of "slowly" and "significantly" membership functions

3.b: "Input1 significantly increases
throughout the last 10 seconds"

Fig. 3. Examples of membership functions for the adverbs of the *Increases* operator
and result on Input 1

The *Increases* operator only differs from the *Decreases* operator because of
the definition domain and the membership function of the adverb fuzzy set.

Figures 2(b) and 3(b) show respectively the result of operators *Decreases*
and *Increases* on the first input whose signal is shown in Fig. 1a.

In a similar way, it is useful to be able to tell that the value of an input
remains stable over time, with an operator like:

$$\text{input varies } \langle adverb \rangle \text{ throughout } S$$

where *adverb* is a fuzzy set which represents, for instance, "fewly" or "highly".
The definition of the *Varies* operator is based on the variance of its signal over
S and on a fuzzy set which defines the *adverb* by characterizing the variance.

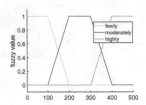 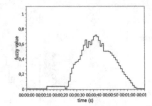 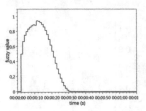

4.a: Examples of "fewly", "moderately" and "highly" membership functions

4.b: "Input1 varies fewly throughout the last 10 seconds"

4.c: "Input1 varies highly throughout the last 10 seconds"

Fig. 4. Examples of membership functions for the adverbs of the *Varies* operator and results on Input 1

The *Varies* operator is defined by:

$$Varies(I, S, \mu_v, t_{now}) = Ratio(\mu_v(Var(I, supp(S))), S, t_{now}) \quad (5)$$

where I is an input of the system whose value changes, Var is the variance of the signal $I(t)$ over S and μ_v is the membership function of the *adverb* fuzzy set.

Figure 4(b) and (c) show the results of the *Varies* operator on the first input (see Fig. 1a), using respectively the adverbs "fewly" and "highly" described in Fig. 4a.

3.2 Comparison

The last family of operators in this article concerns comparison between two input values throughout a scope; one of them can be a fixed value, for instance a threshold. For instance, an expert may want to express that the signal of an input is extremely less than another value:

input1 is ⟨*adverb*⟩ less/greater/close than/to input2 throughout S.

The idea behind these operators is to compare at each time the two values and to characterize the difference between them with a fuzzy set (the adverb). Then, we aggregate the point-to-point comparisons with the *Ratio* operator. Thus, we can define *LessThan*, *GreaterThan*, *CloseTo* as:

$$LessThan(I_1, I_2, S, \mu_{lt}, t_{now}) = Ratio(\mu_{lt}(I_1(t_{now}) - I_2(t_{now})), S, t_{now}) \quad (6)$$

$$GreaterThan(I_1, I_2, S, \mu_{gt}, t_{now}) = Ratio(\mu_{gt}(I_1(t_{now}) - I_2(t_{now})), S, t_{now}) \quad (7)$$

$$ClosteTo(I_1, I_2, S, \mu_{ct}, t_{now}) = Ratio(\mu_{ct}(I_1(t_{now}) - I_2(t_{now})), S, t_{now}) \quad (8)$$

where μ_{lt}, μ_{gt} and μ_{ct} are the membership functions of the adverb fuzzy set which characterizes the difference between the two signals $I_1(t)$ et $I_2(t)$. The operators differ by the definition of the adverb fuzzy set.

Figure 5b shows the application of the *GreaterThan* operator on the input signals shown in Fig. 1(a) and (b) with the adverb "much" (Fig. 5a).

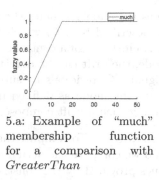 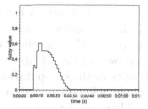

5.a: Example of "much" membership function for a comparison with *GreaterThan*

5.b: "Input1 is much greater than input2 throughout the last 10 seconds"

Fig. 5. Example of membership function for the adverb of the *GreaterThan* operator and result on Input 1 and Input 2

4   Application to a Drift Detection

The goal of the presented work is to apply fuzzy expert systems to predictive maintenance of complex systems. As illustration, we developed a specific software for wind turbines. Figure 6 show some screenshots of our tool. It provides an overview of the system (Fig. 6a) and can locate with a circle the suspected default. The tiles on the left indicate the state of each sub-system of the wind turbine : a green tile indicates it is fully functional while a red tile indicates a critical state. By clicking on a tile, it is possible to access a more detailed view (Fig. 6b) with the signals, the output of the fuzzy expert system, and the rules with their activation which give an explanation of the decision contrarily to other approaches.

In this paper, we focus on the characterization of one of the sub-systems: the rotor-side multicellular converter. It occasionally suffers from drifts which are clues that the energy production is not optimal. It consists of serial cells, each one containing two switches with complementary values. The combination of the

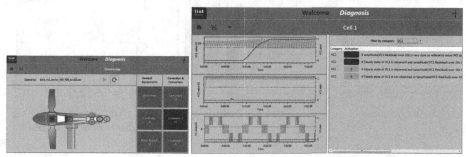

6.a: Screenshot of the system overview

6.b: Screenshot of the detailed view

Fig. 6. Screenshots of the application for windturbines.

values of all the switches in the converter defines a "mode". Among the other variables, the dynamics of the converter is also described by VC_i which is the floating voltage of the capacitors C_i of each cell. An instance of a controlled drift of VC_i is shown in Fig. 7a. According to the mode, the drift can be detected or not. It results in the computation of the new signal VC_i residuals by subtracting the mean reference value to VC_i according to the mode as defined in [13] (Fig. 7b). Then, we defined a rule base for detecting such a drift composed of:

- First, rules for defining the nominal values of the system. To compute that, we wait for a steady state during at least 20 s and we compare the actual values of the amplitude of VC_i residuals to the values provided by the constructor.
- Then, rules for monitoring a drift and, according to its importance and its duration, to yield a suitable level of alarm.

Figure 8 show the membership functions used to compare the amplitude of VC_1 residuals (Fig. 7c) to its reference value and different expressions computed to detect the drift. Once the steady state has been observed, the expression verifying that the amplitude of VC_i residuals is very close to the reference value (Fig. 8b) is associated with a null alert (Fig. 9a), and the one verifying that it is much higher than the reference value (Fig. 8c) is associated with a high alert. The defuzzified value of the alert is shown by a black curve in Fig. 9b. The drift is applied between 40 and 80 s. It begins to be detected after only 15 s which is the delay necessary to compute the *Ratio* operator on the chosen temporal scope. Then, the alert value gradually rises until it reaches its maximum value 40 s afterward.

Monitoring a system with fuzzy temporal rules enables both to estimate a continuous value of the output (an alert here) and to know which rules are activated and led to the results. All the membership functions used as well as temporal scopes have to be chosen according to the application in order to characterize a normal or abnormal behavior of each sub-system. They can be learned when sufficient data of sub-systems are available.

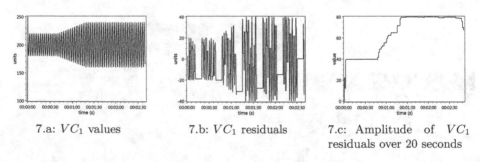

7.a: VC_1 values   7.b: VC_1 residuals   7.c: Amplitude of VC_1 residuals over 20 seconds

Fig. 7. Input example for floating voltage of capacitor C_1

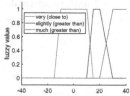

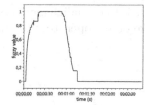

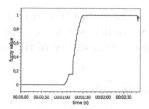

8.a: Membership functions for comparing amplitudes of VC_1 residuals to the reference value

8.b: "Amplitude of VC_1 residuals is very close to reference value during the last 20 seconds"

8.c: "Amplitude of VC_1 residuals is much greater than the reference value during the last 20 seconds"

Fig. 8. Examples of membership functions for comparing the amplitude of VC_1 residuals to the reference value and results of comparison operators

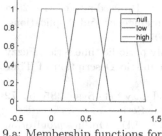

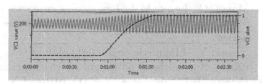

9.a: Membership functions for output "alert"

9.b: VC_1 signal (left-hand axis) and defuzzified value of a drift alert between 0 and 1 (right-hand axis) superimposed

Fig. 9. Example of membership functions for alerting a drift detection and application of detection rules on VC_1 signal

5   Conclusion

In this article, we use a compositional paradigm to build new temporal operators to characterize the kinetics of input values. From simple and intuitive operators like the ratio and the persistence, the temporal aspect is easily handled. These operators can take into account both the temporal uncertainty and the vagueness of the relation between the values.

With such operators, online fuzzy expert systems can play an important role in predictive maintenance or health monitoring. Experts can describe their knowledge about the systems and describe the clues which lead to damage detection from sensors signals. The decision making process can then be justified to the user by tracing activated rules. Moreover, such expert systems are independent of the system on which they are applied, contrary to statistical models which are system-dependent.

The perspectives of our work is to formalize more operators which are suitable for predictive maintenance, like online operators to characterize the seasonality or the periodicity of time-series.

References

1. Barro, S., Bugarín, A., Cariñena, P., Díaz-Hermida, F., Mucientes, M.: Fuzzy temporal rule-based systems: new challenges. In: Actas del XIV Congreso Español sobre Tecnologías y Lógica Fuzzy (ESTYLF), pp. 507–514, Langreo, Spain (2008)
2. Dubois, D., Hadj Ali, A., Prade, H.: Fuzziness and uncertainty in temporal reasoning. J. Univ. Comput. Sci. 9(9), 1168–1194 (2003)
3. Hashemian, H.M., Bean, W.C.: State-of-the-art predictive maintenance techniques. IEEE Trans. Instrum. Measur. 60(10), 3480–3492 (2011)
4. Kobbacy, K.A.H.: Artificial Intelligence in Maintenance, pp. 209–231. Springer, London (2008)
5. Manaf, N.A.A., Beikzadeh, M.R.: Crisp-fuzzy representation of Allen's temporal logic. In: Proceedings of the 25th Conference on Proceedings of the 25th IASTED International Multi-Conference: Artificial Intelligence and Applications, AIAP 2007, pp. 174–179. ACTA Press, Anaheim (2007)
6. Moyse, G., Lesot, M.J.: Linguistic summaries of locally periodic time series. Fuzzy Sets Syst. 285, 94–117 (2016). Special Issue on Linguistic Description of Time Series
7. Novák, V.: Linguistic characterization of time series. Fuzzy Sets Syst. 285, 52–72 (2016). Special Issue on Linguistic Description of Time Series
8. Pereira, R.R., da Silva, V.A.D., Brito, J.N., Nolasco, J.D.: On-line monitoring induction motors by fuzzy logic: a study for predictive maintenance operators. In: 2016 12th International Conference on Natural Computation, Fuzzy Systems and Knowledge Discovery (ICNC-FSKD), pp. 1341–1346, August 2016
9. Poli, J.P., Boudet, L., Mercier, D.: Online temporal reasoning for event and data streams processing. In: FUZZ-IEEE 2016, pp. 2257–2264, July 2016
10. Schockaert, S., De Cock, M., Kerre, E.E.: Fuzzifying Allen's temporal interval relations. Trans. Fuzzy Syst. 16(2), 517–533 (2008)
11. Schockaert, S., Cock, M.D., Kerre, E.: Reasoning About Fuzzy Temporal and Spatial Information from the Web. Intelligent Information Systems, vol. 3. World Scientific, Singapore (2010)
12. da Silva Vicente, S.A., Fujimoto, R.Y., Padovese, L.R.: Rolling bearing fault diagnostic system using fuzzy logic. In: 10th IEEE International Conference on Fuzzy Systems, vol. 2, pp. 816–819, December 2001. vol. 3
13. Toubakh, H., Sayed-Mouchaweh, M.: Hybrid dynamic classifier for drift-like fault diagnosis in a class of hybrid dynamic systems: application to wind turbine converters. Neurocomputing 171, 1496–1516 (2016)

Logics

Complexity of Model Checking
for Cardinality-Based Belief Revision Operators

Nadia Creignou[1](✉), Raïda Ktari[2], and Odile Papini[1]

[1] Aix-Marseille Université, CNRS, LIF, LSIS, Marseille, France
{nadia.creignou,odile.papini}@univ-amu.fr
[2] University of Sfax, ISIMS, Sfax, Tunisia
raida.ktari@isims.usf.tn

Abstract. This paper deals with the complexity of model checking for belief base revision. We extend the study initiated by Liberatore & Schaerf and introduce two new belief base revision operators stemming from consistent subbases maximal with respect to cardinality. We establish the complexity of the model checking problem for various operators within the framework of propositional logic as well as in the Horn fragment.

Keywords: Belief revision · Complexity · Model checking · Maximal cardinality

1 Introduction

Belief revision is an important issue in artificial intelligence, which consists in restoring consistency, keeping new information while modifying as little as possible the agent's initial beliefs. These principles have been formalized in terms of postulates (*AGM postulates*) [1] and numerous operators have been proposed in the literature, which are classified according to two points of view, semantic [12] and syntactic [2,3,10,11,13,18,21]. While early work mainly aimed at defining appropriate semantics for revision, some researchers also investigated the computational complexity of reasoning with the operators introduced in the literature. First works in this direction [4,7,15,16] dealt with the inference problem. Given a set of beliefs, new information and a query (propositional formula), the inference problem consists in deciding whether the query is a logical consequence of the revised set of beliefs. Liberatore and Schaerf [14] proposed to consider another problem, model checking. Given a set of beliefs and new information, it consists in deciding whether an interpretation is a model of the set of revised beliefs. They advocated that studying the complexity of model checking is very important since this problem is the most basic computational task in the setting of model-based knowledge representation and that the succinctness of a knowledge representation formalism is related to the complexity of model checking.

This work has received support from the French Agence Nationale de la Recherche, ASPIQ project reference ANR-12-BS02-0003.

© Springer International Publishing AG 2017
A. Antonucci et al. (Eds.): ECSQARU 2017, LNAI 10369, pp. 387–397, 2017.
DOI: 10.1007/978-3-319-61581-3_35

Liberatore and Schaerf focused on syntactic operators based on consistent sub-bases maximal with respect to set inclusion, in particular Ginsberg's operator [10] and Widtio operator [21]. In this paper we focus on two operators that are similar to Ginsberg's one and Widtio, but which use set cardinality instead of set inclusion as maximality criterion. We also generalize these two operators to stratified belief bases.

We present all these operators within a unified framework. We then study the complexity of the model checking problem for each of them within the propositional framework, as well as within the Horn fragment. We thus obtain a synthetic view of the impact of different strategies and two different maximality criteria (set inclusion and cardinality) on the complexity of the problem, see Table 1. From a computational complexity perspective, maybe surprisingly, the complexity of revision operators based on maximal cardinality can be lower than the complexity of similar revision operators based on set inclusion.

2   Preliminaries

Notations. We assume familiarity with the basics of propositional logic. We remind that a literal is an atom (positive literal) or the negation of an atom (negative literal). Let A be a set of atoms, $\mathrm{Lit}(A)$ denotes the set of literals over A. A clause is a disjunction of literals. We say that a formula is in CNF if it is a conjunction of clauses. A formula is called *Horn* if it is in CNF and in each clause at most one literal is positive. A formula is called *Krom* if it is in CNF and each clause has at most two literals. It is convenient to identify any truth assignment with the set of variables that are true in this assignment. It is thus possible to consider the cardinality of a model of a formula. Let φ be a formula, we denote by $\mathrm{Mod}(\varphi)$ the set of models of φ. A formula ψ is a logical consequence of φ, denoted by $\varphi \models \psi$, if $\mathrm{Mod}(\varphi) \subseteq \mathrm{Mod}(\psi)$, and the two formulas are equivalent, denoted by $\varphi \equiv \psi$, if $\mathrm{Mod}(\varphi) = \mathrm{Mod}(\psi)$.

Let B be a finite set of propositional formulas, $B = \{\varphi_1, \varphi_2, \ldots, \varphi_n\}$ is identified to $\bigwedge B$ the conjunction of its formulas, $\varphi_1 \wedge \varphi_2 \wedge \ldots \wedge \varphi_n$. Given a family of finite sets of formulas $\mathcal{W} = \{B_1, \ldots, B_p\}$, we use $\bigvee_{i=1}^{p} \bigwedge B_i$ for $\bigvee_{i=1}^{p} \bigwedge_{\varphi \in B_i} \varphi$.

Complexity Classes. The classes P and NP are the classes of decision problems solvable in deterministic resp. non deterministic polynomial time. The class coNP is the class of problems whose complementary is in NP [17]. An oracle for a complexity class $\mathcal{C}$ is an entity capable of solving any problem in the class $\mathcal{C}$, it is simply a "black box" that is able to decide for any instance of a given computational problem from $\mathcal{C}$ whether it is a positive instance or not. We write $\mathrm{P}^{\mathcal{C}}$ (resp. $\mathrm{NP}^{\mathcal{C}}$) for the class of decision problems that can be decided by a deterministic (resp. non-deterministic) Turing machine in polynomial time using an oracle for the class $\mathcal{C}$. Stockmeyer inductively defined the polynomial hierarchy, where the first level consists of the P, NP and coNP classes [19]. Within this hierarchy we particularly use the classes of the second level, i.e.

$\Sigma_2 P = NP^{NP}$, $\Delta_2 P = P^{NP}$. We also use the class $\Theta_2 P = P^{NP[\log(n)]}$, introduced by Wagner in [20], which is a sub-class of $\Delta_2 P$. The problems of $\Theta_2 P$ are those of $\Delta_2 P$ that can be solved in polynomial time with only a logarithmic number of calls to an NP-oracle.

3   Belief Base Revision

A formula-based (syntactic) revision operator, denoted by $*$, is a function that takes a belief base B and a formula μ representing new information as input and returns a new belief base $B * \mu$. Many formula-based operators stem from $W(B, \mu)$, the set of maximal subbases of B consistent with μ. They then make use of this set to define the revised belief base according to a given strategy. The maximality criterion as well as the strategy can vary.

In the literature maximality was first considered in terms of set inclusion, and thus the following set was considered:

$$W_\subseteq(B, \mu) = \{B_1 \subseteq B \mid \bigwedge B_1 \not\models \neg\mu \text{ and for all } B_2 \text{ s. t. } B_1 \subset B_2 \subseteq B, \bigwedge B_2 \models \neg\mu\}.$$

We can then consider two different strategies. The first one considers that all maximal subbases are equally plausible. The second one is more drastic and stems from the intersection of consistent maximal subbases, i.e. it only keeps beliefs that are not questioned. Thus, we obtain two well-known operators, namely Ginsberg's operator, $*_G$, [10] and Widtio operator, $*_{wid}$, [21]:

$$B *_G \mu = \bigvee_{B' \in W_\subseteq(B, \mu)} \bigwedge(B' \cup \{\mu\}) \text{ and } B *_{wid} \mu = \bigwedge_{\substack{\\ B' \in W_\subseteq(B, \mu)}} \bigcap (B' \cup \{\mu\}).$$

We focus here on maximality defined in terms of cardinality. So, we consider the set of consistent subbases maximal w.r.t. cardinality:

$$W_{card}(B, \mu) = \{B_1 \subseteq B \mid \bigwedge B_1 \not\models \neg\mu \text{ and for all } B_2 \subseteq B \text{ s. t. } |B_1| < |B_2|, \bigwedge B_2 \models \neg\mu\}.$$

A first operator using this criterion was defined in [2] named RSR [1]. We keep this name and suffix it by G and W to define two operators, RSRG [2] and RSRW, similar respectively to Ginsberg's and Widtio one:

$$B *_{RSRG} \mu = \bigvee_{B' \in W_{card}(B, \mu)} \bigwedge(B' \cup \{\mu\}) \text{ and } B *_{RSRW} \mu = \bigwedge_{\substack{\\ B' \in W_{card}(B, \mu)}} \bigcap B' \cup \{\mu\}.$$

---

[1] The notation RSR comes from the expression *"Removed Sets Revision"*.

Example 1. Consider the belief base $B = \{a \to \neg b, b, b \to c, c \to \neg a, b \to d, d \to \neg a, \neg d \to c\}$, and new information $\mu = a$.

$$\begin{aligned}
\mathcal{W}_{card}(B, \mu) = \{&\{a \to \neg b, b \to c, c \to \neg a, b \to d, d \to \neg a\}; \\
&\{a \to \neg b, b \to c, c \to \neg a, b \to d, \neg d \to c\}; \\
&\{a \to \neg b, b \to c, b \to d, d \to \neg a, \neg d \to c\}\}
\end{aligned}$$

Observe that $\mathcal{W}_{card}(B, \mu) \neq \mathcal{W}_{\subseteq}(B, \mu)$ (since $\mathcal{W}_{\subseteq}(B, \mu)$ contains for instance $\{b, c \to \neg a, d \to \neg a\}$, which is not cardinality-maximal). On the one hand we get

$$\begin{aligned}
B *_{\mathrm{RSRG}} \mu &\equiv (a \wedge \neg b \wedge \neg c \wedge \neg d) \vee (a \wedge \neg b \wedge \neg c \wedge d) \vee (a \wedge \neg b \wedge c \wedge \neg d) \\
&\equiv (a \wedge \neg b) \wedge \neg(c \wedge d).
\end{aligned}$$

On the other hand, since $\bigcap_{B' \in \mathcal{W}_{card}(B,\mu)} B' = \{a \to \neg b, b \to c, b \to d\}$, we have

$$B *_{\mathrm{RSRW}} \mu \equiv a \wedge (a \to \neg b) \wedge (b \to c) \wedge (b \to d) \equiv a \wedge \neg b.$$

The operators RSRG and RSRW may be extended to stratified belief bases. A stratified belief base $B = (S_1, ..., S_n)$ is provided by a partition of the belief base in strata S_i $(1 \leq i \leq n)$ representing priorities between formulas. Let $X \subseteq B$ be a set of formulas, we define $trace(X, B)$ by a tuple of integers as follows: $trace(X, B) = (|X \cap S_1|, ..., |X \cap S_n|)$. The usual lexicographic order over traces, denoted by $\leq_{lex}$, provides a new maximality criterion for consistent subbases.

We define the set of consistent subbases maximal w.r.t. this criterion by $\mathcal{W}_{trace}(B, \mu) = \{B_1 \subseteq B \mid \bigwedge B_1 \not\models \neg\mu$ and for all $B_2 \subseteq B$ s. t. $trace(B_1, B) <_{lex} trace(B_2, B), \bigwedge B_2 \models \neg\mu\}$. Observe that all elements of $\mathcal{W}_{trace}(B, \mu)$ have same trace, denoted by $\mathrm{Tracemax}(B, \mu)$. The operators PRSRG [2,3] and PRSRW are thus defined by:

$$B *_{\mathrm{PRSRG}} \mu = \bigvee_{B' \in \mathcal{W}_{trace}(B,\mu)} \bigwedge (B' \cup \{\mu\}), \quad B *_{\mathrm{PRSRW}} \mu = \bigwedge \bigcap_{B' \in \mathcal{W}_{trace}(B,\mu)} (B' \cup \{\mu\}).$$

Note that all these formula-based operators are sensitive to the syntactic form of the belief base B. While it is well-known that Ginsberg's and Widtio operators satisfy respectively the first seven and the first six AGM postulates, it is proven in [2] that RSRG satisfies all of them when extended to belief set revision.

4 Complexity of Model Checking

In this section we study the computational complexity of the following decision problem according to the considered belief revision operator $*$.

Problem :   MODEL-CHECKING($*$)
Instance : B a belief base, μ a formula, m an interpretation
Question : $m \models B * \mu$?

4.1   Complexity Results for the Operators $*_G$, $*_{RSRG}$ and $*_{PRSRG}$

Complexity of model checking has been studied in [14] by Liberatore and Schaerf for Ginsberg's operator. They proved that MODEL-CHECKING$(*_G)$ is coNP-complete in the general case, and in P when restricted to Horn formulas. So, for Ginsberg's operator restricting instances to Horn formulas makes the model checking problem tractable. On the one hand we prove that MODEL-CHECKING$(*_{RSRG})$ has the same complexity as MODEL-CHECKING$(*_G)$. On the other hand we prove that with this cardinal-maximality criterion, restricting instances to Horn formulas does not make the problem easier.

Theorem 1. MODEL-CHECKING*(*$_{RSRG}$) and MODEL-CHECKING*(*$_{PRSRG}$) are coNP-*complete even if the formulas in the instances are restricted to Horn form.*

Proof. Let (B, μ, m), where B is a stratified belief base, be an instance of MODEL-CHECKING$(*_{PRSRG})$. If m is not a model of μ, then $m \not\models B *_{PRSRG} \mu$. Otherwise, only one subset B' of B is a candidate to have a maximal trace among all subsets of B consistent with μ having m as model: $B' = \{\alpha \in B \mid m \models \alpha\}$. Therefore, to show that $m \not\models B *_{PRSRG} \mu$, we only have to prove that B' is not trace-maximal. For this it is sufficient to guess a set $B_0 \subseteq B$ and an interpretation m_0 such that $m_0 \models \bigwedge(B_0 \cup \{\mu\})$ (and thus $B_0 \cup \{\mu\}$ is consistent) and $trace(B', B) <_{lex} trace(B_0, B)$. All these verifications can be done in polynomial time, thus proving that MODEL-CHECKING$(*_{PRSRG})$ (and a fortiori MODEL-CHECKING$(*_{RSRG})$) is in coNP.

Let us now prove that MODEL-CHECKING$(*_{RSRG})$ (and a fortiori MODEL-CHECKING$(*_{PRSRG})$) is coNP-hard even if the formulas are restricted to Horn form. We use a reduction from the well-known NP-complete problem MAX-INDEPENDENT-SET (see for example [9]) to the complementary of MODEL-CHECKING$(*_{RSRG})$.

Problem :   MAX-INDEPENDENT-SET
Instance : $G = (V, E)$ an undirected graph, k a positive integer.
Question :   Does there exist in G an independent set of size at least k, i.e.,
            $V' \subseteq V$ with $|V'| \geq k$, such that for every two vertices $\{x, y\} \in V'^2$, $\{x, y\} \notin E$?

Let $G = (V, E)$ be an undirected graph and k a positive integer. To every vertex v in G we associate a propositional variable v and let $U = \{u_1, ..., u_{k-1}\}$ be a set of $k - 1$ fresh variables. We then consider the belief base B, the formula μ and the interpretation m defined as follows:

- $B = \{(v_i) \mid v_i \in V\} \cup \{(u_j) \mid u_j \in U\} \cup \{(\neg v_i \vee \neg v_j) \mid \{v_i, v_j\} \in E\} \cup \{(\neg v_i \vee \neg u_j) \mid v_i \in V, u_j \in U\}$
- $\mu = s$ and $m = U \cup \{s\}$ where s is a fresh variable that does not occur in B.

Observe that every consistent subbase of B is consistent with μ, moreover there exists a consistent subbase of B of maximal cardinality, B', that contains all binary clauses. Indeed, if some clause $(\neg v_i \vee \neg w_j)$ is not in B', then by maximality B' contains both v_i and w_j. Since all binary clauses are negative, the set $B' = B \cup \{(\neg v_i \vee \neg w_j)\} \setminus \{v_i\}$ is consistent, has the same cardinality as B' and contains one negative clause more. Finally, such a set cannot contain both a unary clause coming from V and one coming from U.

Suppose that G contains an independent set W with $|W| \geq k$. The set $B' = \{(v_i) \mid v_i \in W\} \cup \{(\neg v_i \vee \neg v_j) \mid \{v_i, v_j\} \in E\} \cup \{(\neg v_i \vee \neg u_j) \mid v_i \in V, u_j \in U\}$ is then consistent (satisfied by the interpretation W). Since $|W| \geq k$, B' shows that the set of clauses from B that are satisfied by m is not cardinality-maximal, which proves that $m \not\models B *_{RSRG} \mu$.

Conversely, according to the observations above, if G does not contain any independent set of size larger than or equal to k, then the set $\{(u_j) \mid u_j \in U\} \cup \{(\neg v_i \vee \neg v_j) \mid \{v_i, v_j\} \in E\} \cup \{(\neg v_i \vee \neg u_j) \mid v_i \in V, u_j \in U\}$ is cardinality-maximal. But this set is exactly the set of clauses from B that are satisfied by m, therefore $m \models B *_{RSRG} \mu$.

4.2   Complexity Results for the Operators $*_{\text{wid}}$, $*_{\text{RSRW}}$ and $*_{\text{PRSRW}}$

Liberatore and Schaerf proved that MODEL-CHECKING($*_{\text{wid}}$) is Σ_2P-complete [14] in the propositional case. As far as we know this is the only known complexity result. In particular the complexity of this problem when restricted to Horn formulas was left open in [14]. We answer this question in proving that the complexity drops by one level in the polynomial hierarchy if the formulas are restricted to Horn.

Theorem 2. MODEL-CHECKING($*_{\text{wid}}$) *is* NP-*complete in the case of Horn formulas.*

Proof. Let us first prove membership. Let B be a set of Horn formulas, μ a Horn formula and m an interpretation. In order to prove that $m \models B *_{\text{wid}} \mu$, one has to show that for every $\alpha \in B$ such that $m \not\models \alpha$, there exists $B'_\alpha \subseteq B$ such that $B'_\alpha \cup \{\mu\}$ is consistent and $B'_\alpha \cup \{\mu\} \cup \{\alpha\}$ is inconsistent (such a set B'_α can be completed until it becomes maximal in terms of set inclusion).

All these consistency checks can be performed in polynomial time since all considered formulas are Horn. Therefore MODEL-CHECKING($*_{\text{wid}}$) is in NP.

Hardness is shown by reduction from the problem PQ-ABDUCTION defined as follows and proven to be NP-complete in [6]:

> Problem :   PQ-ABDUCTION
> Instance :   a Horn formula φ, a set of variables $A = \{x_1, ..., x_n\}$ such that $A \subseteq Var(\varphi)$ and a variable $q \in Var(\varphi) \setminus A$.
> Question :   Does there exist a set $E \subseteq \text{Lit}(A)$ such that $\varphi \wedge \bigwedge E$ is satisfiable and $\varphi \wedge \bigwedge E \wedge \neg q$ is unsatisfiable?

Consider an arbitrary instance (φ, A, q) of PQ-ABDUCTION. Without loss of generality suppose that the Horn formula $\varphi \wedge q$ is satisfiable. Let x_0 be a fresh variable and consider B a set of formulas, μ a formula and m an interpretation defined as follows:

- $B = \{(\varphi \wedge l) \vee \neg x_0 \mid l \in \text{Lit}(A)\} \cup \{\neg q \wedge x_0\}$,
- $\mu = x_0$, and m an interpretation such that $m \models \varphi \wedge x_0 \wedge q$.

First observe that all formulas occurring in B can be rewritten as equivalent Horn formulas. Let us now prove that there exists $E \subseteq \text{Lit}(A)$ such that $\varphi \wedge \bigwedge E$ is consistent and $\varphi \wedge \bigwedge E \wedge \neg q$ is inconsistent if and only if $m \models B *_{\text{wid}} \mu$, that is to say, for all $\alpha \in B$ such that $m \not\models \alpha$, there exists $B'_\alpha \subseteq B$ such that $B'_\alpha \cup \{\mu\}$ is consistent, while $B'_\alpha \cup \{\mu\} \cup \{\alpha\}$ is inconsistent.

First consider $E \subseteq \text{Lit}(A)$ such that $\varphi \wedge \bigwedge E$ is consistent and $\varphi \wedge \bigwedge E \wedge \neg q$ is inconsistent. Let us examine the formulas α in B such that $m \not\models \alpha$. There are two possible cases:

- If $\alpha = (\neg q \wedge x_0)$, then let us consider $B'_\alpha = \{(\varphi \wedge l) \vee \neg x_0 \mid l \in E\}$. We have $\bigwedge(B'_\alpha \cup \{\mu\}) \equiv ((\varphi \wedge \bigwedge E) \vee \neg x_0) \wedge x_0$. On the one hand, this formula is consistent. On the other hand, $\bigwedge(B'_\alpha \cup \{\mu\} \cup \{\alpha\}) \equiv \varphi \wedge \bigwedge E \wedge x_0 \wedge \neg q$ is inconsistent.
- If $\alpha = ((\varphi \wedge l) \vee \neg x_0)$ for some literal l. If $m \not\models \alpha$, then $m \not\models l$ for $m \models \varphi$. Let us then consider $B'_\alpha = \{(\varphi \wedge \neg l) \vee \neg x_0\}$. On the one hand, since $m \models \varphi$ and $m \not\models l$ (and hence $m \models \neg l$), the only formula from B'_α is satisfied by m, and therefore $B'_\alpha \cup \{\mu\}$ is consistent. On the other hand, $B'_\alpha \cup \{\mu\} \cup \{\alpha\} = \{(\varphi \wedge \neg l) \vee \neg x_0\} \cup \{x_0\} \cup \{(\varphi \wedge l) \vee \neg x_0\}$ is inconsistent.

Thus we have proven that if there exists a set $E \subseteq \text{Lit}(A)$ such that $\varphi \wedge \bigwedge E$ is consistent and $\varphi \wedge \bigwedge E \wedge \neg q$ is inconsistent, then $m \models B *_{\text{wid}} \mu$.

Conversely, suppose that $m \models B *_{\text{wid}} \mu$. This means that for every $\alpha \in B$ such that $m \not\models \alpha$ there exists $B'_\alpha \subseteq B$ such that $B'_\alpha \cup \{\mu\}$ is consistent and $B'_\alpha \cup \{\mu\} \cup \{\alpha\}$ is inconsistent. Let us consider $\alpha = \neg q \wedge x_0$ and let us take $E = \{l \mid ((\varphi \wedge l) \vee \neg x_0) \in B'_\alpha\}$. On the one hand, observe that $\varphi \wedge \bigwedge E \wedge x_0 \equiv \bigwedge(B'_\alpha \cup \{\mu\})$, and therefore $\varphi \wedge \bigwedge E$ is consistent. On the other hand, $\varphi \wedge \bigwedge E \wedge x_0 \wedge \neg q \equiv \bigwedge(B'_\alpha \cup \{\mu\} \cup \{\alpha\})$ and therefore $\varphi \wedge \bigwedge E \wedge \neg q$ is inconsistent.

To conclude observe that this reduction is feasible in polynomial time. The only critical point is to find an interpretation m, which is a model of $\varphi \wedge x_0 \wedge q$. This can be done in polynomial time since φ is a Horn formula. So we have finally proven that MODEL-CHECKING($*_{\text{wid}}$) is NP-complete when restricted to Horn formulas.

For this strategy, as stated in the following theorem, considering subbases maximal in terms of cardinality instead of subbases maximal in terms of set inclusion makes the checking problem slightly easier in the context of full propositional logic. However, restricting the instances to Horn formulas is of no help.

Theorem 3. MODEL-CHECKING($*_{\text{RSRW}}$) is $\Theta_2 P$-complete, and MODEL-CHECKING($*_{\text{PRSRW}}$) is in $\Delta_2 P$ and is $\Theta_2 P$-hard. *Hardness results hold even if the formulas occurring in the instances are further restricted to Horn and Krom.*

Proof. Let us first establish membership. We show that MODEL-CHECKING ($*_{\text{RSRW}}$) is in $\Theta_2 P$. Let B be a belief base and μ be a formula. Let k_{max} denote the maximal cardinality of subsets of B that are consistent with μ. In order to decide whether m is a model of the revised belief base $B *_{RSRW} \mu$, we have to check that for every $\alpha \in B$ such that $m \not\models \alpha$, there exists a subset B_α of $B \setminus \{\alpha\}$ consistent with μ such that $|B_\alpha| = k_{max}$.

We proceed by the classical binary search [17] for finding the optimum, asking questions like "Does there exist a subset of B consistent with μ of size least at k?". This requires a logarithmic number of calls to an NP-oracle. When the maximal size k_{max} of all subbases from B that are consistent with μ has thus been computed, we check if for all $\alpha \in B$ such that $m \not\models \alpha$, there exist a subset $B_\alpha \subseteq B \setminus \{\alpha\}$ and an interpretation m_α such that : $m_\alpha \models B_\alpha \cup \{\mu\}$ (and thus $B_\alpha \cup \{\mu\}$ is consistent) and $|B_\alpha| = k_{max}$. This can be done by a single question to an NP-oracle since there are at most $|B|$ formulas to consider and they can be dealt with in parallel. In total, we have then a polynomial algorithm with a logarithmic number of calls to an NP-oracle, thus proving that MODEL-CHECKING($*_{RSRW}$) is in $\Theta_2 P$.

For the problem MODEL-CHECKING($*_{PRSRW}$), the algorithm is similar. The only difference lies in the fact that we have to compute lexicographically maximal trace. This can be done by computing the maximum of each stratum, one after the other. This cannot be done for all strata at once, the oracle calls are adaptative. Therefore the number of calls to an NP-oracle depends linearly on the number of layers of the stratified belief base and logarithmically on the total number of formulas in the belief base. So, the algorithm proves that the problem is in $\Delta_2 P$.

We prove that MODEL-CHECKING($*_{RSRW}$) is $\Theta_2 P$-hard by reduction from the following decision problem:

Problem :   CARDMINSAT
Instance :   Propositional formula φ and an atom x_i.
Question :   Is x_i true in a cardinality-maximal model of φ?

The problem CARDMINSAT is $\Theta_2 P$-hard even if we restrict φ to Krom and moreover the clauses consist of positive literals only (see [5]).

Let (φ, x_i) be an instance of of CARDMINSAT where φ is a conjunction of disjunctions of two positive literals and $X = \{x_1, \ldots, x_n\}$ is the set of variables in φ. We define the following instance of MODEL-CHECKING($*_{RSRW}$):

- $B = \{\varphi \wedge y_j \mid j = 1, \ldots, n+1\} \cup \{(\neg x_i)\} \cup \{(\neg x_j \vee r) \mid j = 1, \ldots, n, j \neq i\}$
- $\mu = \neg r$ and $m = X \cup \{y_1, \ldots y_{n+1}, r\}$ where $y_1, \ldots y_{n+1}, r$ are fresh variables, not occurring in φ.

This reduction is clearly feasible in polynomial time.

On the one hand observe that m satisfies all formulas in B except $(\neg x_i)$. On the other hand, the subbases of B that are consistent with μ and have maximal cardinality necessarily contain $\{\varphi \wedge y_j \mid j = 1, \ldots, n+1\}$.

Let us prove that φ has a cardinality-minimal model containing x_i if and only if there exists $B' \subseteq B$ consistent with μ of maximal cardinality that does not contain $(\neg x_i)$, that is to say if and only if $m \models B *_{RSRW} \mu$.

Suppose that φ has a cardinality-minimal model m' that contains x_i. Consider $B' = \{\varphi \wedge y_j \mid j = 1, \ldots, n+1\} \cup \{(\neg x_j \vee r) \mid x_j \notin m'\}$, $|B'| = n+1+|I'|$ where $I' = \{j \mid x_j \notin m'\}$. It is easy to check that B' is a cardinality-maximal subbase of B consistent with μ that does not contain $(\neg x_i)$.

Conversely, suppose that there exists $B' \subseteq B$ consistent with μ of maximal cardinality that does not contain $(\neg x_i)$. Such a set B' is necessarily of the form $B' = \{\varphi \wedge y_j \mid j = 1, \ldots, n+1\} \cup \{(\neg x_j \vee r) \mid x_j \in I'\}$ where $I' \subseteq \{1, \ldots, n\} \setminus \{i\}$. It is easy to check that $m' = \{x_j \mid j \notin I'\}$ is a cardinality-minimal model of φ that contains x_i.

Observe that all formulas in B, as well as the formula μ are Krom formulas. Therefore, we have proven that MODEL-CHECKING($*_{\text{RSRW}}$) is $\Theta_2 P$-hard even when restricted to Krom formulas. Moreover, each clause in any of these formulas contains at most one negative literal. Hence, in considering the set $\tilde{B}$ and the formula $\tilde{\mu}$ obtained from B and μ by renaming all variables, and the interpretation $\tilde{m} = 1 - m$, we obtain a reduction that proves that MODEL-CHECKING($*_{\text{RSRW}}$) is $\Theta_2 P$-hard even when restricted to formulas that are both Horn and Krom.

Table 1. Complexity results: synthetic summary

Operator	Propositional logic	Horn
Ginsberg	coNP-complete [14, Theorem 1]	P [14, Theorem 17]
Widtio	$\Sigma_2 P$-complete [14, Theorem 2]	NP-complete, Theorem 2
RSRG	coNP-complete, Theorem 1	coNP-complete, Theorem 1
RSRW	$\Theta_2 P$-complete, Theorem 3	$\Theta_2 P$-complete, Theorem 3
PRSRG	coNP-complete, Theorem 1	coNP-complete, Theorem 1
PRSRW	in $\Delta_2 P$, $\Theta_2 P$-hard, Theorem 3	in $\Delta_2 P$, $\Theta_2 P$-hard, Theorem 3

5   Concluding Discussion

It is well known that belief base revision and non-monotonic inference from an inconsistent belief base are the two sides of a same coin [8]. In [4] Cayrol, Lagasquie-Schiex and Schiex present a comparative study of some non-monotonic syntactic inference relations. Let B be a belief base, $<$ be a total pre-order over the formulas of the base, and ϕ be a propositional formula, the inference relations are synthetically defined by $(B, <) \mid\sim^{p,m} \phi$ where $p \in \{T, INCL, LEX\}$ represents the mechanism for selecting consistent subbases, consistent subbases maximal w.r.t. set inclusion or w.r.t. to lexicographic order respectively and $m \in \{\forall, \exists, ARG\}$ represents the inference strategy, universal, existential or argumentative respectively. A comparative study from the complexity perspective is presented in the full propositional case and in the Horn fragment, however it focuses on the inference problem and not on the model checking one. Besides, the strategy stemming from the intersection of maximal consistent subbases is not dealt with. In this paper, on the one hand, we established the complexity of the model checking problem for the Widtio operator in the Horn case, answering to a question left open by Liberatore and Schaerf [14]. On the other hand, we focused on cardinality-based belief revision operators RSRG and RSRW their respective generalization to stratified belief bases. Thus we studied to which extent the use of cardinality as consistent subbases maximality criterion (instead of set inclusion) has an impact on the complexity of the model checking problem. It seems natural to think about the impact of cardinality as maximality criterion for other belief base revision strategies. This will be addressed in a future work.

References

1. Alchourrón, C.E., Gärdenfors, P., Makinson, D.: On the logic of theory change: partial meet contraction and revision functions. J. Symbolic Logic **50**, 510–530 (1985)
2. Benferhat, S., Ben-Naim, J., Papini, O., Würbel, E.: An answer set programming encoding of prioritized removed sets revision: application to GIS. Appl. Intell. **32**(1), 60–87 (2010)
3. Benferhat, S., Cayrol, C., Dubois, D., Lang, J., Prade, H.: Inconsistency management and prioritized syntax-based entailment. In: Proceedings of IJCAI 1993, pp. 640–645 (1993)
4. Cayrol, C., Lagasquie-Schiex, M., Schiex, T.: Nonmonotonic reasoning: from complexity to algorithms. Ann. Math. Artif. Intell. **22**(3–4), 207–236 (1998)
5. Creignou, N., Pichler, R., Woltran, S.: Do hard sat-related reasoning tasks become easier in the Krom fragment? In: IJCAI (2013)
6. Creignou, N., Zanuttini, B.: A complete classification of the complexity of propositional abduction. SIAM J. Comput. **36**, 207–229 (2006)
7. Eiter, T., Gottlob, G.: On the complexity of propositional knowledge base revision, updates, and counterfactuals. Artif. Intell. **57**(2–3), 227–270 (1992)
8. Gärdenfors, P.: Belief revision and nonmonotonic logic: Two sides of the same coin? In: Proceedings of ECAI 1990, pp. 768–773 (1990)
9. Garey, M., Johnson, D.: Computers and Intractability: A Guide to the Theory of NP-Completeness. W.H Freeman, New York (1979)
10. Ginsberg, M.: Counterfactuals. Artif. Intell. **30**, 35–79 (1986)
11. Hansson, S.O.: Revision of belief sets and belief bases. In: Dubois, D., Prade, H. (eds.) Belief Change. Handbook of Defeasible Reasoning and Uncertainty Management Systems, pp. 17–75. Kluwer, Netherlands (1998)
12. Katsuno, H., Mendelzon, A.O.: Propositional knowledge base revision and minimal change. Artif. Intell. **52**, 263–294 (1991)
13. Lehmann, D.: Belief revision, revised. In: Proceedings of IJCAI 1995, pp. 1534–1540 (1995)
14. Liberatore, P., Schaerf, M.: Belief revision and update: complexity of model checking. J. Comput. Syst. Sci. **62**(1), 43–72 (2001)
15. Nebel, B.: Belief revision and default reasoning: syntax-based approaches. In: Proceedings of KR 1991, pp. 417–428 (1991)
16. Nebel, B.: How hard is it to revise a belief base? In: Dubois, D., Prade, H. (eds.) Belief Change. Handbook of Defeasible Reasoning and Uncertainty Management Systems, vol. 3, pp. 77–145. Springer, Netherlands (1998)

17. Papadimitriou, C.H.: Computational Complexity. Addison Wesley, Boston (1994)
18. Papini, O.: A complete revision function in propositional calculus. In: Proceedings of ECAI 1992, pp. 339–343 (1992)
19. Stockmeyer, L.J.: The polynomial-time hierarchy. Theor. Comput. Sci. **3**(1), 1–22 (1976)
20. Wagner, K.: More complicated questions about maxima and minima, and some closures of NP. Theor. Comput. Sci. **51**(1–2), 53–80 (1987)
21. Winslett, M.: Sometimes updates are circumscription. In: Proceedings of IJCAI 1989, pp. 859–863 (1989)

A Two-Tiered Propositional Framework for Handling Multisource Inconsistent Information

Davide Ciucci[1(✉)] and Didier Dubois[2]

[1] DISCo - Università di Milano - Bicocca, Milan, Italy
ciucci@disco.unimib.it
[2] IRIT - CNRS & Université de Toulouse, Toulouse, France
dubois@irit.fr

Abstract. This paper proposes a conceptually simple but expressive framework for handling propositional information stemming from several sources, namely a two-tiered propositional logic augmented with classical modal axioms (BC-logic), a fragment of the non-normal modal logic EMN, whose semantics is expressed in terms of two-valued monotonic set-functions called Boolean capacities. We present a theorem-preserving translation of Belnap logic in this setting. As special cases, we can recover previous translations of three-valued logics such as Kleene and Priest logics. Our translation bridges the gap between Belnap logic, epistemic logic, and theories of uncertainty like possibility theory or belief functions, and paves the way to a unified approach to various inconsistency handling methods.

1 Introduction

A number of works has been published proposing approaches that deal with inconsistent knowledge bases in such a way as to extract useful information from them in a non-explosive way [5,18]. Inconsistency is often due to the presence of multiple sources providing information. Belnap 4-valued logic [4] is one of the earliest approaches to this problem. It is based on a very natural set-up where each source tentatively assigns truth-values to elementary propositions. The sets of truth-values thus collected for these propositions are summarized by so-called epistemic truth-values referring to whether sources are in conflict or not, informed or not. There are 4 such epistemic truth-values, two of which referring to ignorance and conflict. Truth tables for conjunction, disjunction, and negation are used to compute the epistemic status of other complex formulas. This approach underlies both Kleene three-valued logic (when no conflict between sources is observed) and the Priest three-valued logic of paradox [16] (sources are never ignorant and assign truth-values to all elementary propositions).

Besides, inconsistency and incompleteness are present in uncertainty theories as well, using monotonic set-functions called capacities with values in the unit interval, instead of logics. The simplest logical framework for incomplete

© Springer International Publishing AG 2017
A. Antonucci et al. (Eds.): ECSQARU 2017, LNAI 10369, pp. 398–408, 2017.
DOI: 10.1007/978-3-319-61581-3_36

information is the two-tiered propositional logic MEL [3], that accounts for an all-or-nothing view of possibility theory [11], and borrows axioms K and D from modal epistemic logic. Replacing necessity measures by general inclusion-monotonic set-functions can account for the idea of conflicting sources of information. It leads to adopting a fragment of the non-normal modal logic EMN [7] as a general logical framework, which can encompass variants of probabilistic and belief function logics, for instance the logic of risky knowledge [15], where the adjunction rule is not valid. Here, we show that our two-tiered propositional setting related to EMN can encode Belnap 4-valued logic, namely that the four truth-values in Belnap logic are naturally expressed by means of capacities taking values on $\{0, 1\}$. Thus, we construct a bridge between Belnap logic and uncertainty theories. As special cases, we recover our previous translations of Kleene logic for incomplete information [8] and Priest logic of paradox [9]. Showing the possibility of this translation indicates that our logic has potential to support other inconsistency handling approaches as well. The paper is organized as follows. Section 2 presents the propositional logic of Boolean capacities BC and shows its capability to capture the notion of information coming from several sources. Section 3 recalls Belnap 4-valued logic from the point of view of its motivation, its syntactical inference and its semantics. Section 4 contains the main results pertaining to the translation of Belnap logic into BC. Most proofs are omitted due to length constraints.

2   The Logic of Boolean Capacities and Multisource Information Management

In this section, we consider an approach to the handling of pieces of incomplete and conflicting information coming from several sources. We show, following an intuition already suggested in [1, 12] that if we represent each body of information items supplied by each source by means of a set of possible states of affairs, the collective information supplied by the sources can be modelled in a lossless way by a monotonic set function (called a capacity) that takes value on $\{0, 1\}$. These set-functions can serve as natural multisource semantics for a simple flat non-regular modal logic that captures the four Belnap truth-values as already suggested in [10]. This logic looks rather uncommitting for handling multiple source information, while other approaches seem to put additional assumptions.

2.1   Boolean Capacities and Multisource Information

Consider a standard propositional language $\mathcal{L}$ with variables $V = \{a, b, c, \dots\}$ and connectives $\wedge, \vee, \neg$, for conjunction, disjunction and negation, respectively. We denote the propositional formulas of $\mathcal{L}$ by letters $p, q, \dots$. Consider a set of states of affairs Ω which is the set of interpretations of this language.

Definition 1. *A capacity (or fuzzy measure) is a mapping $\gamma : 2^{\Omega} \to [0, 1]$ such that $\gamma(\emptyset) = 0$; $\gamma(\Omega) = 1$; and if $A \subseteq B$ then $\gamma(A) \leq \gamma(B)$.*

The value $\gamma(A)$ can be interpreted as the degree of support of a proposition p represented by the subset $A = [p]$ of its models. A Boolean capacity (B-capacity, for short) is a capacity with values in $\{0,1\}$. It can be defined from a usual capacity and any threshold $\lambda > 0$ as $\beta(A) = 1$ if $\gamma(A) \geq \lambda$ and 0 otherwise.

The useful information in a B-capacity consists of its focal sets. A focal set E is such that $\beta(E) = 1$ and $\beta(E \backslash \{w\}) = 0, \forall w \in E$. Let $\mathcal{F}_\beta$ be the set of focal sets of β. They are minimal sets for inclusion such that $\beta(E) = 1$: we can check that $\beta(A) = 1$ if and only if there is a subset E of A in $\mathcal{F}_\beta$ with $\beta(E) = 1$.

Consider n sources providing information in the form of epistemic states modelled by non-empty sets $E_i \subseteq \Omega$: it is only known from source i that the real state of affairs s should lie in E_i. A capacity β can be built from these pieces of information then viewed as the set $\mathcal{F}_\beta = \{E_1, E_2, \ldots, E_n\}$ of its focal subsets. Then $\beta([p]) = 1$ really means that there is at least one source i that believes that p is true (that is p is true in all states of affairs in E_i). Note that this way of synthetizing information is not destructive: it preserves every initial piece of information. Given a proposition p, there are four epistemic statuses based on the information from sources, that can be described by the capacity:

- *Support of p*: $\beta([p]) = 1$ and $\beta([\neg p]) = 0$. Then p is asserted by at least one source and negated by no other one.
- *Rejection of p*: $\beta([\neg p]) = 1$ and $\beta([p]) = 0$. Then p is negated by at least one source and asserted by no other one.
- *Ignorance about p*: $\beta([p]) = \beta([\neg p]) = 0$. No source supports nor negates p.
- *Conflict about p*: $\beta([p]) = \beta([\neg p]) = 1$. Some sources assert p, some negate it.

Important special cases are
- when β is minitive, i.e., $\beta(A \cap B) = \min(\beta(A), \beta(B))$. It is then a necessity measure and $\mathcal{F}_\beta = \{E\}$. There is only one source and its information is incomplete, but there is no conflict.
- when the focal sets are singletons $\{e_i\}$. Then β is maxitive, i.e., $\beta(A \cup B) = \max(\beta(A), \beta(B))$. All sources have complete information, so there are conflicts, but no ignorance. Letting $E = \{e_1, e_2, \ldots, e_n\}$, then $\beta(A) = 1$ if and only if $A \cap E \neq \emptyset$, formally a possibility measure. But here E is a conjunction of non-mutually exclusive elements, not a possibility distribution.

2.2   The Logic BC

To provide a logical setting to the above situation, we build a higher level propositional language $\mathcal{L}_\square$ on top of $\mathcal{L}$, whose formulas are denoted by Greek letters $\phi, \psi, \ldots$, and defined by: if $p \in \mathcal{L}$ then $\square p \in \mathcal{L}_\square$; if $\phi, \psi \in \mathcal{L}_\square$ then $\neg \phi \in \mathcal{L}_\square$, $\phi \wedge \psi \in \mathcal{L}_\square$. Note that the language $\mathcal{L}$ is not part of $\mathcal{L}_\square$, it is embedded in it, since atomic variables of $\mathcal{L}_\square$ are of the form $\square p, p \in \mathcal{L}$. As usual $\lozenge p$ stands for $\neg \square \neg p$. It defines a very elementary fragment of a modal logic language [3].

A minimal logic for B-capacities has been proposed using the language $\mathcal{L}_\square$ [12]. It is a two-tiered propositional logic plus some modal axioms:

1. All axioms of propositional logics for $\mathcal{L}_\Box$-formulas.
2. The modal axioms:

(RM) $\Box p \to \Box q$ if $\vdash p \to q$ in propositional logic.

  (N) $\Box p$, whenever p is a propositional tautology.

  (P) $\Diamond p$, whenever p is a propositional tautology.

The only rule is modus ponens: If ψ and $\psi \to \phi$ then ϕ. This is a fragment of the non-regular logic EMN [7]. Note that the two dual modalities $\Box$ and $\Diamond$ play the same role. Namely the above axioms remain valid if we exchange $\Box$ and $\Diamond$. So these modalities are not distinguishable. Semantics is usually expressed in terms of neighborhood semantics, but it can be equivalently expressed in terms of B-capacities on the set of interpretations Ω of the language $\mathcal{L}$. We have indicated elsewhere [10] that the set of subsets A such that $\beta(A) = 1$ is a special case of neighborhood family in the sense of neighborhood semantics [7]. This logic can thus be called the logic of Boolean Capacities (BC). A BC-model of an atomic formula $\Box p$ is a B-capacity β. The satisfaction of BC-formulas is defined as:

- $\beta \models \Box p$, if and only if $\beta([p]) = 1$;
- $\beta \models \neg\phi$, $\beta \models \phi \wedge \psi$ in the standard way.

Semantic entailment is defined classically, and syntactic entailment is classical propositional entailment taking RM, N, P as axioms: $\Gamma \vdash_{BC} \phi$ if and only if $\Gamma \cup \{$all instances of $RM, N, P\} \vdash \phi$ (classically defined). It has been proved that BC logic is sound and complete wrt B-capacity models [12]. In fact, axiom RM clearly expresses the monotonicity of capacities, and it is easy to realize that a classical propositional interpretation of $\mathcal{L}_\Box$ that respects the axioms of BC can be precisely viewed as a B-capacity.

As a B-capacity precisely encodes a set of sources each delivering incomplete information items in the form of an n-tuple of focal sets $(E_1, E_2, \ldots, E_n)$, we can see that $\beta \models \Box p$ if and only if $\exists i : E_i \subseteq [p]$, so we may write $(E_1, E_2, \ldots, E_n) \models \varphi$ in place of $\beta \models \varphi$. Denoting by $\Box_i p$ the statement $E_i \subseteq [p]$, the formula $\Box p$ is of the form $\Box_1 p \vee \cdots \vee \Box_n p$ where $\Box_i$ is a standard KD modality in a regular modal logic. Likewise, $\Diamond p = \neg\Box\neg p = \neg\Box_1\neg p \wedge \cdots \wedge \neg\Box_n\neg p = \Diamond_1 p \wedge \cdots \wedge \Diamond_n p$ clearly means that no source is asserting $\neg p$. The four epistemic statuses of propositions in $\mathcal{L}$ can then be expressed by means of modal formulas in $\mathcal{L}_\Box$ as follows [10]:

- *Support of* p: $(E_1, E_2, \ldots, E_n) \models \Box p \wedge \Diamond p$
- *Rejection of* p: $(E_1, E_2, \ldots, E_n) \models \Box\neg p \wedge \Diamond\neg p$
- *Ignorance about* p: $(E_1, E_2, \ldots, E_n) \models \Diamond p \wedge \Diamond\neg p$
- *Conflict about* p: $(E_1, E_2, \ldots, E_n) \models \Box p \wedge \Box\neg p$

Note that this framework is very cautious, in the sense that inferences made are minimal ones one can expect to make from multisource information. If we add axioms K and D of modal logics, then β is forced to be a necessity measure, and the conflict situation disappears: there is only one source with epistemic set E driving β. We get the logic MEL [3], a fragment of the logic KD. In case we restrict to capacities β whose focal sets are singletons, the $\Box$ modality has all properties of a KD possibility modality $\Diamond$. It is a kind of mirror image of logic MEL where conflict is taken into account but there is no ignorance. It can capture Priest logic of paradox [9]. The aim of this paper is to show that the general framework of BC-logic can encode Belnap logic as a special case.

3   Belnap 4-Valued Logic

Belnap [4] considers an artificial information processor, fed from a variety of sources, and capable of answering queries on propositions of interest. The basic assumption is that the computer receives information about atomic propositions in a cumulative way from outside sources, each asserting for each atomic proposition whether it is true, false, or being silent about it. The notion of *epistemic set-up* is defined as an assignment, of one of four values denoted by $\mathbf{T}$, $\mathbf{F}$, $\mathbf{C}$, $\mathbf{U}$, to each atomic proposition $a, b, \ldots$:

1. Assigning $\mathbf{T}$ to a means the computer has only been told that a is true (1) by at least one source, and false (0) by none.
2. Assigning $\mathbf{F}$ to a means the computer has only been told that a is false by at least one source, and true by none.
3. Assigning $\mathbf{C}$ to a means the computer has been told at least that a is true by one source and false by another.
4. Assigning $\mathbf{U}$ to a means the computer has been told nothing about a.

Table 1. Belnap disjunction, conjunction and negation

∨	F	U	C	T
F	F	U	C	T
U	U	U	T	T
C	C	T	C	T
T	T	T	T	T

∧	F	U	C	T
F	F	F	F	F
U	F	U	F	U
C	F	F	C	C
T	F	U	C	T

a	$\neg a$
F	T
U	U
C	C
T	F

If $\{0, 1\}$ is the set of usual truth values (as assigned by the information sources), then the set $\mathbb{V}_4 = \{\mathbf{T}, \mathbf{F}, \mathbf{C}, \mathbf{U}\}$ of epistemic truth values coincides with the power set of $\{0, 1\}$, letting $\mathbf{T} = \{1\}$, $\mathbf{F} = \{0\}$. According to the convention initiated by Dunn [13], $\mathbf{U}$ represents the empty set and corresponds to no information received, while $\mathbf{C} = \{0, 1\}$ represents the presence of conflicting sources, expressing *true and false* at the same time. Belnap's approach relies on two orderings in $\mathbb{V}_4 = \{\mathbf{T}, \mathbf{F}, \mathbf{C}, \mathbf{U}\}$, equipping it with two lattice structures:

- *The information ordering,* $\sqsubseteq$ whose meaning is "less informative than", such that $\mathbf{U} \sqsubseteq \mathbf{T} \sqsubseteq \mathbf{C}; \mathbf{U} \sqsubseteq \mathbf{F} \sqsubseteq \mathbf{C}$. This ordering reflects the inclusion relation of the sets $\emptyset$, $\{0\}$, $\{1\}$, and $\{0, 1\}$. $(\mathbb{V}_4, \sqsubseteq)$ is *the information lattice*.
- *The truth ordering,* $<_t$, representing "more true than" according to which $\mathbf{F} <_t \mathbf{C} <_t \mathbf{T}$ and $\mathbf{F} <_t \mathbf{U} <_t \mathbf{T}$, each chain reflecting the truth-set of Kleene's logic. In other words, ignorance and conflict play the same role with respect to $\mathbf{F}$ and $\mathbf{T}$ according to this ordering. It yields *the logical lattice*, based on

the truth ordering, and the interval extension of standard connectives $\wedge$, $\vee$ and $\neg$ from $\{0, 1\}$ to $2^{\{0,1\}} \backslash \{\emptyset\}$. In this lattice, the maximum of $\mathbf{U}$ and $\mathbf{C}$ is $\mathbf{T}$ and the minimum is $\mathbf{F}$.

The syntax is the one of propositional logic. Connectives of negation, conjunction and disjunction are defined truth-functionally in Belnap 4-valued logic (see Table 1). Belnap 4-valued logic has no tautologies, but it has an inference system. It can be defined only via a set of inference rules, as those that can be found in [14, 17]:

Definition 2. *Let* $a, b, c \in V$. *The inference system of* Belnap 4-valued logic *is defined by no axiom and the following set of rules*

$$(R1): \frac{a \wedge b}{a} \qquad (R2): \frac{a \wedge b}{b} \qquad (R3): \frac{a \quad b}{a \wedge b} \qquad (R4): \frac{a}{a \vee b}$$

$$(R5): \frac{a \vee b}{b \vee a} \qquad (R6): \frac{a \vee a}{a} \qquad (R7): \frac{a \vee (b \vee c)}{(a \vee b) \vee c}$$

$$(R8): \frac{a \vee (b \wedge c)}{(a \vee b) \wedge (a \vee c)} \qquad (R9): \frac{(a \vee b) \wedge (a \vee c)}{a \vee (b \wedge c)} \qquad (R10): \frac{a \vee c}{\neg \neg a \vee c}$$

$$(R11): \frac{\neg(a \vee b) \vee c}{(\neg a \wedge \neg b) \vee c} \qquad (R12): \frac{\neg(a \wedge b) \vee c}{(\neg a \vee \neg b) \vee c} \qquad (R13): \frac{\neg \neg a \vee c}{a \vee c}$$

$$(R14): \frac{(\neg a \wedge \neg b) \vee c}{\neg(a \vee b) \vee c} \qquad (R15): \frac{(\neg a \vee \neg b) \vee c}{\neg(a \wedge b) \vee c}$$

These rules express that conjunction is idempotent and distributes over disjunction, disjunction is idempotent, associative and distributes over conjunction. Negation is involutive and De Morgan Laws are satisfied. It makes clear that the underlying algebra is a De Morgan algebra [17]. Applying these rules, formulas can be put in normal form as a conjunction of clauses, i.e., $p = p_1 \wedge \ldots \wedge p_n$, where the p_i's are disjunctions of literals $l_{ij} = a$ or $\neg a$ where $a \in V$. For the semantics, consider again the four epistemic truth-values forming the set $\mathbb{V}_4 = \{\mathbf{F}, \mathbf{U}, \mathbf{C}, \mathbf{T}\}$. A Belnap valuation is a mapping $vb : \mathcal{L} \mapsto \mathbb{V}_4$. Let $\Gamma \subseteq \mathcal{L}$ and $p \in \mathcal{L}$, then we define the consequence relation by means of the truth ordering $\leq_t$ as

$$\Gamma \vDash_B p \quad \text{iff} \quad \exists p_1, \ldots, p_n \in \Gamma, \; \forall vb \; vb(p_1) \wedge \ldots \wedge vb(p_n) \leq_t vb(p)$$

Now, let us consider the consequence relations $\vDash_U, \vDash_C$ obtained by the designated values $\{\mathbf{U}, \mathbf{T}\}$ or $\{\mathbf{C}, \mathbf{T}\}$, respectively defined as:

$$\Gamma \vDash_U p : \forall vb \quad \text{if} \quad vb(p_i) \in \{\mathbf{U}, \mathbf{T}\}, \forall p_i \in \Gamma, \text{ then } vb(p) \in \{\mathbf{U}, \mathbf{T}\};$$

$$\Gamma \vDash_C p : \forall vb \quad \text{if} \quad vb(p_i) \in \{\mathbf{C}, \mathbf{T}\}, \forall p_i \in \Gamma, \text{ then } vb(p) \in \{\mathbf{C}, \mathbf{T}\}.$$

Font [14] proves the following result: $\Gamma \vDash_B p$ iff $\Gamma \vDash_U p$ and $\Gamma \vDash_C p$. Moreover, due to the symmetric role that $\{\mathbf{U}, \mathbf{T}\}$ and $\{\mathbf{C}, \mathbf{T}\}$ play in Belnap's logic, the two relations $\Gamma \vDash_U p$ and $\Gamma \vDash_C p$ are equivalent: $\Gamma \vDash_U p$ iff $\Gamma \vDash_C p$. The adequacy with the Hilbert-style axiomatization of Belnap logic and the above semantics is proved by Pynko [17] and Font [14]:

Theorem 1. *Belnap logic is sound and complete with respect to Belnap seman-tics, that is $\Gamma \vdash_B p$ iff $\Gamma \vDash_B p$ using the 15 rules $Ri, i = 1, \ldots, 15$.*

Kleene logic has truth tables obtained from Belnap logic's by deleting the truth-value **C**, and has designated truth-value **T**. Priest logic is obtained by deleting the truth-value **U**, keeping **C**, **T** as designated. From a syntactic point of view, Kleene logic has one more inference rule than Belnap 4-valued logic, e.g., $q \wedge \neg q \vdash p \vee \neg p$, while Priest logic is Belnap logic plus one axiom ($p \vee \neg p$, see [14,17]).

4   A Translation of Belnap Logic into BC

The above results, joined with the fact that Kleene logic and Priest logic can be translated into MEL [8,9] strongly suggest that Belnap logic can be expressed in BC. Formulas in BC can be related to Belnap truth-values **T, F, U, C** in an obvi-ous way, provided that we restrict to atomic formulas. Let $\mathcal{T}$ be the translation operation that changes a partial Belnap truth-value assignment $vb(a) \in \Theta \subseteq \mathbb{V}_4$ to an atomic propositional formula a, into a modal formula, indicating its epis-temic status w.r.t a set of sources. In agreement with the multisource semantics of the BC logic, we let $\mathcal{T}(vb(a) \geq_t \mathbf{C}) = \Box a$ and $\mathcal{T}(vb(a) \leq_t \mathbf{C}) = \Box \neg a$. Like-wise $\mathcal{T}(vb(a) \geq_t \mathbf{U}) = \Diamond a, \mathcal{T}(vb(a) \leq_t \mathbf{U}) = \Diamond \neg a$. Hence, we get the modal translation of the four Belnap epistemic values:

$$\mathcal{T}(vb(a) = \mathbf{T}) = \Box a \wedge \Diamond a \qquad \mathcal{T}(vb(a) = \mathbf{F}) = \Box \neg a \wedge \Diamond \neg a$$
$$\mathcal{T}(vb(a) = \mathbf{U}) = \Diamond a \wedge \Diamond \neg a \qquad \mathcal{T}(vb(a) = \mathbf{C}) = \Box a \wedge \Box \neg a$$

In Belnap logic, though, sources provide information only on these elementary propositions, valuations for other propositions being obtained via truth-tables. The translation of Belnap truth-qualified formulas will be carried out using the truth-tables of the logic, which means that in all formulas of $\mathcal{L}_\Box$ that can be reached via the translation, only literals appear in the scope of modalities.

Let us consider the fragment of BC language where we can only put a modal-ity in front of literals: $\mathcal{L}_\Box^\ell = \Box a | \Box \neg a | \neg \phi | \phi \wedge \psi | \phi \vee \psi$. We can proceed to the translation of Belnap truth-tables into BC. First consider negation. It is easy to check that $\mathcal{T}(vb(\neg p) = \mathbf{T}) = \mathcal{T}(vb(p) = \mathbf{F})$, $\mathcal{T}(vb(\neg p) = \mathbf{x}) = \mathcal{T}(vb(p) = \mathbf{x})$, $\mathcal{T}(vb(\neg p) \geq_t \mathbf{x}) = \mathcal{T}(vb(p) \leq_t \mathbf{x})$, $\mathbf{x} \in \{\mathbf{U}, \mathbf{C}\}$. On compound formulas built with conjunction and disjunction, it is clear that $\mathcal{T}(vb(p \wedge q) = \mathbf{T}) = \mathcal{T}(vb(p) = \mathbf{T}) \wedge \mathcal{T}(vb(q) = \mathbf{T})$ but, due to the distributive lattice structure of $\mathbb{V}_4$, we have $\mathcal{T}(vb(p \vee q) = \mathbf{T}) = \mathcal{T}(vb(p) = \mathbf{T}) \vee \mathcal{T}(vb(q) = \mathbf{T}) \vee (\mathcal{T}(vb(p) = \mathbf{U}) \wedge \mathcal{T}(vb(p) = \mathbf{C})) \vee (\mathcal{T}(vb(p) = \mathbf{C}) \wedge \mathcal{T}(vb(p) = \mathbf{U}))$.

For elementary formulas $\neg a, a \vee b, a \wedge b$ of Belnap logic, we get explicit translations using the truth-tables of Belnap logic, for instance:

$$\mathcal{T}(vb(\neg a) = \mathbf{T}) = \mathcal{T}(vb(a) = \mathbf{F})$$
$$\mathcal{T}(vb(\neg a) = \mathbf{U}) = \mathcal{T}(vb(a) = \mathbf{U}); \mathcal{T}(vb(\neg a) = \mathbf{C}) = \mathcal{T}(vb(a) = \mathbf{C})$$
$$\mathcal{T}(vb(\neg a) \geq_t \mathbf{C}) = \mathcal{T}(vb(a) \leq_t \mathbf{C}) = \Box \neg a$$

$$\mathcal{T}(vb(a \wedge b) = \mathbf{T}) = \Box a \wedge \Diamond a \wedge \Box b \wedge \Diamond b$$
$$\mathcal{T}(vb(a \vee b) = \mathbf{T}) = (\Box a \wedge \Diamond a) \vee (\Box b \wedge \Diamond b) \vee (Ca \wedge Ub) \vee (Ua \wedge Cb)$$
$$\mathcal{T}(vb(a \wedge b) \geq_t \mathbf{U}) = \Diamond a \wedge \Diamond b; \quad \mathcal{T}(vb(a \wedge b) \geq_t \mathbf{C}) = \Box a \wedge \Box b$$
$$\mathcal{T}(vb(a \vee b) \geq_t \mathbf{U}) = \Diamond a \vee \Diamond b; \quad \mathcal{T}(vb(a \vee b) \geq_t \mathbf{C}) = \Box a \vee \Box b$$

where $\Diamond a \wedge \Diamond \neg a$ is shortened as Ua and $\Box a \wedge \Box \neg a$ as Ca. Belnap logic has two designated values: $\mathbf{T}$ and $\mathbf{C}$. So, for inference purposes, we use translated semantic expressions $\mathcal{T}(vb(p) \geq_t \mathbf{C})$. According to this translation, Belnap logic reaches the following fragment of BC-language: $\mathcal{L}_\Box^B = \Box a | \Box \neg a | \phi \wedge \psi | \phi \vee \psi$ without negation in front of $\Box$. Conversely, from the fragment $\mathcal{L}_\Box^B$ we can go back to Belnap logic. Namely any formula in $\mathcal{L}_\Box^B$ can be translated into a formula of the propositional logic language as follows: $\Box a$ maps to a and $\Box \neg a$ to $\neg a$; $\theta(\psi \wedge \phi)$ to $\theta(\psi) \wedge \theta(\phi)$ and $\theta(\psi \vee \phi)$ to $\theta(\psi) \vee \theta(\phi)$. We remark that $\Box a \vee \Box \neg a$ is not a tautology in BC, and in general no tautologies can be expressed in the above fragment. This is coherent with the fact that Belnap logic has no theorems.

Theorem 2. *Let ϕ/ψ be any of the 15 inference rules of Belnap logic. Then, the following inference rule is valid in BC:*

$$\frac{\mathcal{T}(vb(\phi) \geq_t \mathbf{C})}{\mathcal{T}(vb(\psi) \geq_t \mathbf{C})}$$

As a consequence we can mimic syntactic inference of Belnap logic in BC, more precisely restricting to formulas in $\mathcal{L}_\Box^B$. The restriction of the scope of modalities to literals also affects the set of B-capacities that can act as a semantic counterpart of the logic. We can check that semantic inference in Belnap logic can be expressed in the modal setting of BC by restricting the capacities that can be used as models of $\mathcal{L}_\Box^B$ formulas. Namely, consider a Belnap set-up where each source i provides a set T_i of atoms considered true by this source, a set F_i of atoms considered false by this source, where $T_i \cap F_i = \emptyset$. It corresponds to a special kind of epistemic state with rectangular shape, namely: $E_i = [(\bigwedge_{a \in T_i} a) \wedge (\bigwedge_{b \in F_i} \neg b)]$.

As there are n sources of this kind, we can restrict to B-capacities β with such rectangular focal sets. In fact, as atoms of $\mathcal{L}_\Box^B$ are of the form $\Box \ell$ where ℓ is a literal, and as we cannot put $\Box$ in front of conjunctions nor disjunctions, it is enough to use capacities whose focal sets are of the form $[a], a \in \cup_{i=1}^n T_i$ and $[\neg b], b \in \cup_{i=1}^n F_i$ to interpret formulas in $\mathcal{L}_\Box^B$. We call such capacities *atomic*. Considering the Belnap valuation vb associated to the information supplied by n sources, there is a one-to-one correspondence between Belnap valuations and atomic B-capacities α induced by this information:

Proposition 1. *For any B-capacity β, there is a single Belnap valuation vb_β such that $\beta \models \phi$ if and only if $vb_\beta(\theta(\phi)) \in \{\mathbf{C}, \mathbf{T}\}$.*

The idea is to let $vb_\beta(a) = \mathbf{T}$ if $\beta([a]) = 1$ and $\beta([\neg a]) = 0$, $vb_\beta(a) = \mathbf{F}$ if $\beta([a]) = 0$ and $\beta([\neg a]) = 1$, etc. In the other way around,

Proposition 2. *For each Belnap valuation vb, there exists a unique atomic B-capacity α_{vb} such that $vb \models p$ if and only if $\alpha_{vb} \models \mathcal{T}(vb(p) \geq_t \mathbf{C})$.*

Indeed, define $T = \{a : vb(a) = \mathbf{T} \text{ or } \mathbf{C}\}$, $F = \{a : vb(a) = \mathbf{F} \text{ or } \mathbf{C}\}$, and let $\alpha([a]) = 1$ if $a \in T$, $\alpha([\neg a]) = 1$ if $a \in F$. However there are several Belnap set-ups inducing a given Belnap valuation vb: for instance only two sources are enough to model the four values [6]. We thus introduce an equivalence relation on the set of B-capacities, whereby two of them are equivalent if they correspond to the same Belnap truth assignment: $\beta \sim_B \beta'$ if and only if $vb_\beta = vb_{\beta'}$.

Proposition 3. *For any B-capacity β, there exists an atomic B-capacity α such that $\beta \sim_B \alpha$.*

Indeed, consider β with focal sets $E_1, \ldots E_n$. Let $T_i = \{a \in V : E_i \subseteq [a]\}$ and $F_i = \{b \in V : E_i \subseteq [\neg b]\}$. The focal sets of α are based on such literals and form the family

$$\mathcal{F}_\alpha = \{[a] : a \in \cup_{i=1}^n T_i\} \cup \{[\neg b] : b \in \cup_{i=1}^n F_i\}.$$

From Proposition 3 we can conclude that for any B-capacity β, there exists an atomic B-capacity $\alpha \sim_B \beta$ such that $\beta \models \phi \in \mathcal{L}_\square^B$ if and only if $\alpha \models \phi$. We then can prove that our translation of Belnap logic into BC is consequence-preserving:

Theorem 3. *Let Γ be a set (conjunction) of formulas in propositional logic interpreted in Belnap logic, and p be another such formula. Then $\Gamma \vdash_B p$ if and only if $\{\mathcal{T}(vb(q) \geq_t \mathbf{C}) : q \in \Gamma\} \vdash_{BC} \mathcal{T}(vb(q) \geq_t \mathbf{C})$.*

Proof. Suppose $\Gamma \vdash_B p$. Then from Theorem 2, all inference rules of Belnap logic become valid inferences in BC using the translations of their premises and conclusions. So the inference can be made in BC. Conversely, by completeness of BC, suppose $\forall \beta$, if $\beta \models \mathcal{T}(vb(q) \geq_t \mathbf{C}), \forall q \in \Gamma$ then $\beta \models \mathcal{T}(vb(p) \geq_t \mathbf{C})$. Using Proposition 3, for all B-capacities β, $\exists \alpha \sim_B \beta$, where α is atomic, such that $\forall q \in \Gamma$, $\alpha \models \mathcal{T}(vb(q) \geq_t \mathbf{C})$ if and only if $\beta \models \mathcal{T}(vb(q) \geq_t \mathbf{C})$ and $\alpha \models \mathcal{T}(vb(p) \geq_t \mathbf{C})$ if and only if $\beta \models \mathcal{T}(vb(p) \geq_t \mathbf{C})$. Then, we have that if $vb(q) \geq_t \mathbf{C}, \forall q \in \Gamma$ then $vb(p) \geq_t \mathbf{C}$ for the Belnap valuation vb associated to α. So $\Gamma \models_B p$. By completeness of Belnap logic, $\Gamma \vdash_B p$ follows. $\square$

We can recover our previous translations of three-valued Kleene logic and the logic of paradox into the logic MEL [8,9], from our translation of Belnap logic into BC, by translating into BC the properties added to Belnap logic to recover these logics. Namely Kleene logic is obtained by adding the inference rule $q \wedge \neg q \vdash p \vee \neg p$ to Belnap logic, which comes down to adding inference rule $(KL) : \quad \square q \wedge \square \neg q \vdash \square p \vee \square \neg p$ to BC. A simpler approach is to add axiom D $(\square p \rightarrow \lozenge p)$ to BC. To recover Priest logic from Belnap's, axiom $p \vee \neg p$ must be added, which means adding to BC the (unusual) axiom $\square p \vee \square \neg p$ [9].

5   Conclusion

In this paper, we have pursued our work regarding a class of many-valued logics dealing with inconsistent or incomplete information processing [8,9]. Just like Kleene logic and Priest's logic of paradox in MEL, we can capture Belnap 4-valued logics in a simple two-tiered propositional logic couched in the language of modal logic EMN involving only depth-1 formulas. The natural semantics for this propositional logic is in terms of all-or-nothing set-functions that model Belnap set-ups and capture both incomplete and inconsistent pieces of information. The use of set-functions clarifies the connection between Belnap 4-valued logic and uncertainty modeling. The use of set-functions beyond possibility and necessity measures is in agreement with the fact that propositions in Belnap 4-valued logics cannot be viewed as S5-like beliefs. The logic BC is cautious enough to be a general setting for modeling incomplete and inconsistent logical information. It subsumes Belnap 4-valued logic, doing away with the restriction to literals, and accounting for generalized Belnap set-ups considered by Avron et al. [2]. It seems that our framework may be used to capture various approaches to inconsistent and incomplete information handling; for instance, the one based on maximal consistent subsets could be obtained by considering B-capacities such that $\beta(A \cap B) = \min(\beta(A), \beta(B))$ if $A \cap B \neq \emptyset$. Moreover it can be extended to handling degrees of support. This is to be explored in the future.

References

1. Assaghir, Z., Napoli, A., Kaytoue, M., Dubois, D., Prade, H.: Numerical information fusion: lattice of answers with supporting arguments. In: Proceedings ICTAI 2011, Boca Raton, FL, USA, pp. 621–628 (2011)
2. Avron, A., Ben-Naim, J., Konikowska, B.: Processing Information from a set of sources. In: Makinson, D., Malinowski, J., Wansing, H. (eds.) Towards Mathematical Philosophy. Trends in Logic, vol. 28, pp. 165–186. Springer, Netherlands (2009)
3. Banerjee, M., Dubois, D.: A simple logic for reasoning about incomplete knowledge. Int. J. Approximate Reasoning **55**, 639–653 (2014)
4. Belnap, N.D.: A useful four-valued logic. In: Dunn, J.M., Epstein, G. (eds.) Modern Uses of Multiple-Valued Logic, pp. 8–37. D. Reidel, Dordrecht (1977)
5. Besnard, P., Hunter, A. (eds.): Reasoning with Actual and Potential Contradictions. The Handbook of Defeasible Reasoning and Uncertain Information, vol. 2. Kluwer, Dordrecht (1998)
6. Carnielli, W., Lima-Marques, M.: Society semantics for multiple-valued logics. In: Advances in Contemporary Logic and Computer Science. Contemporary Mathematics, vol. 235, pp. 33–52. American Mathematical Society (1999)
7. Chellas, B.F.: Modal Logic: An Introduction. Cambridge University Press, Cambridge (1980)
8. Ciucci, D., Dubois, D.: A modal theorem-preserving translation of a class of three-valued logics of incomplete information. J. Appl. Non Classical Logics **23**(4), 321–352 (2013)
9. Ciucci, D., Dubois, D.: From possibility theory to paraconsistency. In: Beziau, J.Y., Chakraborty, M., Dutta, S. (eds.) New Directions in Paraconsistent Logic. Springer Proceedings in Mathematics & Statistics, vol. 152. Springer, New Delhi (2015)
10. Dubois, D.: Reasoning about ignorance and contradiction: many-valued logics versus epistemic logic. Soft Comput. **16**(11), 1817–1831 (2012)
11. Dubois, D., Prade, H.: Possibility theory and its applications: where do we stand? In: Kacprzyk, J., Pedrycz, W. (eds.) Handbook of Computational Intelligence, pp. 31–60. Springer, Heidelberg (2015)

12. Dubois, D., Prade, H., Rico, A.: Representing qualitative capacities as families of possibility measures. Int. J. Approximate Reasoning **58**, 3–24 (2015)
13. Dunn, J.M.: Intuitive semantics for first-degree entailment and coupled trees. Philos. Stud. **29**, 149–168 (1976)
14. Font, J.M.: Belnap's four-valued logic and De Morgan lattices. Logic J. IGPL **5**(3), 1–29 (1997)
15. Kyburg, H.E., Teng, C.-M.: The logic of risky knowledge, reprised. Int. J. Approximate Reasoning **53**(3), 274–285 (2012)
16. Priest, G.: The logic of paradox. J. Philos. Logic **8**, 219–241 (1979)
17. Pynko, A.P.: Characterizing Belnap's logic via De Morgan's laws. Math. Log. Q. **41**, 442–454 (1995)
18. Tanaka, K., Berto, F., Mares, E., Paoli, F. (eds.): Paraconsistency: Logic and Applications, pp. 1–12. Springer, Heidelberg (2013)

Reasoning in Description Logics
with Typicalities and Probabilities of Exceptions

Gian Luca Pozzato$^{(\boxtimes)}$

Dipartimento di Informatica, Università di Torino, Turin, Italy
gianluca.pozzato@unito.it

Abstract. We introduce a nonmonotonic procedure for preferential Description Logics in order to reason about typicality by taking probabilities of exceptions into account. We consider an extension, called $\mathcal{ALC} + \mathbf{T}_{\mathbf{R}}^{\mathsf{P}}$, of the logic of typicality $\mathcal{ALC} + \mathbf{T}_{\mathbf{R}}$ by inclusions of the form $\mathbf{T}(C) \sqsubseteq_p D$, whose intuitive meaning is that "typical Cs are Ds with a probability p". We consider a notion of extension of an ABox containing only some typicality assertions, then we equip each extension with a probability. We then restrict entailment of a query F to those extensions whose probabilities belong to a given and fixed range. We propose a decision procedure for reasoning in $\mathcal{ALC} + \mathbf{T}_{\mathbf{R}}^{\mathsf{P}}$ and we exploit it to show that entailment is ExpTime-complete as for the underlying $\mathcal{ALC}$.

1 Introduction

Nonmonotonic extensions of Description Logics (from now on, DLs for short) have been actively investigated since the early 90s [2–7,16] in order to tackle the problem of representing *prototypical* properties of classes and to reason about *defasible* inheritance. A simple but powerful nonmonotonic extension of DLs is proposed in [8]: in this approach "typical" or "normal" properties can be directly specified by means of a "typicality" operator $\mathbf{T}$ enriching the underlying DL, and a TBox can contain inclusions of the form $\mathbf{T}(C) \sqsubseteq D$ to represent that "typical Cs are also Ds" or "normally, Cs have the property D". The Description Logic so obtained is called $\mathcal{ALC} + \mathbf{T}_{\mathbf{R}}$ and, as a difference with standard DLs, one can consistently express exceptions and reason about defeasible inheritance as well. For instance, a knowledge base can consistently express that "normally, referees do not send-off football managers", whereas "Italian referees usually send-off football managers" (since they usually either protest without justification or kick off water bottles at field side when they become angry) as follows:

$\mathbf{T}(Referee) \sqsubseteq \neg\exists sendoff.FootballManager$
$\mathbf{T}(Referee \sqcap Italian) \sqsubseteq \exists sendoff.FootballManager$

G.L. Pozzato—Partially supported by the project "ExceptionOWL", Università di Torino and Compagnia di San Paolo, call 2014 "Excellent (young) PI", project ID: Torino_call2014_L1_111.

A. Antonucci et al. (Eds.): ECSQARU 2017, LNAI 10369, pp. 409–420, 2017.
DOI: 10.1007/978-3-319-61581-3_37

The semantics of the **T** operator is characterized by the properties of *rational logic* [11], recognized as the core properties of nonmonotonic reasoning. As a consequence, **T** inherits well-established properties like *specificity*: in the example, if one knows that Daniele is a typical Italian referee, then the logic $\mathcal{ALC} + \mathbf{T_R}$ allows us to infer that he usually sends-off football managers, giving preference to the most specific information.

The logic $\mathcal{ALC} + \mathbf{T_R}$ itself is too weak in several application domains. Indeed, although the operator **T** is nonmonotonic ($\mathbf{T}(C) \sqsubseteq E$ does not imply $\mathbf{T}(C \sqcap D) \sqsubseteq E$), the logic $\mathcal{ALC} + \mathbf{T_R}$ is monotonic, in the sense that if the fact F follows from a given knowledge base KB, then F also follows from any KB' $\supseteq$ KB. As a consequence, unless a KB contains explicit assumptions about typicality of individuals, there is no way of inferring defeasible properties about them: in the above example, if KB contains the fact that Mark is a referee, i.e. *Referee*(*mark*) belongs to KB, it is not possible to infer that he does not send-off managers ($\neg\exists sendoff.FootballManager(mark)$). This would be possible only if the KB contained the stronger information that Mark is a *typical* referee, i.e. $\mathbf{T}(Referee)(mark)$ belongs to (or can be inferred from) KB. In order to overwhelm this limit and perform useful inferences, in [10] the authors have introduced a nonmonotonic extension of the logic $\mathcal{ALC} + \mathbf{T_R}$ based on a minimal model semantics, corresponding to a notion of *rational closure* as defined in [11] for propositional logic. Intuitively, the idea is to restrict our consideration to (canonical) models that maximize typical instances of a concept when consistent with the knowledge base. The resulting logic, call it $\mathcal{ALC} + \mathbf{T_R}^{RaCl}$, supports typicality assumptions, so that if one knows that Mark is a referee, one can nonmonotonically assume that he is also a *typical* referee if this is consistent, and therefore that he does not send-off managers. From a semantic point of view, the logic $\mathcal{ALC} + \mathbf{T_R}^{RaCl}$ is based on a preference relation among $\mathcal{ALC} + \mathbf{T_R}$ models and a notion of *minimal entailment* restricted to models that are minimal with respect to such preference relation. However, $\mathcal{ALC} + \mathbf{T_R}^{RaCl}$ imposes to consider *all* consistent typicality assumptions that are consistent with a given KB. This seems to be too strong in several application domains, in particular when the need arises of reasoning about scenarios where exceptional individuals are taken into account.

In this work we introduce a new Description Logic called $\mathcal{ALC} + \mathbf{T_R^P}$, which extends $\mathcal{ALC}$ by means of typicality inclusions equipped by *probabilities of exceptionality* of the form $\mathbf{T}(C) \sqsubseteq_p D$, where $p \in (0, 1)$. The intuitive meaning is that "typical Cs are also Ds with a probability p" or "normally, Cs are Ds and the probability of having exceptional Cs not being Ds is $1 - p$". For instance, we can have

$$\mathbf{T}(Student) \sqsubseteq_{0.3} SportLover \qquad\qquad \mathbf{T}(Student) \sqsubseteq_{0.9} SocialNetworkUser$$

whose intuitive meaning is that being sport lovers and social network users are both typical properties of students, however the probability of having exceptional students not loving sport is higher than the one of finding students not using social networks, in particular we have the evidence that the probability of not having exceptions is 30% and 90%, respectively. As a difference with DLs under the distributed semantics introduced in [13,14], where probabilistic axioms of the form $p :: C \sqsubseteq D$ are used to capture uncertainty in order to represent that Cs are Ds with probability p, in the logic $\mathcal{ALC} + \mathbf{T}_{\mathbf{R}}^{\mathsf{P}}$ we are able to ascribe typical properties to concepts and to reason about probabilities of exceptions to those typicalities. We define different extensions of an ABox containing only some of the "plausible" typicality assertions: each extension represents a scenario having a specific probability. Then, we provide a notion of nonmonotonic entailment restricted to extensions whose probabilities belong to a given and fixed range, in order to reason about scenarios that are not necessarily the most probable. We introduce a decision procedure for checking entailment in $\mathcal{ALC} + \mathbf{T}_{\mathbf{R}}^{\mathsf{P}}$ and we exploit it in order to show that reasoning in $\mathcal{ALC} + \mathbf{T}_{\mathbf{R}}^{\mathsf{P}}$ with probabilities of exceptions is ExpTime complete, therefore we retain the same complexity of the underlying standard $\mathcal{ALC}$.

2 Preferential Description Logics

The logic $\mathcal{ALC} + \mathbf{T}_{\mathbf{R}}$ is obtained by adding to standard $\mathcal{ALC}$ the typicality operator $\mathbf{T}$ [8]. The intuitive idea is that $\mathbf{T}(C)$ selects the *typical* instances of a concept C. We can therefore distinguish between the properties that hold for all instances of concept C ($C \sqsubseteq D$), and those that only hold for the normal or typical instances of C ($\mathbf{T}(C) \sqsubseteq D$).

The semantics of the $\mathbf{T}$ operator can be formulated in terms of *rational models*: a model $\mathcal{M}$ is any structure $\langle \Delta^{\mathcal{I}}, <, \cdot^{\mathcal{I}} \rangle$ where $\Delta^{\mathcal{I}}$ is the domain, $<$ is an irreflexive, transitive, well-founded and modular (for all x, y, z in $\Delta^{\mathcal{I}}$, if $x < y$ then either $x < z$ or $z < y$) relation over $\Delta^{\mathcal{I}}$. In this respect, $x < y$ means that x is "more normal" than y, and that the typical members of a concept C are the minimal elements of C with respect to this relation. An element $x \in \Delta^{\mathcal{I}}$ is a *typical instance* of some concept C if $x \in C^{\mathcal{I}}$ and there is no C-element in $\Delta^{\mathcal{I}}$ *more typical* than x. In detail, $\cdot^{\mathcal{I}}$ is the extension function that maps each concept C to $C^{\mathcal{I}} \subseteq \Delta^{\mathcal{I}}$, and each role R to $R^{\mathcal{I}} \subseteq \Delta^{\mathcal{I}} \times \Delta^{\mathcal{I}}$. For concepts of $\mathcal{ALC}$, $C^{\mathcal{I}}$ is defined as usual. For $\mathbf{T}$, we have $(\mathbf{T}(C))^{\mathcal{I}} = Min_<(C^{\mathcal{I}})$. A model $\mathcal{M}$ can be equivalently defined by postulating the existence of a function $k_{\mathcal{M}} : \Delta^{\mathcal{I}} \longmapsto \mathbb{N}$, where $k_{\mathcal{M}}$ assigns a finite rank to each domain element: $k_{\mathcal{M}}$ and $<$ can be defined from each other by letting $x < y$ if and only if $k_{\mathcal{M}}(x) < k_{\mathcal{M}}(y)$.

Given standard definitions of satisfiability of a KB in a model, we define a notion of entailment in $\mathcal{ALC} + \mathbf{T}_{\mathbf{R}}$. Given a query F (either an inclusion $C \sqsubseteq D$ or an assertion $C(a)$ or an assertion of the form $R(a,b)$), we say that F is entailed from a KB, written KB $\models_{\mathcal{ALC}+\mathbf{T}_{\mathbf{R}}} F$, if F holds in all $\mathcal{ALC} + \mathbf{T}_{\mathbf{R}}$ models satisfying KB.

Even if the typicality operator $\mathbf{T}$ itself is nonmonotonic (i.e. $\mathbf{T}(C) \sqsubseteq E$ does not imply $\mathbf{T}(C \sqcap D) \sqsubseteq E$), what is inferred from a KB can still be inferred from any KB' with KB $\subseteq$ KB', i.e. the logic $\mathcal{ALC} + \mathbf{T}_{\mathbf{R}}$ is monotonic. In order to perform useful nonmonotonic inferences, in [10] the authors have strengthened the above semantics by restricting entailment to a class of minimal models. Intuitively, the idea is to restrict entailment to models that *minimize the untypical instances of a concept.* The resulting logic is called $\mathcal{ALC} + \mathbf{T}_{\mathbf{R}}^{RaCl}$ and it corresponds to a notion of *rational closure* on top of $\mathcal{ALC} + \mathbf{T}_{\mathbf{R}}$. Such a notion is a natural extension of the rational closure construction provided in [11] for the propositional logic.

The nonmonotonic semantics of $\mathcal{ALC} + \mathbf{T}_{\mathbf{R}}^{RaCl}$ relies on minimal rational models that minimize the *rank of domain elements*. Informally, given two models of KB, one in which a given domain element x has rank 2 (because for instance $z < y < x$), and another in which it has rank 1 (because only $y < x$), we prefer the latter, as in this model the element x is assumed to be "more typical" than in the former.

Query entailment is then restricted to minimal *canonical models*. The intuition is that a canonical model contains all the individuals that enjoy properties that are consistent with KB. A model $\mathcal{M}$ is a minimal canonical model of KB if it satisfies KB, it is minimal and it is canonical[1]. A query F is minimally entailed from a KB, written KB $\models_{\mathcal{ALC} + \mathbf{T}_{\mathbf{R}}^{RaCl}} F$, if it holds in all minimal canonical models of KB. In [10] it is shown that query entailment in $\mathcal{ALC} + \mathbf{T}_{\mathbf{R}}^{RaCl}$ is in EXPTIME.

3   Dealing with Probabilities of Exceptions

In this section we define an alternative semantics that allows us to equip a typicality inclusion with the probability of *not* having exceptions for that, and then to reason about such inclusions. In the resulting Description Logic, called $\mathcal{ALC} + \mathbf{T}_{\mathbf{R}}^{\mathsf{P}}$, a typicality inclusion has the form $\mathbf{T}(C) \sqsubseteq_p D$, and its intuitive meaning is "normally, Cs are also Ds with probability p" or, in other words, "typical Cs are also Ds, and the probability of having exceptional Cs not being Ds is $1 - p$". We then define a nonmonotonic procedure whose aim is to describe alternative completions of the ABox obtained by assuming typicality assertions about the individuals explicitly named in the ABox: the basic idea is similar to the one proposed in [8], where a completion of an $\mathcal{ALC}+\mathbf{T}$ ABox is proposed in order to assume that every individual constant of the ABox is a typical element of the most specific concept he belongs to, if this is consistent with the knowledge base. An analogous approach is proposed in [12], where different extensions of the ABox are introduced in order to define plausible but *surprising* scenarios. Here we propose a similar, algorithmic construction in order to compute only *some* assumptions of typicality of individual constants, in order to describe alternative scenarios having different probabilities: different extensions/scenarios are obtained by considering different sets of typicality assumptions of the form $\mathbf{T}(C)(a)$, where a occurs in the ABox.

Definition 1. *We consider an alphabet of concept names $\mathcal{C}$, of role names $\mathcal{R}$, and of individual constants $\mathcal{O}$. Given $A \in \mathcal{C}$ and $R \in \mathcal{R}$, we define:*
$$C := A \mid \top \mid \bot \mid \neg C \mid C \sqcap C \mid C \sqcup C \mid \forall R.C \mid \exists R.C$$

---

[1] In Theorem 10 in [10] the authors have shown that for any consistent KB there exists a finite minimal canonical model of KB.

An $\mathcal{ALC} + \mathbf{T}_{\mathbf{R}}^{\mathsf{P}}$ *knowledge base is a pair* $(\mathcal{T}, \mathcal{A})$. $\mathcal{T}$ *contains axioms of the form either (i)* $C \sqsubseteq C$ *or (ii)* $\mathbf{T}(C) \sqsubseteq_p C$*, where* $p \in \mathbb{R}, p \in (0,1)$. $\mathcal{A}$ *contains assertions of the form* $C(a)$ *and* $R(a,b)$*, where* $a, b \in \mathcal{O}$.

Given an inclusion $\mathbf{T}(C) \sqsubseteq_p D$, the higher the probability p the more the inclusion is "exceptions-free" or, equivalently, the less is the probability of having exceptional Cs not being also Ds. In this respect, the probability p is a real number included in the open interval $(0,1)$: the probability 1 is not allowed, in the sense that an inclusion $\mathbf{T}(C) \sqsubseteq_1 D$ (the probability of having exceptional Cs not being Ds is 0) corresponds to a *strict* inclusion $C \sqsubseteq D$ (all Cs are Ds). Given another inclusion $\mathbf{T}(C') \sqsubseteq_{p'} D'$, with $p' < p$, we assume that this inclusion is less "strict" than the other one, i.e. the probability of having exceptional C's is higher than the one of having exceptional Cs with respect to properties D' and D, respectively. Recalling the example of the Introduction, where KB contains $\mathbf{T}(Student) \sqsubseteq_{0.9} SocialNetworkUser$ and $\mathbf{T}(Student) \sqsubseteq_{0.3} SportLover$, we have that typical students make use of social networks, and that normally they also love sport; however, the second inclusion is less probable with respect to the first one: both are properties of a prototypical student, however there are more exceptions of students not loving sport with respect to those not being active on social networks.

Before introducing formal definitions, we provide an example inspired to Example 1 in [12] in order to give an intuitive idea of what we mean for reasoning in $\mathcal{ALC} + \mathbf{T}_{\mathbf{R}}^{\mathsf{P}}$ with probabilities of exceptions. We will complete it with part 2 in Example 3.

Example 1 (Reasoning in $\mathcal{ALC} + \mathbf{T}_{\mathbf{R}}^{\mathsf{P}}$ *part 1).* Let KB $= (\mathcal{T}, \mathcal{A})$ where $\mathcal{T}$ is as follows:

$$
\begin{aligned}
&AtypicalDepressed \sqsubseteq Depressed \\
&\mathbf{T}(Depressed) \sqsubseteq_{0.85} \neg\exists Symptom.MoodReactivity \\
&\mathbf{T}(AtypicalDepressed) \sqsubseteq_{0.6} \exists Symptom.MoodReactivity \quad\quad (1)\\
&\mathbf{T}(ProstateCancerPatient) \sqsubseteq_{0.5} \exists Symptom.MoodReactivity \\
&\mathbf{T}(ProstateCancerPatient) \sqsubseteq_{0.8} \exists Symptom.Nocturia
\end{aligned}
$$

We have that (2) $\mathbf{T}(Depressed \sqcap Spleenless) \sqsubseteq \neg\exists Symptom.MoodReactivity$ follows[2] from KB, and this is a wanted inference, since undergoing spleen removal is irrelevant with respect to mood reactivity as far as we know. This is a non-monotonic inference that does no longer follow if it is discovered that typical depressed people without their spleen are subject to mood reactivity: given $\mathcal{T}' = \mathcal{T} \cup \{\mathbf{T}(Depressed \sqcap Spleenless) \sqsubseteq \exists Symptom.MoodReactivity\}$, we have that the inclusion (2) does no longer follow from KB with $\mathcal{T}'$ in the logic $\mathcal{ALC} + \mathbf{T}_{\mathbf{R}}^{\mathsf{P}}$. As for rational closure, the set of inclusions that are entailed from a $\mathcal{ALC} + \mathbf{T}_{\mathbf{R}}^{\mathsf{P}}$ KB is closed under the property known as *rational monotonicity*: for instance, from KB and the fact that $\mathbf{T}(Depressed) \sqsubseteq \neg Elder$ is not entailed

---

[2] As mentioned, at this point of the presentation we only want to give an intuition of inferences characterizing $\mathcal{ALC} + \mathbf{T}_{\mathbf{R}}^{\mathsf{P}}$. Technical details and definitions will be provided in Definition 5.

from KB in $\mathcal{ALC} + \mathbf{T_R^P}$, it follows that the inclusion $\mathbf{T}(Depressed \sqcap Elder) \sqsubseteq \neg \exists Symptom.MoodReactivity$ is entailed in $\mathcal{ALC} + \mathbf{T_R^P}$.

Concerning ABox reasoning, if $\mathcal{A} = \{Depressed(jim)\}$, then we can infer that Jim has not mood swings with a probability of 85%, since $\mathbf{T}(Depressed(jim))$ is minimally entailed from KB in $\mathcal{ALC} + \mathbf{T_R^{RaCl}}$ and the inclusion (1) is equipped by a probability of 0.85. If we discover that Jim is an atypical depressed, then $\mathcal{ALC} + \mathbf{T_R^P}$ allows us to retract such inference, whereas the fact that Jim has mood swings ($\exists Symptom.MoodReactivity(jim)$) is entailed and evaluated having probability of 60%.

3.1   Extensions of ABox

Given a KB, we define the finite set $\mathfrak{Tip}$ of concepts occurring in the scope of the typicality operator, i.e. $\mathfrak{Tip} = \{C \mid \mathbf{T}(C) \sqsubseteq_p D \in \text{KB}\}$. Given an individual a explicitly named in the ABox, we define the set of typicality assumptions $\mathbf{T}(C)(a)$ that can be minimally entailed from KB in the nonmonotonic logic $\mathcal{ALC} + \mathbf{T_R^{RaCl}}$, with $C \in \mathfrak{Tip}$. We then consider an ordered set $\mathfrak{Tip}_{\mathcal{A}}$ of pairs (a, C) of all possible assumptions $\mathbf{T}(C)(a)$, for all concepts $C \in \mathfrak{Tip}$ and all individual constants a in the ABox.

Definition 2 (Assumptions in $\mathcal{ALC} + \mathbf{T_R^P}$). *Given an $\mathcal{ALC} + \mathbf{T_R^P}$ KB=$(\mathcal{T}, \mathcal{A})$, let $\mathcal{T}'$ be the set of inclusions of $\mathcal{T}$ without probabilities, namely $\mathcal{T}' = \{\mathbf{T}(C) \sqsubseteq D \mid \mathbf{T}(C) \sqsubseteq_p D \in \mathcal{T}\} \cup \{C \sqsubseteq D \in \mathcal{T}\}$. Given a finite set of concepts $\mathfrak{Tip}$, we define, for each individual name a occurring in $\mathcal{A}$: $\mathfrak{Tip}_a = \{C \in \mathfrak{Tip} \mid (\mathcal{T}', \mathcal{A}) \models_{\mathcal{ALC} + \mathbf{T_R^{RaCl}}} \mathbf{T}(C)(a)\}$. We also define $\mathfrak{Tip}_{\mathcal{A}} = \{(a, C) \mid C \in \mathfrak{Tip}_a$ and a occurs in $\mathcal{A}\}$ and we impose an order on its elements: $\mathfrak{Tip}_{\mathcal{A}} = [(a_1, C_1), (a_2, C_2), \ldots, (a_n, C_n)]$. Furthermore, we define the ordered multiset $\mathcal{P}_{\mathcal{A}} = [p_1, p_2, \ldots, p_n]$, respecting the order imposed on $\mathfrak{Tip}_{\mathcal{A}}$, where $p_i = \prod_{j=1}^{m} p_{ij}$ for all $\mathbf{T}(C_i) \sqsubseteq_{p_{i1}} D_1, \mathbf{T}(C_i) \sqsubseteq_{p_{i2}} D_2, \ldots, \mathbf{T}(C_i) \sqsubseteq_{p_{im}} D_m$ in $\mathcal{T}$.*

The ordered multiset $\mathcal{P}_{\mathcal{A}}$ is a tuple of the form $[p_1, p_2, \ldots, p_n]$, where p_i is the probability of the assumption $\mathbf{T}(C)(a)$, such that $(a, C) \in \mathfrak{Tip}_{\mathcal{A}}$ at position i. p_i is the product of all the probabilities p_{ij} of typicality inclusions $\mathbf{T}(C) \sqsubseteq_{p_{ij}} D$ in the TBox.

Following the basic idea underlying surprising scenarios outlined in [12], we consider different extensions $\widetilde{\mathcal{A}_i}$ of the ABox and we equip them with a probability $\mathbb{P}_i$. Starting from $\mathcal{P}_{\mathcal{A}} = [p_1, p_2, \ldots, p_n]$, the first step is to build all alternative tuples where 0 is used in place of some p_i to represent that the corresponding typicality assertion $\mathbf{T}(C)(a)$ is no longer assumed (Definition 3). Furthermore, we define the *extension* of the ABox corresponding to a string so obtained (Definition 4). In this way, the highest probability is assigned to the extension of the ABox corresponding to $\mathcal{P}_{\mathcal{A}}$, where all typicality assumptions are considered. The probability decreases in the other extensions, where some typicality assumptions are discarded, thus 0 is used in place of the corresponding p_i. The probability of an extension $\widetilde{\mathcal{A}_i}$ corresponding to a string $\mathcal{P}_{\mathcal{A}_i} = [p_{i1}, p_{i2}, \ldots, p_{in}]$

is defined as the product of probabilities p_{ij} when $p_{ij} \neq 0$, i.e. the probability of the corresponding typicality assumption when this is selected for the extension, and $1 - p_j$ when $p_{ij} = 0$, i.e. the corresponding typicality assumption is discarded, that is to say the extension contains an exception to the inclusion.

Definition 3 (Strings of possible assumptions $\mathbb{S}$). *Given a KB = $(\mathcal{T}, \mathcal{A})$, let the set $\mathfrak{Tip}_\mathcal{A}$ and $\mathcal{P}_\mathcal{A} = [p_1, p_2, \ldots, p_n]$ be as in Definition 2. We define the set $\mathbb{S}$ of all the strings of possible assumptions with respect to KB as*

$$\mathbb{S} = \{[s_1, s_2, \ldots, s_n] \mid \forall i = 1, 2, \ldots, n \ either \ s_i = p_i \ or \ s_i = 0\}$$

Definition 4 (Extension of ABox). *Let KB=$(\mathcal{T}, \mathcal{A})$, $\mathcal{P}_\mathcal{A} = [p_1, p_2, \ldots, p_n]$ and $\mathfrak{Tip}_\mathcal{A} = [(a_1, C_1), (a_2, C_2), \ldots, (a_n, C_n)]$ as in Definition 2. Given a string of possible assumptions $[s_1, s_2, \ldots, s_n] \in \mathbb{S}$ of Definition 3, we define the extension $\widetilde{\mathcal{A}}$ of $\mathcal{A}$ w.r.t. $\mathfrak{Tip}_\mathcal{A}$ and $\mathbb{S}$ as:*

$$\widetilde{\mathcal{A}} = \{\mathbf{T}(C_i)(a_i) \mid (a_i, C_i) \in \mathfrak{Tip}_\mathcal{A} \ and \ s_i \neq 0\}$$

We also define the probability of $\widetilde{\mathcal{A}}$ as $\mathbb{P}_{\widetilde{\mathcal{A}}} = \prod_{i=1}^{n} \chi_i$ where $\chi_i = \begin{cases} s_i & if \ s_i \neq 0 \\ 1 - p_i & if \ s_i = 0 \end{cases}$

It can be observed that, in $\mathcal{ALC} + \mathbf{T}_\mathbf{R}^{RaCl}$, the set of typicality assumptions that can be inferred from a KB corresponds to the extension of $\mathcal{A}$ corresponding to the string $\mathcal{P}_\mathcal{A}$ (no element is set to 0): all the typicality assertions of individuals occurring in the ABox, that are consistent with the KB, are assumed. On the contrary, in $\mathcal{ALC} + \mathbf{T}_\mathbf{R}$, no typicality assumptions can be derived from a KB, and this corresponds to extending $\mathcal{A}$ by the assertions corresponding to the string $[0, 0, \ldots, 0]$, i.e. by the empty set.

Example 2. Given a KB=$(\mathcal{T}, \mathcal{A})$, let the only typicality inclusions in $\mathcal{T}$ be $\mathbf{T}(C) \sqsubseteq_{0.6} D$ and $\mathbf{T}(E) \sqsubseteq_{0.85} F$. Let a and b be the only individual constants occurring in $\mathcal{A}$. Suppose also that $\mathbf{T}(C)(a)$, $\mathbf{T}(C)(b)$, and $\mathbf{T}(E)(b)$ are entailed from KB in $\mathcal{ALC} + \mathbf{T}_\mathbf{R}^{RaCl}$. We have that $\mathfrak{Tip}_\mathcal{A} = \{(a, C), (b, C), (b, E)\}$ and $\mathcal{P}_\mathcal{A} = [0.6, 0.6, 0.85]$. All possible strings, corresponding extensions of $\mathcal{A}$ and probabilities are shown in Table 1.

3.2   Reasoning in $\mathcal{ALC} + \mathbf{T}_\mathbf{R}^\mathsf{P}$

We are now ready to provide formal definitions for nonmonotonic entailment in the Description Logic $\mathcal{ALC} + \mathbf{T}_\mathbf{R}^\mathsf{P}$. Intuitively, given KB and a query F, we distinguish two cases: (i) if F is an inclusion $C \sqsubseteq D$, then it is entailed from KB if it is minimally entailed from KB' in the nonmonotonic $\mathcal{ALC} + \mathbf{T}_\mathbf{R}^{RaCl}$, where KB' is obtained from KB by removing probabilities of exceptions, i.e. by replacing each typicality inclusion $\mathbf{T}(C) \sqsubseteq_p D$ with $\mathbf{T}(C) \sqsubseteq D$; (ii) if F is an ABox fact $C(a)$, then it is entailed from KB if it is entailed in the monotonic $\mathcal{ALC} + \mathbf{T}_\mathbf{R}$ from the knowledge bases including the extensions of the ABox of

Table 1. Plausible extensions of the ABox of Example 2.

String	Extension	Probability
[0.6, 0.6, 0.85]	$\widetilde{\mathcal{A}_1} = \{\mathbf{T}(C)(a), \mathbf{T}(C)(b), \mathbf{T}(E)(b)\}$	$\mathbb{P}_{\widetilde{\mathcal{A}_1}} = 0.6 \times 0.6 \times 0.85 = 0.306$
[0, 0, 0.85]	$\widetilde{\mathcal{A}_2} = \{\mathbf{T}(E)(b)\}$	$\mathbb{P}_{\widetilde{\mathcal{A}_2}} = (1-0.6) \times (1-0.6) \times 0.85 = 0.136$
[0, 0.6, 0]	$\widetilde{\mathcal{A}_3} = \{\mathbf{T}(C)(b)\}$	$\mathbb{P}_{\widetilde{\mathcal{A}_3}} = (1-0.6) \times 0.6 \times (1-0.85) = 0.036$
[0.6, 0, 0]	$\widetilde{\mathcal{A}_4} = \{\mathbf{T}(C)(a)\}$	$\mathbb{P}_{\widetilde{\mathcal{A}_4}} = 0.6 \times (1-0.6) \times (1-0.85) = 0.036$
[0, 0.6, 0.85]	$\widetilde{\mathcal{A}_5} = \{\mathbf{T}(C)(b), \mathbf{T}(E)(b)\}$	$\mathbb{P}_{\widetilde{\mathcal{A}_5}} = (1-0.6) \times 0.6 \times 0.85 = 0.204$
[0.6, 0, 0.85]	$\widetilde{\mathcal{A}_6} = \{\mathbf{T}(C)(a), \mathbf{T}(E)(b)\}$	$\mathbb{P}_{\widetilde{\mathcal{A}_6}} = 0.6 \times (1-0.6) \times 0.85 = 0.204$
[0.6, 0.6, 0]	$\widetilde{\mathcal{A}_7} = \{\mathbf{T}(C)(a), \mathbf{T}(C)(b)\}$	$\mathbb{P}_{\widetilde{\mathcal{A}_7}} = 0.6 \times 0.6 \times (1-0.85) = 0.054$
[0, 0, 0]	$\widetilde{\mathcal{A}_8} = \emptyset$	$\mathbb{P}_{\widetilde{\mathcal{A}_8}} = (1-0.6) \times (1-0.6) \times (1-0.85) = 0.024$
$\mathbb{P}_{\widetilde{\mathcal{A}_1}} + \mathbb{P}_{\widetilde{\mathcal{A}_2}} + \cdots + \mathbb{P}_{\widetilde{\mathcal{A}_8}} = 1$		

Definition 4. More in detail, we provide both (i) a notion of entailment restricted to scenarios whose probabilities belong to a given range and (ii), similarly to [13], a notion of probability of the entailment of a query $C(a)$, as the sum of the probabilities of all extensions from which $C(a)$ is so entailed.

Definition 5 (Entailment in $\mathcal{ALC} + \mathbf{T}_{\mathbf{R}}^{\mathbf{P}}$). *Given a KB=$(\mathcal{T}, \mathcal{A})$, given $\mathfrak{Tip}$ a set of concepts, and given $p, q \in (0, 1]$, let $\mathcal{E} = \{\widetilde{\mathcal{A}_1}, \widetilde{\mathcal{A}_2}, \ldots, \widetilde{\mathcal{A}_k}\}$ be the set of extensions of $\mathcal{A}$ of Definition 4 w.r.t. $\mathfrak{Tip}$, whose probabilities are s.t. $p \leq \mathbb{P}_1 \leq q, p \leq \mathbb{P}_2 \leq q, \ldots, p \leq \mathbb{P}_k \leq q$. Let $\mathcal{T}' = \{\mathbf{T}(C) \sqsubseteq D \mid \mathbf{T}(C) \sqsubseteq_r D \in \mathcal{T}\} \cup \{C \sqsubseteq D \in \mathcal{T}\}$. Given a query F, we say that F is entailed from KB in $\mathcal{ALC} + \mathbf{T}_{\mathbf{R}}^{\mathbf{P}}$ in range $\langle p, q \rangle$, written $KB \models_{\mathcal{ALC}+\mathbf{T}_{\mathbf{R}}^{\mathbf{P}}}^{\langle p,q \rangle} F$: (i) if F is a TBox inclusion either $C \sqsubseteq D$ or $\mathbf{T}(C) \sqsubseteq D$, if $(\mathcal{T}', \mathcal{A}) \models_{\mathcal{ALC}+\mathbf{T}_{\mathbf{R}}^{RaCl}} F$; (ii) if F is an ABox fact $C(a)$, where $a \in \mathcal{O}$, if $(\mathcal{T}', \mathcal{A} \cup \widetilde{\mathcal{A}_i}) \models_{\mathcal{ALC}+\mathbf{T}_{\mathbf{R}}} F$ for all $\widetilde{\mathcal{A}_i} \in \mathcal{E}$. We also define the probability of the query as $\mathbb{P}(F) = \sum_{i=1}^{k} \mathbb{P}_i$.*

We conclude by describing a decision procedure for reasoning in the logic $\mathcal{ALC} + \mathbf{T}_{\mathbf{R}}^{\mathbf{P}}$, in order to check whether a query F is entailed from a given KB as in Definition 5. Let $KB = (\mathcal{T}, \mathcal{A})$ be an $\mathcal{ALC} + \mathbf{T}_{\mathbf{R}}^{\mathbf{P}}$ knowledge base. Let $\mathcal{T}'$ be the set of inclusions of $\mathcal{T}$ without probabilities of exceptions: $\mathcal{T}' = \{\mathbf{T}(C) \sqsubseteq D \mid \mathbf{T}(C) \sqsubseteq_r D \in \mathcal{T}\} \cup \{C \sqsubseteq D \in \mathcal{T}\}$, that the procedure will consider in order to reason in $\mathcal{ALC} + \mathbf{T}_{\mathbf{R}}$ and $\mathcal{ALC} + \mathbf{T}_{\mathbf{R}}^{RaCl}$ for checking query entailment and finding all plausible typicality assumptions, respectively. Other inputs of the procedure are the finite set of concepts $\mathfrak{Tip}$, a query F, and two real numbers $p, q \in (0, 1]$ describing a range of probabilities. If F is an inclusion $C \sqsubseteq D$ (where C could be $\mathbf{T}(C')$), we just need to check whether $(\mathcal{T}', \mathcal{A}) \models_{\mathcal{ALC}+\mathbf{T}_{\mathbf{R}}^{RaCl}} C \sqsubseteq D$ in $\mathcal{ALC} + \mathbf{T}_{\mathbf{R}}^{RaCl}$. If F is an ABox formula of the form $C(a)$, we exploit Algorithm 1 in order to check whether $KB \models_{\mathcal{ALC}+\mathbf{T}_{\mathbf{R}}^{\mathbf{P}}}^{\langle p,q \rangle} F$.

We exploit the procedure of Algorithm 1 to show that the problem of entailment in the logic $\mathcal{ALC} + \mathbf{T}_{\mathbf{R}}^{\mathbf{P}}$ is ExpTime complete. This allows us to conclude that reasoning about typicality and defeasible inheritance with probabilities of

---

Algorithm 1. Entailment in $\mathcal{ALC} + \mathbf{T_R^P}$: $\mathrm{KB} \models_{\mathcal{ALC}+\mathbf{T_R^P}}^{\langle p,q \rangle} F$

1: **procedure** ENTAILMENT$((\mathcal{T}, \mathcal{A}), \mathcal{T}', F, \mathfrak{Tip}, p, q)$
2:     $\mathfrak{Tip}_\mathcal{A} \leftarrow \emptyset$ ▷ *build the set* $\mathbb{S}$ *of possible assumptions*
3:     **for** each $C \in \mathfrak{Tip}$ **do**
4:         **for** each individual $a \in \mathcal{A}$ **do**   ▷ *Reasoning in* $\mathcal{ALC} + \mathbf{T_R^{RaCl}}$
5:             **if** $(\mathcal{T}', \mathcal{A}) \models_{\mathcal{ALC}+\mathbf{T_R^{RaCl}}} \mathbf{T}(C)(a)$ **then** $\mathfrak{Tip}_\mathcal{A} \leftarrow \mathfrak{Tip}_\mathcal{A} \cup \{\mathbf{T}(C)(a)\}$

6:     $\mathcal{P}_\mathcal{A} \leftarrow \emptyset$ ▷ *compute the probabilities of Definition 2 given* $\mathcal{T}$ *and* $\mathfrak{Tip}_\mathcal{A}$
7:     **for** each $C \in \mathfrak{Tip}$ **do**
8:         $\Pi_C \leftarrow 1$
9:         **for** each $\mathbf{T}(C) \sqsubseteq_p D \in \mathcal{T}$ **do** $\Pi_C \leftarrow \Pi_C \times p$
10:        $\mathcal{P}_\mathcal{A} \leftarrow \mathcal{P}_\mathcal{A} \cup \Pi_C$
11:    $\mathbb{S} \leftarrow$ build strings of possible assumptions as in Definition 3 given $\mathfrak{Tip}_\mathcal{A}$ and $\mathcal{P}_\mathcal{A}$
12:    $\mathcal{E} \leftarrow \emptyset$ ▷ *build extensions of* $\mathcal{A}$
13:    **for** each $s_i \in \mathbb{S}$ **do**
14:        build the extension $\widetilde{\mathcal{A}_i}$ corresponding to s_i and compute $\mathbb{P}_{\widetilde{\mathcal{A}_i}}$ as in Definition 4
15:        **if** $p \leq \mathbb{P}_{\widetilde{\mathcal{A}_i}} \leq q$ **then** $\mathcal{E} \leftarrow \mathcal{E} \cup \widetilde{\mathcal{A}_i}$ ▷ *select extensions with probability in* $\langle p, q \rangle$
16:    **for** each $\widetilde{\mathcal{A}_i} \in \mathcal{E}$ **do**   ▷ *query entailment in* $\mathcal{ALC} + \mathbf{T_R}$
17:        **if** $(\mathcal{T}', \mathcal{A} \cup \widetilde{\mathcal{A}_i}) \not\models_{\mathcal{ALC}+\mathbf{T_R}} F$ **then return** $\mathrm{KB} \not\models_{\mathcal{ALC}+\mathbf{T_R^P}}^{\langle p,q \rangle} F$

18:    **return** $\mathrm{KB} \models_{\mathcal{ALC}+\mathbf{T_R^P}}^{\langle p,q \rangle} F$ ▷ *F is entailed in all extensions*

---

exceptions is essentially inexpensive, since reasoning retains the same complexity class of the underlying standard $\mathcal{ALC}$, which is known to be ExpTime-complete [1].

Theorem 1 (Complexity of entailment). *Given a KB in* $\mathcal{ALC} + \mathbf{T_R^P}$, *real numbers* $p, q \in (0, 1]$ *and a query* F *whose size is polynomial in the size of KB, the problem of checking whether* $\mathrm{KB} \models_{\mathcal{ALC}+\mathbf{T_R^P}}^{\langle p,q \rangle} F$ *is* ExpTime-*complete.*

Proof (sketch). The algorithm checks, for each concept $C \in \mathfrak{Tip}$ and for each individual name a whether $\mathbf{T}(C)(a)$ is minimally entailed from KB in the non-monotonic logic $\mathcal{ALC} + \mathbf{T_R^{RaCl}}$. Let n be the length of the string representing KB. By definition, the size of $\mathfrak{Tip}$ is $O(n)$. For each $\mathbf{T}(C)(a)$ (they are $O(n^2)$) the algorithm relies on reasoning in $\mathcal{ALC} + \mathbf{T_R^{RaCl}}$, which is in ExpTime [10]. Building $\mathcal{P}_\mathcal{A}$ can be solved with $O(n^2)$ operations. For building the set $\mathbb{S}$ of plausible extensions we have to consider all possible strings obtained by assuming (or not) each typicality assumption $\mathbf{T}(C)(a)$, that are $O(n^2)$: for each s_i, we have two options ($s_i = 0$ or $s_i \neq 0$), then $2 \times 2 \times \cdots \times 2$ different strings, thus $\mathbb{S}$ has exponential size in n. Selecting extensions whose probabilities $\mathbb{P}_{\widetilde{\mathcal{A}_i}}$ are in the range $[p, q]$ can be solved in ExpTime, then the algorithm relies on reasoning in monotonic $\mathcal{ALC} + \mathbf{T_R}$ in order to check whether F is entailed in selected extensions in $\mathcal{E}$, whose size is $O(2^n)$: we have $O(2^n)$ calls to query entailment in $\mathcal{ALC} + \mathbf{T_R}$, which is ExpTime-complete. $\square$

Example 3 (Reasoning in $\mathcal{ALC} + \mathbf{T}_\mathbf{R}^\mathsf{P}$ part 2). We continue Example 1 in the light of definitions provided above. Suppose that the ABox is $\mathcal{A} = \{AtypicalDepressed(john), ProstateCancerPatient(greg)\}$, we can consider two typicality assumptions:

(a) $\mathbf{T}(AtypicalDepressed)(john)$ and   (b) $\mathbf{T}(ProstateCancerPatient)(greg)$

then we can distinguish among four different extensions: (i) both (a) and (b) are assumed: in this scenario, whose probability is $0.6 \times (0.5 \times 0.8) = 0.24$, we can conclude that both John and Greg have mood swings, and that Greg has nocturia. (ii) we assume (b) but not (a): this scenario has probability $(1 - 0.6) \times (0.5 \times 0.8) = 0.16$, and we can only conclude $\exists Symptom.MoodReactivity(greg)$ and $\exists Symptom.Nocturia(greg)$. (iii) we assume (a) and not (b): this scenario, having a probability $0.6 \times (1 - (0.5 \times 0.8)) = 0.36$, allows us to conclude $\exists Symptom.MoodReactivity(john)$. (iv) Neither (a) nor (b) is added to $\mathcal{A}$: here the probability is $(1 - 0.6) \times (1 - (0.5 \times 0.8)) = 0.24$, but we are not able to conclude anything about John and Greg. The probability that John has mood swings is defined as the sum of the probabilities of scenarios where such inference can be performed, namely scenarios (i) and (iii), and it is therefore $0.24 + 0.36 = 0.6$. Similarly, the probability that Greg has nocturia and mood swings is $0.24 + 0.16 = 0.4$. Concerning entailment, we have that, in less predictable scenarios (probability no higher than 20%), Greg has nocturia, i.e. KB $\models_{\mathcal{ALC}+\mathbf{T}_\mathbf{R}^\mathsf{P}}^{\langle 0,0.2\rangle} \exists Symptom.Nocturia(greg)$.

4   Related Works and Conclusions

Several nonmonotonic extensions of DLs have been proposed in the literature in order to reason about inheritance with exceptions, essentially based on the integration of DLs with well established nonmonotonic reasoning mechanisms [2–7,9]. In none of them, probability of exceptions in concept inclusions is taken into account.

Probabilistic extensions of DLs, allowing to label inclusions (and facts) with degrees representing probabilities, have been introduced in [13,14]. In this approach, called DISPONTE, the authors propose the integration of probabilistic information with DLs based on the distribution semantics for probabilistic logic programs [15]. The basic idea is to label inclusions of the TBox as well as facts of the ABox with a real number between 0 and 1, representing their probabilities, assuming that each axiom is independent from each others. The resulting knowledge base defines a probability distribution over *worlds*: roughly speaking, a world is obtained by choosing, for each axiom of the KB, whether it is considered as true of false. The distribution is further extended to queries and the probability of the entailment of a query is obtained by marginalizing the joint distribution of the query and the worlds. There are two main differences between the logic $\mathcal{ALC} + \mathbf{T}_\mathbf{R}^\mathsf{P}$ proposed in this work and probabilistic DLs. On the one hand, as already mentioned in the Introduction, in the logic $\mathcal{ALC} + \mathbf{T}_\mathbf{R}^\mathsf{P}$

probabilities are used in order to express different degrees of admissibility of exceptions with respect to such typicality inclusions. Probabilities are then the basis of different scenarios built by assuming – or not – that individuals are typical instances of a given concept. On the contrary, in DISPONTE probabilities are used to capture a notion of uncertainty about information of the KB, therefore an inclusion $C \sqsubseteq D$ having a very low probability p has a significantly different meaning w.r.t. an inclusion $\mathbf{T}(C) \sqsubseteq_p D$, representing anyway a typical property: normally, Cs are Ds, even if with a high probability of having exceptions to such typical inclusion. On the other hand, in $\mathcal{ALC} + \mathbf{T}_\mathbf{R}^\mathsf{P}$ probabilities are restricted to typicality inclusions only. On the contrary, in DISPONTE probabilities can be associated to concept inclusions as well as to ABox facts. It is worth noticing that the two approaches could be combined in order to describe a probabilistic extension of DLs with typicalities and probabilities of having exceptions: a knowledge base can contain axioms labelled by probabilities that can be interpreted as "epistemic" ones, i.e. as degrees of our belief in those axioms, as in [14], as well as typicality inclusions with probabilities about exceptions. In this respect, an inclusion $p :: \mathbf{T}(C) \sqsubseteq_q D$ represents that we have degree of belief p in the fact that typical Cs are also Ds with a probability q of not having exceptions. Such a further extension will be material for future works.

In [12] a nonmonotonic procedure for reasoning about *surprising* scenarios in DLs has been proposed. In this approach, the Description Logic $\mathcal{ALC} + \mathbf{T}_\mathbf{R}$ is extended by inclusions of the form $\mathbf{T}(C) \sqsubseteq_d D$, where d is a *degree of expectedness*. Similarly to $\mathcal{ALC} + \mathbf{T}_\mathbf{R}^\mathsf{P}$, a notion of extension of an ABox is introduced in order to assume typicality assertions about individuals satisfying cardinality restrictions on concepts, then degrees of expectedness are used in order to define a preference relation among extended ABoxes: entailment of queries is then restricted to ABoxes that are minimal with respect to such preference relations and that represent surprising scenarios. Also in this case, we have two main differences with the approach of the logic $\mathcal{ALC} + \mathbf{T}_\mathbf{R}^\mathsf{P}$: first, in $\mathcal{ALC} + \mathbf{T}_\mathbf{R}^\mathsf{exp}$ degrees of expectedness are non-negative integers used essentially to define a – partial – preference relation among extended ABoxes, whereas they are not used in order to estimate probabilities of typicality inclusions. Second, cardinality restrictions play a fundamental role in order to "filter" extended ABoxes. On the contrary, in the logic $\mathcal{ALC} + \mathbf{T}_\mathbf{R}^\mathsf{P}$, entailment is defined in terms of the probability of a given scenario and can be used to estimate the probability of a given query. In future work we aim at extending the logic $\mathcal{ALC} + \mathbf{T}_\mathbf{R}^\mathsf{P}$ with cardinality restrictions, in order to investigate the precise relation with the approach proposed in [12].

References

1. Baader, F., Calvanese, D., McGuinness, D., Nardi, D., Patel-Schneider, P.: The Description Logic Handbook - Theory, Implementation, and Applications, 2nd edn. Cambridge University Press, Cambridge (2007)
2. Baader, F., Hollunder, B.: Priorities on defaults with prerequisites, and their application in treating specificity in terminological default logic. J. Autom. Reason. **15**(1), 41–68 (1995)
3. Bonatti, P.A., Faella, M., Petrova, I., Sauro, L.: A new semantics for overriding in description logics. Artif. Intell. **222**, 1–48 (2015)
4. Bonatti, P.A., Lutz, C., Wolter, F.: The complexity of circumscription in DLs. J. Artif. Intell. Res. (JAIR) **35**, 717–773 (2009)
5. Casini, G., Straccia, U.: Rational closure for defeasible description logics. In: Janhunen, T., Niemelä, I. (eds.) JELIA 2010. LNCS (LNAI), vol. 6341, pp. 77–90. Springer, Heidelberg (2010). doi:10.1007/978-3-642-15675-5_9
6. Casini, G., Straccia, U.: Defeasible Inheritance-Based Description Logics. J. Artif. Intell. Res. (JAIR) **48**, 415–473 (2013)
7. Donini, F.M., Nardi, D., Rosati, R.: Description logics of minimal knowledge and negation as failure. ACM Trans. Comput. Logics (ToCL) **3**(2), 177–225 (2002)
8. Giordano, L., Gliozzi, V., Olivetti, N., Pozzato, G.L.: ALC+T: a preferential extension of description logics. Fundamenta Informaticae **96**, 341–372 (2009)
9. Giordano, L., Gliozzi, V., Olivetti, N., Pozzato, G.L.: A nonmonotonic description logic for reasoning about typicality. Artif. Intell. **195**, 165–202 (2013)
10. Giordano, L., Gliozzi, V., Olivetti, N., Pozzato, G.L.: Semantic characterization of rational closure: from propositional logic to description logics. Artif. Intell. **226**, 1–33 (2015)
11. Lehmann, D., Magidor, M.: What does a conditional knowledge base entail? Artif. Intell. **55**(1), 1–60 (1992)
12. Pozzato, G.L.: Reasoning about surprising scenarios in description logics of typicality. In: Adorni, G., Cagnoni, S., Gori, M., Maratea, M. (eds.) AI*IA 2016. LNCS (LNAI), vol. 10037, pp. 418–432. Springer, Cham (2016). doi:10.1007/978-3-319-49130-1_31
13. Riguzzi, F., Bellodi, E., Lamma, E., Zese, R.: Probabilistic description logics under the distribution semantics. Semant. Web **6**(5), 477–501 (2015)
14. Riguzzi, F., Bellodi, E., Lamma, E., Zese, R.: Reasoning with probabilistic ontologies. In: Proceedings of IJCAI 2015, Buenos Aires, Argentina, 25–31 July 2015, pp. 4310–4316 (2015)
15. Sato, T.: A statistical learning method for logic programs with distribution semantics. In: Sterling, L. (ed.) Logic Programming, Proceedings of ICLP, pp. 715–729. MIT Press (1995)
16. Straccia, U.: Default inheritance reasoning in hybrid kl-one-style logics. In: Proceedings of IJCAI 1993, pp. 676–681. Morgan Kaufmann (1993)

Orthopairs

Measuring Uncertainty in Orthopairs

Andrea Campagner and Davide Ciucci[✉]

DISCo, University of Milano-Bicocca, Milan, Italy
ciucci@disco.unimib.it

Abstract. In many situations information comes in bipolar form. Orthopairs are a simple tool to represent and study this kind of information, where objects are classified in three different classes: positive, negative and boundary. The scope of this work is to introduce some uncertainty measures on orthopairs. Two main cases are investigated: a single orthopair and a collection of orthopairs. Some ideas are taken from neighbouring disciplines, such as fuzzy sets, intuitionistic fuzzy sets, rough sets and possibility theory.

1 Introduction

Information often comes in bipolar form; that is, positive evidence versus negative one [8]. In order to take into account this bipolarity in a generic and formal way, orthopairs have been introduced and studied [4,5]. An orthopair is just a pair of sets (A, B) with empty intersection, i.e., $A \cap B = \emptyset$. Different meanings can be attached to these two sets, for instance positive and negative examples, affirmed or negated propositional variables, trust and distrust statements, accepted and rejected objects, etc. They can be found at work in several situations in knowledge representation (partial knowledge, borderline cases, consensus) and applications (social network analysis, representing partial or vague knowledge, rough sets, formal concept analysis). Moreover, orthopairs are linked to other paradigms in uncertainty management, in particular they are in bijection with three-valued sets and they can be generalized to obtain Atanassov Intuitionistic Fuzzy Sets (IFS) or possibility distributions.

The two sets A and B usually do not cover the universe, so there is an intrinsic uncertainty in any orthopair. According to the interpretation given to the orthopair, also this uncertainty can be interpreted in several ways. In particular, as a lack of knowledge (we do not have enough evidence to classify all the objects as positive or negative) or as fuzziness (there exist borderline cases which do not belong to either A or B). In any case, it is important to measure this uncertainty and the present work is a preliminary step in this direction. Since, as already mentioned, orthopairs are linked to other paradigms, we will take inspiration from the uncertainty measures already existing on those paradigms and try to cast them on orthopairs. The paper is organized as follows. In Sect. 2, we give the basic definitions concerning orthopairs and formalize their relationship with other paradigms. Then, the uncertainty expressed by a single orthopair is studied in Sect. 3 where it is shown that the E_O measure plays a

© Springer International Publishing AG 2017
A. Antonucci et al. (Eds.): ECSQARU 2017, LNAI 10369, pp. 423–432, 2017.
DOI: 10.1007/978-3-319-61581-3_38

fundamental role. The behaviour of this measure with respect to order relations and aggregation operators is analysed in Sect. 4. Finally, measures of uncertainty on a collection of orthopairs on the same universe are given in Sect. 5.

2   Orthopairs: Basic Definitions

An orthopair on a universe X is a pair of sets (A, B) such that $A \cap B = \emptyset$. Since not necessarily A and B cover the universe, we have to consider also the set $C = X \setminus (A \cup B)$. The sets A, B, C now form a tri-partition of X, thus interesting connections with three-way decision [17] and the theory of opposition [9] can be put forward (see [3,5]). Despite this general definition and the several meanings that can be attached to A, B [5], we will usually interpret them as positive and negative, therefore denoted as (P, N). Moreover, the set C will be named *boundary* and denoted as Bnd. Let $O(X)$ be the collection of all orthopairs on X. It is easy to show that $O(X)$ is in bijection with three-valued sets. Indeed, it is sufficient to fix an orthopair o and consider the (bijective) three-valued function $f_o : X \mapsto V$ with $V = \{a, b, c\}$ defined for all $x \in X$ as: $f_o(x) = a$ if $x \in P$, b if $x \in N$ and c otherwise. In particular, if $V = \{0, \frac{1}{2}, 1\}$ we obtain a fuzzy set with three values. As a first use of this bijection, we define six point-wise orderings, some of which are total and some partial, as schematized in Table 1.

Table 1. A summary of the pointwise order relations on orthopairs.

Order on V	Order on $O(X)$	Symbol	Type
$0 \leq \frac{1}{2} \leq 1$	$P_1 \subseteq P_2,\ N_2 \subseteq N_1$	$\leq_t$	Total
$\frac{1}{2} \leq 1 \leq 0$	$N_1 \subseteq N_2,\ Bnd_2 \subseteq Bnd_1$	$\leq_N$	Total
$\frac{1}{2} \leq 0 \leq 1$	$P_1 \subseteq P_2,\ Bnd_2 \subseteq Bnd_1$	$\leq_P$	Total
$\frac{1}{2} \leq 1, \frac{1}{2} \leq 0$	$P_1 \subseteq P_2,\ N_1 \subseteq N_2$	$\leq_I$	Partial
$0 \leq \frac{1}{2}, 0 \leq 1$	$P_1 \subseteq P_2,\ Bnd_1 \subseteq Bnd_2$	$\leq_{PB}$	Partial
$1 \leq \frac{1}{2}, 1 \leq 0$	$N_1 \subseteq N_2,\ Bnd_1 \subseteq Bnd_2$	$\leq_{NB}$	Partial

In particular, $\leq_t$ and $\leq_I$ are the most used orderings, the first one named *truth ordering* [1] and linked to the logical truth of a statement, whereas the second one is the *knowledge ordering* [1], accounting for the more complete knowledge given by the greater orthopair. Let us notice that some other non-pointwise orderings can be defined [5], but are not considered here.

Clearly, the three total order relations define three lattice structures and hence three different meet and join operations. On orthopairs they read as:

$$(P_1, N_1) \sqcap_t (P_2, N_2) := (P_1 \cap P_2, N_1 \cup N_2)$$
$$(P_1, N_1) \sqcup_t (P_2, N_2) := (P_1 \cup P_2, N_1 \cap N_2)$$
$$(P_1, N_1) \sqcap_N (P_2, N_2) := ((P_1 \cap P_2) \cup [(P_1 \cap N_2) \cup (P_2 \cap N_1)], N_1 \cap N_2))$$
$$(P_1, N_1) \sqcup_N (P_2, N_2) := (P_1 \setminus N_2 \cup P_2 \setminus N_1, N_1 \cup N_2)$$

$$(P_1, N_1) \sqcap_P (P_2, N_2) := (P_1 \cap P_2, (N_1 \cap N_2) \cup [(N_1 \cap P_2) \cup (N_2 \cap P_1)])$$
$$(P_1, N_1) \sqcup_P (P_2, N_2) := (P_1 \cup P_2, N_1 \backslash P_2 \cup N_2 \backslash P_1)$$

When considered on three values, they correspond to strong Kleene (from $\preceq_t$), weak Kleene (the min from $\preceq_P$ and $\preceq_N$) and Sobociński (the max from $\preceq_N$ and $\preceq_P$) conjunction and disjunction. Moreover, it is of interest to consider also the meet operation definable from the knowledge ordering, also known as the *pessimistic combination operator* [12], $(P_1, N_1) \sqcap_I (P_2, N_2) := (P_1 \cap P_2, N_1 \cap N_2)$. On the other hand, the join corresponding to $\leq_I$ (when definable) is the *optimistic combination operator* [12]: $(P_1, N_1) \sqcup_I (P_2, N_2) := (P_1 \cup P_2, N_1 \cup N_2)$.

As a last aggregation operator, let us consider the *consensus* operation $O_1 \odot O_2 := (P_1 \backslash N_2 \cup P_2 \backslash N_1, N_1 \backslash P_2 \cup N_2 \backslash P_1)$ whose aim is to reconcile two orthopairs (for instance representing two agents' opinion), by keeping as positive part only what is not considered negative by the other and dually for the negative part.

Orthopairs can be generalized in several ways [5]. For the scope of the present work it is useful to recall the links with IFS and possibility theory. As the former is concerned, it is enough to consider as P, N two fuzzy (instead of crisp) sets. Indeed, IFSs are pairs of fuzzy sets $f_P, f_N : X \mapsto [0,1]$ such that for all $x \in X$, $f_P(x) + f_N(x) \leq 1$. Hence, orthopairs are particular cases of IFSs. With respect to possibility distributions, orthopairs coincide with the particular class of hyper-rectangular Boolean possibility distributions on the space $\{0,1\}^n$. In detail, given an orthopair (P, N), the associated possibility distribution is the characteristic function of the set of models of the formula $\phi = [\bigwedge_{a \in P} a \wedge \bigwedge_{a \in N} \neg a]$. That is

$$\pi_{(P,N)}(\omega) := \begin{cases} 1 & \text{if } \omega \vDash [\bigwedge_{a \in P} a \wedge \bigwedge_{a \in N} \neg a] \\ 0 & \text{otherwise} \end{cases} \tag{1}$$

So, the partial (i.e., there can exists variables $a \notin \{P, N\}$) truth assignments ω such that $\pi_{(P,N)}(\omega) = 1$ are those compatible with the orthopair (P, N). In order to capture any Boolean possibility distribution π, a set of orthopairs is needed [6]. Specifically, given π we can associate a formula in disjunctive normal form where the disjuncts are mutually exclusive and then associate to each of these conjuncts an orthopair thus obtaining a set of orthopairs.

3   Uncertainty in a Single Orthopair

Let us consider a single orthopair $O = (P, N)$. Of course, the uncertainty contained in O can depend on the interpretation given to the boundary (fuzziness vs lack of information) or if we want to measure a particular facet of uncertainty (for instance, specificity). In the following, we introduce some measures by looking at what happens in generalized theories, such as fuzzy sets and IFS. We will see that a central role is played by a simple counting measure:

$$E_O(O) = \frac{|Bnd|}{|X|} \tag{2}$$

Clearly, E_O counts the number of elements in the boundary region, supposing that they are the elements on which we are uncertain. We notice that E_O is a possible definition of *roughness* in rough set theory if P is interpreted as the lower approximation and N as the exterior region [2].

3.1   Inspired by IFS

Let us consider an orthopair $O = (P, N)$ and denote the characteristic function of P and N as χ_P, χ_N respectively. We consider now O as a crisp version of IFS and apply some IFS uncertainty measures to O, namely: Entropy, Knowledge Measure and Non-Specificity.

Entropy Functions. In [13], Pal et al. define two types of entropy for IFS, in order to distinguish between two different types of uncertainty coexisting in IFS: *Fuzziness* and *Lack of knowledge*. In order to quantify the first type of uncertainty, the authors introduce a system of axioms which are not meaningful when restricted to orthopairs since they constrain an entropy function E to be defined as the constant 0 function. On the other hand, the axioms introduced by Szmidt and Kacprzyk in [16] can be directly applied to orthopairs:

(Ax1) $E(O) = 0$ iff $A \in 2^X$;
(Ax2) $E(O) = 1$ iff $\forall x \in X, \chi_P(x) = \chi_N(x)$;
(Ax1) $E(O_1) \leq E(O_2)$ if $\forall x \in X, \chi_{P_1}(x) \leq \chi_{P_2}(x)$ and $\chi_{N_1}(x) \geq \chi_{N_2}(x)$ for $\chi_{P_2}(x) \leq \chi_{N_2}(x)$, vice versa $\chi_{P_1}(x) \geq \chi_{P_2}(x)$ and $\chi_{N_1}(x) \leq \chi_{N_2}(x)$ for $\chi_{P_2}(x) \geq \chi_{N_2}(x)$;
(Ax4) $E(O) = E(O^c)$

where $(P, N)^c = (N, P)$. In order to quantify the second type of uncertainty the authors in [13] introduce the following set of axioms:

(Ax5) $I(O) = 0$ iff $\forall x \in X, \chi_P(x) + \chi_N(x) = 1$
(Ax6) $I(O) = 1$ iff $\forall x \in X, \chi_P(x) = \chi_N(x) = 0$
(Ax7) $I(O_1) \geq I(O_2)$ if $\forall x \in X, \chi_{P_1}(x) + \chi_{N_1}(x) \leq \chi_{P_2}(x) + \chi_{N_2}(x)$
(Ax8) $I(O) = I(O^c)$

Clearly, on orthopairs the two sets of axioms turn out to be equivalent:

Proposition 1. *Let $E : O(X) \to [0, 1]$ be a function. Then E satisfies axioms 1–4 iff it satisfies axioms 5–8.*

Proof (sketch). Axiom 4 is the same as axiom 8. An orthopair O is a crisp set (i.e. $Bnd_O = \emptyset$) iff $\forall x$, either $\chi_P(x) = 1$ or $\chi_N(x) = 1$, therefore axiom 1 and axiom 5 are equivalent. Axioms 2 and axiom 6 are equivalent since $\chi_P(x) = \chi_N(x)$ iff $\chi_P(x) = \chi_N(x) = 0$. Finally, it is possible to show that from axioms (Ax5)–(Ax8) we get (Ax3) and vice-versa, from axioms (Ax1)–(Ax4) we get (Ax7).

It is easy to observe that E_O satisfies axioms 1–4 (and therefore also axioms 5–8). Furthermore it is the only function, up to constants, to satisfy them. At first, let us recall the following result, adapted to the orthopair case.

Lemma 1. [13] *Let $g : \{0,1\} \to \{0,1\}$ be a function. Then $G : O(X) \to [0,1]$ defined as $G(O) = k\sum_{x \in U} g(\chi_P(x) + \chi_N(x))$ satisfies 5–8 iff $g(1) = 0$ and $g(0) = 1$.*

Proposition 2. *E_O is the only function in the form $k\sum_{x \in X} g(\chi_{P_A}(x) + \chi_{N_A}(x))$ satisfying axioms 1–4 and 5–8, up to a multiplicative constant.*

Proof. First of all, E_O is in the form required by Lemma 1. Indeed, $k = 1/|X|$ and $g'(x) = 1 - (\chi_P(x) + \chi_N(x))$. Thus, we can rewrite $k\sum_{x \in U} g(\chi_P(x) + \chi_N(x)) = k\sum_{x \in P \cup N} g(1) + k\sum_{x \in Bnd} g(0) = k\sum_{x \in Bnd} 1 = k|Bnd|$, which, choosing $k = 1/|X|$, is exactly the definition of E_O, hence the result.

Several definitions of entropy satisfying axioms 1–4 or 5–8 have been given. We consider, in particular, the list of entropies, surveyed by Zhang in [18]. We can divide them in two groups, the first one, clusters all measures that on orthopairs reduce to E_O. They are

- $E_{BB}(O) = \frac{1}{|X|}\sum_{x \in X} \chi_{Bnd_O}(x) = E_O(O)$;
- $E_{SK}(O) = \frac{1}{|X|}\sum_{x \in X} \frac{min(\chi_{P_O}(x),\chi_{N_O}(x))+\chi_{Bnd_O}(x)}{max(\chi_{P_O}(x),\chi_{N_O}(x))+\chi_{Bnd_O}(x)} = E_O(O)$;
- $E_{ZL}(O) = 1 - \frac{1}{|X|}\sum_{x \in X} |\chi_{P_O}(x) - \chi_{N_O}(x)| = E_O(O)$;
- $E_{VS}(O) = -\frac{1}{|X|ln2}\sum_{x \in X} [\chi_{P_O}(x)ln\chi_{P_O}(x) + \chi_{N_O}(x)ln\chi_{N_O}(x) - (1 - \chi_{Bnd_O}(x))ln(1 - \chi_{Bnd_O}(x)) - \chi_{Bnd_O}(x)ln2] = E_O(O)$;
- $E_{Y1}(O) = \frac{1}{|X|}\sum_{x \in X} \{\{sin[\frac{\pi}{4}(1 + \chi_{P_O}(x) - \chi_{N_O}(x))] + sin[\frac{\pi}{4}(1 - \chi_{P_O}(x) + \chi_{N_O}(x))]\} - 1\}\frac{1}{\sqrt{2}-1}\} = E_O(O)$;
- $E_{Y2}(O) = \frac{1}{|X|}\sum_{x \in X} \{\{cos[\frac{\pi}{4}(1 + \chi_{P_O}(x) - \chi_{N_O}(x))] + cos[\frac{\pi}{4}(1 - \chi_{P_O}(x) + \chi_{N_O}(x))]\} - 1\}\frac{1}{\sqrt{2}-1}\} = E_O(O)$;
- $E(O) = 1 - \frac{1}{|X|}\sum_{x \in X} [\sqrt{2(\chi_{P_O}(x) - 0.5)^2 + 2(\chi_{N_O}(x) - 0.5)^2} \ \chi_{Bnd_O}(x)]$;

The second group is made of measures that reduce to the zero constant function:

- $E_{ZJ}(O) = \frac{1}{|X|}\sum_{x \in X} \frac{min(\chi_{P_O}(x),\chi_{N_O}(x))}{max(\chi_{P_O}(x),\chi_{N_O}(x))} = 0$;
- $E_{Z1}(O) = 1 - \sqrt{\frac{2}{|X|}\sum_{x \in X} [(\chi_{P_O}(x) - 0.5)^2 + (\chi_{N_O}(x) - 0.5)^2]} = 0$;
- $E_{Z2}(O) = 1 - \frac{1}{|X|}\sum_{x \in X} [|\chi_{P_O}(x) - 0.5| + |\chi_{N_O}(x) - 0.5|] = 0$;
- $E_{Z3}(O) = 1 - \frac{2}{|X|}\sum_{x \in X} max(|\chi_{P_O}(x) - 0.5|, |\chi_{N_O}(x) - 0.5|) = 0$;
- $E_{Z4}(O) = 1 - \sqrt{\frac{4}{|X|}\sum_{x \in X} max(|\chi_{P_O}(x) - 0.5|^2, |\chi_{N_O}(x) - 0.5|^2)} = 0$;
- $E_{hc}^2(O) = \frac{1}{|X|}\sum_{x \in X} [1 - \chi_{P_O}(x)^2 - \chi_{N_O}(x)^2 - \chi_{Bnd_O}(x)^2] = 0$;
- $E_r^{1/2}(O) = \frac{2}{|X|}\sum_{x \in X} ln[\chi_{P_O}(x)^{1/2} - \chi_{N_O}(x)^{1/2} - \chi_{Bnd_O}(x)^{1/2}] = 0$;

Knowledge Measure. In [10] Guo defines axiomatically a *knowledge measure K* in order to measure the amount of knowledge in an IFS. This notion of knowledge measure is deeply related to the concept of entropy as defined by axioms (Ax1)–(Ax4). Indeed, the set of axioms defining it can be directly obtained by negation of (Ax1)–(Ax4). In particular, Guo defines the following knowledge measure: $K_{AIFS}(O) = 1 - \frac{1}{2n}\sum_{x \in U}(1 - |\chi_{P_O}(x) - \chi_{N_O}(x)|)(1 + \chi_{Bnd_O}(x))$ that on orthopairs gives the negation of E_O: $K_{AIFS}(O) = 1 - E_O(O)$.

Non Specificity. In [15], Song et al. distinguish another type of uncertainty in an IFS, namely *non-specificity*, and define a measure of non-specificity in analogy with the well-known *Hartley measure*[1]. The non-specificity $H_{IFS}(A)$ of an IFS A, once restricted to orthopairs, is defined as follows:

$$H_{IFS}(o) = \begin{cases} \log |U| & \text{if } P = \emptyset \\ \log |P_o \cup Bnd_o| & \text{otherwise} \end{cases}$$

3.2   Inspired by Fuzzy Sets

We can harness the bijective correspondence between three-valued sets and orthopairs in order to translate the measures on fuzzy sets to our context.

Entropy. In [7] De Luca and Termini proposed the well-known non-probabilistic definition of entropy for fuzzy sets based on a set of four axioms. It can be easily shown that, in the case of orthopairs, their axioms are equivalent to axioms (Ax1)–(Ax4). The authors also propose an entropy measure, inspired by Shannon entropy:

$$E_k(O) = k[\sum_{x \in X} \chi_{PO}(x) log(\frac{1}{\chi_{PO}(x)}) + \sum_{x \in X} (1 - \chi_{PO}(x)) log(\frac{1}{1 - \chi_{PO}(x)})].$$

where O is an orthopair on X. We can prove the following result:

Proposition 3. *Restricted to orthopairs* $E_K(O) = E_O(O)$, *with* $k = \frac{1}{|X|}$.

Non Specificity. In [11], Klir defined a measure of non-specificity for fuzzy sets. Considering an orthopair $O = (P, N)$ this measure reads as: $H_{Klir}(O) = \frac{1}{2}log|P \cup Bnd| + \frac{1}{2}log|P|$. Let us remark that when $P = \emptyset$, it is not defined since the term $log|P| = log(0)$ is undefined. We notice that the two measures of non-specificity introduced from IFS and fuzzy sets are different. We have that:

1. Both have value in the range $[0, \log |U|]$;
2. $H_{Klir}(o) \leq H_{IFS}(o)$;
3. $H_{IFS}(o)$ reaches the maximum value $log|U|$, either when $P_o \cup Bnd_o = U$, in this case $H_{Klir}(o) = \frac{1}{2}log|U| + \frac{1}{2}log|P|$, or when $P_o = \emptyset$, in this case $H_{Klir}(o)$ is not defined; $H_{Klir}(o)$ reaches the maximum value whenever $P_o = U$;
4. Both $H_{IFS}(o)$ and $H_{Klir}(o)$ have minimum value 0, when $o = (\{x\}, U \setminus \{x\})$ with $x \in U$.

As a conclusion of this section, we can say that E_O is the unique general measure of uncertainty of a single orthopair. If we want to consider particular aspects such as non-specificity (as above) or proportion between positive and negative (not discussed here) then other measures are possible.

---

[1] We recall that the Hartley measure is defined on crisp sets as: $H_{Hartley}(X) = log|X|$.

4 E_O Measure and Aggregation Operations

In this section, we want to study how uncertainty, measured by E_O, propagates through the various orderings and aggregation operators defined in Sect. 2.

First of all, let us consider the order relations. We can easily observe the following monotonic behaviours.

Proposition 4. *The measure E_O is anti-tonic w.r.t. the orders $\leq_N$, $\leq_P$ and $\leq_I$ and isotonic w.r.t. the orders $\leq_{NB}$, $\leq_{PB}$.*

On the other hand E_O is non-monotonic w.r.t. the order $\leq_t$, as shown in the following example.

Example 1. Consider the orthopairs $O_1 = (\emptyset, U) \leq_t O_2 = (\emptyset, \{1, 2\}) \leq_t O_3 = (\{1\}, \{2\})$ defined on $U = \{1, 2, 3\}$. We have that $E_O(O_1) = 0 \leq E_O(O_2) = \frac{1}{3}$ but $E_O(O_2) = \frac{1}{3} \geq E_O(O_3) = 0$.

Regarding the amount of uncertainty obtained by different binary aggregation operators, the following proposition holds.

Proposition 5. *Let $O_1 = (P_1, N_1)$, $O_2 = (P_2, N_2)$ be two orthopairs defined on universe U. Then, the following properties hold*

1. $E_O(O_1 \sqcap_t O_2) = \frac{|Bnd_1 \cap P_2| + |Bnd_2 \cap P_1| + |Bnd_1 \cap Bnd_2|}{|U|} \leq E_O(O_1) + E_O(O_2)$;
2. $E_O(O_1 \sqcup_t O_2) = \frac{|Bnd_1 \cap N_2| + |Bnd_2 \cap N_1| + |Bnd_1 \cap Bnd_2|}{|U|} \leq E_O(O_1) + E_O(O_2)$;
3. $E_O(O_1), E_O(O_2) \leq E_O(O_1 \sqcap_N O_2) \leq E_O(O_1) + E_O(O_2)$;
4. $E_O(O_1), E_O(O_2) \leq E_O(O_1 \sqcap_P O_2) \leq E_O(O_1) + E_O(O_2)$;
5. $E_O(O_1 \sqcup_N O_2) \leq min(E_O(O_1), E_O(O_2))$;
6. $E_O(O_1 \sqcup_P O_2) \leq min(E_O(O_1), E_O(O_2))$;
7. $E_O(O_1), E_O(O_2) \leq E_O(O_1 \sqcap_I O_2) \leq E_O(O_1) + E_O(O_2) + \frac{|P_1 \cap N_2|}{|U|} + \frac{|P_2 \cap N_1|}{|U|}$;
8. $E_O(O_1 \sqcup_I O_2) \leq min(E_O(O_1), E_O(O_2))$;
9. $E_O(O_1 \odot O_2) = \frac{|P_1 \cap N_2| + |P_2 \cap N_1| + |Bnd_1 \cap Bnd_2|}{|U|}$.

Thus, we can see that aggregating two orthopairs, in some cases the uncertainty diminishes, in some cases it augments, and in others the behaviour is unpredictable, according to the changes occurring in the boundary. Namely, in cases 5,6,8 ($\sqcup_N$, $\sqcup_P$, $\sqcup_I$) the resulting uncertainty is less than the uncertainty of a single orthopair, since these operators have the effect to diminish the boundary. In case of 3,4 ($\sqcap_N$, $\sqcap_P$), it is less than their sum but greater than the single orthopairs' one. In case 1, 2 ($\sqcap_t$, $\sqcup_t$), it can be either greater or lesser than the single orthopairs, depending on their overlapping, but lesser than their sum. In case 7 ($\sqcap_I$) the resulting uncertainty is greater than the single ones, but it can be greater or lesser that their sum, depending if they are in conflict or not; indeed, in the case that O_1, O_2 are not in conflict we have that $(E_O((P_1, N_1) \sqcap_I (P_2, N_2)) \leq E_O(O_1) + E_O(O_2))$. Finally, the uncertainty of case 9 ($\odot$) is in general incomparable to both $E_O(O_1)$ and $E_O(O_2)$, the reason is that $O_1 \odot O_2$ puts in the boundary all the elements that put O_1 and O_2 in conflict.

5   Uncertainty in a Collection of Orthopairs

In the following section, we will consider contexts (in particular rough sets, possibility distributions) in which we have a collection of orthopairs instead of a single one. A generic approach to generalize a measure of uncertainty to a collection $\mathcal{O}$ of orthopairs is to associate a probability distribution $P_\mathcal{O}$ to the collection $\mathcal{O}$ and then define the global uncertainty $E(\mathcal{O})$ of the collection as a weighted sum:

$$E(\mathcal{O}) = \sum_{O_i \in \mathcal{O}} P(O_i) E_O(O_i) \tag{3}$$

The exact value of the probability and the interpretation associated to it and to the entropy clearly depend on the context. In general, if we have no reasons to assume that an orthopair in the collection is more probable than others, we can assume $P_\mathcal{O}$ as the uniform distribution. So, if we have n orthopairs defined on the same universe U, the associated entropy is $E(\mathcal{O}) = \frac{1}{n|U|} \sum_{i=1}^{n} |Bnd_i|$.

5.1   Rough Sets

We suppose that the collection of orthopairs represents the rough sets definable on a universe U based on a partition π. That is, given an approximation space (U, π), we collect all the lower-exterior approximation pairs $(l(A), e(A))$ of subsets A of U, defined in the standard manner [14].

Now, given (U, π), we can, as suggested by Zhu and Wen [19], associate to any rough approximation $R(A) = (l(A), e(A))$ induced by π the probability $P_i(R(A)) = \frac{r_i(A)}{2^{|U|}}$, where $r_i(A)$ is defined as the number of subsets of U represented by $R(A)$, that is $r_i(A) = |\{B \subseteq U | R(B) = R(A)\}|$. We can then define the global entropy of the partition π as the weighted sum

$$E(\pi) = \sum_{i=1}^{m} P_i(R(A)) \cdot E_O(R(A)) \tag{4}$$

where m is the number of different approximation pairs induced by π.

Given the standard ordering $\pi_1 \leq \pi_2$ between partitions (i.e., $\pi_1 \leq \pi_2$ iff $\forall C \in \pi_1 \; \exists D \in \pi_2 : C \subseteq D$), we can verify the following result:

Proposition 6. E_O *is isotonic w.r.t. the standard ordering between partitions, i.e.,* $\pi \leq \sigma \rightarrow E(\pi) \leq E(\sigma)$.

Thus, in case of exact knowledge, where the partition is made by singletons, we have that $r_i(A) = 1$ for all A, and the entropy assumes the minimum value.

5.2   Possibility Distribution

We now suppose that the collection of orthopairs represents a possibility distribution. Indeed, as observed in Sect. 2 we can associate to any Boolean possibility distribution π a set of orthopairs $\mathcal{O}_\pi^*$.

In order to apply Eq. (3) to this particular case, we can assign to each orthopair $(P_i, N_i) = O_i \in \mathcal{O}_\pi^*$ a probability measure $P(O_i) = \frac{|\pi_{O_i}|}{|\pi_\mathcal{O}|}$ where $|\pi_{O_i}|$ is the number of valuation functions $\omega : X \mapsto [0,1]$ such that $\omega \models [\bigwedge_{x \in P_i} x \wedge \bigwedge_{x \in N_i} \neg x]$ and $|\pi_\mathcal{O}|$ is the cardinality of the set of valuations of which $\pi_\mathcal{O}$ is the characteristic function. Thus we obtain the global uncertainty of $\mathcal{O}_\pi^*$ as:

$$E(\mathcal{O}_\pi^*) = \sum_{O_i \in \mathcal{O}_\pi^*} P(O_i) E_O(O_i) = \frac{1}{|X||\pi_\mathcal{O}|} \sum_{O_i \in \mathcal{O}_\pi} |\pi_{O_i}||Bnd(O_i)| \qquad (5)$$

More in general given a set of orthopairs $\mathcal{O} = \{O_1, ..., O_n\}$ (not necessarily obtained from a possibility distribution) we can associate to each of these orthopairs a probability measure $P(O_i) = \frac{m(O_i)}{|\pi_\mathcal{O}|}$ where $m(O_i)$ is defined, as suggested in [2], as $m(O_i) = \sum_{\omega \in \pi_\mathcal{O}} \frac{\chi_{\pi_{O_i}}(\omega)}{\sum_{O_j \in \mathcal{O}} \chi_{\pi_{O_j}}(\omega)}$. We can then define the global entropy as

$$E(\mathcal{O}) = \sum_{O_i \in \mathcal{O}} \frac{m(O_i)}{|\pi_\mathcal{O}|} E_O(O_i). \qquad (6)$$

Besides these definitions based on the weighted sum of Eq. (3), we can give another definition based on the non-specificity of the associated Boolean possibility distribution. At first, given a single orthopair O we can associate to it π_O and measure its non-specificity with the Hartley measure, $H(\pi_O) = log_2(|\pi_O|) = log_2(2^{|Bnd_O|}) = |Bnd_O|$. Once normalized over $|X|$ we get back the counting measure E_O of Eq. (2). Then, if $\mathcal{O}_\pi^*$ is a set of orthopairs representing a Boolean possibility distribution in disjunctive normal form, we can easily observe that $E_O(\mathcal{O}_\pi^*) = \frac{log(\sum_{O \in \mathcal{O}_\pi^*} 2^{E_O(O)*|X|})}{|X|}$. In case of a generic set $\mathcal{O} = \{O_1, ..., O_n\}$ of orthopairs, this measure is an upper bound since the π_{O_i} have a non-empty intersection, that is in general, $E_O(\mathcal{O}) \leq \frac{log(\sum_{i=1}^n 2^{E_O(O_i)*|X|})}{|X|}$.

6 Conclusions

A preliminary study on uncertainty measures for orthopairs has been put forward. In case of a single orthopair we have seen that a prominent role is played by the counting measure E_O. In case of a collection of orthopairs a generic entropy can be defined (see Eq. (3)). This measure can then be instantiated in particular collections and the cases of rough sets and possibility distribution have been investigated. However, the picture is far from being complete. Particular facets of uncertainty can be studied, and we have just seen the non-specificity case on a single orthopair. The case of conflict/agreement among orthopairs is another one worth consideration, with the possibility to combine them in a consistent orthopair. Further, we concentrated on the boundary region, another possibility is to analyze the balance between positive and negative. Finally, in a forthcoming work, we will introduce a generalized notion of partition on the space of all orthopairs, introduce a mutual information on this partition and use it for clustering.

References

1. Belnap, N.: A useful four-valued logic. In: Dunn, M., Epstein, G. (eds.) Modern Uses of Multiple Valued Logics, pp. 8–37. D. Reidel (1977)
2. Bianucci, D., Cattaneo, G.: Information entropy and granulation co-entropy of partitions and coverings: a summary. In: Peters, J.F., Skowron, A., Wolski, M., Chakraborty, M.K., Wu, W.-Z. (eds.) Transactions on Rough Sets X. LNCS, vol. 5656, pp. 15–66. Springer, Heidelberg (2009). doi:10.1007/978-3-642-03281-3_2
3. Ciucci, D., Yao, Y.: Special issue on three-way decisions, orthopairs and square of opposition. Int. J. Approximate Reasoning (2017). http://www.sciencedirect.com/science/journal/0888613X/vsi/10WGD1MNXV4
4. Ciucci, D.: Orthopairs: a simple and widely used way to model uncertainty. Fundam. Inf. 108(3–4), 287–304 (2011)
5. Ciucci, D.: Orthopairs and granular computing. Granular Comput. 1, 159–170 (2016)
6. Ciucci, D., Dubois, D., Lawry, J.: Borderline vs. unknown: comparing three-valued representations of imperfect information. Int. J. Approximate Reasoning 55(9), 1866–1889 (2014)
7. De Luca, A., Termini, S.: A definition of a nonprobabilistic entropy in the setting of fuzzy sets theory. Inf. Control 20(4), 301–312 (1972)
8. Dubois, D., Prade, H.: An introduction to bipolar representations of information and preference. Int. J. Intell. Syst. 23(3), 866–877 (2008)
9. Dubois, D., Prade, H.: From Blanché's hexagonal organization of concepts to formal concept analysis and possibility theory. Log. Univers. 6, 149–169 (2012)
10. Guo, K.: Knowledge measure for Atanassov's intuitionistic fuzzy sets. IEEE Trans. Fuzzy Syst. 24(5), 1072–1078 (2016)
11. Klir, G.J.: Generalized information theory: aims, results, and open problems. Reliab. Eng. Syst. Saf. 85(1–3), 21–38 (2004)
12. Lawry, J., Dubois, D.: A bipolar framework for combining beliefs about vague propositions. In: Brewka, G., Eiter, T., McIlraith, S.A. (eds.) Principles of Knowledge Representation and Reasoning: Proceedings of the Thirteenth International Conference, pp. 530–540. AAAI Press (2012)
13. Pal, N., Bustince, H., Pagola, M., Mukherjee, U., Goswami, D., Beliakov, G.: Uncertainties with Atanassov's intuitionistic fuzzy sets: fuzziness and lack of knowledge. Inf. Sci. 228, 61–74 (2013)
14. Skowron, A., Jankowski, A., Swiniarski, R.W.: Foundations of rough sets. In: Kacprzyk, J., Pedrycz, W. (eds.) Springer Handbook of Computational Intelligence, pp. 331–348. Springer, Heidelberg (2015)
15. Song, Y., Wang, X., Yu, X., Zhang, H., Lei, L.: How to measure non-specificity of intuitionistic fuzzy sets. J. Int. Fuzzy Syst. 29(5), 2087–2097 (2015)
16. Szmidt, E., Kacprzyk, J.: Entropy for intuitionistic fuzzy sets. Fuzzy Sets Syst. 118(3), 467–477 (2001)
17. Yao, Y.: An outline of a theory of three-way decisions. In: Yao, J.T., Yang, Y., Słowiński, R., Greco, S., Li, H., Mitra, S., Polkowski, L. (eds.) RSCTC 2012. LNCS (LNAI), vol. 7413, pp. 1–17. Springer, Heidelberg (2012). doi:10.1007/978-3-642-32115-3_1
18. Zhang, H.: Entropy for intuitionistic fuzzy sets based on distance and intuitionistic index. Int. J. Uncertainty Fuzziness Knowl. Based Syst. 21(01), 139–155 (2013)
19. Zhu, P., Wen, Q.: Information-theoretic measures associated with rough set approximations. Inf. Sci. 212, 33–43 (2012)

Possibilistic Networks

Possibilistic MDL: A New Possibilistic Likelihood Based Score Function for Imprecise Data

Maroua Haddad[1,2](✉), Philippe Leray[2], and Nahla Ben Amor[1]

[1] LARODEC Laboratory ISG, Université de Tunis, Tunis, Tunisia
nahla.benamor@gmx.fr
[2] LINA-UMR CNRS 6241, Université de Nantes, Nantes, France
maroua.haddad@gmail.com, philippe.leray@univ-nantes.fr

Abstract. Recent years have seen a surge of interest in methods for representing and reasoning with imprecise data. In this paper, we propose a new possibilistic likelihood function handling this particular form of data based on the interpretation of a possibility distribution as a contour function of a random set. The proposed function can serve as the foundation for inferring several possibilistic models. In this paper, we apply it to define a new scoring function to learn possibilistic network structure. Experimental study showing the efficiency of the proposed score is also presented.

1 Introduction

In statistics, likelihood functions are generally viewed as adequateness functions of a probability distribution w.r.t a set of data. They play a key role in model inference, especially in estimating model parameters given observed data. Most of research endeavors elaborated in this context are defined in the probabilistic framework which, for a long time, has been considered as the unique normative model manipulating uncertain but *precise* information. Nevertheless, probability theory, as good as it is, does not remain the best alternative where imprecision is inherent in the studied domain or where we are faced to incomplete information. Thereby, over the last five decades, a lot of effort has been put into developing new non-classical uncertainty theories and proposing methods handling such imperfect data. This paper rigorously fits this context by proposing a new likelihood function in the possibilistic framework [7], one of the non-classical uncertainty theories that has gained considerable interest in recent years. The proposed likelihood function handles imprecise data i.e. set-valued data based on the acknowledged interpretation of a possibility distribution as a contour function of a random set [17]. Since this form of imperfect data may arise in many real world applications, the proposed likelihood function could be used to infer multiple types of possibilistic/random sets models. In this paper, it represents the key concept of proposing a new scoring function to learn possibilistic network structure from imprecise data.

© Springer International Publishing AG 2017
A. Antonucci et al. (Eds.): ECSQARU 2017, LNAI 10369, pp. 435–445, 2017.
DOI: 10.1007/978-3-319-61581-3_39

This paper is organized as follows: Sect. 2 briefly introduces the probabilistic likelihood function. In Sect. 3, we propose a new possibilistic likelihood function exploring the link between possibility theory and random sets theory. Section 4 defines a new likelihood based scoring function. In Sect. 5, experimental study showing the efficiency of the proposed score in the context of structure learning of possibilistic networks is proposed.

2   Probabilistic Likelihood

The likelihood function is the probability of the joint occurrence of all the given data for a specified value of the parameter defined by a probability distribution. More formally, let us consider a set of data $\mathcal{D}_i = \{d_i^{(1)}, d_i^{(2)}, ..., d_i^{(l)}\}$ relative to a variable X_i defined on D_i and let x_{ik} be an instance of X_i. Let $p_{i1}, p_{i2}, ..., p_{ir_i}$ be the parameter values of p_i relative to X_i. The most used likelihood function is the logarithmic-based likelihood (log-likelihood) defined as follows:

$$LL(p_i, \mathcal{D}_i) = \sum_{o=1}^{l} \log(p(d_i^{(o)})) \tag{1}$$

If we name N_{ik} the number of occurrences of each x_{ik} in $\mathcal{D}_i$ i.e. the number of times x_{ik} appears in $\mathcal{D}_i$: $N_{ik} = |\{o \text{ s.t. } x_{ik} = d_i^{(o)}\}|$, the log-likelihood function could be re-written as follows:

$$LL(p_i, \mathcal{D}_i) = \sum_{k=1}^{r_i} N_{ik} \log(p_{ik}) \tag{2}$$

where $r_i = |D_i|$. Note that this likelihood function has been widely used in many domains to infer probabilistic models from precise data. Despite the fact that most of researches of the last decades have proved that imperfection, be it uncertainty or imprecision, is unavoidable in real world applications and must be incorporated in information systems, only one attempt [6] has been made to adapt the probabilistic likelihood function to imperfect data, in particular, evidential data. This adaptation combines aleatory uncertainty captured by a parametric statistical model with epistemic uncertainty induced by an imperfect observation process and represented by belief functions. A recent work [5] has shown that this evidential adaptation of likelihood function is limited and unsound in some situations especially, when used to estimate model parameters. In what follows, we propose a possibilistic likelihood function which handles a more common form of imperfect data i.e. imprecise data. This form of imperfection has gained increasing interest in recent years, in particular, in the context of learning models from data. Note that in this paper, we do not consider the issue of imprecision due to a small number of precise observations (see for instance [16]).

3   Possibilistic Likelihood

The main objective of this work is to propose a likelihood function that represents data as they have been collected, i.e. including imprecision due to the physical measurement, in the possibilistic framework [7]. The latter is able to offer a natural and simple formal framework representing imprecise and uncertain information. In fact, it refers to the study of maxitive and minitive set-functions and can be interpreted as an approximation of upper and lower frequentist set probabilities in the presence of imprecise observations. The basic building block of possibility theory is the notion of possibility distribution π which corresponds to a mapping from the universe of discourse D_i to the unit interval $[0, 1]$. For any state $x_{ik} \in D_i$, $\pi(x_{ik}) = 1$ means that the realization of x_{ik} is totally possible. $\pi(x_{ik}) = 0$ means that x_{ik} is an impossible state. It is generally assumed that at least one state x_{ik} is totally possible and π is then said to be normalized. The particularity of the possibilistic scale is that it can be interpreted in two ways: (i) in an ordinal manner which means that possibility degrees reflect only a specific order between possible values (ii) in a numerical manner which means that possibility degrees make sense in the ranking scale. Given a possibility distribution π, we can define for any subset $A \subseteq D_i$ two dual measures: possibility measure $\Pi(A) = \max_{x_{ik} \in A} \pi(x_{ik})$ and necessity measure $N(A) = 1 - \Pi(\bar{A})$. The definition of a possibility distribution could be generalized to a set of variables $V = \{X_1, X_2, ..., X_n\}$ defined on the universe of discourse $\Omega = D_1 \times ... \times D_n$ encoded by π and $\omega \in \Omega$ is called interpretation or event.

The formulation of our likelihood function is made in two steps: first, we propose an extension of the probabilistic likelihood function based on random sets since their definition supports naturally set-valued data. Then, we propose its approximation in the possibilistic framework using the interpretation of a possibility distribution π on X_i as a *contour function* (CF$_{m \to \pi}$) of a random set [17] of D_i. A random set of D_i is a pair $(\mathcal{F}, m)$ where $\mathcal{F}$ is the family of all focal sets i.e. $A_{ik} \subseteq D_i$ such that $m(A_{ik}) > 0$. m is called basic probability assignment or mass function and corresponds to a mapping $m : 2^{|D_i|} \longmapsto [0, 1]$ such that $\sum_{A_{ik} \subseteq D_i}(m(A_{ik})) = 1$ and $m(\emptyset) = 0$. So, a possibility distribution could be derived using the following equation:

$$\text{CF}_{m \to \pi}(x_{ik}) = \pi(x_{ik}) = \sum_{A_{ik} | x_{ik} \in A_{ik}} m(A_{ik}) \tag{3}$$

The extension of the probabilistic log-likelihood (Eq. 2), named random set likelihood, consists in replacing the probability distribution by mass functions. More formally, let us consider a set of imprecise data $\mathcal{D}_i = \{d_i^{(1)}, d_i^{(2)}, ..., d_i^{(l)}\}$ relative to a variable X_i defined on D_i and let A_{ik} be a focal set belonging to D_i, i.e. $A_{ik} \subseteq D_i$. We name N_{ik} the number of occurrences of each A_{ik} in $\mathcal{D}_i$ i.e. the number of times A_{ik} appears in $\mathcal{D}_i$: $N_{ik} = |\{l \text{ s.t. } x_{ik} = d_i^{(l)}\}|$. Let $m_{i1}, m_{i2}, ..., m_{irs_i}$ be the parameter values of m_i relative to X_i, we define random set log-likelihood function as follows:

$$mLL(m_i, \mathcal{D}_i) = \sum_{k=1}^{rs_i} N_{ik} \log(m_{ik}) \tag{4}$$

where $rs_i = 2^{|D_i|}$. It should be noted that computing the random set log-likelihood function is computationally expensive. In fact, a random set relative to a variable X_i is defined on $2^{|D_i|}$ and its cardinality grows exponentially with the number of values in D_i. To alleviate this complexity, we propose to investigate the link between possibility theory and random sets theory expressed in Eq. 3 and we replace mass functions by possibility distributions. More formally, let us consider a set of imprecise data $\mathcal{D}_i = \{d_i^{(1)}, d_i^{(2)}, ..., d_i^{(l)}\}$ relative to a variable X_i. We name N_{ik} the number of occurrences of each x_{ik} in $\mathcal{D}_i$ i.e. the number of times x_{ik} appears in $\mathcal{D}_i$: $N_{ik} = |\{l \text{ s.t. } x_{ik} \subseteq d_i^{(l)}\}|$. Let $\pi_{i1}, \pi_{i2}, ..., \pi_{ir_i}$ be the parameter values of π_i relative to X_i, we express the possibilistic likelihood function as follows:

$$\pi LL(\pi_i, \mathcal{D}_i) = \sum_{k=1}^{r_i} N_{ik} \log(\pi_{ik}) \tag{5}$$

where $r_i = |D_i|$. It is evident that random set likelihood and possibilistic likelihood functions are not equivalent. However, possibility distributions obtained by transforming mass functions, obtained by maximizing random sets likelihood, using $\text{CF}_{m \to \pi}$ leads to the ones obtained by directly maximizing possibilistic likelihood in Eq. 5 (for more details about maximizing possibilistic and random set likelihood functions, see [12]). More formally,

Proposition 1. $argmax(\pi LL(\pi_{ik}, \mathcal{D}_i)) = CF_{m \to \pi}(argmax(mLL(m_{ik}, \mathcal{D}_i)))$.

Proof. $argmax(\pi LL(\pi_{ik}, \mathcal{D}_i)) = \dfrac{N_{ik}}{N} = \sum\limits_{A_{ik}|x_{ik} \in A_i} \dfrac{N_{A_{jk}}}{N} = \text{CF}_{m \to \pi}(\dfrac{N_{A_{jk}}}{N}) = CF_{m \to \pi}(argmax(mLL(m_{ik}, \mathcal{D}_i)))$.

4   New Possibilistic-Likelihood-Based Score

πLL could be used in many research fields such as pattern recognition and classification. We propose here to illustrate it in an ill-explored area of research that is learning possibilistic network structure. Before detailing its application, we briefly introduce these models and present existing learning methods.

4.1   Learning Possibilistic Networks from Data

Possibilistic networks [8] represent the possibilistic counterpart of Bayesian networks [13] having similarly two components: a *graphical component* composed of a DAG which encodes a set of independence relations (i.e. each variable $X_i \in V$ is conditionally independent of its non-descendant given its parents) and a *numerical component* corresponding to the set of conditional possibility distributions relative to each node $X_i \in V$ in the context of its parents, denoted by $Pa(X_i)$, i.e. $\pi(X_i|Pa(X_i))$. The two interpretations of the possibilistic scale lead naturally to two different ways to define possibilistic networks: *product-based* possibilistic networks and *min-based* possibilistic networks. In this paper, we are interested in *product-based* possibilistic networks, defined in the numerical interpretation, using the product-based conditioning expressed by:

$$\pi(\omega|\Phi) = \begin{cases} \frac{\pi(\omega)}{\Pi(\Phi)} & \text{if } \omega \in \Phi \\ 0 & \text{otherwise.} \end{cases} \tag{6}$$

where $\Phi \subseteq \Omega$. The joint distribution relative to *product-based* possibilistic networks can be computed via the following product-based chain rule

$$\pi(X_1, ..., X_n) = \prod_{i=1..n} \pi(X_i \mid Pa(X_i)) \tag{7}$$

Contrarily to Bayesian networks, learning possibilistic networks from data has not been deeply studied. In fact, for the last years, learning Bayesian networks has been widely studied and various approaches were proposed to learn both DAG structure and parameters. More precisely, structure learning algorithms can be classified into three families: constraint-based approaches, score-based approaches and hybrid methods. Regarding possibilistic networks, few attempts address the problem of their learning and existing ones [2, 15] are direct adaptations of Bayesian networks learning methods without any awareness of specificities of the possibilistic framework and of advances made concerning possibilistic networks as models of independence [1]. In fact, Sangüesa et al. [15] have proposed two hybrid methods handling precise data: the first one learns trees and the second one learns the more general structure of DAGs. Borgelt et al. [2] have adapted two methods initially proposed to learn Bayesian networks: K2 [4] and maximum weight spanning tree (MWST) [3] to learn possibilistic networks from imprecise data. These two algorithms are based on local scores to guide the search in graph candidates. In the current work, we retain two scores, namely, *possibilistic mutual information* and *possibilistic χ^2 measure* which are direct adaptations of probabilistic independence tests mutual information and χ^2. In fact, Borgelt et al. have shown that these adaptations lead to good structures [2]. Given two variables X_i and X_j in V, then:

- *Possibilistic mutual information* is expressed by:

$$d_{mi}(X_i, X_j) = - \sum_{\substack{x_{ik} \in D_i \\ x_{jl} \in D_j}} \frac{N_{ik,jl}}{N}.log_2 \frac{N_{ik,jl}}{\min(N_{ik}, N_{jl})} \tag{8}$$

- *Possibilistic χ^2 measure* is expressed by:

$$d_{\chi^2}(X_i, X_j) = \sum_{\substack{x_{ik} \in D_i \\ x_{jl} \in D_j}} \frac{(min(N_{ik}, N_{jl}) - N_{ik,jl})^2}{min(N_{ik}, N_{jl})} \tag{9}$$

Note that these scores are computed in a binary manner which fit well with MWST that generates trees. For K2 algorithm, a generalization of these scores to more than two attributes is made as follows: given a variable X_i, all its parents could be ombined into one pseudo-variable representing the Cartesian product of their domains. It should be noted that none of these works is theoretically sound and every proposed score lacks an explanation of its use and its contributions regarding the others (for more details, see [11]). Moreover, contrarily to the probabilistic case, none of these scores is based on likelihood and assesses the adequateness between the learned possibilistic networks and the training dataset. In what follows, we investigate the use of the possibilistic likelihood expressed by Eq. 5 to propose a new scoring function based on *minimum description length* (MDL) principle [14].

4.2   Possibilistic MDL

MDL principle is based on the following insight: any regularity in a given set of data can be used to compress the data, i.e. to describe it using fewer symbols than needed to describe the data literally [9]. More explicitly, the underlying idea of MDL principle is that the model that best represents a data set is the one that minimizes the sum of two terms : (i) the coding length of the model and (ii) the data coding length when this model is used to represent this data. This principle has been applied to define the probabilistic scoring function, named MDL [14], to learn Bayesian networks. MDL establishes an appropriate trade-off between complexity and precision and is based on the following insight: the model to be selected is the one that is best balanced in terms of simplicity and fitness of given data. In fact, it includes two terms: likelihood function to quantify fitness between the graph and the data and complexity computed via the dimension of the graph. The latter corresponds to the sum over variables X_i of the number of parameters required to represent $p(X_i|Pa(X_i))$. So, by analogy to the probabilistic case, our possibilistic adaptation of MDL, named πMDL, includes the likelihood function of the possibilistic network given data and its dimension. We define these two quantities as follows:

Definition 1. *The dimension of a possibilistic network G denoted by $Dim(G)$ is the number of parameters required to represent its conditional possibility distributions and is expressed by:*

$$Dim(G) = \sum_{i=1}^{n} dim(X_i, G) \tag{10}$$

where $dim(X_i, G) = |D_i| * \prod_{X_j \in Pa(X_i)} |D_j|$.

Definition 2. *Let G be a DAG and $\{\pi_1, \pi_2, ..., \pi_n\}$ be the parameters relative to $\{X_1, X_2, ..., X_n\}$ to be estimated and $\mathcal{D}_{ij} = \{d_{ij}^{(l)}\}$ be a dataset relative to a variable X_i and its parents $Pa(X_i) = x_j$, $d_{ij}^{(l)} \subseteq D_{ij}$. The number of occurrences of each $x_{ik} \in D_i$ such that such that $Pa(X_i) = x_j$, denoted by N_{ijk}, is the number of times x_{ijk} appears in $\mathcal{D}_{ij}$: $N_{ijk} = |\{l \ s.t. \ x_{ijk} \subseteq d_{ij}^{(l)}\}|$. We express the possibilistic likelihood by:*

$$\pi LL(\pi, G, \mathcal{D}) = \sum_{i=1}^{n} \sum_{j=1}^{q_i} \sum_{k=1}^{r_i} N_{ijk} \log \pi_{ijk} \tag{11}$$

where for each X_i, $q_i = |Pa(X_i)|$, $r_i = |D_i|$, π_{ijk} is the parameter to be estimated when $X_i = x_{ik}$ and $Pa(X_i) = x_j$.

So, we define πMDL as follows:

$$\pi MDL(G|\mathcal{D}) = \pi LL(\pi, G, \mathcal{D}) - Dim(G) \tag{12}$$

where the conditional possibility distributions are inferred from data by maximizing πLL (for more details see [12]) and computed as follows:

$$\hat{\pi}(X = x_{ik}|Pa(X_i) = x_j) = \frac{N_{ijk}}{\sum_{k=1}^{r_i} N_{ijk}} \tag{13}$$

So, MDL could be re-written as follows:

$$\pi MDL(G|\mathcal{D}) = \sum_{i=1}^{n} \sum_{j=1}^{q_i} \sum_{k=1}^{r_i} N_{ijk} \log \hat{\pi}(X = x_{ik}|Pa(X_i) = x_j) - Dim(G) \tag{14}$$

4.3   Property Analysis

Most of Bayesian scores proposed in the literature satisfy two criteria: *decomposability* and *Markov equivalence*. In what follows, we will check if these two criteria are satisfied by πMDL.

Decomposability: A score S is said to be decomposable if it can be described in terms (generally a sum) of local scores i.e. depending only on a node and all its parents. πMDL is decomposable as follows:

$$\pi MDL(G|\mathcal{D}) = \sum_{i=1}^{n} \pi mdl(X_i|Pa(X_i)) \tag{15}$$

where $\pi mdl(X_i|Pa(X_i)) = \pi LL(X_i|Pa(X_i), \mathcal{D}) - Dim(X_i, G)$.

Markov equivalence: A score is said to be Markov equivalent if it assigns the same value to two equivalent graphs. Two DAGs are equivalent if and only if they have the same skeleton (the skeleton of a directed graph is the same underlying undirected graph) and the same v-structures ($X_i \rightarrow X_j \leftarrow X_k$). Contrarily to the case of the probabilistic MDL, πMDL does not satisfy Markov equivalence as shown by the following example:

Table 1. Example of an imprecise dataset

X_1	X_2	nb of occurrences
x_{12}	x_{22}	3
x_{12}	x_{21}, x_{22}	3
x_{11}, x_{12}	x_{22}	3
x_{11}	x_{21}, x_{22}	1

Example 1. Let us consider the imprecise dataset $\mathcal{D}$ in Table 1, the two Markov equivalent graphs composed of two variables X_1 and X_2 (G_1: $X_1 \rightarrow X_2$ and G_2: $X_1 \leftarrow X_2$).

$\pi MDL(G_1|\mathcal{D}) = 4\log(0.4) + 9\log(0.9) + \log(0.25) + 4\log(1) + 3\log(\frac{1}{3}) + 9\log(1) - 6 = -10.03.$

$\pi MDL(G_2|\mathcal{D}) = 4\log(0.4) + 10\log(1) + \log(0.25) + 3\log(0.75) + 4\log(0.4) + 9\log(0.9) - 6 = -11.10.$

Now, to learn possibilistic network structure, we propose to adapt greedy search algorithm initially proposed to learn Bayesian network structure using πMDL. Greedy search algorithm is an iterative method which given an initial DAG (an empty DAG, randomly generated network or the tree obtained by MWST algorithm), generates all neighbor structures obtained after performing one of elementary operation, i.e., adding, deleting or reversing an edge. Then, it computes obtained neighbors structures scores using πMDL and picks the operation that leads to the structure having the highest score. This process is repeated until the already obtained structure has a higher score than DAGs in the list of neighbors. Note that the decomposability property satisfied by our score πMDL allows us to efficiently evaluate the elementary operators performed by greedy search. In fact, it reduces the number of calculations by locally estimating the change in the score between two neighboring structures, instead of recalculating entirely to the new structure. Note also that the non satisfaction of Markov equivalence property deprives to perform greedy search in an efficient way, i.e., by reducing the search space from DAGs space to CPDAGs (graph representative of Markov equivalence class) space.

5   Experimental Study

To evaluate πMDL, we use the evaluation strategy proposed for product-based possibilistic networks learning algorithms [10] described in Fig. 1. More precisely, we generate 20 possibilistic networks (10 with 10 variables and 10 with 20 variables) to derive 60 synthetic datasets containing 1000 imprecise observations. We also vary the maximum number of parents between 2 and 4 and the maximum number of variable domain cardinality between 2 and 5[1]. Using the

---

[1] Used benchmarks are publicly available in: https://sites.google.com/site/karimtabiasite/mappos.

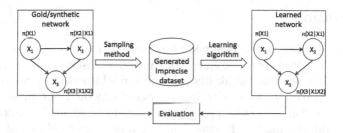

Fig. 1. Evaluation process of possibilistic networks learning algorithms

generated networks, we apply existing possibilistic learning structure algorithms which handle imprecise data, i.e. the possibilistic adaptation of k2 (πK2) [2][2], the Maximum Weight Spanning Tree (πMWST) [2] and our approach which combines greedy search (πGS) and πMDL. In order to better appreciate the impact of πMDL, we also propose to test πGS with other scores i.e. sum over V of the local scores d_{χ^2} and d_{mi}, denoted respectively by $\sum_{d_{\chi^2}}$ and $\sum_{d_{mi}}$. Then, we compare the learned and the initial possibilistic networks using the editing distance [18] which corresponds to the number of operations required to transform a learned possibilistic network DAG into the initial one (add, reverse or delete an edge increases the editing distance by 1). Table 2 presents the average of obtained results. Unsurprisingly, πGS combined with πMDL outperforms the remaining learning algorithms (i.e. πK2 and πMWST). Indeed, this result reaffirms the quality of greedy search already observed in the probabilistic framework in learning Bayesian networks. Moreover, using πGS with πMDL is better than combining it with $\sum_{d_{\chi^2}}$ and $\sum_{d_{mi}}$ and this result is in coherence with a previous work on parameters learning in possibilistic networks [12] using πLL from which we derive πMDL.

Table 2. Editing distance between initial and learned networks

n	10	20
πGS + πMDL	**19.77 +/− 1.5**	**31.55 +/− 2.92**
πGS + $\sum_{d_{\chi^2}}$	28.83 +/− 2.32	51.66 +/− 1.33
πGS + $\sum_{d_{mi}}$	35.66 +/− 2.06	49.55 +/− 1.41-
πMWST + d_{χ^2}	23.44 +/− 1.63	47.33 +/− 0.88
πMWST + d_{mi}	22.77 +/− 1.6	47.55 +/− 1.41
π K2 + d_{χ^2}	27.44 +/− 2.95	42.22 +/− 6.87
πK2 + d_{mi}	28.38 +/− 4.53	42.77 +/− 5.66

---

[2] Since πK2 handles variables in a predefined order, so, we generate 5 orders in each experiment and we retain the best structure.

6   Conclusion

In this paper, we propose two likelihood functions, namely, random set likelihood function which represents an extension of the probabilistic one and handles set-valued data, and possibilistic likelihood function which is an approximation of the first one based on the interpretation of a possibility distribution as a countour function of a random set. This approximation could be applied to infer different types of possibilistic models. In this study, it represents the basis of defining a new possibilistic score πMDL to learn possibilistic network structure. Proposed experimental study shows that πMDL combined with greedy search outperforms existing learning algorithms. Such results are preliminary and clearly deserve more investigations but encouraging. A further comparative study on a large number of benchmarks and problems using other evaluation measures will be needed to really evaluate the efficiency of the proposed score. Moreover, it will be interesting to evaluate the impact of non-satisfaction of Markov equivalence property on the learned possibilistic network structure quality.

References

1. Ben Amor, N., Benferhat, S.: Graphoid properties of qualitative possibilistic independence relations. Int. J. Uncertainty, Fuzziness Knowl.-Based Syst. **13**(01), 59–96 (2005)
2. Borgelt, C., Kruse, R.: Operations and evaluation measures for learning possibilistic graphical models. Artif. Intell. **148**(1), 385–418 (2003)
3. Chow, C., Liu, C.: Approximating discrete probability distributions with dependence trees. IEEE Trans. Inf. Theory **14**(3), 462–467 (1968)
4. Cooper, G.F., Herskovits, E.: A Bayesian method for the induction of probabilistic networks from data. Mach. Learn. **9**(4), 309–347 (1992)
5. Couso, I., Dubois, D.: Maximum likelihood under incomplete information: toward a comparison of criteria. In: Ferraro, M.B., Giordani, P., Vantaggi, B., Gagolewski, M., Gil, M.Á., Grzegorzewski, P., Hryniewicz, O. (eds.) Soft Methods for Data Science. AISC, vol. 456, pp. 141–148. Springer, Cham (2017). doi:10.1007/978-3-319-42972-4_18
6. Denoeux, T.: Maximum likelihood estimation from uncertain data in the belief function framework. IEEE Trans. knowl. data Eng. **25**(1), 119–130 (2013)
7. Dubois, D., Prade, H.: Possibility Theory. Springer, Berlin (1988)
8. Fonck, P.: Propagating uncertainty in a directed acyclic graph. In: Proceedings of the Fourth Information Processing and Management of Uncertainty Conference, **92**, 17–20 (1992)
9. Grünwald, P.D.: Theory and applications: advances in minimum description length, Mdl tutorial (2005)
10. Haddad, M., Leray, P., Amor, N.B.: Evaluating product-based possibilistic networks learning algorithms. In: Proceedings of Symbolic and Quantitative Approaches to Reasoning with Uncertainty, pp. 312–321 (2015)
11. Haddad, M., Leray, P., Amor, N.B.: Learning possibilistic networks from data : a survey. In: 16th World Congress of the International Fuzzy Systems Association and the 9th Conference of the European Society for Fuzzy Logic and Technology, pp. 194–201 (2015)

12. Haddad, M., Leray, P., Levray, A., Tabia, K.: Possibilistic networks parameter learning: Preliminary empirical comparison. In: 8èmes journées francophones de réseaux bayésiens (JFRB 2016), (2016)
13. Pearl, J.: Probabilistic Reasoning in Intelligent Systems: Networks of Plausible Inference. Morgan Kaufmann, San Francisco (1988)
14. Rissanen, J.: Modeling by shortest data description. Automatica **14**(5), 465–471 (1978)
15. Sangüesa, R., Cabós, J., Cortes, U.: Possibilistic conditional independence: A similarity-based measure and its application to causal network learning. Int. J. Approximate Reasoning **18**(1), 145–167 (1998)
16. Serrurier, M., Prade, H.: An informational distance for estimating the faithfulness of a possibility distribution, viewed as a family of probability distributions, with respect to data. Int. J. Approximate Reasoning **54**(7), 919–933 (2013)
17. Shafer, G.: A mathematical Theory of Evidence, vol. 1. Princeton University Press, Princeton (1976)
18. Shapiro, L.G., Haralick, R.M.: A metric for comparing relational descriptions. IEEE Trans. Pattern Anal. Mach. Intell. **1**(1), 90–94 (1985)

Probabilistic Logics, Probabilistic Reasoning

The Complexity of Inferences and Explanations in Probabilistic Logic Programming

Fabio G. Cozman[1]([✉]) and Denis D. Mauá[2]

[1] Escola Politécnica, Universidade de São Paulo, São Paulo, Brazil
fgcozman@usp.br
[2] Instituto de Matemática e Estatística, Universidade de São Paulo,
São Paulo, Brazil

Abstract. A popular family of probabilistic logic programming languages combines logic programs with independent probabilistic facts. We study the complexity of marginal inference, most probable explanations, and maximum a posteriori calculations for propositional/relational probabilistic logic programs that are acyclic/definite/stratified/normal/ disjunctive. We show that complexity classes Σ_k and PP^{Σ_k} (for various values of k) and NP^{PP} are all reached by such computations.

1 Introduction

The goal of this paper is to shed light on the computational complexity of inference for probabilistic logic programs interpreted in the spirit of Sato's distribution semantics [25]; that is, we have logic programs where some facts are annotated with probabilities, so as to define probability distributions over models. This framework has been shown to be quite useful in modeling practical problems [15,24].

The distribution defined by a probabilistic logic program can be used to answer many queries of interest. Two common queries are to compute the probability of some ground atom given evidence (inference), and to find a (partial) interpretation that maximizes probability while being consistent with evidence (MPE/MAP).

We present results on the complexity of *acyclic*, *definite*, *stratified*, *normal* and *disjunctive* probabilistic logic programs; these results are summarized by Table 1. While most semantics agree on stratified programs, there is less consensus on non-stratified programs. Here we examine two semantics: the *credal semantics*, based on stable models, and the *well-founded semantics*.

We start in Sects. 2 and 3 by reviewing relevant background on probabilistic logic programs and on complexity theory. Our contributions appear in Sect. 4. These results are further discussed in the concluding Sect. 5.

2 Background

The results in this paper depend on an understanding of logic and answer set programming; the topic is dense and cannot be described in detail in the space

© Springer International Publishing AG 2017
A. Antonucci et al. (Eds.): ECSQARU 2017, LNAI 10369, pp. 449–458, 2017.
DOI: 10.1007/978-3-319-61581-3_40

Table 1. Summary of complexity results presented in this paper (all entries indicate completeness with respect to many-one reductions). Entries containing known results have orange background (grey background if printed in black-and-white) [6–8].

	Propositional			Bounded arity		
	Inferential	MPE	MAP	Inferential	MPE	MAP
Acyclic normal	PP	NP	NP^{PP}	PP^{NP}	Σ_2^P	NP^{PP}
Definite (positive query)	PP	NP	NP^{PP}	PP^{NP}	Σ_2^P	NP^{PP}
Stratified normal	PP	NP	NP^{PP}	PP^{NP}	Σ_2^P	NP^{PP}
Normal, credal	PP^{NP}	Σ_2^P	NP^{PP}	PP$^{\Sigma_2^P}$	Σ_3^P	NP^{PP}
Normal, well-founded	PP	NP	NP^{PP}	PP^{NP}	Σ_2^P	NP^{PP}
Disjunctive, credal	PP$^{\Sigma_2^P}$	Σ_3^P	NP^{PP}	PP$^{\Sigma_3^P}$	Σ_4^P	NP^{PP}

we have here. We just mention the main concepts, and refer the reader to any in-depth presentation in the literature [9,14].

We have a vocabulary consisting of *predicates*, *constants*, and *logical variables*; a *term* is a constant or logical variable, and an *atom* is a predicate of arity k associated with k terms; a *ground atom* is an atom without logical variables. We often resort to *grounding* to produce ground atoms. A *disjunctive logic program* (DLP) consists of a set of rules written as

$$A_1 \vee \cdots \vee A_h : -B_1, \ldots, B_{b'}, \textbf{not } B_{b'+1}, \ldots, \textbf{not } B_b.$$

where each A_i and B_i is an atom. The lefthand side is the *head* of the rule; the remainder is its *body*. A rule without disjunction (i.e., $h = 1$) and with empty body, written simply as A_1., is a *fact*. A program without disjunction is a *normal logic program*; a normal logic program without negation is a *definite program*; finally, a program without variables is a *propositional program*. The *dependency graph* of a program is the graph where each atom is a vertex and there are arcs from atoms in the bodies to atoms in the heads of a same rule; a normal logic program is *acyclic* when the dependency graph of its grounding is acyclic. A normal logic program is (locally) *stratified* when the dependency graph of its grounding has no cycles containing an arc involving a negated literal.

The *Herbrand base* of a program is the set of all ground atoms built from constants and predicates in the program. An *interpretation* is a set of ground literals mentioning exactly once each atom in the Herbrand base. A *model* is an interpretation that *satisfies* every grounding of a rule (a rule is satisfied iff the interpretation contains all of $B_1, \ldots, B_{b'}$, none of $B_{b'+1}, \ldots, B_b$, and some of $A_1, \ldots, A_h$). A *minimal model* minimizes the number of non-negated literals.

The most common semantics for DLPs is the *stable model* semantics. Given a program **P** and an interpretation $\mathcal{I}$, define their *reduct* to be program obtained by removing from **P** every rule whose body is not satisfied by $\mathcal{I}$. An interpretation is a *stable model* if it is a minimal model of its reduct. A normal program may have zero, one or several stable models. *Brave reasoning* asks whether there is a stable model containing a specific literal (possibly returning one). *Cautious reasoning* asks whether a specific literal appears in all stable models (possibly listing all).

Here is an example: Several robots can perform, each, one of three operations, called "red", "green", "yellow". A robot placed in a site also covers adjacent sites, so there is no need to place same-color robots on adjacent sites. There is a list of robots and a list of (one-way) connections between sites; the goal is to distribute the robots and verify whether the sites are connected. This disguised 3-coloring problem can be encoded as [14]:

$$\text{color}(X, \text{red}) \vee \text{color}(X, \text{green}) \vee \text{color}(X, \text{yellow}) : -\text{site}(X).$$
$$\text{clash} : -\textbf{not}\ \text{clash}, \text{edge}(X, Y), \text{color}(X, C), \text{color}(Y, C).$$
$$\text{path}(X, Y) : -\text{edge}(X, Y). \qquad \text{path}(X, Y) : -\text{edge}(X, Z), \text{path}(Z, Y).$$

We might have a database of facts, consisting of a list of sites and their connections, say $\text{site}(s_1), \text{site}(s_2), \ldots, \text{edge}(s_1, s_4)$, and so on. Each stable model of this program is a possible placement (a 3-coloring) and a list of paths between sites.

An alternative semantics for normal logic programs is the *well-founded semantics*; a model under this semantics might fix only a truth-value for some of the atoms (leaving the remaining atoms undefined) [30]. One way to define the well-founded semantics is as follows [4]. Write $\text{LFT}_{\mathbf{P}}(\mathcal{I})$ to mean the least fixpoint of $\mathbb{T}_{\mathbf{P}\mathcal{I}}$, where $\mathbb{T}_{\mathbf{P}}$ is a transformation such that: atom A is in $\mathbb{T}_{\mathbf{P}}(\mathcal{I})$ iff there is grounded rule with head A with the whole body true in interpretation $\mathcal{I}$. Then the well-founded semantics of $\mathbf{P}$ consists of those atoms A that are in the least fixpoint of $\text{LFT}_{\mathbf{P}}(\text{LFT}_{\mathbf{P}}(\cdot))$ plus the literals $\neg A$ for those atoms A that are *not* in the greatest fixpoint of $\text{LFT}_{\mathbf{P}}(\text{LFT}_{\mathbf{P}}(\cdot))$.

We also need standard concepts from complexity theory: languages (sets of strings), decision problems (deciding whether input is in language), complexity classes (sets of languages), many-one reductions [22]. We use well-known complexity classes such as P, NP, PP. We also consider oracle machines and corresponding complexity classes such as Π_i^{P} and Σ_i^{P} (the so-called *polynomial hierarchy*). We also use Wagner's *polynomial counting hierarchy* defined as the smallest set of classes containing P and, recursively, for any class C in the polynomial counting hierarchy, the classes PP^{C}, NP^{C}, and $\mathsf{coNP}^{\mathsf{C}}$ [29,31].

3   Probabilistic Logic Programming

In this paper we focus on a particularly simple combination of logic programming and probabilities [23,25]. A *probabilistic disjunctive logic program*, abbreviated PDLP, is a pair $\langle \mathbf{P}, \mathbf{PF} \rangle$ consisting of a disjunctive logic program $\mathbf{P}$ and a set of *probabilistic facts* $\mathbf{PF}$. A probabilistic fact is a pair consisting of an atom A and a probability value α, written as $\alpha :: A$. [15]. Note that we allow a probabilistic fact to contain logical variables. As an instance of PDLP, take our running example on robots and sites: to generate a random graph over a set of five sites add the rules: $0.5 :: \text{edge}(X, Y).$, $\text{site}(s_1).$, $\text{site}(s_2).$, $\text{site}(s_3).$, $\text{site}(s_4).$, $\text{site}(s_5)..$ If $\mathbf{P}$ is a normal logic program, we just write *probabilistic logic program*, abbreviated PLP. If $\mathbf{P}$ is normal and acyclic/definite/stratified, we say the PLP is acyclic/definite/stratified.

To build the semantics of a PDLP, we first take its grounding. From a PDLP with n ground probabilistic facts, we can generate 2^n DLPs: for each probabilistic fact $\alpha :: A.$, either keep fact $A.$ with probability α, or erase $A.$ with probability $1 - \alpha$. A *total choice* θ is a subset of the set of ground probabilistic facts that is selected to be kept (other grounded probabilistic facts are discarded). For any total choice θ we obtain a DLP $\mathbf{P} \cup \mathbf{PF}^{\downarrow\theta}$ with probability $\prod_{A_i \in \theta} \alpha_i \prod_{A_i \notin \theta} (1 - \alpha_i)$. The distribution over total choices induces a distribution over DLPs.

We first define a semantics proposed by Lukasiewicz [18,19]. A *probability model* for a PDLP $\langle \mathbf{P}, \mathbf{PF} \rangle$ is a probability measure $\mathbb{P}$ over interpretations, such that (i) every interpretation $\mathcal{I}$ with $\mathbb{P}(\mathcal{I}) > 0$ is a stable model of $\mathbf{P} \cup \mathbf{PF}^{\downarrow\theta}$ for the total choice θ that agrees with $\mathcal{I}$ on the probabilistic facts; and (ii) the probability of a total choice θ is $\mathbb{P}(\theta) = \prod_{A_i \in \theta} \alpha_i \prod_{A_i \notin \theta} (1 - \alpha_i)$. The set of all probability models for a PDLP is the semantics of the program; note that if a PDLP does not have stable models for some total choice, there is no semantics for it (the program is *inconsistent*). Because a set of probability measures is often called a *credal set* [2]; we adopt the term *credal semantics*.

If $\mathbf{P}$ is definite, then $\mathbf{P} \cup \mathbf{PF}^{\downarrow\theta}$ is definite for any θ, and $\mathbf{P} \cup \mathbf{PF}^{\downarrow\theta}$ has a unique minimal model that is also its unique stable/well-founded model. Thus the distribution over total choices induces a single probability model. This is Sato's *distribution semantics* [25]. Similarly, suppose that $\mathbf{P}$ is acyclic or stratified; then $\mathbf{P} \cup \mathbf{PF}^{\downarrow\theta}$ is respectively acyclic or stratified for any θ, and $\mathbf{P} \cup \mathbf{PF}^{\downarrow\theta}$ has a unique stable model that is also its unique well-founded model [1].

Given a consistent PDLP whose credal semantics is the credal set $\mathbb{K}$, we may be interested in computing *lower conditional probabilities*, defined as $\underline{\mathbb{P}}(\mathbf{Q}|\mathbf{E}) = \inf_{\mathbb{P} \in \mathbb{K}: \mathbb{P}(\mathbf{E}) > 0} \mathbb{P}(\mathbf{Q}|\mathbf{E})$ or *upper conditional probabilities*, defined as $\overline{\mathbb{P}}(\mathbf{Q}|\mathbf{E}) = \sup_{\mathbb{P} \in \mathbb{K}: \mathbb{P}(\mathbf{E}) > 0} \mathbb{P}(\mathbf{Q}|\mathbf{E})$, where $\mathbf{Q}$ and $\mathbf{E}$ are consistent set of literals. Note that we leave conditional lower/upper probabilities undefined when $\overline{\mathbb{P}}(\mathbf{E}) = 0$ (that is, when $\mathbb{P}(\mathbf{E}) = 0$ for every probability model).

Consider again our running example. Suppose we have a graph over five sites, with edges (to save space, e means edge):
0.5 :: e(s_4, s_5).   e(s_1, s_3).   e(s_1, s_4).   e(s_2, s_1).   e(s_2, s_4).   e(s_3, s_5).   e(s_4, s_3)..
That is, we have an edge e(s_4, s_5) which appears with probability 0.5. If this edge is kept, there are 6 stable models; if it is discarded, there are 12 stable models. If additional facts color(s_2, red). and color(s_5, green). are given, then there is a single stable model if edge(s_4, s_5) is kept, and 2 stable models if it is discarded. Then, $\underline{\mathbb{P}}(\text{color}(s_4, \text{green})) = 0$ and $\overline{\mathbb{P}}(\text{color}(s_4, \text{green})) = 1/2$.

A different semantics is defined by Hadjichristodoulou and Warren [17] for *normal* PLPs: they allow probabilities directly over well-founded models, thus allowing probabilities over atoms that are undefined. That is, given a PLP $\langle \mathbf{P}, \mathbf{PF} \rangle$, associate to each total choice θ the unique well-founded model of $\mathbf{P} \cup \mathbf{PF}^{\downarrow\theta}$ to θ; the unique distribution over total choices induces a unique distribution over well-founded models. To conclude, we note that one can find other semantics in the literature that serve deserve further study [3,5,18,21,26].

4   Complexity Results

We consider three different problems in this paper: (marginal) inference, most probable explanation (MPE), and maximum a posteriori (MAP).

In the following problem definitions a PDLP $\langle \mathbf{P}, \mathbf{PF} \rangle$ is always specified using rational numbers as probability values, and with a bound on the arity of predicates (so the Herbrand base is always polynomial in the input size). A query $(\mathbf{Q}, \mathbf{E})$ is always a pair of sets of consistent literals (consistent here means that the set does not contain both a literal and its negation). The set $\mathbf{E}$ is called *evidence*. The symbol $\mathbf{M}$ denotes a set of atoms in the Herbrand base of the union of the program $\mathbf{P}$ and all the facts in $\mathbf{PF}$. The symbol γ is always a rational number in $[0, 1]$.

The *inferential complexity* of a class of PDLPs is the complexity of the following decision problem: with **input** equal to a PDLP $\langle \mathbf{P}, \mathbf{PF} \rangle$, a query $(\mathbf{Q}, \mathbf{E})$, and a number γ, the **output** is whether or not $\mathbb{P}(\mathbf{Q}|\mathbf{E}) > \gamma$; by convention, the input is rejected if $\mathbb{P}(\mathbf{E}) = 0$.

The *MPE complexity* of a class of PDLPs is the complexity of the following decision problem: with **input** equal to a PDLP $\langle \mathbf{P}, \mathbf{PF} \rangle$, evidence $\mathbf{E}$, and a number γ, the **output** is whether or not there is an interpretation $\mathcal{I}$ that agrees with $\mathbf{E}$ and satisfies $\mathbb{P}(\mathcal{I}) > \gamma$.

The *MAP complexity* of a class of PDLPs is the complexity of the following decision problem: with **input** equal to a PDLP $\langle \mathbf{P}, \mathbf{PF} \rangle$, a set $\mathbf{M}$, and a number γ, the **output** is whether or not there is a consistent set of literals $\mathbf{Q}$ mentioning all atoms in $\mathbf{M}$ such that $\mathbb{P}(\mathbf{Q}|\mathbf{E}) > \gamma$; by convention, the input is rejected if $\mathbb{P}(\mathbf{E}) = 0$.

The contributions of this paper are summarized by Table 1. Darker entries are already known [6–8], and some entries on acyclic PLPs can also be found in work by Ceylan et al. [5]. All entries in this table indicate completeness with respect to many-one reductions. In this section we prove these facts through a series of results.[1] In all proofs the argument for membership depends on the complexity of logical reasoning on logic programs that are obtained by fixing all total choices; this suffices to even make decisions concerning conditional probabilities (using for instance techniques by Park [10, Theorem 11.5]).

Theorem 1. *The MPE complexity of acyclic propositional PLPs is* NP-*hard, and of stratified propositional PLPs is in* NP. *The MPE complexity of acyclic PLPs is* Σ_2^P-*hard, and of stratified PLPs is in* Σ_2^P.

Proof. Membership for stratified propositional PLPs: "guess" a polynomial-sized interpretation consistent with evidence, and then decide if its probability exceeds a given threshold in polynomial time (by checking the stable model [12, Table 4]). Hardness: use acyclic propositional PLPs to encode Bayesian networks [11].

---

[1] IMPORTANT NOTE: Due to space restrictions; we only present proof sketches; the reader can find the complete proofs at http://sites.poli.usp.br/p/fabio.cozman/ Publications/Article/cozman-maua-ecsqaru2017.pdf.

To prove membership for stratified PLPs, note that deciding whether a given interpretation is a stable model of a stratified logic program can be reduced to an instance of cautious reasoning. Hence, we can "guess" an interpretation consistent with evidence, then decide whether it is a stable model using a $\mathsf{P^{NP}}$ oracle [12, Table 5]. To obtain Σ_2^P-hardness, use an encoding employed by Eiter et al. [12]. Suppose we have formula $\phi = \exists \mathbf{X} : \neg \exists \mathbf{Y} : \varphi(\mathbf{X}, \mathbf{Y})$, where $\varphi(\mathbf{X}, \mathbf{Y})$ is a propositional formula in 3CNF with sets $\mathbf{X}$ and $\mathbf{Y}$ of propositional variables. Deciding satisfiability of such formulas is a Σ_2^P-complete problem [27]. Introduce $0.5 :: \mathsf{x}$. with predicate x for each x in $\mathbf{X}$. A clause c in φ contains $k \in \{0, \dots, 3\}$ propositional variables from $\mathbf{Y}$. Introduce a predicate c of arity k, and for each predicate c introduce a set of rules: For each one of the 2^k groundings $\mathbf{y}'$ of the logical variables $\mathbf{Y}'$ in c, if $\mathbf{y}'$ satisfies c (for all assignments of $\mathbf{X}$), introduce a fact $\mathsf{c}(\mathbf{y}')$.; if $\mathbf{y}'$ does not satisfy c (for some assignment of $\mathbf{X}$), introduce $3 - k$ rules of the form $\mathsf{c}(\mathbf{y}') : -[\mathbf{not}] \mathsf{x}$., where $\mathbf{not}$ appears in the rule depending on whether x is preceded by negation or not in the clause c. The formula φ is encoded by the rule $\mathsf{cnf} : -\mathsf{c}_1, \mathsf{c}_2, \dots$. (conjunction extends over all clauses).

Then the MPE with evidence $\{\neg \mathsf{cnf}\}$ and threshold $\gamma = 0$ decides whether ϕ is satisfiable.   □

As definite programs are stratified, they are already covered by previous results. However, it makes sense to assume that any query with respect to such a program will also be *positive* in the sense that it only contains non-negated literals. Even then we have the same complexity as stratified programs:

Theorem 2. *Assume all queries are positive. The inferential complexity of definite propositional* PLP*s is* PP*-complete, and of definite* PLP*s is* $\mathsf{PP^{NP}}$*-complete. The MPE complexity of definite propositional* PLP*s is* NP*-complete, and of definite* PLP*s is* Σ_2^P*-complete.*

Proof. Membership follows from results for stratified PLPs [7]. To show hardness for propositional programs, consider a 3CNF formula φ over variables $x_1, \dots, x_n$, and obtain a new monotone formula $\tilde{\varphi}$ by replacing every literal $\neg x_i$ by a fresh variable y_i. Now the formula φ has M satisfying ssigments iff the formula $\tilde{\varphi} \wedge (\bigwedge_i x_i \vee y_i) \vee \bigvee_i (x_i \wedge y_i)$ has $M + 2^{2n} - 3^n$ satisfying assignments [16, Proposition 4]. The latter formula is monotone (i.e., contains no negated variables), so we can encode it as a definite program using probabilistic facts $0.5 :: \mathsf{x}$. to represent each logical variable, c_i to represent clauses, and cnf to represent the value of the formula. To decide whether the number of solutions of φ exceeds M, verify whether $\mathbb{P}(\mathsf{cnf}) > (2^{2n} - 3^n + M)/2^{2n}$. To decide if there is a solution, decide the MPE with evidence $\{\mathsf{cnf}\}$ and threshold $2^{2n} - 3^n$. The same reasoning applies to definite PLPs, by building existential quantification over part of the variables as in the proof of Theorem 1.   □

Non-stratified programs climb one step up in the polynomial hierarchy:

Theorem 3. *Assume the credal semantics for* PLP*s. The MPE complexity of propositional* PLP*s is* Σ_2^P*-complete, and of* PLP*s is* Σ_3^P*-complete.*

Proof. Adapt the proof of Theorem 1, as follows. For the propositional case, membership obtains as cautious reasoning is coNP-complete [12, Table 2], and Σ_2^P-hardness obtains by encoding a formula $\phi = \exists \mathbf{X} : \forall \mathbf{Y} : \varphi(\mathbf{X}, \mathbf{Y})$, where $\varphi(\mathbf{X}, \mathbf{Y})$ is a propositional formula in 3DNF. Encode each $x \in \mathbf{X}$ as before, and introduce y_i and ny_i for each y_i in $\mathbf{Y}$, together with rules $y_i : -\mathsf{not}\ ny_i.$ and $ny_i : -\mathsf{not}\ y_i.$. Then encode φ by rules $\mathsf{dnf} := d_j$ (where each d_j is a conjunct of φ); the MPE of this program with evidence $\{\mathsf{dnf}\}$ and threshold 0 decides whether ϕ is satisfiable (the "inner" universal quantifier is "produced" by going over all stable models when doing cautious reasoning). For the arity-bounded case, membership obtains as cautious reasoning is Π_2^P-complete [12, Table 5], and Σ_3^P-hardness obtains by a combination of strategies used in the proof of Theorem 1 and in the proof for propositional PLPs. That is, encode a formula $\phi = \exists \mathbf{X} : \forall \mathbf{Y} : \exists \mathbf{Z} : \varphi(\mathbf{X}, \mathbf{Y}, \mathbf{Z})$, where φ is in CNF. $\square$

Now consider inferential complexity under the credal semantics, for PDLPs. The credal semantics of a PDLP is a credal set that dominates an infinite monotone Choquet capacity [6]. This result is important because it implies that $\mathbb{P}(\mathcal{M}) = \sum_{\theta \in \Theta : \Gamma(\theta) \subseteq \mathcal{M}} \mathbb{P}(\theta)$ and $\overline{\mathbb{P}}(\mathcal{M}) = \sum_{\theta \in \Theta : \Gamma(\theta) \cap \mathcal{M} \neq \emptyset} \mathbb{P}(\theta)$, where Θ is the set of total choices and Γ maps a total choice to the set of resulting stable models. Also, we have that $\mathbb{P}(A|B) = \mathbb{P}(A \cap B) / (\mathbb{P}(A \cap B) + \overline{\mathbb{P}}(A^c \cap B))$ whenever $\mathbb{P}(A \cap B) + \overline{\mathbb{P}}(A^c \cap B) > 0$; otherwise, either $\mathbb{P}(A|B) = 1$ when $\mathbb{P}(A \cap B) + \overline{\mathbb{P}}(A^c \cap B) = 0$ and $\overline{\mathbb{P}}(A \cap B) > 0$, or $\mathbb{P}(A|B)$ is undefined. Similarly, $\overline{\mathbb{P}}(A|B) = \overline{\mathbb{P}}(A \cap B) / (\overline{\mathbb{P}}(A \cap B) + \underline{\mathbb{P}}(A^c \cap B))$, when $\overline{\mathbb{P}}(A \cap B) + \underline{\mathbb{P}}(A^c \cap B) > 0$, with similar special cases. Using these results we see that computing lower and upper probabilities can be reduced to going through the total choices and running brave/cautious inference for each total choice [6].

Theorem 4. *Assume the credal semantics for PDLPs. The inferential complexity of propositional PDLPs is* $\mathrm{PP}^{\Sigma_2^P}$*-complete, and of PDLPs is* $\mathrm{PP}^{\Sigma_3^P}$*-complete.*

Proof. To prove membership for propositional PDLPs, note that once a total choice is guessed, the cost of checking whether a set of literals holds under cautious reasoning is in Π_2^P [12, Table 2]. To obtain $\mathrm{PP}^{\Sigma_2^P}$-hardness, consider a formula $\phi(\mathbf{X}) = \forall \mathbf{Y} : \neg \forall \mathbf{Z} : \varphi(\mathbf{X}, \mathbf{Y}, \mathbf{Z})$, where $\varphi(\mathbf{X}, \mathbf{Y}, \mathbf{Z})$ is a propositional formula in 3DNF with conjuncts d_j and sets of propositional variables $\mathbf{X}$, $\mathbf{Y}$, and $\mathbf{Z}$. Deciding whether the number of truth assignments to $\mathbf{X}$ that satisfy the formula is strictly larger than an integer M is a $\mathrm{PP}^{\Sigma_2^P}$-complete problem [31, Theorem 7]. To emulate counting, introduce a predicate x_i for each propositional variable x_i in $\mathbf{X}$, associated with a probabilistic fact $0.5 :: x_i.$. And to encode $\phi(\mathbf{X})$, we combine the use of stable sets as in the proof of Theorem 3 with an adapted version of a proof by Eiter and Gottlob [13] on disjunctive programming. We focus on the latter construction here. Introduce predicates z_i and nz_i for each propositional variable z_i in $\mathbf{Z}$, and auxiliary predicate w, together with the rules $z_i \vee nz_i.$, $z_i : -w.$, and $nz_i : -w.$ for each z_i in $\mathbf{Z}$, plus the rule $w : -L_j^1, L_j^2, L_j^3.$ for each conjunct d_j, where L_j^r is obtained from L, the rth literal of d_j, as follows: (1) if $L = z_i$, then $L_j^r = z_i$; (2) if $L = \neg z_i$, then $L_j^r = nz_i$; (3) if $L = x_i$, then

$L_j^r = \mathsf{x}_i$; (4) if $L = \neg x_i$, then $L_j^r = \mathbf{not}\ \mathsf{x}_i$; (5) if $L = y_i$, then $L_j^r = \mathsf{y}_i$; (6) if $L = \neg y_i$, then $L_j^r = \mathbf{not}\ \mathsf{y}_i$. Finally, introduce nw : $-\mathbf{not}\ \mathsf{w}$.. Now reason as follows. To decide whether $\mathbb{P}(\mathsf{nw} = \mathsf{true}) > \gamma$, we must go through all possible total choices; each one of them has probability 2^{-n} where n is the length of $\mathbf{X}$. For each total choice, we must run cautious inference; this is done by going through all stable models, and verifying whether nw is true in all of them. For each truth assignment of $\mathbf{Y}$, the program has a stable model where w is true iff for all truth assignments of $\mathbf{Z}$ we have that φ holds [13, Theorem 3.2]. Hence for fixed $\mathbf{y}$, resulting stable models have nw as true iff $\forall \mathbf{Z} : \varphi(\mathbf{x}, \mathbf{y}, \mathbf{Z})$ is false ($\mathbf{x}$ is fixed by the selected total choice). Thus if we take $\gamma = M/2^{-n}$, we obtain that $\mathbb{P}(\mathsf{nw} = \mathsf{true}) > \gamma$ decides whether $\phi(\mathbf{X})$ has a number of satisfying truth assignments of $\mathbf{X}$ that is strictly larger than M.

To prove membership for PDLPs, note that once a total choice is guessed, the cost of checking whether a set of literals holds under cautious reasoning is in Π_3^P [12, Table 5]. To obtain $\mathsf{PP}^{\Sigma_3^P}$-hardness, we use a combination of strategies used in the proof of Theorem 1 and in the proof for propositional PLPs (previous paragraph) to encode a formula $\phi(\mathbf{X}) = \forall \mathbf{Y} : \neg \forall \mathbf{Z} : \exists \mathbf{V} : \varphi(\mathbf{V}, \mathbf{X}, \mathbf{Y}, \mathbf{Z})$. $\square$

Theorem 5. *Assume the credal semantics for PDLPs. The MPE complexity of propositional PDLPs is Σ_3^P-complete, and of PDLPs is Σ_4^P-complete.*

Proof. Membership and hardness are proved by adapting arguments in the proofs of Theorems 3 and 4, using the complexity of cautious reasoning [12, Tables 2 and 5], and encodings of DNF and CNF formulas as before. $\square$

Theorem 6. *Assume the well-founded semantics for PLPs. The MPE complexity of propositional PLPs is NP-complete, and of PLPs is Σ_2^P-complete.*

Proof. Hardness follows from Theorem 1 (in both cases). Membership: use the argument in the proof of Theorem 1, employing complexity of logical inference of propositional case [9] and the bounded-arity case [8] as appropriate. $\square$

Theorem 7. *Assume the credal semantics both for PLPs and for PDLPs. The MAP complexity of propositional PLPs is $\mathsf{NP}^{\mathsf{PP}}$-hard, and of PDLPs is in $\mathsf{NP}^{\mathsf{PP}}$.*

Proof. Hardness: a propositional PLP can encode a Bayesian network with binary variables, and MAP in such networks is $\mathsf{NP}^{\mathsf{PP}}$-complete [11]. To show membership, reason in two steps: one can solve MAP by first guessing literals for the MAP-predicates that are not fixed by evidence, and then running inference in an $\mathsf{PP}^{\Sigma_3^P}$ oracle. That is, the decision problem is in $\mathsf{NP}^{\mathsf{PP}^{\Sigma_3^P}}$. Now resort to a theorem by Toda and Watanabe [28] that shows that, for any k, $\mathsf{PP}^{\Sigma_k^P}$ collapses to $\mathsf{PP}^{\mathsf{PP}}$, to note that the decision problem of interest is in $\mathsf{NP}^{\mathsf{PP}}$. $\square$

Theorem 8. *Assume the credal semantics for PLPs and positive queries. The MAP complexity of definite propositional PLPs is $\mathsf{NP}^{\mathsf{PP}}$-hard, and of PLPs is in $\mathsf{NP}^{\mathsf{PP}}$.*

Proof. Membership: Theorem 7. Hardness: MAP for Bayesian networks described without negation is already $\mathsf{NP}^{\mathsf{PP}}$-hard [20, Theorem 5]. $\square$

5   Conclusion

As conveyed by Table 1, we have presented several novel results concerning the complexity of probabilistic logic programming, both analyzing conditional probabilities (inferences) and explanations (MPE and MAP)—we note that previous work has not examined these latter problems in the context of probabilistic logic programming. These computations go up several layers within the counting hierarchy, reaching some interesting complexity classes that are rarely visited. In particular, MAP is *always* $\mathsf{NP}^{\mathsf{PP}}$-complete, a rather interesting result.

A future step is to obtain the complexity of relational programs without bounds on arity (exponential complexity is sure to appear), and perhaps the complexity of programs with functions (with suitable restrictions to guarantee decidability). The complexity of other constructs, such as aggregates, should also be explored. Future work should look at "query" complexity; that is, the complexity of computing inferences when the program is fixed and the query varies—this is akin to data complexity as studied in database theory [7].

Acknowledgements. The first author is partially supported by CNPq, grant 308433/2014-9. The second author received support from the São Paulo Research Foundation (FAPESP), grant 2016/01055-1. The work reported in this paper was partially funded by FAPESP grant #2015/21880-4 (project Proverbs).

References

1. Apt, K.R., Bezem, M.: Acyclic programs. New Gener. Comput. **9**, 335–363 (1991)
2. Augustin, T., Coolen, F.P.A., de Cooman, G., Troffaes, M.C.M.: Introduction to Imprecise Probabilities. Wiley, USA (2014)
3. Baral, C., Gelfond, M., Rushton, N.: Probabilistic reasoning with answer sets. Theor. Pract. Logic Program. **9**(1), 57–144 (2009)
4. Baral, C., Subrahmanian, V.: Dualities between alternative semantics for logic programming and nonmonotonic reasoning. J. Autom. Reason. **10**(3), 399–420 (1993)
5. Ceylan, Í.Í, Lukasiewicz, T., Peñaloza, R.: Complexity results for probabilistic Datalog$^{\pm}$. In: European Conference on Artificial Intelligence, pp. 1414–1422 (2016)
6. Cozman, F.G., Mauá, D.D.: The structure and complexity of credal semantics. In: Workshop on Probabilistic Logic Programming, pp. 3–14 (2016)
7. Cozman, F.G., Mauá, D.D.: Probabilistic graphical models specified by probabilistic logic programs: semantics and complexity. In: Conference on Probabilistic Graphical Models – JMLR Proceedings, vol. 52, pp. 110–121 (2016)
8. Cozman, F.G., Mauá. D.D.: The well-founded semantics of cyclic probabilistic logic programs: meaning and complexity. In: Encontro Nacional de Inteligência Artificial e Computacional, pp. 1–12 (2016)
9. Dantsin, E., Eiter, T., Voronkov, A.: Complexity and expressive power of logic programming. ACM Comput. Surv. **33**(3), 374–425 (2001)
10. Darwiche, A.: Modeling and Reasoning with Bayesian Networks, Cambridge (2009)
11. Polpo de Campos, C., Cozman, F.G.: The inferential complexity of Bayesian and credal networks. In: IJCAI, pp. 1313–1318 (2005)
12. Eiter, T., Faber, W., Fink, M., Woltran, S.: Complexity results for answer set programming with bounded predicate arities and implications. Ann. Math. Artif. Intell. **5**, 123–165 (2007)

13. Eiter, T., Gottlob, G.: On the computational cost of disjunctive logic programming: propositional case. Ann. Math. Artif. Intell. **15**, 289–323 (1995)
14. Eiter, T., Ianni, G., Krennwallner, T.: Answer set programming: a primer. In: Tessaris, S., Franconi, E., Eiter, T., Gutierrez, C., Handschuh, S., Rousset, M.-C., Schmidt, R.A. (eds.) Reasoning Web 2009. LNCS, vol. 5689, pp. 40–110. Springer, Heidelberg (2009). doi:10.1007/978-3-642-03754-2_2
15. Fierens, D., Van den Broeck, G., Renkens, J., Shrerionov, D., Gutmann, B., Janssens, G., de Raedt, L.: Inference and learning in probabilistic logic programs using weighted Boolean formulas. Theor. Pract. Logic Program. **15**(3), 358–401 (2014)
16. Goldsmith, J., Hagen, M., Mundhenk, M.: Complexity of DNF minimization and isomorphism testing for monotone formulas. Inf. Comput. **206**(6), 760–775 (2008)
17. Hadjichristodoulou, S., Warren, D.S.: Probabilistic logic programming with well-founded negation. In: International Symposium on Multiple-Valued Logic, pp. 232–237 (2012)
18. Lukasiewicz, T.: Probabilistic description logic programs. In: Conference on Symbolic and Quantitative Approaches to Reasoning with Uncertainty, pp. 737–749 (2005)
19. Lukasiewicz, T.: Probabilistic description logic programs. Int. J. Approx. Reason. **45**(2), 288–307 (2007)
20. Mauá, D.D., Polpo de Campos, C., Cozman, F.G.: The complexity of MAP inference in Bayesian networks specified through logical languages. In: International Joint Conference on Artificial Intelligence (IJCAI), pp. 889–895 (2015)
21. Michels, S., Hommersom, A., Lucas, P.J.F., Velikova, M.: A new probabilistic constraint logic programming language based on a generalised distribution semantics. Artif. Intell. J. **228**, 1–44 (2015)
22. Papadimitriou, C.H.: Computational Complexity. Addison-Wesley, Longman (1994)
23. Poole, D.: Probabilistic Horn abduction and Bayesian networks. Artif. Intell. **64**, 81–129 (1993)
24. Poole, D.: The independent choice logic and beyond. In: Raedt, L., Frasconi, P., Kersting, K., Muggleton, S. (eds.) Probabilistic Inductive Logic Programming. LNCS, vol. 4911, pp. 222–243. Springer, Heidelberg (2008). doi:10.1007/978-3-540-78652-8_8
25. Sato, T.: A statistical learning method for logic programs with distribution semantics. In: International Conference on Logic Programming, pp. 715–729 (1995)
26. Sato, T., Kameya, Y., Zhou, N.-F.: Generative modeling with failure in PRISM. In: International Joint Conference on Artificial Intelligence, pp. 847–852 (2005)
27. Stockmeyer, L.J.: The polynomial-time hierarchy. Theor. Comput. Sci. **3**(1), 1–22 (1976)
28. Toda, S., Watanabe, O.: Polynomial-time 1-Turing reductions from #PH to #P. Theor. Comput. Sci. **100**, 205–221 (1992)
29. Tóran, J.: Complexity classes defined by counting quantifiers. J. ACM **38**(3), 753–774 (1991)
30. van Gelder, A., Ross, J.A., Schlipf, J.S.: The well-founded semantics for general logic programs. J. Assoc. Comput. Mach. **38**(3), 620–650 (1991)
31. Wagner, K.W.: The complexity of combinatorial problems with succinct input representation. Acta Informatica **23**, 325–356 (1986)

Count Queries in Probabilistic Spatio-Temporal Knowledge Bases with Capacity Constraints

John Grant[1], Cristian Molinaro[2]([⊠]), and Francesco Parisi[2]

[1] University of Maryland, College Park, USA
grant@cs.umd.edu
[2] DIMES Department, Università della Calabria, Rende, Italy
{cmolinaro,fparisi}@dimes.unical.it

Abstract. The problem of managing spatio-temporal data arises in many applications, such as location-based services, environment monitoring, geographic information system, and many others. In real life, this kind of data is often uncertain. The SPOT framework has been proposed for the representation and processing of probabilistic spatio-temporal data where probability is represented as an interval because the exact value is unknown.

In this paper, we enhance the SPOT framework with *capacity constraints*, which allow users to better model many real-world scenarios. The resulting formalization is called PST knowledge base. We study the computational complexity of consistency checking, a central problem in this setting. Specifically, we show that the problem is NP-complete and also identify tractable cases. We then consider a relevant kind of queries to reason on PST knowledge bases, namely *count queries*, which ask for how many objects are in a region at a certain time point. We investigate the computational complexity of answering count queries, and show cases for which consistency checking can be exploited for query answering.

1 Introduction

Tracking moving objects is fundamental for all those applications that provide location-based and context-aware services, such as emergency call-out assistance and live traffic reports [2,14]. Such innovative services are becoming so widely diffused that MarketsandMarkets forecasts that the location-based services market will grow from \$15.04 billion in 2016 to \$77.84 billion in 2021 [16].

An important aspect of systems providing location-based and context-aware services is that they need to manage spatial and temporal data together. For this reason, researchers have investigated the representation and processing of spatio-temporal data, both in AI [5,8,31,32] and databases [1,26]. However, in many cases the location of objects is uncertain: such cases can be handled by using probabilities [25,30]. Sometimes the probabilities themselves are not known exactly. Indeed, the position of an object at a given time is estimated by means of location estimation methods such as proximity (where the location of an object is derived from its vicinity to antennas), fingerprinting (where radio signal strength

© Springer International Publishing AG 2017
A. Antonucci et al. (Eds.): ECSQARU 2017, LNAI 10369, pp. 459–469, 2017.
DOI: 10.1007/978-3-319-61581-3_41

measurements produced by a moving object are matched against a radio map built before the system is working), and dead reckoning (where the position of an object is derived from the last known position, assuming that the direction of motion and either the speed or the travelled distance are known) [2,14]. However, since location estimation methods have limited accuracy and precision, what can be asserted is that an object is at a given position at a given time with a probability belonging to an interval. The SPOT (Spatial PrObabilistic Temporal) framework was introduced in [24] to provide a declarative language for the representation and processing of probabilistic spatio-temporal data with probabilities that are not known exactly.

The SPOT framework allows statements of the form "object id is/was/will be inside region r at time t with probability in the interval $[\ell, u]$". This allows the representation of information concerning moving objects in several domains. For instance, a cell phone provider is interested in knowing which cell phones will be in the range of some towers at a given time and with what probability [4]. A transportation company is interested in predicting the vehicles that will be on a given road at a given time (and with what probability) in order to avoid congestion [13]. A retailer is interested in knowing the positions of people moving in a shopping mall in order to offer customized coupons [15].

Past work on the SPOT framework included a formal syntax and semantics as well as checking for consistency. Additional research focused on the efficient processing of selection queries [20,22], database updates [10], and aggregate count queries [9].

In particular, count queries ask how many objects are in a certain region at a given time. Answering this kind of query is useful in several applications. The cell phone provider, the transportation company, and the retailer mentioned earlier are mostly interested in knowing the number of cell phones, vehicles, and shoppers, respectively, in the regions of interest. Moreover, counting how many people or vehicles are in a given region is important during natural disasters or extreme weather in order to arrange evacuation operations and help injured people. As a recent example, during Hurricane Matthew a massive sheltering operation was launched with thousands of people seeking refuge in evacuation shelters. A system providing information on safe places to ride out the dangerous storm has to deal with (i) uncertainty on the actual positions of people; and (ii) the maximum capacity of the roads that can be used to reach shelters, the capacity of the shelters and other critical locations such as first aid stations. Analogously, for the applications mentioned earlier, exploiting knowledge on the capacities of the regions that can be occupied by objects would help interpret available data.

Although the SPOT framework can manage the uncertainty on the positions of moving objects, it does not allow us to represent knowledge on the capacities of regions.

In this paper we make the following contributions. We extend the SPOT framework to include capacity constraints enabling us to express lower- and/or upper-bounds on the number of objects that can be in a certain region. In particular, we provide a formal syntax and semantics for probabilistic spatio-temporal (PST) knowledgebases (KBs) consisting of atomic statements, such as those representable in the SPOT framework, and capacity constraints. We investigate the computational complexity of the problem of deciding whether a PST KB is consistent and, after showing that the problem is NP-complete in general, we identify restricted classes of PST KBs for which the problem is tractable. In fact, deciding consistency is fundamental before answering queries. We introduce the formal semantics of count queries in PST KBs, investigate the computational complexity of answering such kind of queries, and show how checking consistency can be exploited for query answering.

Due to space constraints, proofs are omitted and will be provided in an extended version of the paper.

2   The PST Framework

This section introduces the syntax and semantics of PST KBs, augmenting the framework with capacity constraints.

2.1   Syntax

We assume the existence of a finite set ID of *object ids* and a finite set *Space* of *spatial points*. A non-empty subset of *Space* is called a *region*. We also assume an arbitrarily large but fixed size window of time $T = [0, 1, \ldots, tmax]$. Time point $tmax$ can be as large as a developer needs, and the user may choose the granularity of time according to her needs.

A *spatio-temporal atom* (*st-atom*, for short) is an expression of the form $loc(id, r, t)$, where $id \in ID$, $\emptyset \subsetneq r \subseteq Space$, and $t \in T$. The intuitive meaning of $loc(id, r, t)$ is that object id is/was/will be inside region r at time t.

Definition 1 (PST atom). *A PST atom is an st-atom $loc(id, r, t)$ annotated with a probability interval $[\ell, u] \subseteq [0, 1]$ (with ℓ and u rational numbers), and denoted as $loc(id, r, t)[\ell, u]$.*

Intuitively, the PST atom $loc(id, r, t)[\ell, u]$ says that object id is/was/will be inside region r at time t with probability in the interval $[\ell, u]$. Hence, PST atoms can represent information about the past and the present, but also information about the future, such as when data are obtained from methods for predicting the destination of moving objects [12,17,29], or from querying predictive databases [3,21].

In our framework, *Space* is an arbitrary set of points and a region is any non-empty subset of *Space*. For convenience, we use rectangular regions in our running example where each point of *Space* is represented as an ordered pair (x, y) with integer coordinates. We identify a region r with its bottom-left (x_1, y_1) and top-right (x_2, y_2) endpoints.

Example 1. Figure 1(a) shows a map for our running example concerning a delivery company that has trucks going to various addresses in a city. Here,

$Space = \{(x,y) \mid x,y \in \mathbb{N} \text{ and } 0 \leq x,y \leq 8\}$. The map consists of the company warehouse (coloured in orange), streets (coloured in grey), a lake (coloured in lightblue), and a botanic park (coloured in green). Eight regions are shown, each of them represented by a rectangle containing the region's points and with the region's name in the top-left corner of the rectangle. For instance r_1 consists of the points $(0,7),(1,7),(0,8),(1,8)$. The bottom-left (x_1,y_1) and top-right (x_2,y_2) endpoints of each region are reported in Fig. 1(b).

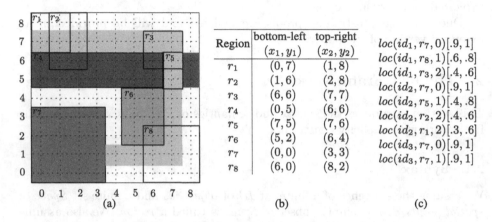

Region	bottom-left (x_1, y_1)	top-right (x_2, y_2)
r_1	$(0,7)$	$(1,8)$
r_2	$(1,6)$	$(2,8)$
r_3	$(6,6)$	$(7,7)$
r_4	$(0,5)$	$(6,6)$
r_5	$(7,5)$	$(7,6)$
r_6	$(5,2)$	$(6,4)$
r_7	$(0,0)$	$(3,3)$
r_8	$(6,0)$	$(8,2)$

$loc(id_1, r_7, 0)[.9, 1]$
$loc(id_1, r_8, 1)[.6, .8]$
$loc(id_1, r_3, 2)[.4, .6]$
$loc(id_2, r_7, 0)[.9, 1]$
$loc(id_2, r_5, 1)[.4, .8]$
$loc(id_2, r_2, 2)[.4, .6]$
$loc(id_2, r_1, 2)[.3, .6]$
$loc(id_3, r_7, 0)[.9, 1]$
$loc(id_3, r_7, 1)[.9, 1]$

(a)   (b)   (c)

Fig. 1. (a) Running example's map (best viewed in color); (b) regions' bottom-left and top-right endpoints; (c) PST atoms. (Color figure online)

Some regions denote specific places: r_4 represents a part of the lake; r_5 represents a bridge linking the street on the bottom to that on the top; r_6 represents the botanic park; and r_7 represents the company warehouse. Other regions are meant to represent areas where trucks were detected by sensors (e.g., GPS).

The information of the positions of the trucks is represented using the set of PST atoms in Fig. 1(c). For instance, the PST atom $loc(id_1, r_7, 0)[.9, 1]$ says that the truck id_1 was in region r_7 (i.e., the company warehouse) at time 0 with probability in the interval $[.9, 1]$ (the high-accuracy sensors used inside the company warehouse entail a narrow probability interval with upper bound equal to 1). Atom $loc(id_1, r_3, 2)[.4, .6]$ says that id_1 was recognized in r_3 by a roadside sensor at the later time 2 with probability in $[.4, .6]$.   □

Although PST atoms express much useful information, they cannot express additional knowledge such as constraints on how many objects are allowed in a region, which can be stated with *capacity constraints*, introduced below. Indeed, in many real-life applications, it is useful to express a lower- and/or upper-bound on the number of objects that can be in a certain region. For instance, the number of trucks that can be in a company warehouse, or on a bridge, is clearly bounded by some constant. Also, at some time points, as for instance during non-working hours, all the trucks of the company must be in the warehouse.

Definition 2 (Capacity constraint). *A capacity constraint is an expression of the form capacity(r, k_1, k_2, t), where r is a region, k_1 and k_2 are two integers such that $0 \leq k_1 \leq k_2 \leq |ID|$, and t is a time point in T.*

For convenience, we write a constraint *capacity*(r, k, k, t) simply as *capacity*(r, k, t).

Example 2. In our running example, the constraint *"there cannot be more than one truck on the bridge (region r_5) at any time between 0 and 2"* can be expressed by the capacity constraints $\kappa_{1,t} = capacity(r_5, 0, 1, t)$ with $t \in [0, 2]$. Similarly, $\kappa_{2,t} = capacity(r_7, 1, 3, t)$, with $t \in [0, 1]$, says that *"the number of trucks in the company warehouse is between* 1 *and* 3 *at any time between* 0 *and* 1*".* Herein, lower bound 1 can be derived for instance from the fact that one truck was damaged in that period. Moreover, $\kappa_{3,t} = capacity(r_4, 0, 0, t)$ and $\kappa_{4,t} = capacity(r_6, 0, 0, t)$, with $t \in [0, 2]$, say that no truck can be in the lake or the botanic park at any time point (we are assuming $tmax = 2$).   □

Definition 3 (PST knowledge base). *A PST knowledge base is a pair $\langle A, C \rangle$, where A is a finite set of PST atoms and C is a finite set of capacity constraints.*

Example 3. In our running example, $ID = \{id_1, id_2, id_3\}$, $T = [0, 2]$, *Space* is the set of points (x, y) such that $0 \leq x \leq 8$ and $0 \leq y \leq 8$, and PST KB $\mathcal{K}_{ex}$ is the pair $\langle A_{ex}, C_{ex} \rangle$, where A_{ex} is the set consisting of the PST atoms in Fig. 1(c), and C_{ex} is the set of capacity constraints defined in Example 2.   □

Throughout the paper we use $\mathcal{K}$ to denote an arbitrary PST knowledge base.

2.2 Semantics

The set of all object ids appearing in $\mathcal{K}$ is denoted by $ID(\mathcal{K})$. The semantics of a PST KB is defined through the concept of *worlds*.

Definition 4 (World). *A world w for $\mathcal{K}$ is a function $w : ID(\mathcal{K}) \times T \to Space$.*

Basically, a world w specifies a trajectory for each $id \in ID(\mathcal{K})$. That is, for each $id \in ID(\mathcal{K})$, w says where object id was/is/will be in *Space* at each time point $t \in T$. This means that an object can be in only one location at a time. However, a location may contain multiple objects. We use $\mathcal{W}(\mathcal{K})$ to denote the set of all worlds for $\mathcal{K}$.

Example 4. World w_1 describing the positions of id_1, id_2 and id_3 for time points in $[0, 2]$ is such that $w_1(id_1, 0) = (1, 1)$, $w_1(id_1, 1) = (7, 2)$, $w_1(id_1, 2) = (7, 6)$, $w_1(id_2, 0) = (2, 1)$, $w_1(id_2, 1) = (7, 5)$, $w_1(id_2, 2) = (1, 7)$, $w_1(id_3, 0) = (1, 2)$, $w_1(id_3, 1) = (1, 2)$, $w_1(id_3, 2) = (6, 1)$.   □

Definition 5 (Satisfaction). *A world w satisfies an st-atom $a = loc(id, r, t)$, denoted $w \models a$, iff $w(id, t) \in r$. Moreover, w satisfies a capacity constraint $\kappa = capacity(r, k_1, k_2, t)$, denoted $w \models \kappa$, iff $k_1 \leq |\{id \in ID(\mathcal{K}) \mid w(id, t) \in r\}| \leq k_2$.*

Example 5. World w_1 of Example 4 satisfies the st-atom $loc(id_1, r_7, 0)$, as $w_1(id_1, 0) = (1, 1)$ belongs to region r_7 (see Fig. 1(a)). Moreover, for each $t \in [0, 2]$, w_1 satisfies $\kappa_{1,t}$, as $\{id \in ID(\mathcal{K}) \mid w_1(id, 0) \in r_5\} = \emptyset$, $\{id \in ID(\mathcal{K}) \mid w_1(id, 1) \in r_5\} = \{id_2\}$, $\{id \in ID(\mathcal{K}) \mid w_1(id, 2) \in r_5\} = \{id_1\}$, and the cardinalities of these sets are all in the interval $[0, 1]$, as required by $\kappa_{1,t}$.   □

Definition 6 (Interpretation). *An* interpretation I *for* $\mathcal{K}$ *is a probability distribution function (PDF) over* $\mathcal{W}(\mathcal{K})$, *that is,* $I : \mathcal{W}(\mathcal{K}) \rightarrow [0, 1]$ *and* $\sum_{w \in \mathcal{W}(\mathcal{K})} I(w) = 1$.

Intuitively, $I(w)$ is the probability that w describes the actual trajectories of all objects.

Definition 7 (Model). *A* model M *for* $\mathcal{K} = \langle \mathcal{A}, \mathcal{C} \rangle$ *is an interpretation for* $\mathcal{K}$ *such that:*

1. $\forall\ loc(id, r, t)[\ell, u] \in \mathcal{A},\ \left(\sum_{w \mid w \models loc(id, r, t)} M(w) \right) \in [\ell, u];$
2. $\forall\ \kappa \in \mathcal{C},\ \sum_{w \mid w \not\models \kappa} M(w) = 0.$

The first condition in the definition above says that, for each PST atom $loc(id, r, t)[\ell, u] \in \mathcal{A}$, the sum of the probabilities assigned by M to the worlds satisfying the st-atom $loc(id, r, t)$ must belong to the probability interval $[\ell, u]$. The second condition says that M must assign probability 0 to every world not satisfying all constraints $\kappa \in \mathcal{C}$.

Example 6. Let w_1 be the world introduced in Example 4. Let w_2 be the same as w_1 except that $w_2(id_1, 2) = (5, 7)$ and $w_2(id_2, 2) = (4, 7)$. Let w_3 be the same as w_2 except that $w_3(id_1, 1) = (7, 3)$ and $w_3(id_2, 1) = (7, 4)$. Let interpretation M be such that $M(w_1) = .6$, $M(w_2) = .2$, $M(w_3) = .2$, and $M(w) = 0$ for all other worlds in $\mathcal{W}(\mathcal{K}_{ex})$. It can be checked that M satisfies both conditions of Definition 7 for the PST KB $\mathcal{K}_{ex}$ of our running example. For instance, for $loc(id_1, r_8, 1)[.6, .8] \in \mathcal{A}_{ex}$, we have $\sum_{w \mid w \models loc(id_1, r_8, 1)} M(w) = M(w_1) + M(w_2) = 0.6 + 0.2 \in [.6, .8]$. Also, it is easy to check that w_1, w_2, w_3 satisfy every capacity constraint in $\mathcal{C}_{ex}$. In particular, we have shown in Example 5 that $w_1 \models \kappa_{1,t}$. It is easy to see that $w_2 \models \kappa_{1,t}$ and $w_3 \models \kappa_{1,t}$ as well, as w_2 and w_3 coincide with w_1 for time point 0, and for time point 1 (resp. 2) only id_2 (resp. id_1) is in r_5. Thus, M is a model for $\mathcal{K}_{ex}$.

We say that $\mathcal{K}$ is *consistent* iff there exists a model for it. The set of models for $\mathcal{K}$ will be denoted as $\mathbf{M}(\mathcal{K})$. PST KB $\mathcal{K}_{ex}$ of our running example is consistent, because the interpretation M of Example 6 is a model for it.

3   Consistency Checking

In this section, we study the complexity of deciding whether a PST KB is consistent. Specifically, we show that the problem is NP-complete. As checking the consistency of PST KBs without capacity constraints is in PTIME [24], this shows that capacity constraints increase the complexity. Then, we identify restricted classes of PST KBs for which the problem is in PTIME. These restrictions are fairly strong; however, they are also quite different from one another. Thus, if we have a PST KB that does not satisfy any of these classes, we may still be able to make small modifications, such as deleting some capacity constraints, so that the modified KB is in one of the classes whose consistency we can check efficiently.

Theorem 1. *Deciding whether a* PST *KB* $\mathcal{K}$ *is consistent is NP-complete.*

The first restricted class we consider is when the capacity constraints allow no objects in some regions. For instance, there cannot be trucks in the lake.

Theorem 2. *Let* $\mathcal{K} = \langle \mathcal{A}, \mathcal{C} \rangle$ *be a* PST *KB. If* $\mathcal{C}$ *consists of capacity constraints of the form* $capacity(r, 0, t)$, *then checking whether* $\mathcal{K}$ *is consistent is in PTIME.*

We now propose sound but possibly incomplete ways of checking consistency, that is, techniques that when the answer is 'yes', the KB is consistent, and when the answer is 'no' we cannot conclude anything about the consistency of the KB.

Below we consider the case where the upper bounds of all PST atoms is 1 and where we can divide *Space* into small enough regions so that regions in different capacity constraints are disjoint.

Theorem 3. *Let* $\mathcal{K} = \langle \mathcal{A}, \mathcal{C} \rangle$ *be a* PST *KB that satisfies the following conditions:*

- $\mathcal{A}$ *consists of* PST *atoms of the form* $loc(id, r, t)[\ell, 1]$ *and there are no two distinct* PST *atoms in* $\mathcal{A}$ *for the same object id and time point, and*
- *for every time point* t, *every pair of distinct capacity constraints* $capacity(r, k_1, k_2, t)$ *and* $capacity(r', k_1', k_2', t)$ *in* $\mathcal{C}$ *is such that* $r \cap r' = \emptyset$.

Deciding if there exists a world $w \in \mathcal{W}(\mathcal{K})$ *s.t. (i)* $w \models \mathcal{C}$ *and (ii)* $w(id, t) \in r$ *for every* $loc(id, r, t)[\ell, 1]$ *in* $\mathcal{A}$ *with* $\ell > 0$, *is in PTIME. If such a world exists, then* $\mathcal{K}$ *is consistent.*

A PST KB $\langle \mathcal{A}, \mathcal{C} \rangle$ is called *simple* iff for every time point $t \in T$, there is at most one capacity constraint of the form $capacity(r, k_1, k_2, t)$ in $\mathcal{C}$.

Theorem 4. *Let* $\mathcal{K} = \langle \mathcal{A}, \mathcal{C} \rangle$ *be a simple* PST *KB. If* $\langle \mathcal{A}, \emptyset \rangle$ *is consistent and, for every* $capacity(r, k_1, k_2, t) \in \mathcal{C}$, $[z, Z] \subseteq [k_1, k_2]$, *where*

$$z = \min_{M \in \mathbf{M}(\langle \mathcal{A}, \emptyset \rangle)} |\{id \mid id \in ID \wedge \left(\sum_{w \mid w(id,t) \in r} M(w) \right) = 1\}|,$$

$$Z = \max_{M \in \mathbf{M}(\langle \mathcal{A}, \emptyset \rangle)} |\{id \mid id \in ID \wedge \left(\sum_{w \mid w(id,t) \in r} M(w) \right) \neq 0\}|,$$

then $\mathcal{K}$ *is consistent. Checking consistency under such conditions is in PTIME.*

The reverse implication does not hold. As an example, the PST KB containing only the PST atom $loc(id, Space, 0)[0, 1]$ and the capacity constraint $capacity(\{p\}, 1, 2, 0)$ is consistent (here p is an arbitrary point in *Space*), but $[z, Z] = [0, 1] \not\subseteq [k_1, k_2] = [1, 2]$.

4   Count Queries

In this section, we consider the problem of answering count queries over PST KBs. We first define the syntax and semantics of count queries, and then analyze the complexity of query answering. *Throughout this section* PST *KBs are assumed to be consistent.*

Definition 8 (Count query). *A* count query *is an expression of the form* $Count(q,t)$, *where* $q \subseteq Space$ *and* $t \in T$.

The count query $Count(q,t)$ asks: "How many objects are inside region q at time t?". Before defining the semantics of count queries, we need to introduce the following auxiliary definition, which introduces the probability that exactly i objects are in a region q at a time point t according to a given model M.

Definition 9. *Let* M *be a model for* $\mathcal{K}$. *For* $0 \leq i \leq |ID|$, *the probability of having exactly* i *objects in a region* q *at a time point* t *w.r.t.* M *is as follows:*

$$Prob_M(q,i,t) = \sum\nolimits_{w|w \models capacity(q,i,t)} M(w)$$

Next, we define the ranking answer $Q(\mathcal{K})$ to a count query $Count(q,t)$ as the set of pairs of the form $\langle i, [\ell_i, u_i] \rangle$ (with $0 \leq i \leq |ID|$) where i is the number of objects that may be in the given region q at the given time point t, and ℓ_i and u_i are the minimum and maximum probabilities of having exactly i objects in q at a time point t over all models.

Definition 10. *The ranking answer to a count query* $Q = Count(q,t)$ *w.r.t.* $\mathcal{K}$ *is:*

$$Q(\mathcal{K}) = \{\langle i, [\ell_i, u_i] \rangle \mid 0 \leq i \leq |ID| \wedge \ell_i = \min_{M \in \mathbf{M}(\mathcal{K})} Prob_M(q,i,t) \wedge$$

$$u_i = \max_{M \in \mathbf{M}(\mathcal{K})} Prob_M(q,i,t)\}.$$

Example 7. Continuing our running example, one may be interested in knowing the number of trucks that are at time 2 in the region $q = \{(x,y) \in Space \mid (6 \leq x \leq 8) \wedge (6 \leq y \leq 8)\}$ (this region includes the whole region r_3, a portion of r_5, and some other points). This can be expressed by the count query $Q = Count(q,2)$. With a little effort, the reader can check that $Q(\mathcal{K}) = \{\langle 0, [.4, .6] \rangle, \langle 1, [.4, 1] \rangle, \langle 2, [0, .3] \rangle, \langle 3, [0, .1] \rangle\}$. $\square$

Our definition of ranking answer generalizes that introduced in [9] where PST KBs of the form $\langle \mathcal{A}, \emptyset \rangle$ (using our notation) were considered and the ranking answer was defined by assuming the independence of the events involving the locations of different objects. This does not hold for PST KBs where the presence of capacity constraints entails that the events that different objects occupy a given region are implicitly correlated.

Theorem 5. *Computing* $Q(\mathcal{K})$ *is* $FP^{NP[\log n]}$-*hard.*

Our final result relates portions of the ranking answer to a consistency check, showing that solving some particular instances of the consistency check problem allows us to answer some specific count queries.

Proposition 1. *Let* $Q = Count(q, t)$ *and* $\mathcal{K} = \langle \mathcal{A}, \mathcal{C} \rangle$.

- *If* $\mathcal{K}' = \langle \mathcal{A}, \mathcal{C} \cup \{capacity(q, k_1, k_2, t)\} \rangle$ *is consistent, then* $\ell_i = 0$ *in* $Q(\mathcal{K})$ *for all* i *such that* $i < k_1$ *or* $i > k_2$.
- *If* $\mathcal{K}' = \langle \mathcal{A}, \mathcal{C} \cup \{capacity(Space \setminus q, k_1, k_2, t)\} \rangle$ *is consistent, then* $u_i = 1$ *in* $Q(\mathcal{K})$ *for all* $i \in [|ID| - k_2, |ID| - k_1]$.

5   Summary and Outlook

The SPOT framework is a declarative language suitable in many current applications dealing with uncertain spatio-temporal data. We have enhanced the SPOT formalism with capacity constraints, which enable users to model semantic information commonly arising in practice. Furthermore, we considered count queries, a relevant kind of queries in several real-world scenarios. We have investigated the computational complexity of the consistency checking and query answering problems, showing that they are intractable and proposed tractable approaches for restricted cases.

Different frameworks have been proposed in the literature to handle spatial information [1,5,8,26,30 32], with some of them being able to model uncertainty too. However, to the best of our knowledge, the SPOT framework is the only one that allows probabilities to be uncertain— in different real-world applications, it is often the case that a precise estimation of the single-value probabilities cannot be obtained. Different frameworks to reason about time have been proposed in [7,27,28], but no spatial information is taken into account.

A full logic (including negation, disjunction and quantifiers) for managing SPOT data was proposed in [6]. Grant et al. [11] is a comprehensive survey of the results on the SPOT framework. The original SPOT framework introduced in [24] has been extended with different kinds of constraints. Grant et al. [10] and Parker et al. [23] considered *reachability constraints* on moving objects—e.g., an object in a given location cannot reach another location within a time point. Recently, [18] extended the SPOT framework with a general form of spatio-temporal denial constraints, which allow us to state that some movements are not allowed. However, the extensions previously introduced were not able to express all the capacity constraints we considered.

As directions for future work, we plan to integrate the framework proposed in this paper with the one proposed in [18] into a unified approach that allows for a wide range of constraints to be expressed. We also plan to investigate count queries and other kinds of aggregate queries in the unified framework. Finally, following [19], where the problem of restoring consistency of PST KBs without integrity constraints has been explored, we plan to address the problems of repairing and querying inconsistent PST KBs with constraints.

References

1. Agarwal, P.K., Arge, L., Erickson, J.: Indexing moving points. J. Comput. Syst. Sci. **66**(1), 207–243 (2003)
2. Ahson, S.A., Ilyas, M.: Location-Based Services Handbook: Applications, Technologies, and Security. CRC Press, Hoboken (2010)
3. Akdere, M., Cetintemel, U., Riondato, M., Upfal, E., Zdonik, S.B.: The case for predictive database systems: opportunities and challenges. In: Proceedings of the 5th Biennial Conference on Innovative Data Systems Research (CIDR), pp. 167–174 (2011)
4. Bayir, M.A., Demirbas, M., Eagle, N.: Mobility profiler: a framework for discovering mobility profiles of cell phone users. Pervasive Mob. Comput. **6**(4), 435–454 (2010)
5. Cohn, A.G., Hazarika, S.M.: Qualitative spatial representation and reasoning: an overview. Fundamenta Informaticae **46**(1–2), 1–29 (2001)
6. Doder, D., Grant, J., Ognjanović, Z.: Probabilistic logics for objects located in space and time. J. Logic Comput. **23**(3), 487–515 (2013)
7. Dousson, C., Maigat, P.L.: Chronicle recognition improvement using temporal focusing and hierarchization. In: Proceedings of International Joint Conference on Artificial Intelligence (IJCAI), pp. 324–329 (2007)
8. Gabelaia, D., Kontchakov, R., Kurucz, Á., Wolter, F., Zakharyaschev, M.: Combining spatial and temporal logics: expressiveness vs. complexity. J. Artif. Intell. Res. **23**, 167–243 (2005)
9. Grant, J., Molinaro, C., Parisi, F.: Aggregate count queries in probabilistic spatio-temporal databases. In: Liu, W., Subrahmanian, V.S., Wijsen, J. (eds.) SUM 2013. LNCS (LNAI), vol. 8078, pp. 255–268. Springer, Heidelberg (2013). doi:10.1007/978-3-642-40381-1_20
10. Grant, J., Parisi, F., Parker, A., Subrahmanian, V.S.: An AGM-style belief revision mechanism for probabilistic spatio-temporal logics. Artif. Intell. **174**(1), 72–104 (2010)
11. Grant, J., Parisi, F., Subrahmanian, V.S.: Research in probabilistic spatiotemporal databases: the SPOT framework. In: Ma, Z., Yan, L. (eds.) Advances in Probabilistic Databases for Uncertain Information Management, Studies in Fuzziness and Soft Computing, vol. 304, pp. 1–22. Springer, Heidelberg (2013)
12. Hammel, T., Rogers, T.J., Yetso, B.: Fusing live sensor data into situational multimedia views. In: Proceedings of International Workshop on Multimedia Information Systems, pp. 145–156 (2003)
13. Karbassi, A., Barth, M.: Vehicle route prediction and time of arrival estimation techniques for improved transportation system management. In: Proceedings of the 2013 IEEE Intelligent Vehicles Symposium, pp. 511–516 (2003)
14. Karimi, H.A.: Advanced Location-Based Technologies and Services. CRC Press, Boca Raton (2013)
15. Kurkovsky, S., Harihar, K.: Using ubiquitous computing in interactive mobile marketing. Pers. Ubiquit. Comput. **10**(4), 227–240 (2006)
16. MarketsandMarkets (2016). http://www.marketsandmarkets.com/Market-Reports/location-based-service-market-96994431.html
17. Mittu, R., Ross, R.: Building upon the coalitions agent experiment (CoAX) - integration of multimedia information in GCCS-M using impact. In: Proceedings of International Workshop on Multimedia Information Systems (MIS), pp. 35–44 (2003)

18. Parisi, F., Grant, J.: Knowledge representation in probabilistic spatio-temporal knowledge bases. J. Artif. Intell. Res. (JAIR) **55**, 743–798 (2016)
19. Parisi, F., Grant, J.: On repairing and querying inconsistent probabilistic spatio-temporal databases. Int. J. Approx. Reason. (IJAR) **84**, 41–74 (2017)
20. Parisi, F., Parker, A., Grant, J., Subrahmanian, V.S.: Scaling cautious selection in spatial probabilistic temporal databases. In: Jeansoulin, R., Papini, O., Prade, H., Schockaert, S. (eds.) Methods for Handling Imperfect Spatial Information, Studies in Fuzziness and Soft Computing, vol. 256, pp. 307–340. Springer, Heidelberg (2010)
21. Parisi, F., Sliva, A., Subrahmanian, V.S.: A temporal database forecasting algebra. Int. J. Approx. Reason. **54**(7), 827–860 (2013)
22. Parker, A., Infantes, G., Grant, J., Subrahmanian, V.S.: SPOT databases: efficient consistency checking and optimistic selection in probabilistic spatial databases. IEEE Trans. Knowl. Data Eng. (TKDE) **21**(1), 92–107 (2009)
23. Parker, A., Infantes, G., Subrahmanian, V.S., Grant, J.: An AGM-based belief revision mechanism for probabilistic spatio-temporal logics. In: Proceedings of AAAI Conference on Artificial Intelligence (AAAI), pp. 511–516 (2008)
24. Parker, A., Subrahmanian, V.S., Grant, J.: A logical formulation of probabilistic spatial databases. IEEE Trans. Knowl. Data Eng. (TKDE) **19**(11), 1541–1556 (2007)
25. Parker, A., Yaman, F., Nau, D.S., Subrahmanian, V.S.: Probabilistic go theories. In: Proceedings of International Joint Conference on Artificial Intelligence (IJCAI), pp. 501–506 (2007)
26. Pelanis, M., Saltenis, S., Jensen, C.S.: Indexing the past, present, and anticipated future positions of moving objects. ACM Trans. Database Syst. **31**(1), 255–298 (2006)
27. Saint-Cyr, F.D., Lang, J.: Reasoning about unpredicted change and explicit time. In: Gabbay, D.M., Kruse, R., Nonnengart, A., Ohlbach, H.J. (eds.) ECSQARU/FAPR -1997. LNCS, vol. 1244, pp. 223–236. Springer, Heidelberg (1997). doi:10.1007/BFb0035625
28. de Saint-Cyr, F.D., Lang, J.: Belief extrapolation (or how to reason about observations and unpredicted change). Artif. Intell. **175**(2), 760–790 (2011)
29. Southey, F., Loh, W., Wilkinson, D.F.: Inferring complex agent motions from partial trajectory observations. In: Proceedings of International Joint Conference on Artificial Intelligence (IJCAI), pp. 2631–2637 (2007)
30. Tao, Y., Cheng, R., Xiao, X., Ngai, W.K., Kao, B., Prabhakar, S.: Indexing multi-dimensional uncertain data with arbitrary probability density functions. In: Proceedings of International Conference on Very Large Data Bases (VLDB), pp. 922–933 (2005)
31. Yaman, F., Nau, D.S., Subrahmanian, V.S.: A logic of motion. In: Proceedings of International Conference on Principles of Knowledge Representation and Reasoning (KR), pp. 85–94 (2004)
32. Yaman, F., Nau, D.S., Subrahmanian, V.S.: Going far, logically. In: Proceedings of International Joint Conference on Artificial Intelligence (IJCAI), pp. 615–620 (2005)

RankPL: A Qualitative Probabilistic Programming Language

Tjitze Rienstra[(✉)]

Computer Science and Communication, University of Luxembourg,
Luxembourg City, Luxembourg
`tjitze@gmail.com`

Abstract. In this paper we introduce *RankPL*, a modeling language
that can be thought of as a qualitative variant of a probabilistic programming language with a semantics based on Spohn's ranking theory.
Broadly speaking, RankPL can be used to represent and reason about
processes that exhibit uncertainty expressible by distinguishing "normal"
from "surprising" events. RankPL allows (iterated) revision of rankings
over alternative program states and supports various types of reasoning,
including abduction and causal inference. We present the language, its
denotational semantics, and a number of practical examples. We also
discuss an implementation of RankPL that is available for download.

1 Introduction

Probabilistic programming languages (PPLs) are programming languages
extended with statements to (1) draw values at random from a given probability
distribution, and (2) perform conditioning due to observation. Probabilistic programs yield, instead of a deterministic outcome, a probability distribution over
possible outcomes. PPLs greatly simplify representation of, and reasoning with
rich probabilistic models. Interest in PPLs has increased in recent years, mainly
in the context of Bayesian machine learning. Examples of modern PPLs include
Church, Venture and *Figaro* [4,8,10], while early work goes back to Kozen [6].

Ranking theory is a qualitative abstraction of probability theory in which
events receive discrete degrees of surprise called *ranks* [11]. That is, events are
ranked 0 (not surprising), 1 (surprising), 2 (very surprising), and so on, or ∞ if
impossible. Apart from being computationally simpler, ranking theory permits
meaningful inference without requiring precise probabilities. Still, it provides
analogues to powerful notions known from probability theory, like conditioning
and independence. Ranking theory has been applied in logic-based AI (e.g. belief
revision and non-monotonic reasoning [1,3]) as well as formal epistemology [11].

In this paper we develop a language called *RankPL*. Semantically, the language draws a parallel with probabilistic programming in terms of ranking theory. We start with a minimal imperative programming language (**if-then-else**,
while, etc.) and extend it with statements to (1) draw choices at random from
a given ranking function and (2) perform ranking-theoretic conditioning due to
observation. Analogous to probabilistic programs, a RankPL programs yields,
instead of a deterministic outcome, a ranking function over possible outcomes.

© Springer International Publishing AG 2017
A. Antonucci et al. (Eds.): ECSQARU 2017, LNAI 10369, pp. 470–479, 2017.
DOI: 10.1007/978-3-319-61581-3_42

Broadly speaking, RankPL can be used to represent and reason about processes whose input or behavior exhibits uncertainty expressible by distinguishing normal (rank 0) from surprising (rank > 0) events. Conditioning in RankPL amounts to the (iterated) revision of rankings over alternative program states. This is a form of revision consistent with the well-known AGM and DP postulates for (iterated) revision [1,2]. Various types of reasoning can be modeled, including abduction and causal inference. Like with PPLs, these reasoning tasks can be modeled without having to write inference-specific code.

The overview of this paper is as follows. Section 2 deals with the basics of ranking theory. In Sect. 3 we introduce RankPL and present its syntax and formal semantics. In Sect. 4 we discuss two generalized conditioning schemes (L-conditioning and J-conditioning) and show how they can be implemented in RankPL. All the above will be demonstrated by practical examples. In Sect. 5 we discuss our RankPL implementation. We conclude in Sect. 6.

2 Ranking Theory

Here we present the necessary basics of ranking theory, all of which is due to Spohn [11]. The definition of a *ranking function* presupposes a finite set Ω of possibilities and a boolean algebra $\mathcal{A}$ over subsets of Ω, which we call *events*.

Definition 1. *A ranking function is a function* $\kappa : \Omega \to \mathbb{N} \cup \{\infty\}$ *that associates every possibiltiy with a* rank. κ *is extended to a function over events by defining* $\kappa(\emptyset) = \infty$ *and* $\kappa(A) = min(\{\kappa(w) \mid w \in A\})$ *for each* $A \in \mathcal{A} \backslash \emptyset$. *A ranking function must satisfy* $\kappa(\Omega) = 0$.

As mentioned in the introduction, ranks can be understood as degrees of surprise or, alternatively, as inverse degrees of plausibility. The requirement that $\kappa(\Omega) = 0$ is equivalent to the condition that at least one $w \in \Omega$ receives a rank of 0. We sometimes work with functions $\lambda : \Omega \to \mathbb{N} \cup \{\infty\}$ that violate this condition. The *normalization* of λ is a ranking function denoted by $||\lambda||$ and defined by $||\lambda||(w) = \lambda(w) - \lambda(\Omega)$. Conditional ranks are defined as follows.

Definition 2. *Given a ranking function* κ, *the rank of* A *conditional on* B *(denoted* $\kappa(A \mid B)$ *is defined by*

$$\kappa(A \mid B) = \begin{cases} \kappa(A \cap B) - \kappa(B) & \text{if } \kappa(B) \neq \infty, \\ \infty & \text{otherwise.} \end{cases}$$

We denote by κ_B *the ranking function defined by* $\kappa_B(A) = \kappa(A \mid B)$.

In words, the effect of conditioning on B is that the rank of B is shifted down to zero (keeping the relative ranks of the possibilities in B constant) while the rank of its complement is shifted up to ∞.

How do ranks compare to probabilities? An important difference is that ranks of events do not add up as probabilities do. That is, if A and B are disjoint, then $\kappa(A \cup B) = min(\kappa(A), \kappa(B))$, while $P(A \cup B) = P(A) + P(B)$. This is, however, consistent with the interpretation of ranks as degrees of surprise (i.e., $A \cup B$ is no less surprising than A or B). Furthermore, ranks provide deductively closed beliefs, whereas probabilities do not. More precisely, if we say that A is believed with firmness x (for some $x > 0$) with respect to κ iff $\kappa(\overline{A}) > x$, then if A and B are believed with firmness x then so is $A \cap B$. A similar notion of belief does not exist for probabilities, as is demonstrated by the Lottery paradox [7].

Finally, note that ∞ and 0 in ranking theory can be thought of as playing the role of 0 and 1 in probability, while min, $-$ and $+$ play the role, respectively, of $+$, $\div$ and $\times$. Recall, for example, the definition of conditional probability, and compare it with Definition 2. This correspondence also underlies notions such as (conditional) independence and ranking nets (the ranking-based counterpart of Bayesian networks) that have been defined in terms of rankings [11].

3   RankPL

We start with a brief overview of the features of RankPL. The basis is a minimal imperative language consisting of integer-typed variables, an **if-then-else** statement and a **while-do** construct. We extend it with the two special statements mentioned in the introduction. We call the first one *ranked choice*. It has the form $\{s_1\} \langle e \rangle \{s_2\}$. Intuitively, it states that either s_1 or s_2 is executed, where the former is a normal (rank 0) event and the latter a typically surprising event whose rank is the value of the expression e. Put differently, it represents a draw of a statement to be executed, at random, from a ranking function over two choices. Note that we can set e to zero to represent a draw from two equally likely choices, and that larger sets of choices can be represented through nesting.

The second special statement is called the *observe* statement **observe** b. It states that the condition b is observed to hold. Its semantics corresponds to ranking-theoretic conditioning. To illustrate, consider the program

$$\text{x} := 10; \ \{\text{y} := 1\} \ \langle 1 \rangle \ \{\{\text{y} := 2\} \ \langle 1 \rangle \ \{\text{y} := 3\}\}; \ \text{x} := \text{x} \times \text{y};$$

This program has three possible outcomes: $\text{x} = 10$, $\text{x} = 20$ and $\text{x} = 30$, ranked 0, 1 and 2, respectively. Now suppose we extend the program as follows:

$$\text{x} := 10; \ \{\text{y} := 1\} \ \langle 1 \rangle \ \{\{\text{y} := 2\} \ \langle 1 \rangle \ \{\text{y} := 3\}\}; \textbf{observe} \ \text{y} > 1; \ \text{x} := \text{x} \times \text{y};$$

Here, the observation rules out the event $\text{y} = 1$, and the ranks of the remaining possibilities are shifted down, resulting in two outcomes $\text{x} = 20$ and $\text{x} = 30$, ranked 0 and 1, respectively.

A third special construct is the *rank expression* **rank** b., which evaluates to the rank of the boolean expression b. Its use will be demonstrated later.

3.1   Syntax

We fix a set *Vars* of variables (ranged over by x) and denote by *Val* the set of integers including ∞ (ranged over by n). We use e, b and s to range over the numerical expressions, boolean expressions, and statements. They are defined by the following BNF rules:

$$e := n \mid x \mid \mathbf{rank}\ b \mid (e_1 \Diamond e_2)\ (\text{for } \Diamond \in \{-, +, \times, \div\})$$
$$b := \neg b \mid (b_1 \vee b_2) \mid (e_1 \blacklozenge e_2)\ (\text{for } \blacklozenge \in \{=, <\})$$
$$s := \{s_0; s_1\} \mid x := e \mid \mathbf{if}\ b\ \mathbf{then}\ \{s_1\}\ \mathbf{else}\ \{s_2\} \mid$$
$$\mathbf{while}\ b\ \mathbf{do}\ \{s\} \mid \{s_1\}\ \langle e \rangle\ \{s_2\} \mid \mathbf{observe}\ b \mid \mathbf{skip}$$

We omit parentheses and curly brackets when possible and define $\wedge$ in terms of $\vee$ and $\neg$. We write **if** b **then** $\{s\}$ instead of **if** b **then** $\{s\}$ **else** $\{\mathbf{skip}\}$, and abbreviate statements of the form $\{x := e_1\}\ \langle e \rangle\ \{x := e_2\}$ to $x := e_1\ \langle e \rangle\ e_2$. Note that the **skip** statement does nothing and is added for technical convenience.

3.2   Semantics

The denotational semantics of RankPL defines the meaning of a statement s as a function $D[\![s]\!]$ that maps prior rankings into posterior rankings. The subjects of these rankings are program states represented by *valuations*, i.e., functions that assign values to all variables. The *initial valuation*, denoted by σ_0, sets all variables to 0. The *initial ranking*, denoted by κ_0, assigns 0 to σ_0 and ∞ to others. We denote by $\sigma[x \rightarrow n]$ the valuation equivalent to σ except for assigning n to x.

From now on we associate Ω with the set of valuations and denote the set of rankings over Ω by K. Intuitively, if $\kappa(\sigma)$ is the degree of surprise that σ is the actual valuation *before* executing s, then $D[\![s]\!](\kappa)(\sigma)$ is the degree of surprise that σ is the actual valuation *after* executing s. If we refer to the result of running the *program* s, we refer to the ranking $D[\![s]\!](\kappa_0)$. Because s might not execute successfully, $D[\![s]\!]$ is not a total function over K. There are two issues to deal with. First of all, non-termination of a loop leads to an undefined outcome. Therefore $D[\![s]\!]$ is a partial function whose value $D[\![s]\!](\kappa)$ is defined only if s terminates given κ. Secondly, observe statements may rule out all possibilities. A program whose outcome is empty because of this is said to *fail*. We denote failure with a special ranking κ_∞ that assigns ∞ to all valuations. Since $\kappa_\infty \notin K$, we define the range of $D[\![s]\!]$ by $K^* = K \cup \{\kappa_\infty\}$. Thus, the semantics of a statement s is defined by a partial function $D[\![s]\!]$ from K^* to K^*.

But first, we define the semantics of expressions. A numerical expression is evaluated w.r.t. both a ranking function (to determine values of **rank** expressions) and a valuation (to determine values of variables). Boolean expressions may also contain **rank** expressions and therefore also depend on a ranking function. Given a valuation σ and ranking κ, we denote by $\sigma_\kappa(e)$ the value of the numerical expression e w.r.t. σ and κ, and by $[b]_\kappa$ the set of valuations satisfying the boolean expression b w.r.t. κ. These functions are defined as follows.[1]

$$\sigma_\kappa(n) = n$$
$$\sigma_\kappa(x) = \sigma(x)$$
$$\sigma_\kappa(\mathbf{rank}\ b) = \kappa([b]_\kappa)$$
$$\sigma_\kappa(a_1 \Diamond a_2) = \sigma_\kappa(a_1) \Diamond \sigma_\kappa(a_2)$$

$$[\neg b]_\kappa = \Omega \setminus [b]_\kappa$$
$$[b_1 \vee b_2]_\kappa = [b_1]_\kappa \cup [b_2]_\kappa$$
$$[a_1 \blacklozenge a_2]_\kappa = \{\sigma \in \Omega \mid \sigma_\kappa(a_1) \blacklozenge \sigma_\kappa(a_2)\}$$

---

[1] We omit explicit treatment of undefined operations (i.e. division by zero and some operations involving ∞). They lead to program termination.

Given a boolean expression b we will write $\kappa(b)$ as shorthand for $\kappa([b]_\kappa)$ and κ_b as shorthand for $\kappa_{[b]_\kappa}$. We are now ready to define the semantics of statements. It is captured by seven rules, numbered (**D1**) to (**D7**). The first deals with the **skip** statement, which does nothing and therefore maps to the identity function.

$$D[\![\mathbf{skip}]\!](\kappa) = \kappa. \tag{D1}$$

The meaning of $s_1; s_2$ is the composition of $D[\![s_1]\!]$ and $D[\![s_2]\!]$.

$$D[\![s_1; s_2]\!](\kappa) = D[\![s_2]\!](D[\![s_1]\!](\kappa)) \tag{D2}$$

The rank of a valuation σ after executing an assignment $x := e$ is the minimum of all ranks of valuations that equal σ after assigning the value of e to x.

$$D[\![x := e]\!](\kappa)(\sigma) = \kappa(\{\sigma' \in \Omega \mid \sigma = \sigma'[x \to \sigma'_\kappa(e)]\}) \tag{D3}$$

To execute **if** b **then** $\{s_1\}$ **else** $\{s_2\}$ we first execute s_1 and s_2 conditional on b and $\neg b$, yielding the rankings $D[\![s_1]\!](\kappa_b)$ and $D[\![s_2]\!](\kappa_{\neg b})$. These are adjusted by adding the prior ranks of b and $\neg b$ and combined by taking the minimum of the two. The result is normalized to account for the case where one branch fails.

$$D[\![\mathbf{if}\ e\ \mathbf{then}\ \{s_1\}\ \mathbf{else}\ \{s_2\}]\!](\kappa) = \|\lambda\|,$$
$$\text{where } \lambda(\sigma) = min \begin{pmatrix} D[\![s_1]\!](\kappa_b)(\sigma) + \kappa(b), \\ D[\![s_2]\!](\kappa_{\neg b})(\sigma) + \kappa(\neg b) \end{pmatrix} \tag{D4}$$

Given a prior κ, the rank of a valuation after executing $s_1 \langle e \rangle s_2$ is the minimum of the ranks assigned by $D[\![s_1]\!](\kappa)$ and $D[\![s_2]\!](\kappa)$, where the latter is increased by e. The result is normalized to account for the case where one branch fails.

$$D[\![\{s_1\} \langle e \rangle \{s_2\}]\!](\kappa) = \|\lambda\|, \text{ where } \lambda(\sigma) = min \begin{pmatrix} D[\![s_1]\!](\kappa)(\sigma), \\ D[\![s_2]\!](\kappa)(\sigma) + \sigma_\kappa(e) \end{pmatrix} \tag{D5}$$

The semantics of **observe** b corresponds to conditioning on the set of valuations satisfying b, unless the rank of this set is ∞ or the prior ranking equals κ_∞.

$$D[\![\mathbf{observe}\ b]\!](\kappa) = \begin{cases} \kappa_\infty & \text{if } \kappa = \kappa_\infty \text{ or } \kappa(b) = \infty, \text{ or} \\ \kappa_b & \text{otherwise.} \end{cases} \tag{D6}$$

We define the semantics of **while** b **do** $\{s\}$ as the iterative execution of **if** b **then** $\{s\}$ **else** $\{\mathbf{skip}\}$ until the rank of b is ∞ (the loop terminates normally) or the result is undefined (s does not terminate). If neither of these conditions is ever met (i.e., if the **while** statement loops endlessly) then the result is undefined.

$$D[\![\mathbf{while}\ b\ \mathbf{do}\{s\}]\!](\kappa) = \begin{cases} F_{b,s}^n(\kappa) & \text{for the first } n \text{ s.t. } F_{b,s}^n(\kappa)(b) = \infty, \text{ or} \\ \text{undef.} & \text{if there is no such } n, \end{cases} \tag{D7}$$

where $F_{b,s} : K_\perp \to K_\perp$ is defined by $F_{b,s}(\kappa) = D[\![\mathbf{if}\ b\ \mathbf{then}\ \{s\}\ \mathbf{else}\{\mathbf{skip}\}]\!](\kappa)$.

Some remarks. Firstly, the semantics of RankPL can be thought of as a ranking-based variation of the Kozen's semantics of probabistic programs [6] (i.e., replacing × with + and + with *min*). Secondly, a RankPL implementation does not need to compute complete rankings. Our implementation discussed in Sect. 5 follows a *most-plausible-first* execution strategy: different alternatives are explored in ascending order w.r.t. rank, and higher-ranked alternatives need not be explored if knowing the lowest-ranked outcomes is enough, as is often the case.

Example. Consider the *two-bit full adder* circuit shown in Fig. 1. It contains two *XOR* gates X_1, X_2, two *AND* gates A_1, A_2 and an *OR* gate O_1. The function of this circuit is to generate a binary representation (b_1, b_2) of the number of inputs among a_1, a_2, a_3 that are high. The *circuit diagnosis problem* is about explaining observed incorrect behavior by finding minimal sets of gates that, if faulty, cause this behavior.[2]

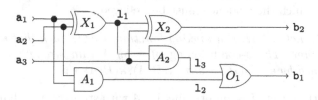

Fig. 1. A two-bit full adder

The listing below shows a RankPL solution. On line 1 we set the constants L and H (representing a *low* and *high* signal); and OK and FAIL (to represent the state of a gate). Line 2 encodes the space of possible inputs (L or H, equally likely). The failure variables fa_1, fa_2, fo_1, fx_2 and fx_2 represent the events of individual gates failing and are set on line 3. Here, we assume that failure is surprising to degree 1. The circuit's logic is encoded on lines 4–8, where the output of a failing gate is arbitrarily set to L or H. Note that ⊕ stands for XOR. At the end we observe ϕ.

```
1 L := 0; H := 1; OK := 0; FAIL := 1;
2 a₁ := (L ⟨0⟩ H); a₂ := (L ⟨0⟩ H); a₃ := (L ⟨0⟩ H);
3 fx₁ := (OK ⟨1⟩ FAIL); fx₂ := (OK ⟨1⟩ FAIL); fa₁ := (OK ⟨1⟩ FAIL);
       fa₂ := (OK ⟨1⟩ FAIL); fo₁ := (OK ⟨1⟩ FAIL);
4 if fx₁ = OK then l₁ := a₁ ⊕ a₂ else l₁ := (L ⟨0⟩ H);
5 if fa₁ = OK then l₂ := a₁ ∧ a₂ else l₂ := (L ⟨0⟩ H);
6 if fa₂ = OK then l₃ := l₁ ∧ a₃ else l₃ := (L ⟨0⟩ H);
7 if fx₂ = OK then b₂ := l₁ ⊕ a₃ else b₂ := (L ⟨0⟩ H);
8 if fo₁ = OK then b₁ := l₃ ∨ l₂ else b₁ := (L ⟨0⟩ H);
9 observe φ;
```

The different valuations of the failure variables produced by this program represent explanations for the observation ϕ, ranked according to plausibility. Suppose we observe that the input (a_1, a_2, a_3) is valued (L, L, H) while the

---

[2] See Halpern [5, Chap. 9] for a similar treatment of this example.

output (b_1, b_2) is incorrectly valued (H,L) instead of (L,H). Thus, we set ϕ to $a_1 = L \wedge a_2 = L \wedge a_3 = H \wedge b_1 = H \wedge b_2 = L$. The program then produces one outcome ranked 0, namely $(\mathtt{fa_1}, \mathtt{fa_2}, \mathtt{fo_1}, \mathtt{fx_2}, \mathtt{fx_2}) = (\mathtt{OK}, \mathtt{OK}, \mathtt{OK}, \mathtt{FAIL}, \mathtt{OK})$. That is, ϕ is most plausibly explained by failure of X_1. Other outcomes are ranked higher than 0 and represent explanations involving more than one faulty gate.

4   Noisy Observation and Iterated Revision

Conditioning by A means that A becomes believed with infinite firmness. This is undesirable if we have to deal with iterated belief change or noisy or untrustworthy observations, since we cannot, after conditioning on A, condition on events inconsistent with A. *J-conditioning* [3] is a more general form of conditioning that addresses this problem. It is parametrized by a rank x that indicates the firmness with which the evidence must be believed.

Definition 3. *Let $A \in \Omega$, κ a ranking function over Ω such that $\kappa(A), \kappa(\overline{A}) < \infty$, and x a rank. The J-conditioning of κ by A with strength x, denoted by $\kappa_{A \to x}$, is defined by $\kappa_{A \to x}(B) = min(\kappa(B|A), \kappa(B|\overline{A}) + x)$.*

In words, the effect of J-conditioning by A with strength x is that A becomes believed with firmness x. This permits iterated belief change, because the rank of $\overline{A}$ is shifted up only by a finite number of ranks and hence can be shifted down again afterwards. Instead of introducing a special statement representing J-conditioning, we show that we can already express it in RankPL, using ranked choice and observation as basic building blocks. Below we write $\kappa_{b \to x}$ as shorthand for $\kappa_{[b]_\kappa \to x}$. Proofs are omitted due to space constraints.

Theorem 1. *Let b be a boolean expression and κ a ranking function s.t. $\kappa(b) < \infty$ and $\kappa(\neg b) < \infty$. We then have $\kappa_{b \to x} = D[\![\{\mathbf{observe}\ b\}\ \langle x \rangle\ \{\mathbf{observe}\ \neg b\}]\!](\kappa)\}$.*

L-conditioning [3] is another kind of generalized conditioning. Here, the parameter x characterizes the 'impact' of the evidence.

Definition 4. *Let $A \in \Omega$, κ a ranking function over Ω such that $\kappa(A), \kappa(\overline{A}) < \infty$, and x a rank. The L-conditioning of κ is denoted by $\kappa_{A \uparrow x}$ and is defined by $\kappa_{A \uparrow x}(B) = min(\kappa(A \cap B) - y, \kappa(\neg A \cap B) + x - y)$, where $y = min(\kappa(A), x)$.*

Thus, L-conditioning by A with strength x means that A improves by x ranks w.r.t. the rank of $\neg A$. Unlike J-conditioning, L-conditioning satisfies two properties that are desirable for modeling noisy observation: *reversibility* $((\kappa_{A \uparrow x})_{\overline{A} \uparrow x} = \kappa)$ and *commutativity* $((\kappa_{A \uparrow x})_{B \uparrow x} = (\kappa_{B \uparrow x})_{A \uparrow x})$ [11]. We can expression L-conditioning in RankPL using ranked choice, observation and the rank expression as basic building blocks. Like before, we write $\kappa_{b \uparrow x}$ to denote $\kappa_{[b]_\kappa \uparrow x}$.

Theorem 2. *Let b be a boolean expression, κ a ranking function over Ω such that $\kappa(b), \kappa(\neg b) < \infty$, and x a rank. We then have:*

$$\kappa_{b\uparrow x} = D \left[\!\!\left[\begin{array}{c} \textbf{if } (\textbf{rank}(b) \leq x) \textbf{ then} \\ \{\textbf{observe } b\} \ \langle x - \textbf{rank}(b) + \textbf{rank}(\neg b)\rangle \ \{\textbf{observe } \neg b\} \\ \textbf{else} \\ \{\textbf{observe } \neg b\} \ \langle \textbf{rank}(b) - x\rangle \ \{\textbf{observe } b\} \end{array} \right]\!\!\right] \quad (\kappa)$$

In what follows we use the statement $\textbf{observe}_L(b,x)$ as shorthand for the statement that represents L-conditioning as defined in Theorem 2.

Example. This example involves both iterated revision and noisy observation. A robot navigates a grid world and has to determine its location using a map and two distance sensors. Figure 2 depicts the map that we use. Gray cells represent walls and other cells are empty (ignoring, for now, the red cells and dots). The sensors (oriented north and south) measure the distance to the nearest wall or obstacle. To complicate matters, the sensor readings might not match the map. For example, the X in Fig. 2 marks an obstacle that affects sensor readings, but as far as the robot knows, this cell is empty.

The listing below shows a RankPL solution. The program takes as input: (1) A map, held by an array map of size m × n, storing the content of each cell (0 = empty, 1 = wall); (2) An array mv (length k) of movements (N/E/S/W for north/east/south/west) at given time points; and (3) Two arrays ns and ss (length k) with distances read by the north and south sensor at given time points. Note that, semantically, arrays are just indexed variables.

```
Input: k: number of steps to simulate
Input: mv: array (size > k) of movements (N/E/S/W)
Input: ns and ss: arrays (size ≥ k) of north and south sensor readings
Input: map: 2D array (size m × n) encoding the map
 1  t := 0;  x := 0 ⟨0⟩ {1 ⟨0⟩ {2 ⟨0⟩ ...m}};  y := 0 ⟨0⟩ {1 ⟨0⟩ {2 ⟨0⟩ ...n}};
 2  while (t < k) do {
 3      if (mv[t] = N) then y := y + 1
 4      else if (mv[t] = S) then y := y − 1
 5      else if (mv[t] = W) then x := x − 1
 6      else if (mv[t] = E) then x := x + 1 else skip;
 7      nd := 0;  while map[x][y + nd + 1] = 0 do nd := nd + 1;
 8      observe_L(1, nd = ns[t]);
 9      sd := 0;  while map[x][y − sd − 1] = 0 do sd := sd + 1;
10      observe_L(1, sd = ss[t]);
11      t := t + 1;
12  }
```

The program works as follows. On line 1 the time point t is set to 0 and the robot's location (x, y) is set to a randomly chosen coordinate (all equally likely) using nested ranked choice statements. Inside the **while** loop, which iterates over t, we first process the movement mv[t] (lines 3–6). We then process (lines 7–8) the north sensor reading, by counting empty cells between the robot and nearest wall, the result of which is observed to equal ns[t]—and likewise for the

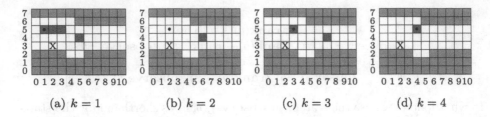

(a) $k = 1$　　　　(b) $k = 2$　　　　(c) $k = 3$　　　　(d) $k = 4$

Fig. 2. Most plausible inferred locations during four iterations (Color figure online)

south sensor (lines 9–10). We use L-conditioning with strength 1 to account for possible incorrect observations. On line 11 we update t.

Suppose we want to simulate a movement from $(0, 5)$ to $(4, 5)$. Thus, we use the inputs mv $= \{E, E, E, E\}$, ns $= \{1, 1, 1, 1\}$, ss $= \{2, 1, 2, 3\}$, while map is set as shown in Fig. 2 (i.e., 1 for every cell containing a wall, 0 for every other cell). Note that the observed distances stored in ns and ss are consistent with the distances observed along this path, where, at $t = 1$, the south sensor reads a distance of 1 instead of 2, due to the obstacle X.

The different values of x, y generated by this program encode possible locations of the robot, ranked according to plausibility. The dots in Fig. 2 show the actual locations, while the red cells represent the inferred most plausible (i.e., rank zero) locations generated by the program. Terminating after $t = 0$ (i.e., setting k to 1) yields four locations, all consistent with the observed distances 1 (north) and 2 (south). If we terminate after $t = 1$, the robot wrongly believes to be at (6, 4), due to having observed the obstacle. However, if we terminate after $t = 3$, the program produces the actual location.

Note that using L-conditioning here is essential. Regular conditioning would cause failure after the third iteration. We could also have used J-conditioning, which gives different rankings of intermediate results.

5   Implementation

A RankPL interpreter written in Java can be found at http://github.com/tjitze/ RankPL. It runs programs written using the syntax described in this paper, or constructed using Java classes that map to this syntax. The latter makes it possible to embed RankPL programs inside Java code and to make it interact with and use classes and methods written Java. The interpreter is faithful to the semantics described in Sect. 3 and implements the *most-plausible-first* execution strategy discussed in Sect. 3.2. All examples discussed in this paper are included, as well as a number of additional examples.

6   Conclusion and Future Work

We have introduced RankPL, a language semantically similar to probabilistic programming, but based on Spohn's ranking theory, and demonstrated its utility using examples involving abduction and iterated revision. We believe that the approach has great potential for applications where PPLs are too demanding due to their computational complexity and dependence on precise probability values. Moreover, we hope that our approach will generate a broader and more practical scope for the topics of ranking theory and belief revision which, in the past, have been studied mostly from purely theoretical perspectives.

A number of aspects were not touched upon and will be addressed in future work. This includes a more fine grained characterization of termination and a discussion of the relationship with nondeterministic programming, which is a special case of RankPL. Furthermore, we have omitted examples to show that RankPL subsumes ranking networks and can be used to reason about causal rules and actions [3]. We also did not contrast our approach with default reasoning formalisms that use ranking theory as a semantic foundation (see, e.g., [9]).

Even though we demonstrated that RankPL is expressive enough to solve fairly complex tasks in a compact manner, it is a very basic language that is best regarded as proof of concept. In principle, the approach can be applied to any programming language, whether object-oriented, functional, or LISP-like. Doing so would make it possible to reason about ranking-based models expressed using, for example, recursion and complex data structures. These features are also supported by PPLs such as Church [4], Venture [8] and Figaro [10].

References

1. Darwiche, A., Pearl, J.: On the logic of iterated belief revision. Artif. Intell. **89**(1–2), 1 29 (1996)
2. Gärdenfors, P., Rott, H.: Belief revision. In: Gabbay, D.M., Hogger, C.J., Robinson, J.A. (eds.) Handbook of Logic in Artificial Intelligence and Logic Programming, vol. 4, pp. 35–132. Oxford University Press, Oxford (1995)
3. Goldszmidt, M., Pearl, J.: Qualitative probabilities for default reasoning, belief revision, and causal modeling. Artif. Intell. **84**(1), 57–112 (1996)
4. Goodman, N.D., Mansinghka, V.K., Roy, D.M., Bonawitz, K., Tenenbaum, J.B.: Church: a language for generative models. In: McAllester, D.A., Myllymäki, P. (eds.) UAI 2008, Proceedings of the 24th Conference in Uncertainty in Artificial Intelligence, Helsinki, Finland, 9–12 July 2008, pp. 220–229. AUAI Press, Helsinki (2008)
5. Halpern, J.Y.: Reasoning about Uncertainty. MIT Press, Cambridge (2005)
6. Kozen, D.: Semantics of probabilistic programs. J. Comput. Syst. Sci. **22**(3), 328–350 (1981)
7. Kyburg Jr., H.E.: Probability and the Logic of Rational Belief (1961)
8. Mansinghka, V.K., Selsam, D., Perov, Y.N.: Venture: a higher-order probabilistic programming platform with programmable inference (2014). CoRR abs/1404.0099
9. Pearl, J.: System Z: a natural ordering of defaults with tractable applications to nonmonotonic reasoning. In: Parikh, R. (ed.) Proceedings of the 3rd Conference on Theoretical Aspects of Reasoning about Knowledge, Pacific Grove, CA, March 1990, pp. 121–135. Morgan Kaufmann, San Mateo (1990)
10. Pfeffer, A.: Figaro: an object-oriented probabilistic programming language. Charles River Analytics Technical Report 137 (2009)
11. Spohn, W.: The Laws of Belief - Ranking Theory and Its Philosophical Applications. Oxford University Press, Oxford (2014)

Generalized Probabilistic Modus Ponens

Giuseppe Sanfilippo[1(✉)], Niki Pfeifer[2], and Angelo Gilio[3]

[1] Department of Mathematics and Computer Science,
University of Palermo, Palermo, Italy
giuseppe.sanfilippo@unipa.it
[2] Munich Center for Mathematical Philosophy, LMU Munich, Munich, Germany
niki.pfeifer@lmu.de
[3] Department SBAI, University of Rome "La Sapienza", Rome, Italy
angelo.gilio@sbai.uniroma1.it

Abstract. Modus ponens (*from A and "if A then C" infer C*) is one of the most basic inference rules. The probabilistic modus ponens allows for managing uncertainty by transmitting assigned uncertainties from the premises to the conclusion (i.e., *from $P(A)$ and $P(C|A)$ infer $P(C)$*). In this paper, we generalize the probabilistic modus ponens by replacing A by the conditional event $A|H$. The resulting inference rule involves iterated conditionals (formalized by conditional random quantities) and propagates previsions from the premises to the conclusion. Interestingly, the propagation rules for the lower and the upper bounds on the conclusion of the generalized probabilistic modus ponens coincide with the respective bounds on the conclusion for the (non-nested) probabilistic modus ponens.

Keywords: Coherence · Conditional random quantities · Conjoined conditionals · Iterated conditionals · Modus ponens · Prevision

1 Introduction

There is a long ongoing interest in combining logic and probability (see, e.g., [7,13,27,31]). In this paper we use coherence-based probability logic, which is characterized by properly managing conditioning events of zero probability and by using arbitrary families of conditional events (see, e.g., [4,9,10,17,21]). Specifically, we investigate and generalize the modus ponens (*from A and "if A then C" infer C*) which is one of the most basic and important inference rules. By instantiating the antecedent of a conditional it allows for detaching the consequent of the conclusion. It is well-known that modus ponens is logically valid (i.e., it is impossible that A and $\overline{A} \vee C$ are true while C is false, where the event $\overline{A} \vee C$ denotes the material conditional as defined in classical logic). It is also well-known that there

G. Sanfilippo—Partially supported by INdAM–GNAMPA Project 2016 Grant U 2016/000391

N. Pfeifer—Supported by his DFG project PF 740/2-2 (within the SPP1516)

A. Gilio—Retired

© Springer International Publishing AG 2017
A. Antonucci et al. (Eds.): ECSQARU 2017, LNAI 10369, pp. 480–490, 2017.
DOI: 10.1007/978-3-319-61581-3_43

are philosophical arguments [2,14] and psychological arguments [15,35] in favor of the hypothesis that a conditional *if A, then C* is best represented by a suitable conditional probability assertion $P(C|A)$ and not by a probability of a corresponding material conditional $P(\overline{A} \vee C)$. Consequently, coherence-based probability logic generalizes the classical modus ponens probabilistically by propagating assigned probabilities from the premises to the conclusion as follows (see, e.g., [36,37,41]):

Probabilistic modus ponens. *From* $P(A) = x$ (probabilistic categorical premise) *and* $P(C|A) = y$ (probabilistic conditional premise) *infer* $xy \leq P(C) \leq xy + 1 - x$ (probabilistic conclusion).

In our paper, $P(C|A)$ is the probability of the *conditional event* $C|A$ (see, e.g., [10–12,21,29,38,39]). The probabilistic modus ponens is *p*-valid, which means that the premise set $\{A, C|A\}$ *p*-entails the conclusion C (i.e., if $P(A) = 1$ and $P(C|A) = 1$, then $P(C) = 1$); moreover, it is probabilistically informative ([19,21,24,37]). However, we recall that some other classical logic inference rules are not *p*-valid, for instance *transitivity* is in general not *p*-valid, i.e., the premise set $\{A|H, C|A\}$ does not *p*-entail the conclusion $C|H$ ([21]). We observe that the lower and upper bounds for the conclusion of the CUT rule coincide with the lower and upper bounds for the conclusion of the modus ponens; indeed, if $P(A|H) = x$ and $P(C|A \wedge H) = y$ it holds that $xy \leq P(C|H) \leq xy + 1 - x$ ([17]). The CUT rule is also *p*-valid and reduces to modus ponens when H is the sure event.

In this paper we generalize the probabilistic modus ponens by replacing the categorical premise (i.e., A) and the antecedent of the conditional premise (i.e., A in "*if A then C*") by the conditional event $A|H$. The resulting inference rule involves the prevision $\mathbb{P}(C|(A|H))$ of the iterated conditional $C|(A|H)$ (formalized by a suitable conditional random quantity, see [20,22,23,26]) and propagates the uncertainty from the premises to the conclusion:

Generalized probabilistic modus ponens. *From* $P(A|H)$ (generalized categorical premise) *and* $\mathbb{P}(C|(A|H))$ (generalized conditional premise) *infer* $P(C)$ (conclusion).

The conditional event $A|H$ is interpreted as a conditional random quantity, with $\mathbb{P}(A|H) = P(A|H)$ (see below). As mentioned above, modus ponens instantiates the antecedent of a conditional and governs the detachment of the consequent of the conclusion. In our generalization, we study the case where the unconditional event A is replaced by the conditional event $(A|H)$ and the conditional event $C|A$ is replaced by the iterated conditional $C|(A|H)$. The following instantiation is an example of our generalization (see, e.g., [16, p. 237]):

$$\overbrace{The\ cup\ breaks\ if\ dropped.}^{A|H}$$

$$\text{If } \overbrace{the\ cup\ breaks\ if\ dropped}^{A|H}, \text{ then } \overbrace{the\ cup\ is\ fragile}^{C}.$$

$$\text{Therefore, } \overbrace{the\ cup\ is\ fragile}^{C}.$$

In what follows, we study how to interpret the uncertainty of the premises and how to propagate the uncertainty from the premises to the conclusion. The outline of the paper is as follows. In Sect. 2 we first recall basic notions and results on coherence and previsions of conditional random quantities. Then, we illustrate the notions of conjunction and of iterated conditional involving conditional events, by recalling some results. In Sect. 3 we prove a generalized decomposition formula for conditional events, with other results on compounded and iterated conditionals. Then, we propagate the previsions from the premises of the generalized probabilistic modus ponens to the conclusion. We observe that this propagation rule coincides with the probability propagation rule for the (non-nested) probabilistic modus ponens (where $H = \Omega$) [37]. Section 4 concludes the paper with an outlook for future work.

2   Preliminary Notions

In this section we recall some basic notions and results on coherence for conditional prevision assessments. In our approach an event A represents an uncertain fact described by a (non-ambiguous) logical entity, where A is two-valued and can be true (T), or false (F). The indicator of A, denoted by the same symbol, is a two-valued numerical quantity which is 1, or 0, according to whether A is true, or false, respectively. The sure event is denoted by Ω and the impossible event is denoted by $\emptyset$. Moreover, we denote by $A \wedge B$, or simply AB, (resp., $A \vee B$) the logical conjunction (resp., logical disjunction). The negation of A is denoted by $\overline{A}$. Given any events A and B, we simply write $A \subseteq B$ to denote that A logically implies B, that is, $A\overline{B}$ is the impossible event $\emptyset$. We recall that n events are logically independent when the number m of constituents, or possible worlds, generated by them is 2^n (in general $m \leq 2^n$). Given two events A and H, with $H \neq \emptyset$, the *conditional event* $A|H$ is defined as a three-valued logical entity which is *true* if AH is true, *false* if $\overline{A}H$ is true, and *void* if H is false. In terms of random quantities, the symbol $A|H$ also denotes the random quantity $AH + p\overline{H} \in \{1, 0, p\}$, where $p = P(A|H)$. Moreover, the negation of $A|H$ is defined as $\overline{A|H} = 1 - A|H = \overline{A}|H$. Notice that, in the approach of de Finetti, the usual terms random variable and expected value are replaced by random quantity and prevision, respectively.

2.1   Coherent Conditional Prevision

We recall the notion of coherent conditional prevision (see, e.g., [5,6,8,10,21,26, 34]). Let $\mathcal{K}$ be an arbitrary family of conditional random quantities with finite sets of possible values. Moreover, let $\mathbb{P}$ be a prevision function defined on $\mathcal{K}$. Consider a finite subfamily $\mathcal{F}_n = \{X_i|H_i, i \in J_n\} \subseteq \mathcal{K}$, where $J_n = \{1, \ldots, n\}$, and the vector $\mathcal{M}_n = (\mu_i, i \in J_n)$, where $\mu_i = \mathbb{P}(X_i|H_i)$ is the assessed prevision for the conditional random quantity $X_i|H_i$. With the pair $(\mathcal{F}_n, \mathcal{M}_n)$ we associate the random gain $G = \sum_{i \in J_n} s_i H_i(X_i - \mu_i)$; moreover, we set $\mathcal{H}_n = H_1 \vee \cdots \vee H_n$ and we denote by $\mathcal{G}_{\mathcal{H}_n}$ the set of values of G restricted to $\mathcal{H}_n$. Then, using the *betting scheme* of de Finetti, we stipulate

Definition 1. The function $\mathbb{P}$ defined on $\mathcal{K}$ is coherent if and only if, $\forall n \geq 1$, $\forall \mathcal{F}_n \subseteq \mathcal{K}$, $\forall s_1, \ldots, s_n \in \mathbb{R}$, it holds that: $min \; \mathcal{G}_{\mathcal{H}_n} \leq 0 \leq max \; \mathcal{G}_{\mathcal{H}_n}$.

Given a family $\mathcal{F}_n = \{X_1|H_1, \ldots, X_n|H_n\}$, for each $i \in J_n$ we denote by $\{x_{i1}, \ldots, x_{ir_i}\}$ the set of possible values of the random quantity X_i when the conditioning event H_i is true (that is, $\{x_{i1}, \ldots, x_{ir_i}\}$ is the set of possible values for the restriction of X_i to H_i); then, for each $i \in J_n$ and $j = 1, \ldots, r_i$, we set $A_{ij} = (X_i = x_{ij})$. Of course, for each $i \in J_n$, the family $\{\overline{H}_i, A_{ij}H_i, \; j = 1, \ldots, r_i\}$ is a partition of the sure event Ω, with $A_{ij}H_i = A_{ij}$, $\bigvee_{j=1}^{r_i} A_{ij} = H_i$. Then, the constituents generated by the family $\mathcal{F}_n$ are (the elements of the partition of Ω) obtained by expanding the expression $\bigwedge_{i \in J_n}(A_{i1} \vee \cdots \vee A_{ir_i} \vee \overline{H}_i)$. We set $C_0 = \overline{H}_1 \cdots \overline{H}_n$ (it may be $C_0 = \emptyset$); moreover, we denote by $C_1, \ldots, C_m$ the constituents contained in $\mathcal{H}_n = H_1 \vee \cdots \vee H_n$. Hence $\bigwedge_{i \in J_n}(A_{i1} \vee \cdots \vee A_{ir_i} \vee \overline{H}_i) = \bigvee_{h=0}^{m} C_h$. With each C_h, $h \in J_m$, we associate a vector $Q_h = (q_{h1}, \ldots, q_{hn})$, where $q_{hi} = x_{ij}$ if $C_h \subseteq A_{ij}$, $j = 1, \ldots, r_i$, while $q_{hi} = \mu_i$ if $C_h \subseteq \overline{H}_i$; C_0 is associated with $Q_0 = \mathcal{M}_n = (\mu_1, \ldots, \mu_n)$. Denoting by $\mathcal{I}_n$ the convex hull of $Q_1, \ldots, Q_m$, the condition $\mathcal{M}_n \in \mathcal{I}_n$ amounts to the existence of a vector $(\lambda_1, \ldots, \lambda_m)$ such that: $\sum_{h \in J_m} \lambda_h Q_h = \mathcal{M}_n$, $\sum_{h \in J_m} \lambda_h = 1$, $\lambda_h \geq 0$, $\forall h$; in other words, $\mathcal{M}_n \in \mathcal{I}_n$ is equivalent to the solvability of the system (Σ_n), associated with $(\mathcal{F}_n, \mathcal{M}_n)$,

$$(\Sigma_n) \quad \sum_{h \in J_m} \lambda_h q_{hi} = \mu_i, \; i \in J_n; \; \sum_{h \in J_m} \lambda_h = 1; \; \lambda_h \geq 0, \; h \subset J_m. \quad (1)$$

Given the assessment $\mathcal{M}_n = (\mu_1, \ldots, \mu_n)$ on $\mathcal{F}_n = \{X_1|H_1, \ldots, X_n|H_n\}$, let S be the set of solutions $\Lambda = (\lambda_1, \ldots, \lambda_m)$ of system (Σ_n) defined in (1). Then, the following characterization theorem for coherent assessments on finite families of conditional events can be proved ([5]).

Theorem 1 [*Characterization of coherence*]. Given a family of n conditional random quantities $\mathcal{F}_n = \{X_1|H_1, \ldots, X_n|H_n\}$, with finite sets of possible values, and a vector $\mathcal{M}_n = (\mu_1, \ldots, \mu_n)$, the conditional prevision assessment $\mathbb{P}(X_1|H_1) = \mu_1, \ldots, \mathbb{P}(X_n|H_n) = \mu_n$ is coherent if and only if, for every subset $J \subseteq J_n$, defining $\mathcal{F}_J = \{X_i|H_i, i \in J\}$, $\mathcal{M}_J = (\mu_i, i \in J)$, the system (Σ_J) associated with the pair $(\mathcal{F}_J, \mathcal{M}_J)$ is solvable.

We point out that the solvability of system (Σ_n) (i.e., the condition $\mathcal{M}_n \in \mathcal{I}_n$) is a necessary (but not sufficient) condition for coherence of $\mathcal{M}_n$ on $\mathcal{F}_n$.

Coherence can be also characterized in terms of proper scoring rules ([6]), which can be related to the notion of entropy in information theory ([30]).

2.2   Conjunction and Iterated Conditional

Given a random quantity X and a non impossible event H, the conditional random quantity $X|H$ can be seen as the random quantity $XH + \mu\overline{H}$, where $\mu = \mathbb{P}(X|H)$ ([22,23,26]). Before introducing the notion of conjunction (see also [28] for related work), we observe that given any pair of conditional events $A|H$ and $B|K$, with $P(A|H) = x$, $P(B|K) = y$, in numerical terms they coincide with the random quantities $AH + x\overline{H}$ and $BK + y\overline{K}$, respectively; then, the symbol $min\{A|H, B|K\}$ denotes the random quantity $min\{AH + x\overline{H}, BK + y\overline{K}\}$.

Definition 2. Given any pair of conditional events $A|H$ and $B|K$, with $P(A|H) = x$, $P(B|K) = y$, we define their conjunction as the conditional random quantity $(A|H) \wedge (B|K) = Z \,|\, (H \vee K)$, where $Z = \min\{A|H, B|K\}$.

Of course, the random quantity Z, as a function of the random quantities $A|H$ and $B|K$, is defined on the set of possible values of the random vector $(A|H, B|K)$. Based on the betting scheme, the compound conditional $(A|H) \wedge (B|K)$ coincides with $1 \cdot AHBK + x \cdot \overline{H}BK + y \cdot AH\overline{K} + z \cdot \overline{H}\,\overline{K}$, where z is the *prevision* of the random quantity $(A|H) \wedge (B|K)$, denoted by $\mathbb{P}[(A|H) \wedge (B|K)]$. Notice that z represents the amount you agree to pay, with the proviso that you will receive the random quantity $(A|H) \wedge (B|K)$. In other words, you agree to pay z with the proviso that you will receive: 1, if both conditional events are true; 0, if at least one of the conditional events is false; z, if both conditional events are void; the probability of that conditional event which is void (i.e., either x or y), otherwise. Notice that this notion of conjunction, with positive probabilities for the conditioning events, has been already proposed in [33].

A well-known notion of conjunction among conditional events, which plays an important role in nonmonotonic reasoning, is the quasi conjunction [1, 3, 24], i.e., the following conditional event: $QC(A|H, B|K) = (AH \vee \overline{H}) \wedge (BK \vee \overline{K}) | (H \vee K)$. In numerical terms, since $AH \vee \overline{H} = AH + \overline{H}$ and $BK \vee \overline{K} = BK + \overline{K}$, the quasi conjunction is the following conditional random quantity: $QC(A|H, B|K) = \min\{AH + \overline{H}, BK + \overline{K}\} | (H \vee K)$. As it is well known, $AH \vee \overline{H}$ is the material conditional associated with the conditional "if H then A". Then, the quasi conjunction is defined by taking the minimum of the material conditionals given $H \vee K$. However, we define the conjunction by taking the minimum of the conditional events given $H \vee K$. Our conjunction is (in general) a conditional random quantity, whereas the quasi conjunction is a conditional event. Moreover, as given in Theorem 2 ([26]), classical results concerning lower and upper bounds for the conjunction of unconditional events still hold for our notion of conjunction, which do not hold for the upper bound of the quasi conjunction ([18, 25]):

Theorem 2. Let A, H, B, K be logically independent events with $H \neq \emptyset, K \neq \emptyset$. Given any coherent assessment $P(A|H) = x$ and $P(B|K) = y$, then the extension $z = \mathbb{P}[(A|H) \wedge (B|K)]$ is coherent if and only if the Fréchet-Hoeffding bounds are satisfied: $max\{x + y - 1, 0\} \leq z \leq min\{x, y\}$.

Now, we recall the notion of iterated conditioning.

Definition 3 (Iterated conditioning). Given any pair of conditional events $A|H$ and $B|K$, the iterated conditional $(B|K)|(A|H)$ is the conditional random quantity $(B|K)|(A|H) = (B|K) \wedge (A|H) + \mu \overline{A}|H$, where $\mu = \mathbb{P}[(B|K)|(A|H)]$.

Notice that, in the context of betting scheme, $\mu = \mathbb{P}[(B|K)|(A|H)]$ represents the amount you agree to pay, with the proviso that you will receive the quantity

$$
(B|K)|(A|H) = \begin{cases} 1, & \text{if } AHBK \text{ true,} \\ 0, & \text{if } AH\overline{B}K \text{ true,} \\ y, & \text{if } AH\overline{K} \text{ true,} \\ \mu, & \text{if } \overline{A}H \text{ true,} \\ x + \mu(1 - x), & \text{if } \overline{H}BK \text{ true,} \\ \mu(1 - x), & \text{if } \overline{H}\overline{B}K \text{ true,} \\ z + \mu(1 - x), & \text{if } \overline{H}\overline{K} \text{ true.} \end{cases} \tag{2}
$$

By applying Definition 3, with $K = \Omega$, the iterated conditional becomes $B|(A|H) = B \wedge (A|H) + \mu \overline{A}|H$, where $\mu = \mathbb{P}[B|(A|H)]$. Moreover, the values of $B|(A|H)$ are 1, 0, μ, $x + \mu(1 - x)$, and $\mu(1 - x)$, associated with the constituents $AHB, AH\overline{B}, \overline{A}H, \overline{H}B$, and $\overline{H}\overline{B}$, respectively.

We recall that if a conditional random quantity is a conditional event, then its prevision ($\mathbb{P}$) coincides with its probability (P). By linearity of prevision ([26]) it holds that $\mathbb{P}[(B|K)|(A|H)] = \mu = \mathbb{P}[(B|K) \wedge (A|H)] + \mu P(\overline{A}|H) = z + \mu(1 - x)$, from which it follows $z = \mu \cdot x$, that is $\mathbb{P}[(B|K) \wedge (A|H)] = \mathbb{P}[(B|K)|(A|H)]P(A|H)$. More formally (see also [23]),

Theorem 3 (Product formula). Given any assessment $x = P(A|H), \mu = \mathbb{P}[(B|K)|(A|H)]$, $z = \mathbb{P}[(B|K) \wedge (A|H)]$, if (x, μ, z) is coherent, then $z = \mu \cdot x$.

Concerning Definition 3, it may seem strange that, in the context of betting scheme, the bet is called off when $A|H$ is false and not when $A|H$ is void. However, we point out that our definition is reasonable because it preserves the product formula.

We recall that coherence requires that $(x, y, \mu, z) \in [0, 1]^4$ (see, e.g., [20]).

Remark 1. We note that in our approach the iterated conditional $(A|H)|K$ does not coincide with the conditional event $A|HK$ ([26]). Moreover, in our approach $(A|H)|K$ is (not a conditional event but) a conditional random quantity. Therefore, the Import-Export Principle ([33]) does not hold in our approach (like in [1,28]). Thus, as shown in [26] we avoid the counter-intuitive consequences related to well-known first triviality result by Lewis ([32]).

Remark 2. Given any random quantity X and any events H, K, with $H \subseteq K$, $H \neq \emptyset$, it holds that (see [26, Sect. 3.3]): $(X|H)|K = X|HK = X|H$. In particular, given any events A, H, K, with $H \neq \emptyset$, it holds that: $(A|H)|(H \vee K) = A|H$.

3   Generalized Modus Ponens

In this section we present a decomposition formula, by also considering a particular case. Then, we give a result on the coherence of a prevision assessment on $\mathcal{F} = \{A|H, C|(A|H), C|(\overline{A}|H)\}$ which will be used to obtain the generalized modus ponens.

Proposition 1. *Let $A|H, B|K$ be two conditional events. Then*

$$B|K = (A|H) \wedge (B|K) + (\overline{A}|H) \wedge (B|K). \tag{3}$$

Proof. Let (x, y, z_1, z_2) be a (coherent) prevision on $(A|H, B|K, (A|H) \wedge (B|K), (\overline{A}|H) \wedge (B|K))$. Of course, coherence requires that $P(\overline{A}|H) = 1 - x$. By Definition 2 it holds that $(A|H) \wedge (B|K) = AHBK + x\overline{H}BK + y\overline{K}AH + z_1\overline{H}\overline{K}$ and $(\overline{A}|H) \wedge (B|K) = \overline{A}HBK + (1 - x)\overline{H}BK + y\overline{K}\overline{A}H + z_2\overline{H}\overline{K}$. Then,

$$(A|H) \wedge (B|K) + (\overline{A}|H) \wedge (B|K) = HBK + \overline{H}BK + y\overline{K}H + z_1\overline{H}\overline{K} + z_2\overline{H}\overline{K}$$
$$= BK + y\overline{K}H + (z_1 + z_2)\overline{H}\overline{K}. \tag{4}$$

Moreover,

$$B|K = BK + y\overline{K} = BK + y\overline{K}H + y\overline{H}\overline{K}. \tag{5}$$

From (4) and (5), when $H \vee K$ is true, it holds that

$$(A|H) \wedge (B|K) + (\overline{A}|H) \wedge (B|K) = BK + y\overline{K}H = B|K.$$

Then, the difference $[(A|H) \wedge (B|K) + (\overline{A}|H) \wedge (B|K)] - B|K$ is zero when $H \vee K$ is true. Thus, $\mathbb{P}[((A|H) \wedge (B|K) + (\overline{A}|H) \wedge (B|K) - B|K)|(H \vee K)] = \mathbb{P}[((A|H) \wedge (B|K))|(H \vee K)] + \mathbb{P}[((\overline{A}|H) \wedge (B|K))|(H \vee K)] - \mathbb{P}[(B|K)|(H \vee K)] = 0$. By Remark 2 it holds that $[(A|H) \wedge (B|K)]|(H \vee K)$, $[(\overline{A}|H) \wedge (B|K)]|(H \vee K)$, and $(B|K)|(H \vee K)$ coincide with $(A|H) \wedge (B|K)$, $(\overline{A}|H) \wedge (B|K)$, and $B|K$, respectively. Then,

$$\mathbb{P}[((A|H) \wedge (B|K) + (\overline{A}|H) \wedge (B|K) - B|K)|(H \vee K)] =$$
$$\mathbb{P}[(A|H) \wedge (B|K)] + \mathbb{P}[(\overline{A}|H) \wedge (B|K)] - P(B|K) = z_1 + z_2 - y = 0.$$

Therefore, $(A|H) \wedge (B|K) + (\overline{A}|H) \wedge (B|K)$ and $B|K$ also coincide when $H \vee K$ is false. Thus, $(A|H) \wedge (B|K) + (\overline{A}|H) \wedge (B|K) = B|K$. $\qquad \blacksquare$

From Proposition 1, by the linearity of prevision, and by the product formula, we obtain $P(B|K) = \mathbb{P}[(B|K)|(A|H)]P(A|H) + \mathbb{P}[(B|K)|(\overline{A}|H)]P(\overline{A}|H)$.

Remark 3. Notice that Proposition 1 also holds when there are some logical relations among the events A, B, H, K, provided that $H \neq \emptyset$ and $K \neq \emptyset$. In particular, if $K = \Omega$ the proof of Proposition 1 is simpler because, by Definition 2,

$$(A|H) \wedge B = AHB + x\overline{H}B, \quad (\overline{A}|H) \wedge B = \overline{A}HB + (1 - x)\overline{H}B,$$

hence

$$(A|H) \wedge B + (\overline{A}|H) \wedge B = HB + \overline{H}B = B. \tag{6}$$

The next result shows that, assuming logical independence, every point $(x, y, z) \in [0, 1]^3$ is a coherent assessment on $\{A|H, C|(A|H), C|(\overline{A}|H)\}$.

Theorem 4. *Let three logically independent events A, C, H be given, with $A \neq \emptyset$, $H \neq \emptyset$. The set of all coherent assessments $\mathcal{M} = (x, y, z)$ on $\mathcal{F} = \{A|H, C|(A|H), C|(\overline{A}|H)\}$ is the unit cube $[0, 1]^3$.*

Due to the lack of space we omit the proof of Theorem 4. A detailed proof of this theorem is available in [40]: by exploiting Theorem 1 the coherence of any assessment $\mathcal{M} = (x, y, z) \in [0, 1]^3$ on $\mathcal{F} = \{A|H, C|(A|H), C|(\overline{A}|H)\}$ is proved by showing that for each subset $J \subseteq \{1, 2, 3\}$ the respective system (Σ_J) is solvable.

We now generalize the *modus ponens* to the case where the first premise A is replaced by the conditional event $A|H$.

Theorem 5. *Given any coherent assessment* (x, y) *on* $\{A|H, C|(A|H)\}$, *with* A, C, H *logically independent, with* $A \neq \emptyset$ *and* $H \neq \emptyset$, *the extension* $z = P(C)$ *is coherent if and only if* $z \in [z', z'']$, *where*

$$z' = xy \quad and \quad z'' = xy + 1 - x. \tag{7}$$

Proof. We recall that (Theorem 4) the assessment (x, y) on $\{A|H, C|(A|H)\}$ is coherent for every $(x, y) \in [0, 1]^2$. From (6), by the linearity of prevision, and by Theorem 3, we obtain

$$z = P(C) = \mathbb{P}[(A|H) \wedge C + (\overline{A}|H) \wedge C] = \mathbb{P}[(A|H) \wedge C] + \mathbb{P}[(\overline{A}|H) \wedge C]$$
$$= P(A|H)\mathbb{P}[C|(A|H)] + P(\overline{A}|H)\mathbb{P}[C|(\overline{A}|H)] = xy + (1 - x)\mathbb{P}[C|(\overline{A}|H)].$$

From Theorem 4, given any coherent assessment (x, y) on $\{A|H, C|(A|H)\}$, the extension $t = \mathbb{P}[C|(\overline{A}|H)]$ on $C|(\overline{A}|H)$ is coherent for every $t \in [0, 1]$. Then, as $z = xy + (1 - x)t$, it follows that $z' = xy$ and $z'' = xy + 1 - x$. □

We notice that Theorem 5 can be rewritten as

Theorem 5'. *Given any logically independent events* A, C, H, *with* $A \neq \emptyset$ *and* $H \neq \emptyset$, *the set* Π *of all coherent assessments* (x, y, z) *on* $\{A|H, C|(A|H), C\}$ *is*

$$\Pi = \{(x, y, z) \in [0, 1]^3 : (x, y) \in [0, 1]^2, z \in [xy, xy + 1 - x]\}. \tag{8}$$

Finally, we observe that the propagation rules for the lower and the upper bounds on the conclusion of the generalized probabilistic modus ponens given in Theorem 5 coincide with the respective bounds on the conclusion for the (non-nested) probabilistic modus ponens.

4   Concluding Remarks

We generalized the probabilistic modus ponens in terms of conditional random quantities in the setting of coherence. Specifically, we replaced the categorical premise A and the antecedent A of the conditional premise $C|A$ by the conditional event $A|H$. We proved a generalized decomposition formula for conditional events and we gave some results on compound conditionals and iterated conditionals. We propagated the previsions from the premises of the generalized probabilistic modus ponens to the conclusion. Interestingly, the lower and the upper bounds on the conclusion of the generalized probabilistic modus ponens coincide with the respective bounds on the conclusion for the (non-nested) probabilistic modus ponens. We obtained our results by also avoiding Lewis' triviality results. Thereby, we aim to provide a unified methodology for investigating compound conditionals under uncertainty. In future work we will study other instantiations to obtain further generalizations of modus ponens, e.g., by also replacing the consequent C of the conditional premise $C|A$ and the conclusion C by a conditional event $C|K$: from $\{A|H, (C|K)|(A|H)\}$ infer $C|K$. Moreover, we will focus on similar generalizations (also involving imprecision) of other argument forms like the probabilistic modus tollens.

Acknowledgments. We thank three anonymous referees for their useful comments and suggestions. We thank *DFG*, *FMSH*, and *Villa Vigoni* for supporting joint meetings at Villa Vigoni where parts of this work originated (Project: "Human Rationality: Probabilistic Points of View").

References

1. Adams, E.W.: The Logic of Conditionals. Reidel, Dordrecht (1975)
2. Adams, E.W.: A Primer of Probability Logic. CSLI, Stanford (1998)
3. Benferhat, S., Dubois, D., Prade, H.: Nonmonotonic reasoning, conditional objects and possibility theory. Artif. Intell. **92**, 259–276 (1997)
4. Biazzo, V., Gilio, A., Lukasiewicz, T., Sanfilippo, G.: Probabilistic logic under coherence: complexity and algorithms. Ann. Math. Artif. Intell. **45**(1–2), 35–81 (2005)
5. Biazzo, V., Gilio, A., Sanfilippo, G.: Generalized coherence and connection property of imprecise conditional previsions. In: Proceedings of IPMU 2008, Malaga, Spain, 22–27 June, pp. 907–914 (2008)
6. Biazzo, V., Gilio, A., Sanfilippo, G.: Coherent conditional previsions and proper scoring rules. In: Greco, S., Bouchon-Meunier, B., Coletti, G., Fedrizzi, M., Matarazzo, B., Yager, R.R. (eds.) IPMU 2012. CCIS, vol. 300, pp. 146–156. Springer, Heidelberg (2012). doi:10.1007/978-3-642-31724-8_16
7. Boole, G.: An Investigation of the Laws of Thought, On Which are Founded the Mathematical Theories of Logic and Probabilities. Walton and Maberly, London (1854)
8. Capotorti, A., Lad, F., Sanfilippo, G.: Reassessing accuracy rates of median decisions. Am. Stat. **61**(2), 132–138 (2007)
9. Capotorti, A., Vantaggi, B.: Locally strong coherence in inference processes. Ann. Math. Artif. Intell. **35**(1), 125–149 (2002)
10. Coletti, G., Scozzafava, R.: Probabilistic Logic in a Coherent Setting. Kluwer, Dordrecht (2002)
11. de Finetti, B.: The logic of probability. Philos. Stud. **77**, 181–190 (1995). Originally published in 1936
12. de Finetti, B.: Foresight: its logical laws, its subjective sources. In: Kyburg, Jr. H., Smokler, H.E. (eds.) Studies in Subjective Probability, pp. 55–118. Robert E. Krieger Publishing Company, Huntington (1980). Originally published in 1937
13. De Morgan, A.: Formal Logic: Or, the Calculus of Inference, Necessary and Probable. Taylor and Walton, London (1847). Reprinted 2002 by Eliborn Classics series
14. Edginton, D.: Indicative conditionals. In: Zalta, E.N. (ed.) The Stanford Encyclopedia of Philosophy (Winter 2014 Edition). https://plato.stanford.edu/archives/win2014/entries/conditionals

15. Evans, J.S.B.T., Handley, S.J., Over, D.E.: Conditionals and conditional probability. J. Exp. Psychol.: Learn. Mem. Cogn. **29**(2), 321–355 (2003)
16. Gibbard, A.: Two recent theories of conditionals. In: Harper, W.L., Stalnaker, R., Pearce, G. (eds.) Ifs, pp. 221–247. Reidel, Dordrecht (1981)
17. Gilio, A.: Probabilistic reasoning under coherence in System P. Ann. Math. Artif. Intell. **34**, 5–34 (2002)
18. Gilio, A.: Generalizing inference rules in a coherence-based probabilistic default reasoning. Int. J. Approximate Reasoning **53**(3), 413–434 (2012)
19. Gilio, A., Over, D.E., Pfeifer, N., Sanfilippo, G.: Centering with conjoined and iterated conditionals under coherence. https://arxiv.org/abs/1701.07785
20. Gilio, A., Over, D.E., Pfeifer, N., Sanfilippo, G.: Centering and compound conditionals under coherence. In: Ferraro, M., et al. (eds.) Soft Methods for Data Science. AISC, vol. 456, pp. 253–260. Springer, Heidelberg (2017)
21. Gilio, A., Pfeifer, N., Sanfilippo, G.: Transitivity in coherence-based probability logic. J. Appl. Logic **14**, 46–64 (2016)
22. Gilio, A., Sanfilippo, G.: Conditional random quantities and iterated conditioning in the setting of coherence. In: van der Gaag, L.C. (ed.) ECSQARU 2013. LNCS (LNAI), vol. 7958, pp. 218–229. Springer, Heidelberg (2013). doi:10.1007/978-3-642-39091-3_19
23. Gilio, A., Sanfilippo, G.: Conjunction, disjunction and iterated conditioning of conditional events. In: Kruse, R., Berthold, M., Moewes, C., Gil, M., Grzegorzewski, P., Hryniewicz, O. (eds.) Synergies of Soft Computing and Statistics for Intelligent Data Analysis. AISC, vol. 190, pp. 399–407. Springer, Berlin (2013)
24. Gilio, A., Sanfilippo, G.: Probabilistic entailment in the setting of coherence: the role of quasi conjunction and inclusion relation. Int. J. Approximate Reasoning **54**(4), 513–525 (2013)
25. Gilio, A., Sanfilippo, G.: Quasi conjunction, quasi disjunction, t-norms and t-conorms: probabilistic aspects. Inf. Sci. **245**, 146–167 (2013)
26. Gilio, A., Sanfilippo, G.: Conditional random quantities and compounds of conditionals. Stud. Logica **102**(4), 709–729 (2014)
27. Hailperin, T.: Probability semantics for quantifier logic. J. Philos. Logic **29**, 207–239 (2000)
28. Kaufmann, S.: Conditionals right and left: probabilities for the whole family. J. Philos. Logic **38**, 1–53 (2009)
29. Lad, F.: Operational Subjective Statistical Methods: A Mathematical, Philosophical, and Historical Introduction. Wiley, New York (1996)
30. Lad, F., Sanfilippo, G., Agró, G.: Extropy: complementary dual of entropy. Stat. Sci. **30**(1), 40–58 (2015)
31. Lambert, J.H.: Neues Organon oder Gedanken über die Erforschung und Bezeichnung des Wahren und dessen Unterscheidung vom Irrthum und Schein. Wendler, Leipzig (1764)
32. Lewis, D.: Probabilities of conditionals and conditional probabilities. Philos. Rev. **85**, 297–315 (1976)
33. McGee, V.: Conditional probabilities and compounds of conditionals. Philos. Rev. **98**(4), 485–541 (1989)
34. Petturiti, D., Vantaggi, B.: Envelopes of conditional probabilities extending a strategy and a prior probability. Int. J. Approximate Reasoning **81**, 160–182 (2017)
35. Pfeifer, N.: The new psychology of reasoning: a mental probability logical perspective. Thinking Reasoning **19**(3–4), 329–345 (2013)
36. Pfeifer, N., Kleiter, G.D.: Inference in conditional probability logic. Kybernetika **42**, 391–404 (2006)

37. Pfeifer, N., Kleiter, G.D.: Framing human inference by coherence based probability logic. J. Appl. Logic **7**(2), 206–217 (2009)
38. Pfeifer, N., Sanfilippo, G.: Probabilistic squares and hexagons of opposition under coherence. Int. J. Approximate Reasoning. doi:10.1016/j.ijar.2017.05.014. (in press)
39. Pfeifer, N., Sanfilippo, G.: Square of opposition under coherence. In: Ferraro, M., et al. (eds.) Soft Methods for Data Science. AISC, vol. 456, pp. 407–414. Springer, Berlin (2017)
40. Sanfilippo, G., Gilio, A., Pfeifer, N.: A Generalized Probabilistic Version of Modus Ponens. http://arxiv.org/abs/1705.00385 (2017)
41. Wagner, C.: Modus Tollens probabilized. Br. J. Philos. Sci. **55**, 747–753 (2004)

A First-Order Logic for Reasoning About Higher-Order Upper and Lower Probabilities

Nenad Savić[1]([✉]), Dragan Doder[2], and Zoran Ognjanović[3]

[1] Institute of Computer Science, University of Bern, Bern, Switzerland
savic@inf.unibe.ch
[2] Faculty of Mechanical Engineering, Belgrade, Serbia
dragan.doder@gmail.com
[3] Mathematical Institute of SASA, Belgrade, Serbia
zorano@mi.sanu.ac.rs

Abstract. We present a first-order probabilistic logic for reasoning about the uncertainty of events modeled by sets of probability measures. In our language, we have formulas that essentially say that "according to agent Ag, for all x, formula $\alpha(x)$ holds with the lower probability at least $\frac{1}{3}$". Also, the language is powerful enough to allow reasoning about higher order upper and lower probabilities. We provide corresponding Kripke-style semantics, axiomatize the logic and prove that the axiomatization is sound and strongly complete (every satisfiable set of formulas is consistent).

Keywords: Probabilistic logic · Uncertainty · Axiomatization · Strong completeness

1 Introduction

Reasoning with uncertainty has gained an important role in computer science, artificial intelligence and cognitive science. These applications require the development of formal models which could capture reasoning through probability [3,4,6–9,11,13,17,19].

We investigate a probabilistic logic approach, considering the situation when there is also uncertainty about probabilities. In this case, the uncertainty is often described using the two boundaries, called *upper probability* and *lower probability* [14,15]. Those probabilities are previously formalized in logics developed in [12, 20,21]. Halpern and Pucella [12] give the following example: a bag contains 100 marbles, 30 of them are red and the remaining 70 are either blue or yellow, but we do not know their exact proportion. Thus, we can assign exact probability 0.3 to the event that a randomly picked ball from the bag is red, while for each possible probability p for picking a blue ball, we know that the remaining probability for yellow one is $0.7 - p$. For the set of possible probability measures obtained in that way, we can assign a pair of functions, the upper and lower probability measure, that assign the supremum and the infimum the probability of an event according to the probability measures in the set.

© Springer International Publishing AG 2017
A. Antonucci et al. (Eds.): ECSQARU 2017, LNAI 10369, pp. 491–500, 2017.
DOI: 10.1007/978-3-319-61581-3_44

We use the papers [12, 20, 21] as a starting point and generalize them in two ways:

- We want to reason not only about lower and upper probabilities an agent assigns to a certain event, but also about her uncertain belief about other agent's imprecise probabilities. Thus, we introduce separate lower and upper probability operators for different agents, and we allow nesting of the operators, similarly as it has been done in [6], in the case of simple probabilities[1]. Suppose that an agent a is planning a visit to the city C based on the weather reports from several sources, and she decides to take an action if probability of rain is at most $\frac{1}{10}$, according to all reports she considers. Since she wishes to go together with b, she should be sure with probability at least $\frac{9}{10}$ that b (who might consult different weather reports) has the same conclusion about possibility of rain. In our language, it can be formalized as

$$U^a_{\leq \frac{1}{10}} Rain(C) \wedge L^a_{\geq \frac{9}{10}} (U^b_{\leq \frac{1}{10}} Rain(C)).$$

- We extend both [6, 12, 20, 21] by allowing reasoning about events expressible in a first-order language. The papers [12, 20] deal with propositional reasoning, while [21] introduces a logic whose syntax allows only Boolean combinations of formulas in which lower and upper probability operators are applied to first order sentences. On the other hand, here we use the most general approach, allowing arbitrary combination of probability operators and quantifiers, so we can express the statement like "according to the agent a, the lower probability of rain in all cities is at least $\frac{1}{3}$" ($L^a_{\geq \frac{1}{3}} \forall x Rain(x)$), but also "There exists a city in which it will surely not rain" $((\exists x) U^a_{=0} Rain(x))$.

Formally, if the uncertainty about probabilities is modeled by a set of probability measures P defined on a given algebra H, then the lower probability measure $P_\star$ and the upper probability measure $P^\star$ are defined by $P_\star(X) = \inf\{\mu(X) \mid \mu \in P\}$ and $P^\star(X) = \sup\{\mu(X) \mid \mu \in P\}$, for every $X \in H$. Those two functions are related by the formula $P_\star(X) = 1 - P^\star(X^c)$.

In this paper, we logically formalize such situations using a generalization of Kripke models – for each agent, every world is equipped with a probabilistic space which consists of the accessible worlds, algebra of subsets, and a set of measures. We denote our logic by $\mathcal{L}_{lu}$.

We propose a sound and strongly complete axiomatization of the logic. Since we use different completion technique than the one used in [6, 12], we did not have to incorporate the arithmetical operations in the language. Instead, we use unary operators for upper and lower probability, following [20]. Since, like the other real-valued probabilistic logics, $\mathcal{L}_{lu}$ is not compact, any finitary axiomatic system would be incomplete [22]. In order to achieve completeness, we use two infinitary rules of inference, with countably many premises and one conclusion.

---

[1] For a discussion on higher-order probabilities we refer the reader to [10].

2   The Logic $\mathcal{L}_{lu}$ – Syntax and Semantics

Let $S = \mathbb{Q} \cap [0,1]$, $Var = \{x, y, z, \ldots\}$ be a denumerable set of variables and let $\Sigma = \{a, b, \ldots\}$ be a finite, non-empty set of agents. The language of the logic $\mathcal{L}_{lu}$ consists of:

- the elements of set Var,
- classical propositional connectives $\neg$ and $\wedge$,
- universal quantifier $\forall$,
- for every integer $k \geq 0$, denumerably many function symbols $F_0^k, F_1^k, \ldots$ of arity k,
- for every integer $k \geq 0$, denumerably many relation symbols $P_0^k, P_1^k, \ldots$ of arity k,
- the list of upper probability operators $U_{\geq s}^a$, for every $s \in S$,
- the list of lower probability operators $L_{\geq s}^a$, for every $s \in S$,
- comma, parentheses.

Functions of arity 0 will be called constants.

Note that we use conjunction and negation as primitive connectives, while $\vee$, $\rightarrow$, $\leftrightarrow$ and $\exists$ are introduced in the usual way. The notions of a term, atomic formula, bound and free variables, sentence and a term free for a variable in formula, can be defined as usual.

Definition 1 (Formula). *The set $For_{\mathcal{L}_{lu}}$ of formulas is the smallest set containing atomic formulas and that is closed under following formation rules: if α, β are formulas, then $L_{\geq s}^a \alpha$, $U_{\geq s}^a \alpha$, $\neg \alpha$, $\alpha \wedge \beta$, $(\forall x)\alpha$ are formulas as well. The formulas from $For_{\mathcal{L}_{lu}}$ will be denoted by $\alpha, \beta, \ldots$*

We use the following abbreviations to introduce other types of inequalities:

- $U_{<s}^a \alpha$ is $\neg U_{\geq s}^a \alpha$, $U_{\leq s}^a \alpha$ is $L_{\geq 1-s}^a \neg \alpha$, $U_{=s}^a \alpha$ is $U_{\leq s}^a \alpha \wedge U_{\geq s}^a \alpha$, $U_{>s}^a \alpha$ is $\neg U_{\leq s}^a \alpha$,
- $L_{<s}^a \alpha$ is $\neg L_{\geq s}^a \alpha$, $L_{\leq s}^{\bar{a}} \alpha$ is $U_{\geq 1-s}^{\bar{a}} \neg \alpha$, $L_{=s}^a \alpha$ is $L_{\leq s}^{\bar{a}} \alpha \wedge L_{\geq s}^{\bar{a}} \alpha$, $L_{>s}^a \alpha$ is $\neg L_{\leq s}^{\bar{a}} \alpha$.

We also denote $\alpha \vee \neg \alpha$ by $\top$, and $\alpha \wedge \neg \alpha$ by $\bot$.

The semantics for the logic $\mathcal{L}_{lu}$ is based on the possible-world approach.

Definition 2 ($\mathcal{L}_{lu}$-structure). *An $\mathcal{L}_{lu}$-structure is a tuple $\mathcal{M} = \langle W, D, I, LUP \rangle$, where:*

- *W is a nonempty set of worlds,*
- *D associates a non-empty domain $D(w)$ with every world $w \in W$,*
- *I associates an interpretation $I(w)$ with every world $w \in W$ such that:*
  - *$I(w)(F_i^k) : D(w)^k \rightarrow D(w)$, for all i and k,*
  - *$I(w)(P_i^k) \subseteq D(w)^k$, for all i and k,*
- *LUP assigns, to every $w \in W$ and every agent $a \in \Sigma$, a space, such that $LUP(a, w) = \langle W(a, w), H(a, w), P(a, w) \rangle$, where:*
  - *$\emptyset \neq W(a, w) \subseteq W$,*

- $H(a, w)$ is an algebra of subsets of $W(a, w)$, i.e. a set of subsets of $W(a, w)$ such that:
    - $W(a, w) \in H(a, w)$,
    - if $A, B \in H(a, w)$, then $W(a, w) \setminus A \in H(a, w)$ and $A \cup B \in H(a, w)$,
- $P(a, w)$ is a set of finitely additive probability measures defined on $H(a, w)$, i.e. for every $\mu(a, w) \in P(a, w)$, $\mu(a, w) : H(a, w) \longrightarrow [0, 1]$ and the following conditions hold:
    - $\mu(a, w)(W(a, w)) = 1$,
    - $\mu(a, w)(A \cup B) = \mu(a, w)(A) + \mu(a, w)(B)$, whenever $A \cap B = \emptyset$.

Definition 3 (Variable valuation). *Let* $\mathcal{M} = \langle W, D, I, LUP \rangle$ *be an* $\mathcal{L}_{lu}$-*structure. A variable valuation* v *assigns to every variable some element of the corresponding domain to every world* $w \in W$, *i.e.* $v(w)(x) \in D(w)$. *For* v, $w \in W$ *and* $d \in D(w)$ *we define* $v_w[d/x]$ *is a valuation same as* v *except that* $v_w[d/x](w)(x) = d$.

Definition 4. *Let* $\mathcal{M} = \langle W, D, I, LUP \rangle$ *be an* $\mathcal{L}_{lu}$-*structure and* t *a term. The value of a term* t, *denoted by* $I(w)(t)_v$ *is defined as follows:*

- *if* t *is a variable* x, *then* $I(w)(x)_v = v(w)(x)$, *and*
- *if* $t = F_i^m(t_1, \ldots, t_m)$, *then*

$$I(w)(t)_v = I(w)(F_i^m)(I(w)(t_1)_v, \ldots, I(w)(t_m)_v).$$

Now we define satisfiability of the formulas from $For_{\mathcal{L}_{lu}}$ in the worlds of $\mathcal{L}_{lu}$-structures.

Definition 5. *The truth value of a formula* α *in a world* $w \in W$ *of a model* $\mathcal{M} = \langle W, D, I, LUP \rangle$ *for a given valuation* v, *denoted by* $I(w)(\alpha)_v$ *is defined as follows:*

- *if* $\alpha = P_i^m(t_1, \ldots, t_m)$, *then* $I(w)(\alpha)_v = true$ *if* $\langle I(w)(t_1)_v, \ldots, I(w)(t_m)_v \rangle \in I(w)(P_i^m)$, *otherwise* $I(w)(\alpha)_v = false$,
- *if* $\alpha = \neg\beta$, *then* $I(w)(\alpha)_v = true$ *if* $I(w)(\beta)_v = false$, *otherwise* $I(w)(\alpha)_v = false$,
- *if* $\alpha = \beta \wedge \gamma$, *then* $I(w)(\alpha)_v = true$ *if* $I(w)(\beta)_v = true$ *and* $I(w)(\gamma)_v = true$,
- *if* $\alpha = U_{\geq s}^a \beta$, *then* $I(w)(\alpha)_v = true$ *if* $P^\star(w, a)\{u \in W(w, a) \mid I(u)(\beta)_v = true\} \geq s$, *otherwise* $I(w)(\alpha)_v = false$,
- *if* $\alpha = L_{\geq s}^a \beta$, *then* $I(w)(\alpha)_v = true$ *if* $P_\star(w, a)\{u \in W(w, a) \mid I(u)(\beta)_v = true\} \geq s$, *otherwise* $I(w)(\alpha)_v = false$,
- *if* $\alpha = (\forall x)\beta$, *then* $I(w)(\alpha)_v = true$ *if for every* $d \in D(w)$, $I(w)(\beta)_{v_w[d/x]} = true$, *otherwise* $I(w)(\alpha)_v = false$.

Recall that $P_\star(w, a)\{u \in W(w, a) \mid I(u)(\beta)_v = true\} = \inf\{\mu(w, a)(\{u \in W(w, a) \mid I(u)(\beta)_v = true\}) \mid \mu(w, a) \in P(w, a)\}$, and $P^\star(w, a)\{u \in W(w, a) \mid I(u)(\beta)_v = true\} = \sup\{\mu(w, a)(\{u \in W(w, a) \mid I(u)(\beta)_v = true\}) \mid \mu(w, a) \in P(w, a)\}$.

Definition 6. *A formula α holds in a world w from a model $\mathcal{M} = \langle W, D, I, LUP \rangle$, denoted by $\mathcal{M}, w \models \alpha$, if for every valuation v, $I(w)(\alpha)_v = true$. If $d \in D(w)$, we will use $\mathcal{M}, w \models \alpha(d)$ to denote that $I(w)(\alpha(x))_{v_w[d/x]} = true$, for every valuation v.*

A sentence α is satisfiable if there is a world w in an $\mathcal{L}_{lu}$-model $\mathcal{M}$ such that $\mathcal{M}, w \models \alpha$. A sentence α is valid if it is satisfied in every world in every $\mathcal{L}_{lu}$-model $\mathcal{M}$. A set of sentences T is satisfiable if there is a world w in an $\mathcal{L}_{lu}$-model $\mathcal{M}$ such that $\mathcal{M}, w \models \alpha$ for every $\alpha \in T$.

We will consider a class of $\mathcal{L}_{lu}$ models that satisfy:

- all the worlds from a model have the same domain, i.e., for all $v, w \in W$, $D(v) = D(w)$,
- for every sentence α, for every agent $a \in \Sigma$ and every world w from a model $\mathcal{M}$, the set $\{u \in W(w, a) \mid I(u)(\alpha)_v = true\}$ of all worlds from $W(w, a)$ that satisfy α is measurable,
- the terms are rigid, i.e., for every model their meanings are the same in all the worlds.

We will use the notation $[\alpha]_w^a$ for the set $\{u \in W(w, a) \mid I(u)(\alpha)_v = true\}$, and also $\mathcal{L}_{lu_{Meas}}$ to denote the class of all fixed domain measurable models with rigid terms.

The following example shows that Compactness theorem does not hold for the logic $\mathcal{L}_{lu}$, i.e. we can construct a set T such that every finite subset of a set T is satisfiable, but T itself is not.

Example 1. *Consider the set of formulas*

$$T = \{\neg U_{=0}^a \alpha\} \cup \{U_{<\frac{1}{n}}^a \alpha \mid n \quad is \quad a \quad positive \quad integer\}.$$

It is clear that every finite subset of T is $\mathcal{L}_{lu_{Meas}}$-satisfiable, but the set T is not.

3 The Axiomatization $Ax_{\mathcal{L}_{lu}}$

In this section we introduce an axiomatic system for the logic $\mathcal{L}_{lu}$. That system will be denoted by $Ax_{\mathcal{L}_{lu}}$. In order to axiomatize upper and lower probabilities, we need to completely characterize them with a small number of properties. There are many complete characterizations in the literature, and the earliest appears to be by Lorentz [16]. We use the characterization result by Anger and Lembcke [1]. It uses the notion of (n, k)-cover.

Definition 7 ((n, k)-cover). *A set A is said to be covered n times by a multiset $\{\{A_1, \ldots, A_m\}\}$ of sets if every element of A appears in at least n sets from $A_1, \ldots, A_m$, i.e., for all $x \in A$, there exists $i_1, \ldots, i_n$ in $\{1, \ldots, m\}$ such that for all $j \leq n$, $x \in A_{i_j}$. An (n, k)-cover of (A, W) is a multiset $\{\{A_1, \ldots, A_m\}\}$ that covers W k times and covers A $n + k$ times.*

Theorem 1 ([1]). *Let W be a set, H an algebra of subsets of W, and f a function $f : H \longrightarrow [0,1]$. There exists a set P of probability measures such that $f = P^*$ iff f satisfies the following three properties:*

(1) $f(\emptyset) = 0$,
(2) $f(W) = 1$,
(3) for all natural numbers m, n, k and all subsets $A_1, \ldots, A_m$ in H, if the multiset $\{\{A_1, \ldots, A_m\}\}$ is an (n,k)-cover of (A, W), then $k + nf(A) \leq \sum_{i=1}^{m} f(A_i)$.

This theorem is also used in the Halpern and Pucella's paper on the logical formalization of upper and lower probabilities [12].

Axiom Schemes

(1) all instances of the classical propositional tautologies
(2) $(\forall x)(\alpha \rightarrow \beta) \rightarrow (\alpha \rightarrow (\forall x)\beta)$, where the variable x does not occur free in α
(3) $(\forall x)\alpha(x) \rightarrow \alpha(t)$, where $\alpha(t)$ is obtained by substitution of all free occurrences of x in the first-order formula $\alpha(x)$ by the term t which is free for x in $\alpha(x)$
(4) $U^a_{\leq 1}\alpha \wedge L^a_{\leq 1}\alpha$
(5) $U^a_{\leq r}\alpha \rightarrow U^a_{<s}\alpha$, $s > r$
(6) $U^a_{<s}\alpha \rightarrow U^a_{\leq s}\alpha$
(7) $(U^a_{\leq r_1}\alpha_1 \wedge \cdots \wedge U^a_{\leq r_m}\alpha_m) \rightarrow U^a_{\leq r}\alpha$, if $\alpha \rightarrow \bigvee_{J \subseteq \{1,\ldots,m\}, |J|=k+n} \bigwedge_{j \in J} \alpha_j$ and $\bigvee_{J \subseteq \{1,\ldots,m\}, |J|=k} \bigwedge_{j \in J} \alpha_j$ are tautologies, where $r = \frac{\sum_{i=1}^{m} r_i - k}{n}$, $n \neq 0$
(8) $\neg(U^a_{\leq r_1}\alpha_1 \wedge \cdots \wedge U^a_{\leq r_m}\alpha_m)$, if $\bigvee_{J \subseteq \{1,\ldots,m\}, |J|=k} \bigwedge_{j \in J} \alpha_j$ is a tautology and $\sum_{i=1}^{m} r_i < k$
(9) $L^a_{=1}(\alpha \rightarrow \beta) \rightarrow (U^a_{\geq s}\alpha \rightarrow U^a_{\geq s}\beta)$

Inference Rules

(1) From α and $\alpha \rightarrow \beta$ infer β
(2) From α infer $(\forall x)\alpha$
(3) From α infer $L^a_{\geq 1}\alpha$
(4) From the set of premises

$$\{\alpha \rightarrow U^a_{\geq s - \frac{1}{k}}\beta \mid k \geq \frac{1}{s}\}$$

  infer $\alpha \rightarrow U^a_{\geq s}\beta$
(5) From the set of premises

$$\{\alpha \rightarrow L^a_{\geq s - \frac{1}{k}}\beta \mid k \geq \frac{1}{s}\}$$

  infer $\alpha \rightarrow L^a_{\geq s}\beta$.

The axioms 7 and 8 together capture the condition (3) from the Theorem 1. Indeed, note that $\{\{A_1, \ldots, A_m\}\}$ covers a set A n times iff

$$A \subseteq \bigcup_{J \subseteq \{1, \ldots, m\}, |J| = n} \bigcap_{j \in J} A_j.$$

Hence, the condition that a formula $\alpha \to \bigvee_{J \subseteq \{1, \ldots, m\}, |J| = k+n} \bigwedge_{j \in J} \alpha_j$ is a tautology gives us that, for every $a \in \Sigma$ and $w \in W$, $[\alpha]_w^a$ is covered $n + k$ times by a multiset $\{\{[\alpha_1]_w^a, \ldots, [\alpha_m]_w^a\}\}$, while the condition that $\bigvee_{J \subseteq \{1, \ldots, m\}, |J| = k} \bigwedge_{j \in J} \alpha_j$ is a tautology ensures that, for every $a \in \Sigma$ and $w \in W$, $W(w, a) = [\top]_w^a$ is covered k times by a multiset $\{\{[\alpha_1]_w^a, \ldots, [\alpha_m]_w^a\}\}$.

Rule 4 and Rule 5 are infinitary rules of inference and intuitively says that if upper/lower probability is arbitrary close to a rational number s then it is at least s.

Definition 8 (Inference relation).

- $\vdash \alpha$ (α is a theorem) iff there is an at most denumerable sequence of formulas $\alpha_1, \alpha_2, \ldots, \alpha$, such that every α_i is an axiom or it is derived from the preceding formulas by an inference rule;
- $T \vdash \alpha$ (α is derivable from T) if there is an at most denumerable sequence of formulas $\alpha_1, \alpha_2, \ldots, \alpha$, such that every α_i is an axiom or a formula from the set T, or it is derived from the preceding formulas by an inference rule, with the exception that Inference Rule 3 can be applied only to the theorems;
- T is consistent if there is at least one formula $\alpha \in For_{\mathcal{L}_{lu}}$ that is not deducible from T, otherwise T is inconsistent;
- T is maximally consistent set if it is consistent and for every $\alpha \in For_{\mathcal{L}_{lu}}$, either $\alpha \in T$ or $\neg \alpha \in T$;
- T is deductively closed if for every $\alpha \in For_{\mathcal{L}_{lu}}$, if $T \vdash \alpha$, then $\alpha \in T$;
- T is saturated if it is maximally consistent and satisfies: if $\neg (\forall x) \alpha(x) \in T$, then for some term t, $\neg \alpha(t) \in T$.

Note that T is inconsistent iff $T \vdash \bot$. Also, it is easy to check that every maximally consistent set is deductively closed.

It is straightforward to prove that our axiomatic system is sound with respect to the class of $\mathcal{L}_{lu_{Meas}}$-models.

4 Completeness

Deduction theorem holds for $Ax_{\mathcal{L}_{lu}}$: if T is a set of formulas and α a sentence, then $T \cup \{\alpha\} \vdash \beta$ iff $T \vdash \alpha \to \beta$. This theorem can be proved using the facts that our infinitary inference rules have implicative form, and that the application of Rule 3 is restricted to theorems only.

Now, we show how to extend an arbitrary consistent set of formulas T to a saturated set of formulas T^*. In the end the canonical model $\mathcal{M}_{Can}$ is constructed and after that, it is proved that for every world w and every formula

α, $\alpha \in w$ iff $w \models \alpha$, so the proof of the completeness theorem is an easy consequence.

Theorem 2 (Lindenbaum's theorem). *Every consistent set of formulas can be extended to a saturated set.*

Sketch of the proof. Consider a consistent set T and let $\alpha_0, \alpha_1, \ldots$ be an enumeration of all formulas from $For_{\mathcal{L}_{lu}}$. A sequence of sets T_i, $i = 0, 1, 2, \ldots$ is defined as follows:

(1) $T_0 = T$,
(2) for every $i \geq 0$,
    (a) if $T_i \cup \{\alpha_i\}$ is consistent, then $T_{i+1} = T_i \cup \{\alpha_i\}$, otherwise
    (b) if α_i is of the form $\beta \rightarrow U_{\geq s}^a \alpha$, then $T_{i+1} = T_i \cup \{\neg\alpha_i, \beta \rightarrow \neg U_{\geq s - \frac{1}{n}}^a \alpha\}$,
        for some positive integer n, so that T_{i+1} is consistent, otherwise
    (c) if α_i is of the form $\beta \rightarrow L_{\geq s}^a \alpha$, then $T_{i+1} = T_i \cup \{\neg\alpha_i, \beta \rightarrow \neg L_{\geq s - \frac{1}{n}}^a \alpha\}$,
        for some positive integer n, so that T_{i+1} is consistent, otherwise
    (d) if the set T_{i+1} is obtained by adding a formula of the form $\neg(\forall x)\beta(x)$
        to the set T_i, then for some $c \in C$ (C is a countably infinite set of new
        constant symbols), $\neg\beta(c)$ is also added to T_{i+1}, so that T_{i+1} is consistent,
        otherwise
    (e) $T_{i+1} = T_i \cup \{\neg\alpha_i\}$.
(3) $T^\star = \bigcup_{i=0}^{\infty} T_i$.

Obviously, the set T_0 is consistent. Natural numbers (n), from the steps 2(b) and 2(c) of the construction exist (this is a direct consequence of the Deduction Theorem), and each T_i is consistent. The maximality of $T^\star$ (either $\alpha \in T$ or $\neg\alpha \in T$) is ensured by the steps (1) and (2) of the above construction. It is clear that $T^\star$ does not contain all the formulas because for a formula $\alpha \in For_{\mathcal{L}_{lu}}$, the set $T^\star$ does not contain both $\alpha = \alpha_i$ and $\neg\alpha = \alpha_j$, since the set $T_{\max\{i,j\}+1}$ is consistent.

It only remains to prove that $T^\star$ is deductively closed. Let $\alpha \in For_{\mathcal{L}_{lu}}$. We will prove by the induction on the length of the inference that if $T^\star \vdash \alpha$, then $\alpha \in T^\star$. Consider the infinitary Rule 5. Let $\alpha_i = \beta \rightarrow L_{\geq s}^a \gamma$ be obtained from the set of premises $\{\alpha_i^k = \beta \rightarrow L_{\geq s_k}^a \gamma \mid s_k \in S\}$. Using the induction hypothesis, we conclude that $\alpha_i^k \in T^\star$, for every k. If $\alpha_i \notin T^\star$, by step (2)(c) of the construction, there must be some l and j such that $\neg(\beta \rightarrow L_{\geq s}^a \gamma)$, $\beta \rightarrow \neg L_{\geq s - \frac{1}{t}}^a \gamma \in T_j$. Hence, we have that for some $j' \geq j$: $\beta \wedge \neg L_{\geq s}^a \gamma \in T_{j'}$; $\beta \in T_{j'}$; $\neg L_{\geq s - \frac{1}{t}}^a \gamma$, $L_{\geq s - \frac{1}{t}}^a \gamma \in T_{j'}$.

Therefore, we have that $T^\star$ is deductively closed set, and $T^\star$ does not contain all the formulas, so it is consistent.

The step (2)(d) of the construction guarantees that $T^\star$ is saturated. $\quad\square$

Now we define a canonical model, using the saturated sets of formulas.

Definition 9 (Canonical model). *A canonical model $\mathcal{M}_{Can} = \langle W, D, I, LUP \rangle$ is a tuple such that:*

- W is the set of all saturated sets of formulas,
- D is the set of all variable-free terms,
- for every $w \in W$, $I(w)$ is an interpretation such that:
  - for every function symbol F_i^m, $I(w)(F_i^m) : D^m \to D$ such that for all variable-free terms $t_1, \ldots, t_m$, $I(w)(F_i^m) : \langle t_1, \ldots, t_m \rangle \mapsto F_i^m(t_1, \ldots, t_m)$,
  - for every relation symbol P_i^m, $I(w)(P_i^m) = \{\langle t_1, \ldots, t_m \rangle \mid P_i^m(t_1, \ldots, t_m) \in w\}$, for all variable-free terms $t_1, \ldots, t_m$,
- for $a \in \Sigma$ and $w \in W$, $LUP(w, a) = \langle W(w, a), H(w, a), P(w, a) \rangle$ is defined:
  - $W(w, a) = W$,
  - $H(w, a) = \{\{u \mid u \in W(w, a), \alpha \in u\} \mid \alpha \in For_{\mathcal{L}_{lu}}\}$,
  - $P(w, a)$ is any set of probability measures such that $P^\star(w, a)(\{u \mid u \in W(w, a), \alpha \in u\}) = \sup\{s \mid U_{\geq s}^a \alpha \in w\}$.

Lemma 1. *For every formula α and every $w \in W$, $\alpha \in w$ iff $w \models \alpha$.*

Theorem 3 (Strong completeness). *Every consistent set of formulas T is $\mathcal{L}_{lu_{Meas}} - satisfiable$.*

Sketch of the proof. Let T be a consistent set of formulas and let $\mathcal{M}_{Can} = \langle W, D, I, LUP \rangle$ be a canonical model. It can be shown that $\mathcal{M}_{Can}$ is a well defined measurable structure. Furthermore, from Lemma 1 we obtain that for every formula α, and every $w \in W$, $w \models \alpha$ iff $\alpha \in w$. Finally, using Theorem 2, we can extend T to a saturated set T^*, and since $T^* \in W$, we obtain $\mathcal{M}_{Can}, T^* \models T$. $\qquad\Box$

5 Conclusion

In this paper we present the proof-theoretical analysis of a logic which allows making statements about upper and lower probabilities of formulas according to some agent. We combine the approaches from [6,12,20] and generalize them to an expressive modal language $\mathcal{L}_{lu}$ which extend first-order logic with the unary operators $U_{\geq r}^a$ and $L_{\geq r}^a$, where r ranges over the unit interval of rational numbers. The corresponding semantics $\mathcal{L}_{lu_{Meas}}$ consists of the measurable Kripke models with a set of finitely additive probability measures attached to each possible world. For a given world of a model, every probability form the corresponding set of probabilities is defined on the same algebra of a chosen sets of worlds. We prove that the proposed axiomatic system $Ax_{\mathcal{L}_{lu}}$ is strongly complete with respect to the class of $\mathcal{L}_{lu_{Meas}}$-models. Since the logic is not compact, the axiomatization contains infinitary rules of inference.

Finally, upper and lower probabilities are just one approach in development of imprecise probability models [2,5,18,23,24]. In the future work, we also wish to logically formalize different approaches to imprecise probabilities.

Acknowledgments. This work was supported by the SNSF project 200021_165549 Justifications and non-classical reasoning, and by the Serbian Ministry of Education and Science through projects ON174026, III44006 and ON174008.

References

1. Anger, B., Lembcke, J.: Infnitely subadditive capacities as upper envelopes of measures. Z. Wahrscheinlichkeitstheorie Verwandte Gebiete **68**, 403–414 (1985)
2. de Cooman, G., Hermans, F.: Imprecise probability trees: bridging two theories of imprecise probability. Artif. Intell. **172**(11), 1400–1427 (2008)
3. Doder, D.: A logic with big-stepped probabilities that can model nonmonotonic reasoning of system P. Pub. Inst. Math. **90**(104), 13–22 (2011)
4. Doder, D., Ognjanović, Z.: Probabilistic logics with independence and confirmation. Stud. Logica (2017). doi:10.1007/s11225-017-9718-z
5. Dubois, D., Prade, H.: Possibility Theory. Plenum Press, New York (1988)
6. Fagin, R., Halpern, J.: Reasoning about knowledge and probability. J. ACM **41**(2), 340–367 (1994)
7. Fagin, R., Halpern, J., Megiddo, N.: A logic for reasoning about probabilities. Inf. Comput. **87**(1–2), 78–128 (1990)
8. Fattorosi-Barnaba, M., Amati, G.: Modal operators with probabilistic interpretations. Stud. Logica **46**(4), 383–393 (1989)
9. Frisch, A., Haddawy, P.: Anytime deduction for probabilistic logic. Artif. Intell. **69**, 93–122 (1994)
10. Gaifman, H., Haddawy, P.: A theory of higher order probabilities. Causation, Chance and Credence, vol. 41, pp. 191–219. Springer, Netherlands (1988)
11. Halpern, J.Y.: An analysis of first-order logics of probability. Artif. Intell. **46**, 311–350 (1990)
12. Halpern, J.Y., Pucella, R.: A logic for reasoning about upper probabilities. J. Artif. Intell. Res. **17**, 57–81 (2002)
13. Heifetz, A., Mongin, P.: Probability logic for type spaces. Games Econ. Behav. **35**, 31–53 (2001)
14. Huber, P.J.: Robust Statistics. Wiley, New York (1981)
15. Kyburg, H.E.: Probability and the Logic of Rational Belief. Wesleyan University Press, Middletown (1961)
16. Lorentz, G.G.: Multiply subadditive functions. Can. J. Math. **4**(4), 455–462 (1952)
17. Meier, M.: An infinitary probability logic for type spaces. Isr. J. Math. **192**(1), 1–58 (2012)
18. Miranda, E.: A survey of the theory of coherent lower previsions. Int. J. Approximate Reasoning **48**(2), 628–658 (2008)
19. Ognjanović, Z., Rašković, M.: Some first-order probability logics. Theor. Comput. Sci. **247**(1–2), 191–212 (2000)
20. Savić, N., Doder, D., Ognjanović, Z.: A logic with upper and lower probability operators. In: Proceedings of the 9th International Symposium on Imprecise Probability: Theories and Applications, pp. 267–276, Pescara, Italy (2015)
21. Savić, N., Doder, D., Ognjanović, Z.: Logics with lower and upper probability operators. Int. J. Approximate Reasoning (2017). doi:10.1016/j.ijar.2017.05.013
22. van der Hoek, W.: Some consideration on the logics $P_F D$. J. Appl. Non-Classical Logics **7**(3), 287–307 (1997)
23. Walley, P.: Statistical Reasoning with Imprecise Probabilities. Chapman and Hall, London (1991)
24. Walley, P.: Towards a unified theory of imprecise probability. Int. J. Approximate Reasoning **24**(2–3), 125–148 (2000)

Author Index

Abassi, Lina 159
Abdelkhalek, Raoua 169
Aguzzoli, Stefano 353
Amgoud, Leila 25
Antonucci, Alessandro 282
Augustin, Thomas 329

Beierle, Christoph 225, 236
Ben Amor, Nahla 295, 306, 435
Ben-Naim, Jonathan 25
Bianchi, Matteo 353
Bolt, Janneke H. 83
Bonesana, Claudio 282
Boudet, Laurence 375
Boudjani, Nadira 36
Boukhris, Imen 159, 169, 201
Bronevich, Andrey G. 271

Campagner, Andrea 423
Ciucci, Davide 398, 423
Coletti, Giulianella 364
Cornez, Laurence 375
Cozman, Fabio G. 93, 449
Creignou, Nadia 387

Dao, Tien-Tuan 135
Destercke, Sébastien 179
Doder, Dragan 491
Dubois, Didier 398

Eichhorn, Christian 236, 257
EL khalfi, Zeineb 306
Elouedi, Zied 169, 201, 212
Espinosa, Bruno 375
Essghaier, Fatma 295

Fargier, Hélène 295, 306
Flaminio, Tommaso 246
Fragnito, Giulia 318

Gabarro, Joaquim 318
Gerla, Brunella 353
Gilio, Angelo 480

Godo, Lluis 246
Gouaich, Abdelkader 36
Grabisch, Michel 340
Grant, John 459
Guyet, Thomas 135

Haddad, Maroua 435
Helal, Nathalie 190
Hoang, Tuan Nha 135
Hosni, Hykel 246
Hunter, Anthony 46

Jansen, Christoph 329
Järvisalo, Matti 57
Jensen, Frank 115
Jeppesen, Nicolaj Søndberg 115

Kaci, Souhila 36
Kern-Isberner, Gabriele 236, 257
Kratochvíl, Václav 146
Ktari, Raïda 387
Kutsch, Steven 225

Labreuche, Christophe 340
Lefèvre, Éric 190, 201, 212
Lehtonen, Tuomo 57
Leray, Philippe 435
Lohse, Niels 115
Lopatatzidis, Stavros 104

Madsen, Anders L. 115
Mallek, Sabrine 201
Mangili, Francesca 282
Mauá, Denis D. 93, 449
Mercier, David 190
Molinaro, Cristian 459
Moser, Ulrich 115

Negrevergne, Benjamin 135
Neto, Luis 115
Niland, Richard 257

Ognjanović, Zoran 491

Papini, Odile   387
Parisi, Francesco   459
Petturiti, Davide   364
Pfeifer, Niki   480
Pichon, Frédéric   190
Plajner, Martin   125
Poli, Jean-Philippe   375
Porumbel, Daniel   190
Potyka, Nico   46
Pozzato, Gian Luca   409
Prade, Henri   3, 10
Prakken, Henry   69

Reis, Joao   115
Renooij, Silja   83
Richard, Gilles   3, 10
Ridaoui, Mustapha   340
Rienstra, Tjitze   470
Rozenberg, Igor N.   271

Sabaddin, Régis   306
Samet, Ahmed   135
Sanfilippo, Giuseppe   480
Savić, Nenad   491
Sayed, Mohamed S.   115
Schollmeyer, Georg   329
Serna, Maria   318

Tho, Marie Christine Ho Ba   135
Trabelsi, Asma   212

Valota, Diego   353
van der Gaag, Linda C.   104
Vantaggi, Barbara   364
Vomlel, Jiří   125, 146

Wallner, Johannes P.   57
Wilhelm, Marco   257

Printed in the United States
By Bookmasters